S0-EPA-976

FISH MORPHOLOGY
HORIZON OF NEW RESEARCH

Fish Morphology
Horizon of New Research

Editors

Dr. J.S. Datta Munshi
*Professor Emeritus of Zoology
Postgraduate Department
Bhagalpur University
Bhagalpur 812 007
INDIA*

and

Dr. Hiran M. Dutta
*Professor of Biological Sciences
Kent State University
Kent OH 44242
U.S.A.*

Science Publishers, Inc.
U.S.A.

Science Publishers, Inc.
P.O. Box 699
Enfield, N H 03748
United States of America

© 1996, Copyright Reserved

ISBN 1-886106-31-2

Printed at Baba Barkha Nath Printers, New Delhi 110 015.

Acknowledgments

Our sincere thanks to Mrs. Linda L. Matz of the Department of Biological Sciences, Kent State University, Kent, Ohio, USA for her untiring efforts in word processing several drafts of different chapters of this book. While doing it she also formated several chapters in accordance with the specifications of the *American Zoologist*. Dr. Ashok K. Dutt, Professor of Geography/Planning and Urban Studies of the University of Akron is thanked for his help in initiation and editing of the book. We gratefully acknowledge the benefit derived from the Smithsonian Institution grant, which enabled the editors to initiate, discuss and finalize the drafts of the book.

J.S. Datta Munshi
and
Hiran M. Dutta
Editors

Contents

Acknowledgments v

List of contributors xiii

An Introduction to Fish Research xviii
J. S. Datta Munshi and Hiran M. Dutta

1. Transformation Morphology on Structures in the Head of Cichlid Fishes. 1
   G.C. Anker and P. Dullemeijer

   The Eye of Cichlid Fishes 2
   Morphological Connection of the Eye with Other Functional Components. 11

2. Catfish Morphology: A Reappraisal 21
   C.B.L. Srivastava

   Pineal Window and Carotid Labyrinth 25
   Electroreceptors 26
   Pseudobranchial Neurosecretory System 27

3. Scanning Electron Microscopy of the Fish Gill 31
   Kenneth R. Olson

   Gill Anatomy 32
      Pavement cells 33
      Chloride cells 35
      Mucous cells 37
      Pathophysiology and other applications 38
   Gill Vessels 39
   X-Ray Analysis 41
   Conclusion 41

4. Vascular Organization of Lungfish, a Landmark in Ontogeny and Phylogeny of Air-breathers 47
   Pierre Laurent

   Circulation in Lungfish 48
      Structure and control of the ductus arteriosus 49
      Comparison of ductus in lungfish and mammals 51
      Pulmonary artery vasomotor segments 52
      Gill shunts 52

  From morphology to physiology of lungfish circulation 53
  Significance of the Lungfish Model 54
   Ontogeny 54
   Phylogeny 55
  Conclusion 56

5. **Phylogeny, Ontogeny, Structure and Function of Digestive Tract Appendages (Caeca) in Teleost Fish** 59
  Amjad M. Hossain and Hiran M. Dutta

  Phylogenetic History of Digestive Tract Appendages 60
  Morphological Relationship among Stomach, Caeca and Intestine ... 63
  Era of Confusion Regarding Teleost Digestive Tract Appendages 66
  Ontogeny of Fish Caeca 66
  Functional Morphology, Gross and Microscopic Anatomy of Caeca .. 68
  Physiology: Study of Caecal Enzymes and Microorganisms 71
  Environmental Effect 73

6. **The Structure and Function of Fish Liver** 77
  Jacques Bruslé and Gemma Gonzàlez i Anadon

  Gross Anatomy 78
   Shape 78
   Color 78
   Vascularization 79
   Hepatosomatic index 79
    Sex 80
    Sexual maturation 80
    Feeding 80
    Season and photoperiod 80
    Stress 80
  Light Microscopy 80
   Organization of hepatic parenchyma 80
   Hepatic architecture 81
  Electron Microscopy 83
   Hepatocytes 83
   Sinusoids and their associated cells 84
   Ito cells or fat-storing cells 85
   Macrophages and Kupffer cells 85
   Bile canaliculi 85
  Physiological Hepatocyte Polymorphism 85
   Species variability 87
   Age changes 87
   Sex 87
   Temperature 88
   Feeding 88
   Complex relationship 88

Highlights on Fish Liver Morphology and Suggestions for Further Investigation	88
7. Ultrastructural Diversity of the Biliary Tract and the Gallbladder in Fish *J. Gilloteaux, C.K. Oldham and S. Biagianti-Risbourg*	95
Morphological Interrelationship between Liver, the Biliary System, and the Gallbladder: Development and Histology	96
Agnatha: class Cyclostoma, Myxinidae and Petromyzontidae	96
Gnatha: class Chondrichtyes, subclass Elasmobranchii	96
Gnatha: Osteichthyes	96
Biliary System	97
Ultrastructural Characteristics and Nomenclature	97
Gallbladder Ultrastructure	99
Agnatha: class Cyclostoma	99
Elasmobranch gallbladders	100
Osteichthyes gallbladders, class Teleost	101
Examples of Teleost Gallbladder Ultrastructure	101
Gallbladder Parasites	103
Example of gallbladder parasite ultrastructure	105
Other Peculiar Structures: the "Rodlet Cells"	105
Effect of Natural and Man-made Toxins on the Liver, Biliary Tract, and Gallbladder of Fishes	106
Conclusion	106
8. Recent Advances in the Functional Morphology of Follicular Wall, Egg-Surface Components, and Micropyle in the Fish Ovary *Sardul S. Guraya*	111
Follicular Epithelium	111
Morphology and histochemistry	111
Functions	114
Protein and lipid synthesis and their transport into the oocyte	114
Steroid hormone synthesis	116
Theca and Surface Epithelium	119
Morphology and histochemistry	119
Thecal layer	119
Steroid hormone synthesis	121
Zona Pellucida (or Chorion)	123
Structure	123
Microvilli	123
Zona material	124
Origin	126
Chemistry	128
Surface Structures of Eggs	130
Micropyle(s)	133
Summary	133

9. Atretic Follicles and Corpora Lutea in the Ovaries of Fishes: Structure-Function Correlations and Significance 147
   Bhagyashri A. Shanbhag and Srinivas K. Saidapur

   Atretic Follicles 147
      Process of atresia: Elasmobranchs 148
      Atresia of previtellogenic follicles: Bony fishes 150
      Atresia of vitellogenic (yolky) follicles 151
      Causes of follicular atresia 152
      Origin of interstitial gland cells 153
      Significance of atresia 154
   Corpora Lutea 155
      Luteogenesis and luteolysis 155
         Elasmobranchs 155
      Steroidogenic potential of corpus luteum 158
         Bony fishes 158
      Steroidogenic potential of corpus luteum 160
      Significance of corpora lutea 161
   Terminology 162
   Conclusions and Future Directions 162

10. Structure of Fish Locomotory Muscle 169
    Seth M. Kisia

    Gross Structure 169
    Fine Structure of Muscle 170
       Muscle/fiber types 171
       Fiber differences 173
       Mitochondria 173
       Oxygen diffusion distances 174
       Blood supply 175
    Innervation of Muscle 175
    Conclusion 176

11. Morphology of the Swim (Air) Bladder of a Cichlid Teleost: *Oreochromis alcalicus grahami* (Trewavas, 1983), A Fish Adapted to a Hyperosmotic, Alkaline, and Hypoxic Environment: A Brief Outline of the Structure and Function of the Swimbladder 179
    J.N. Maina, C.M. Wood, A. Narahara, H.L. Bergman, P. Laurent, and P. Walsh

    Introduction 179
    Materials and Methods 181
       Gross dissection 181
          Latex casting of the gills and the airbladder 181
          Transmission electron microscopy 182
       Scanning electron microscopy 182
    Results 182

Discussion	184
Conclusion	188

12. Cephalic Sensory Canal System of Some Cyprinodont Fishes in Relation to their Habitat — 193
P.K. Ray and N.C. Datta

Material and Methods	194
Observations	195
Discussion	199

13. Morphometrics of the Respiratory System of Air-breathing Fishes of India — 203
P.K. Roy and J.S. Datta Munshi

Methodology and Principles Involved in Morphometrics	203
Gills	203
Accessory respiratory organs	206
Types of grids	206
Measurement of diffusing capacity	207
Morphological Analysis	210
Gills	210
Accessory respiratory organs	212
Modification of pharyngeal chambers	212
Modification of branchial chamber	213
Modification of opercular chambers	216
Modification of gastrointestinal tract	216
Modification of swimbladder	216
Dimensional Analysis	218
Scaling of gill and its parameters	218
Juvenile and adult groups	220
Scaling of air-breathing organs	223
Considering entire fish group as one	223
Juvenile and adult groups	224
Diffusing capacity	225
Morphometric estimation of oxygen uptake	229
Future Perspectives in morphometrics	229
Summary	230

14. Effects of Gill Dimension on Respiration — 235
Brian A. Hills

Respiration	235
Gas exchange	235
Gill dimensions	236
Basics of mass transfer	238
Steady supply	238
Steady state	239

Diffusion	239
Diffusivities	240
Gas Exchange in the Gill	240
Laminar flow	241
Boundary layer	242
Film coefficient	242
Types of flow	242
Reynolds number	243
Transitional flow	244
Quantification	244
Flow within tubes	244
Reynolds analogy	245
The Gill	245
Boundary layer analysis	246
Water velocity	246
Energy considerations	246

15. A Composite Approach for Evaluation of the Effects of Pesticides on Fish   249
Hiran M. Dutta

Overview	249
Author's Research	255
Electrophoresis	255
Acetylcholinesterase activity	258
Optomotor behavior	261
Histopathology	265
Gills	265
Liver	266
Ovary	266
Electron microscopy study	267
Scanning electron microscopic study of gills	267
Treated gill	267
Transmission electron microscopic study of normal and treated gills	268
Towards a Composite Model	269

List of Contributors

1. G.C. ANKER
   INSTITUTE OF EVOLUTIONARY AND
      ECOLOGICAL SCIENCES
   LEIDEN UNIVERSITY
   P.O. BOX 9516, 2300A LEIDEN
   THE NETHERLANDS

2. H.L. BERGMAN
   DEPARTMENT OF ZOOLOGY AND PHYSIOLOGY
   UNIVERSITY OF WYOMING
   LARAMIE
   WYOMING 82071
   USA

3. S. BIAGIANTI-RISBOURG
   LABORATOIRE DE ZOOLOGIE ET DES
      SCIENCES DE l' ENVIRONNEMENT
   UNIVERSITE DE REIMS-CHAMPAGNE ARDENNE
   MOULIN DE LA HOUSSE
   51062 REIMS CEDEX
   FRANCE

4. JACQUES BRUSLE
   LABORATOIRE DE BIOLOGIE MARINE
   UNIVERSITE DE PERPIGNAN
   52 AVENUE VILLENEUVE (66860)
   PERPIGNAN CEDEX
   FRANCE

5. N.C. DATTA
   FISHERY AND ECOLOGY RESEARCH UNIT
   DEPARTMENT OF ZOOLOGY
   UNIVERSITY OF CALCUTTA
   35 BALLYGUNGE CIRCULAR ROAD
   CALCUTTA 700 019
   INDIA

6. J.S. DATTA MUNSHI
   POSTGRADUATE DEPARTMENT OF ZOOLOGY
   BHAGALPUR UNIVERSITY
   BHAGALPUR 812 007
   INDIA

7. P. DULLEMEIJER
   INSTITUTE OF EVOLUTIONARY AND
      ECOLOGICAL SCIENCES
   LEIDEN UNIVERSITY
   P.O. BOX 9516, 2300RA
   LEIDEN
   THE NETHERLANDS

8. HIRAN M. DUTTA
   DEPARTMENT OF BIOLOGICAL SCIENCES
   KENT STATE UNIVERSITY
   KENT, OHIO 44242
   USA

9. GEMMA GONZALEZ I ANADON
   LABORATOIRE DE BIOLOGIE MARINE
   UNIVERSITE DE PERPIGNAN
   52 AVENUE VILLENEUVE (66860)
   PERPIGNAN CEDEX
   FRANCE

10. J. GILLOTEAUX
    NORTHEASTERN OHIO UNIVERSITIES
       COLLEGE OF MEDICINE
    ROOTSTOWN
    OHIO 44272
    USA

11. SARDUL S. GURAYA
    DEPARTMENT OF ZOOLOGY
    PUNJAB AGRICULTURAL UNIVERSITY
    LUDHIANA 141 004
    INDIA

12. BRIAN A. HILLS
    THE PEADIATRIC RESPIRATORY RESEARCH CENTRE
    MATER CHILDREN'S HOSPITAL
    SOUTH BRISBANE
    QUEENSLAND 4101
    AUSTRALIA

13. A.M. HOSSAIN
    DIVISION OF REPRODUCTIVE/
    ENDOCRINOLOGY
    DEPARTMENT OF OB/GYN
    UNIVERSITY OF SOUTH ALABAMA
    MOBILE
    ALABAMA 36688
    USA

14. SETH M. KISIA
    DEPARTMENT OF VETERINARY ANATOMY
    UNIVERSITY OF NAIROBI
    P.O. BOX 30197
    NAIROBI
    KENYA

15. PIERRE LAURENT
    LABORATOIRE DE MORPHOLOGIE
    FUNCTIONNELLE ET ULTRASTRUCTURALE
      DES ADAPTATIONS
    CENTRE D'ECOLOGIE ET DE PHYSIOLOGIE
    ENERGETIQUE, CNRS BP 20CR
    F-67037, STRASBOURG
    CEDEX, FRANCE

16. J.N. MAINA
    DEPARTMENT OF VETERINARY ANATOMY
    UNIVERSITY OF NAIROBI
    P.O. BOX 30197
    NAIROBI, KENYA

17. A. NARAHARA
    DEPARTMENT OF ZOOLOGY AND PHYSIOLOGY
    UNIVERSITY OF WYOMING
    LARAMIE
    WYOMING 82071, USA

18. C.K. OLDHAM
    NORTHEASTERN OHIO UNIVERSITIES
      COLLEGE OF MEDICINE
    ROOTSTOWN
    OHIO 44272, USA

19. KENNETH R. OLSON
    CENTRE FOR MEDICAL EDUCATION
    B-19 HAGGAR HALL
    UNIVERSITY OF NOTRE DAME
    NOTRE DAME
    INDIANA 46635, USA

20. P.K. RAY
    FISHERY AND ECOLOGY RESEARCH UNIT
    DEPARTMENT OF ZOOLOGY
    UNIVERSITY OF CALCUTTA
    35 BALLYGUNGE CIRCULAR ROAD
    CALCUTTA 700 019
    INDIA

21. P.K. ROY
    POSTGRADUATE DEPARTMENT OF ZOOLOGY
    BHAGALPUR UNIVERSITY
    BHAGALPUR 812 007
    INDIA

22. SRINIVAS K. SAIDAPUR
    DEPARTMENT OF ZOOLOGY
    KARNATAK UNIVERSITY
    DHARWAD 580 003
    INDIA

23. BHAGYASHRI A. SHANBHAG
    DEPARTMENT OF ZOOLOGY
    KARNATAK UNIVERSITY
    DHARWAD 580 003
    INDIA

24. C.B.L. SRIVASTAVA
    ZOOLOGY DEPARTMENT
    UNIVERSITY OF ALLAHABAD
    ALLAHABAD 211 001
    U.P.
    INDIA

25. P. WALSH
    DIVISION OF MARINE BIOLOGY
       AND FISHERIES
    ROSENTHEL SCHOOL OF MARINE
       AND ATMOSPHERIC SCIENCES
    MIAMI
    FLORIDA 33149
    USA

26. C.M. WOOD
    DEPARTMENT OF BIOLOGY
    MCMASTER UNIVERSITY
    HAMILTON
    ONTARIO
    CANADA L85 4K1

An Introduction to Fish Research

J.S. Datta Munshi and Hiran M. Dutta

Explanation of structural configurations in the functional context constitutes functional morphology. Research in morphology has recently trended toward establishing a relationship between functions and forms, not only between hard and the soft structures, but also among soft structures themselves. This relationship also exists at the cellular level. Functional morphology emphasizes the mechanical and spatial arrangements of parts and structural features in accordance with their function or performance. When hard tissues (bones and ligaments) and soft tissues (muscles, aponeuroses) are involved in the realization of functions, a mechanical demand influences the shape, size, and location of these structures, although the environmental impact cannot be ignored. Chapter 1 by Anker and Dullemeijer extensively examines this relationship, designated as ecological morphology.

Modern biologists are showing renewed interest in understanding the various facets of evolutionary morphology, encompassing developmental biology, and paleontology in relation to phylogeny. Morphological features are frequently used to establish phylogenetic relationship among species and therefore morphology has a multifaceted purpose. Knowledge of the vascular organization of the fish lung can help us to achieve a better understanding of the phylogeny of air-breathers. The function of various structures can be explained in the framework of systems. Such explanations include the vascular system, gills, lungs, swimbladder, digestive organs, gallbladder, liver, gonads, locomotory organs (muscles) and the eyes of fish—all of which are discussed in this book.

Recently, morphometrics has been widely used to correlate form and function. This relationship can be applied to any structure of the body but accessory respiratory organs and lungs are considered to be the most suitable. Study of the relationship between the morphometric characteristics and the growth patterns of animals provides an in-depth understanding of the structures and their differential function in relation to growth.

Morphometric analysis can also be applied to assess the environmental impact on structures. The impact of environmental pollutants on the form-function relationship can be severe. Pollutants such as pesticides can easily disrupt the fragile thread which keeps this relationship intact. Indiscriminate use of pesticides can be considered one of several factors which bring changes in the environment, inducing several imbalances in the ecosystems, especially the denizens of the aquatic environment. Fish are one of the most important aquatic animals. Therefore, a composite measure from different disciplines should be taken into account when formulating the tolerance limit at which ecologically significant changes in structures, functions and behavior occur. Fish is a staple food for human consumption and occupies an important place in the food chain. Moreover, the morphology of fish can provide base-line knowledge which can be applied to the morphology of higher forms of animals, including human beings.

ABOUT THE BOOK

This book is an outcome of contributions by reputed scholars from five continents. New research on fish morphology conducted by the contributors has been synthesized in a broad perspective. Each chapter of this book reflects a review of not only the contributors' research, but synthesizes the ongoing research by others in a much broader framework. A précis of each chapter is given below.

The authors of Chapter 1, Anker and Dullemeijer, explain the broad aspects of functional morphology by using the eye structure of cichlid fishes in relation to their niche, behavior, and food intake. This aspect of the scientific approach has been designated ecological morphology. The performance of the vision system is important for the animal's survival. Inductive and the deductive explanations can be used to establish the relationship between features or for the derivation of one feature from another, such as form from function. The facial part, the eye as an example of a cichlid fish, has been used to show the importance and limitation of the deductive method. Ecological parameters can induce derivation of the necessary behavior. Among the ecological factors, the wavelength and the amount of light become essential for observing food, partners and enemies. As observation is distance dependent, the resolution and sensitivity of the eye is related to the accommodation mechanism and structure of the retina. Four species of *Haplochromis* are compared in relation to their variable habitat, which includes water depth, day, and night. These variables are correlated with the distribution and density of the retinal cones. There is a morphological connection of the eye with other functional components. Further, the eye size, shape, and its location are related to the demand of necessary vision. As in most vertebrates, the shape of the eye can be derived from laws of optics and the demand of smooth rotational movement, although it may be affected by other factors.

At the end of the twentieth century, fish morphologists stand on the threshold of a new view regarding the status of morphological characteristics of various groups of fish in order to determine their phylogenetic relationships. In Chapter 2, Srivastava analyzes the autapomorphic, synapomorphic, and plesiomorphic characters of catfish and raises questions about their proper phylogenetic position. The cladistic approach of phyletic schemes of classification is particularly important for teleosteans as a fossil record is not readily available for them. Teleostei are considered to be advanced bony fishes, very competitive and successful. Among teleosts, the ostariophysian forms are considered by the author to be far older than previously believed. Living catfish are considered to be primitive among ostariophysian forms, recalling similarities with even lower actinopterygians. The autapomorphies of the catfish group reveal that they can no longer be considered a sister group to cyprinids and other ostariophysians. The recent addition to catfish autapomorphies suggests an independent evolution of the catfish group as a phylogenetic lineage. Further, the phylogenetic and systematic status of the catfish should be subjected to revision and future researchers of the twenty-first century should throw light on this and similar issues related to the understanding of fish morphology.

Fish gills as respiratory organs have fascinated biologists. In spite of intense work done on physiological and biochemical aspects of osmoregulation in fishes, there is limited morphometric data on the chloride cells of gills. These cells are found on the arch support epithelium, lamellar portion of the filament, and medial margin of the lamellae. In Chapter 3 the varied functions of mucous glands in gills of fishes are discussed by Olson. The variations in microridge organization of pavement epithelial cells of filaments and lamellae are remarkable and must have some phylogenetic and physiological significance. Olson has worked on the cellular organization of gill structure of fishes in relation to their vasculature using modern techniques. The most significant use of SEM is illustrated in the analysis of corrosion replicas of the complex gill microcirculation. Orientation of endothelial cells distinguishes between arterial and venous vessels respectively. The interlamellar system has a definite organization and is present in all fishes. More work is required in order to clarify the origin and relationship of interlamellar and nutrient pathways in the filament. The use of x-ray in conjunction with SEM offers a quantitative elemental analysis of gill tissue and can be valuable in the studies of physiology and toxicology.

The functional morphology of Dipnoan lungfish has attracted the attention of biologists for more than two centuries. In Chapter 4 Laurent shows how the vascular organization of lungfish helps us to understand the phylogeny of air-breathers. Lungfish living in a tropical swampy environment have the capacity to withstand drought conditions by living in encysted cocoons buried in dry mud for several months and even years. During aestivation the urea level in the body rises considerably. At this time the lungfish relies for both O_2 and CO_2 exchanges on the "snorkel" which connects the inside of the cocoon with the surface of the soil, thereby supplying fresh air to the lung. The circulatory system of the lungfish is configured to operate in relation to its specialized gas exchangers. The lungfish retains the capability of shifting its mode of respiration from one medium to another (water/air) throughout its life. This cardiorespiratory innovation evolved in this group of fishes several million years ago, in the Paleozoic era, and is responsible for their survival to the present day. Laurent emphasizes the structure and function of the *ductus arteriosus* in the circulatory system, since it resembles the same structure in the postfetal stage of a newborn mammal. A new functional concept of the system is proposed, taking lessons from a comprehensive study of the circulatory system of *Protopterus aethiopicus* and *Lepidosiren paradoxa*. The main features of their circulatory system are: i) persistence of the *ductus arteriosus* throughout life, ii) development of a septum dividing the atria, iii) partial division of the ventricle, and iv) presence of a spiral valve in the conus, besides the six aortic arches supporting blood supply to the gills and lungs. The morphological basis of nervous control of the *ductus arteriosus* is much like that of higher vertebrates. The existence of anatomical gill shunts in dipnoan and teleostean fishes (see Olson this volume) has been established. These shunts consist of large bore vessels connecting the afferent and efferent gill arteries at the base of the gill filaments. In amphibians these shunts are under neurohormonal control, but this has not been demonstrated in fish. The lungfish might be a good experimental model for studying some of the problems of prenatal medicine.

The digestive system appendages, such as caeca, have long attracted the attention of biologists interested in fishery. In Chapter 5 Hossain and Hiran Dutta make an important contribution to unraveling the ontogeny, phylogeny, and function of these appendages. The phylogenetic history of the caeca has been explicitly explained. A correlation was established between the shape of the stomach and the absence or presence and number of caeca. There is a relationship between feeding groups and number of caeca. A detailed explanation of the morphological relationship among stomach, caeca, and intestine is given. The functional morphological approach with computer assisted microscopic analysis of cross sections of caeca and intestine is described. The unit area of intestine is about twice that of caeca. A cross section was used to measure some of the structural components, namely the amount of muscle, mucosa, and lumen. This light microscopic study is supported by an electron microscopic study. The same major cell types, columnar epithelial (absorbent cell), goblet (secretory) cell, and endocrine cells, are found in both the caecal and intestinal mucosa. However, the absorbent cells of the intestine are larger than those of the caeca. Movement of food in the digestive tract was monitored by using fish food mixed with a special dye or by radiography. It was confirmed that stomach contents enter the caeca in the same way that food enters the intestine. Considerable attention was paid to detection of enzymatic activities and microorganisms of the caeca. The lipase and alkaline phosphatase activities are quite high in the intestinal caeca. Different levels of response of the caeca and intestine to environmental hazards occur. The caeca retain food during starvation.

The fish liver is a key organ which controls many life functions and plays a prominent role in fish physiology, both in anabolism and catabolism. It also plays an important role in vitellogenesis. Fish are especially susceptible to environmental variations and respond more sensitively to pollutants than mammals. In Chapter 6 Bruslé and Anadon show the versatile nature of the fish liver and demonstrate its hepatic polymorphic nature. The fish liver is shown to be a very interesting model for the study of interactions between environmental factors and hepatic structure and function.

The gallbladder in fish is essentially an accessory organ of the digestive system that stores and secretes concentrated bile. Several functions, such as facilitating several digestive functions, eliminating conjugated metabolites in the liver, and enterohepatic bile circulation are assigned to the gallbladder. In Chapter 7 Gilloteaux, Oldham and Biagianti-Risbourg review the work done on the biliary tract and gallbladder of various groups of fish. The ontogeny of the gallbladder, biliary system, and liver in fish differs in different taxa from cyclostomes, elasmobranchs, and teleostean groups of fishes. Ultrastructural investigation of the gallbladder and biliary system in fish has only recently been done and comparisons with the mammalian system made. In this chapter the ultrastructure of the gallbladder of several teleostean fish, e.g. *Triglosporam lastoviza*, *Uranoscopus* sp., *Sorpena scrofa*, is compared with that of elasmobranchs and other groups of fishes. The gallbladder in fish is often infected by Myxosporidian protozoa, e.g. *Cnidospora*. Myxosporidia can be detected as diffuse infiltrates in tissues and can form voluminous tumors. Some species live free in the lumen of the gallbladder and other viscera, such as the urinary bladder, testes and ovary, but they most frequently infect the gallbladder. The presence of "rodlet cells" in the gallbladder of fish seems to be a universal phenomenon. They are present in cyclostomes, elasmobranchs, and teleosts. Their nature is still controversial; they have been considered parasitic but this interpretation is not widely accepted. The authors also deal with the effects of natural and man-made toxins on the liver, biliary tract, and gallbladder in fishes. *Microcystis aeroginosa*, a bluegreen alga, was recently shown to produce hepatotoxin which affects the liver and biliary structures. Some experiments have also been conducted on the effects of disodium arsenate heptahydrate on juvenile rainbow trout.

A great deal of work has been done on fish ovaries in the last three decades. In Chapter 8 Guraya has made significant contributions to our understanding of the functional morphology of the follicular wall, egg surface components, and micropyle in the fish ovary. He not only reviews the work done in this field, but also describes the functional components of the fish ovary. The great diversity in their development, structure, and function in various fish species is revealed. The possible synthesis of lipids and steroid hormones, e.g. estrogen is discussed. The isolated specialized cells lying close to capillaries in the thecal layer synthesize estrogen precursors which are converted to 17β-estradiol in the follicular epithelium. There exist possible mechanisms of transport of substances across their cellular membranes. The surface pattern of eggs and their micropyles form important characteristic features for the identification of different species of teleosts.

In Chapter 9, Shanbhag and Saidapur review the work on the structure and function of atretic follicles and corpora lutea in the ovaries of fish. The endocrine capacity of the corpora lutea, their role in gestation and other reproductive functions, and the development, structure, and function of atretic follicles, their degeneration, and the possible hormonal and environmental factors affecting and governing follicular atresia are discussed. The study of various histological, ultrastructural, histochemical, and biochemical changes that occur in the atretic follicles and corpora lutea in both cartilaginous and bony fish indicate that in all viviparous and some oviparous species, the granulosa cells undergo luteinization and synthesize steroids. The corpus luteum degenerates very rapidly in most oviparous fish and is long lived in all viviparous forms with ovarian gestation. Experiments involving hypophyseal hormones, ovariectomy, or deluteinization are needed to understand the precise role of corpora lutea in fish gestation. Some researchers are of the opinion that atretic follicles are merely abortive while others relate them with steroidogenic potential.

The structure and function of the locomotory muscles of fishes have been studied by many zoologists throughout the world. These muscles are likely to vary structurally and physiologically due to the great diversity in form and habitat of fish. In Chapter 10 Kisia presents some aspects of the gross and fine structure of the locomotory muscles of fish. Fish display great diversity in their body movements. Most possess a streamlined body and vary in their swimming behavior: surface swimmers (needlefishes, halfbeaks, and tapminnows) display dart movements, ocean flying fish (South American freshwater flying fish) can swim above the water, midwater swimmers (trouts and tuna) are fast swimmers, while those in quiet waters (sunfish and freshwater angelfish) are not. The variation in demands placed on fish by the environment in relation to swimming behavior has brought about differences in the structure and physiology of fish locomotory muscles. These differences include different muscle fiber types (red, pink and white) and their dimensions and proportions in different fish, blood supply, innervation and densities of muscle fiber components such as mitochondria. The variation in mitochondrial volume densities is related to species of fish, temperature, fiber type, activity of fish, and environment. Oxygen diffusion distances are important in determining the rate at which oxygen diffuses from red blood cells in capillaries to muscle fibers and metabolic wastes diffuse back to the capillaries from the muscle fibers. Phylogenetic changes are also evident in the gross and fine structure of the locomotory muscles among different orders of fish.

In Chapter 11 Maina and his collaborators show how the specialized swim (air) bladder has enabled the cichlid fish to adapt to a hyperosmotic, alkaline, and hypoxic environment. The swimbladder in fish has received sustained scientific interest throughout the ages. The principal aspects of interest are related to their evolutionary origin, structure, and function. The variety of functions carried out by this airbladder in various species is astounding. Maina and others investigated the structure of the airbladder in cichlid fish by gross dissection and latex rubber casting to study its topographic anatomy and that of its basic structural components with transmission and scanning electron microscopy. The airbladder of *Oreochromis alcalicus grahami* is designed to play both the roles of respiration in extreme conditions of anoxia and buoyancy control by gas secretion below the water surface when feeding.

The organization of the sensory apparatus in fishes has been a subject of special interest to zoologists for many years because of its structural and functional specialization. In Chapter 12. Dutta and Ray picture the cephalic sensory canal system of cyprinodont fishes in relation to their habitat. There are interesting variations in the structure and disposition of the neuromasts in the canals, which show a close relationship with the habitat of the animal.

In Chapter 13 Roy and Datta Munshi review the structural organization of gills and accessory respiratory organs and analyze their morphometric characteristics in relation to growth. Effect of body size on scaling of respiratory organs is discussed. The relationships among total respiratory surface area, gill area, and accessory respiratory surface areas (ARS) of juvenile and adult air-breathing fishes are analyzed with respect to body weight and compared with water-breathing fishes. The inflection of regression lines is related to age and maturity in various species and their hypoxic environment. The gill areas of adult *Heteropneustes fossilis*, *Clarias batrachus*, *Channa punctatus*, and *Channa gachua* increase with higher exponent values than those of juveniles. In obligate air-breathers, *Anabas testudinus* and *Channa striat*, the slope values of adult fish are insignificantly lower than those of juveniles, as in water-breathing fishes, but the exponent values of air-breathing surface areas of adult fish are significantly higher than those of juveniles. This suggests that after the onset of maturity the adults require more accessory respiratory surface area for respiration. In water-breathers, the regression line remains more or less straight throughout their postlarval life history, whereas in air-breathing fish it shows a diphasic allometry. The diffusing capacities of different respiratory organs vary from species to species.

Gill dimensions play an important role in gas exchange as well as in resistance to fluid flow in the gill. It is suggested by Hills in Chapter 14 that in all transport processes the rate of transfer is determined by two basic factors—driving force and resistance. In respiratory gas exchange the driving force is oxygen tension (P_{O_2}) whether for providing a gradient to force O_2 into solution or to displace the position of chemical equilibrium which occurs in the conversion of hemoglobin to oxyhemoglobin. The secondary lamellae of the gill filament are bordered by an epithelial surface and the central capillaries, which run lengthwise, are separated only by the basement membrane to give a mean blood-to-water distance. Hills also suggests that fish invoke the countercurrent principle, whereby blood flows in the capillaries in a direction opposite to water in the adjacent channels. Some fundamental principles of movement of solutes in solution are presented. The concept of steady supply applicable in this situation suggests that if a concentration of solute molecules at one surface of the medium is maintained from an external source, then diffusion continues until the rest of the medium reaches the same concentration, provided other boundaries are impermeable. Since the diffusion coefficient is low in biological creatures, they circumvent this problem by increasing surface area and having a thin membrane. Gas exchange in the gill involves the exchange of oxygen and carbon dioxide in the secondary lamellae, in which Fick's law plays an important role. In order to explain the exchange, the author has incorporated the theory of laminar flow, boundary layer, and film coefficient. The parallel nature of the movement of different laminations of fluid elements characterizes laminar flow. A water radiator is an appropriate analogy for laminar flow. In the case of flow state, if the surface is flat but the velocity greatly increased, the parallel and well-ordered flow will break up and lead to a haphazard movement of the elements, i.e. turbulence. Turbulence in the flow is related to the roughness of the surface, constriction in pipes (or branching as indicated in Reynolds number), and also to length (as expressed by Schlichting, 1960). The quantification of mass transfer involves mathematical analysis, flow within tube, and Reynolds analogy. Reynolds analogy with heat transfer has been applied to "gilled systems" to drive the flux of respiratory gases across the boundary layer of water adjacent to the gill membrane. Mass transfer is appreciably more effective at higher velocities but turbulence induction raises the work load. The energy expenditure and the pressure gradient are also related to width of the channel.

Indiscriminate use, careless handling, accidental spillage, or discharge of pesticides or untreated effluents into natural waterways have harmful effects on the fish population. In Chapter 15 the purpose of Hiran Dutta's study is to determine the effects of sublethal doses of malathion and diazinon on microscopic ultrastructures, serum proteins, behavior and acetylcholinesterase activity of certain teleosts. SDS-polyacrylamide gel electrophoresis was used to detect the quantitative and qualitative changes in the different serum protein fractions. A quantitative estimation of serum protein fractions provides a good index of stress from malathion and diazinon, or any other chemical present in the environment. A qualitative estimation provides an index of a new protein which may form due to the breakdown of cellular components. Changed acetylcholinesterase (AChE) activity and optomotor behavior responses are useful as a valuable tool for assessing the pesticide pollution. Pavlo et al. (1992) found that exposure of *Abramis bremia* to the organophosphorus pesticide DDVP resulted in decreased food consumption and inhibited AChE activity. AChE is not only essential for normal behavior, it has an important role in embryonic development (Hanneman, 1992). The optomotor response is widespread throughout the animal kingdom and is considered to be extremely important for maintaining position within the habitat and for schooling in fish. Any type of imbalance in the rheotropic response brought on by pollution will certainly be detrimental to the existence of the fish. Histopathology appears to be a very sensitive parameter and is crucial in determining cellular changes that may occur in the target organs such as gills, liver, and ovary. Light microscopy reveals only the superficial changes that occur in the structures of gill, liver, and ovary. Scanning electron microscopy is an important modern tool revealing changes in the surface ultrastructures of the aforementioned organs. Transmission electron microscopy provides a clear, in-depth understanding of the normal ultrastructure as well as changes in the ultrastructure of gills, liver, and ovary. A descriptive model showing changes in structure, physiology, and behavior in fish exposed to pesticides was constructed. By correlating data from the four different techniques, it is possible to detect environmental pollution at an early stage and to prevent unwanted effects on nontarget species such as fish.

HORIZON OF NEW RESEARCH

Although research in anatomy has a long history, the study of functional anatomy has evolved gradually, mainly since the nineteenth century. The concept, theory, and philosophy of functional morphology was nurtured in the Department of Zoology, Leiden University, the Netherlands. Prof. Vanderklaauw pioneered the concept of modern functional morphology. Prof. P. Dullemeijer, a pupil of Prof. Vanderklaauw, not only perpetuated that concept and philosophy, but gave a new and in-depth theoretical foundation to functional morphology. Several scholars trained in the department have made notable contributions in this field: G. Anker, H.A. Akster, C.D.N. Barel, H. Berkhoudt, J.L. Dubbledam, Hiran M. Dutta, H.S. DeJong, J.L. van Leeuwen, J.W.M. Osse, E. Otten, M.J.P. van Oijen, W.A. Weijs, F. Witte, F.W. Wuijens, G.A. Zweers among others.

Apart from the contributing authors of this book, several other outstanding scholars who have made significant contributions in the research of fish functional morphology, namely: V.O. Adanina, Sechenov, Institute of Evolutionary Physiology and Biochemistry, Russia; P.E. Ahlberg, University of Oxford, United Kingdom; R. McN. Alexander, University of Leeds, United Kingdom; V. Banerjee, Patna University, India; C.R. Braekevelt, University of Manitoba, Canada; M.M. Bryden, University of Sydney, Australia; C.J. DeVre, University of Antwerp, Belgium; H. Evans, Cornell University, USA; T.P. Evgenjeva, Institute of Evolutionary Animal Morphology and Ecology, Moscow, Russia; S. Gemballa, University of Tubingen, Germany; G.M. Hughes, Bristol University, United Kingdom; A. Huysseune, University of Ghent, Belgium; M. Jakubowski, Jagiellonian University, Poland; S.B. Lall, Udaipur University, India; G.V. Lauder, University of California at Irvine, USA; K.F. Liem, Harvard University, USA; P.J. Motta, University of South Florida, USA; G.B. Müller, University of Vienna, Austria; R.G. Northcult, University of California, USA; B.R. Singh, Bihar University, India; A.K. Mittal, Banaras Hindu University, India, P.E. Wainwright, Florida State University, USA: G.P. Wagner, Yale University, USA; E. Zeiska, University of Hamburg, Germany.

Today, functional morphological research has advanced considerably. Two aspects have helped development of research—technique and concept. They build upon each other, forging the discipline ahead. The use of light microscope, electron microscope, x-rays, moving x-rays, electromyography, and cinematography-electromyography in combination, as well as analysis of blood serum and muscle fibers chemically, histochemically, and physiologically, have helped functional morphologists to understand not only the gross anatomy, but also its functions.

Though Greek and Roman ancients knew the property of the crude lens, it was only in the thirteenth century that eyeglasses were used. The precision required for a simple microscope was attained in 1590 by Hans and Zacharias Janssen, who were eyeglass makers in Middleburg, The Netherlands. Thereafter, advancement both in the structure and function of the simple microscope took place and compound microscopes have evolved since 1870. Microscopic research provided a detailed visual basis for revealing many priorly unknown functions of animals and human beings.

Greater precision in microscopic study became necessary, however, and free electron beams were identified by physicists. Based on the laws of electron optics, Max Knoll and Ernst Ruska of Berlin constructed the first two-stage electron microscope in 1931. This electron microscope magnified smaller organic materials several thousand times and thus provided the base for many functional anatomical studies that continues today. The x-ray, discovered by Prof. Wilhelm Conrad Roentgen, Royal University of Wurzburg, Germany in 1895, not only revolutionized diagnostic procedures in the field of medicine, but unfolded the actual, pictorial basis of vivid animal anatomy. When later, in the twentieth century, x-ray and cinematography were combined to produce x-ray cinematography, this technique ushered in a new horizon for functional anatomy. While these discoveries and techniques added greater understanding of animal form and function, others augmented understanding the functions of blood, muscle fibers, bones, cartilages, and nerves. Today, near the end of the twentieth century, we are endowed with modern scientifically based tools that unravel the structures, functions and form/function relationships that were impossible to visualize when the century began. Moreover, the use of computers for statistical analysis, configuration measurement, and functional simulation have further advanced the field of research.

The concepts of functional morphology have also evolved to include not only the techniques and ideas derived from biological researchers, but those taken increasingly from other disciplines. Thus the science of functional morphology has become an interdisciplinary one. As the twenty-first century unfolds, such interdisciplinary aspects of functional morphology will be further enhanced. With this, interest and research will further develop in the applied branches of functional morphology, such as the study of structural changes in diseased and contaminated situations. Considering the rapid advancements in technology, in medical and space research, techniques and their applications in functional morphology will also continue to advance. Thus the twenty-first century awaits a much higher level of research in functional morphology, some for which foundations have already been laid in this book.

References

Hanneman, E.H. 1992. Diisopropylfluorophosphate inhibits acetylcholinesterase activity and disrupts somitogenesis in the zebrafish. *J. Exp. Zool*. 263(1):41–52.

Pavlo, D.D., G.M. Chuiko, Y.V. Serrassimov, and V.D. Tonkoply. 1992. Feeding behavior and brain acetylcholinesterase activity in brean (*Abramis brama* L.) as affected by DDVP, an organophosphorus insecticide. *Comp. Biochem. Physiol*. 103c:563–568.

Schlichting, H. 1960. *Boundary Layer Theory*. McGraw Hill, New York (4th ed.).

1

Transformation Morphology on Structures in the Head of Cichlid Fishes

G.C. Anker and P. Dullemeijer

Induction and deduction are the two methods for establishing and explaining the relationship between features or for deriving one feature from another, such as form from function. Out of the two methods deduction is the strongest, while induction is used mainly in a narrative way for collecting and classifying data and, in functional morphology, for describing correlations between form and function. This description asks subsequently for a tested explanation. Deduction in functional morphology, and also in constructional and ecological morphology, aims to explain animal form and features from function, behavior, and environmental conditions. This methodological reasoning follows a reverse course. Ecological parameters are measured from which a possible or necessary behavior is derived and from this in turn a function followed by a structure. The structures in the deductively obtained model are tested by comparing them with the actual structures (Dullemeijer and Barel, 1977).

In the following example, the structures in the facial part of cichlid fishes, we shall show how deduction can be used, where it fails or falls short, and consequently what kind of new information or models are required to be obtained in further investigations.

We shall start the procedure by presenting the course of research for the explanation of one functional component, viz. the eye, and from there proceed by adding more components to explain structures in the facial part of the head in relation to the parameters of the niche of the fish.

We shall show that structures can be explained by deduction, but also that we have often to return to induction and even have to admit that as yet the structures cannot be explained. In particular it will be demonstrated that deduction has its limitations for the following reasons. The niche parameters cannot be defined accurately, the functional capacities of the structure are insufficiently known, constraints and thus limitations occur when structures are combined, nonfunctional structures or functional for only part of the lifecycle have to be taken into account and, last but not least, some basic structural properties can only be explained historically, and thus have to be taken as given boundary conditions in the design of the model. The latter condition should force us to start with a description of structures constituting the basic plan. However, we trust that the reader will have some knowledge of the general body plan and building material of a percoid fish.

THE EYE OF CICHLID FISHES

Most cichlids have normal fish eyes, i.e., nearly hemispherical in shape with a spherical lens of which the radius (r) is about 2.55 times smaller than the radius of the retinal sphere. The lens is the only refractive element with a focal distance of 2.3 r (Fig. 1). It is situated in a myopic position to the caudal part of the retina and nearly at focal distance to the rostral part. A retractor lentis muscle can shift the lens caudally with respect to the retina.

As in most vertebrates, the shape of the eye can be derived from laws of optics and the demand of smooth rotational movement. Although the shape can be affected by other factors (see below), for the model we shall start with a hemispherical one.

It is obvious that for proper vision the eye must have a position at the surface of the head, but its rostrocaudal and dorsoventral situation can vary in relation to constructional demands or required visual fields.

Conceiving the eye as a functional component, which is equal in shape and basic construction and has similar optical properties in the fishes compared, only differences in size and retinal composition will be considered in relation to functional differences. The latter are related to behavior and environment.

Among the ecological factors, wavelength and the amount of light stand out for observing food, partners, and enemies. As observation is distance dependent, we shall first discuss the accommodation mechanism and subsequently the structure of the retina in relation to resolution and sensitivity.

The distance from which objects are to be observed may affect the balance between demands of retinal resolution and sensitivity. Nearby objects form larger images on the retina and from them more light enters the eye than from similar objects far away. Also, accurate focusing on nearby objects asks for more precise adjustment. Therefore, catching prey items at a distance less than one cm in front of the snout demands a perfect accommodation mechanism of the lens in a rostrad-caudad direction. This is provided by the retractor lentis muscle and an antagonistic force produced by the elasticity of the lens-suspension tissue inside the eye. For such nearby vision, a high resolution is possible because in that position of the prey, the most image-forming light is available. Therefore, the highest resolution can be expected in the caudal part of the retina (sharp, myopic vision).

Retinal resolution is inversely related to the separable angle and consequently proportional to the density of photoreceptors and to eye size (Fig. 2). With the same optimal packing of receptors, and ignoring differences in neuronal convergence, sensitivity depends only on the size of the photoreceptors because the amount of photons arriving at the retina per retinal square unit and per unit time is independent of eye size (Van der Meer and Anker, 1986). Maximal packing of receptors in the retinal plane depends on their shape and configuration but not on the size of the receptors (Van der Meer, 1992). It may be noted that the receptor convergence, which is the ratio between the number of receptors and the number of ganglion cells, is positively related to sensitivity and negatively to resolution.

Per vision angle more equal receptors or the same number of larger receptors can be placed in a bigger eye (Fig. 3). This simple model formulates a possible increase of retinal resolution or an increase of sensitivity or an increase of both in bigger eyes (this indicates

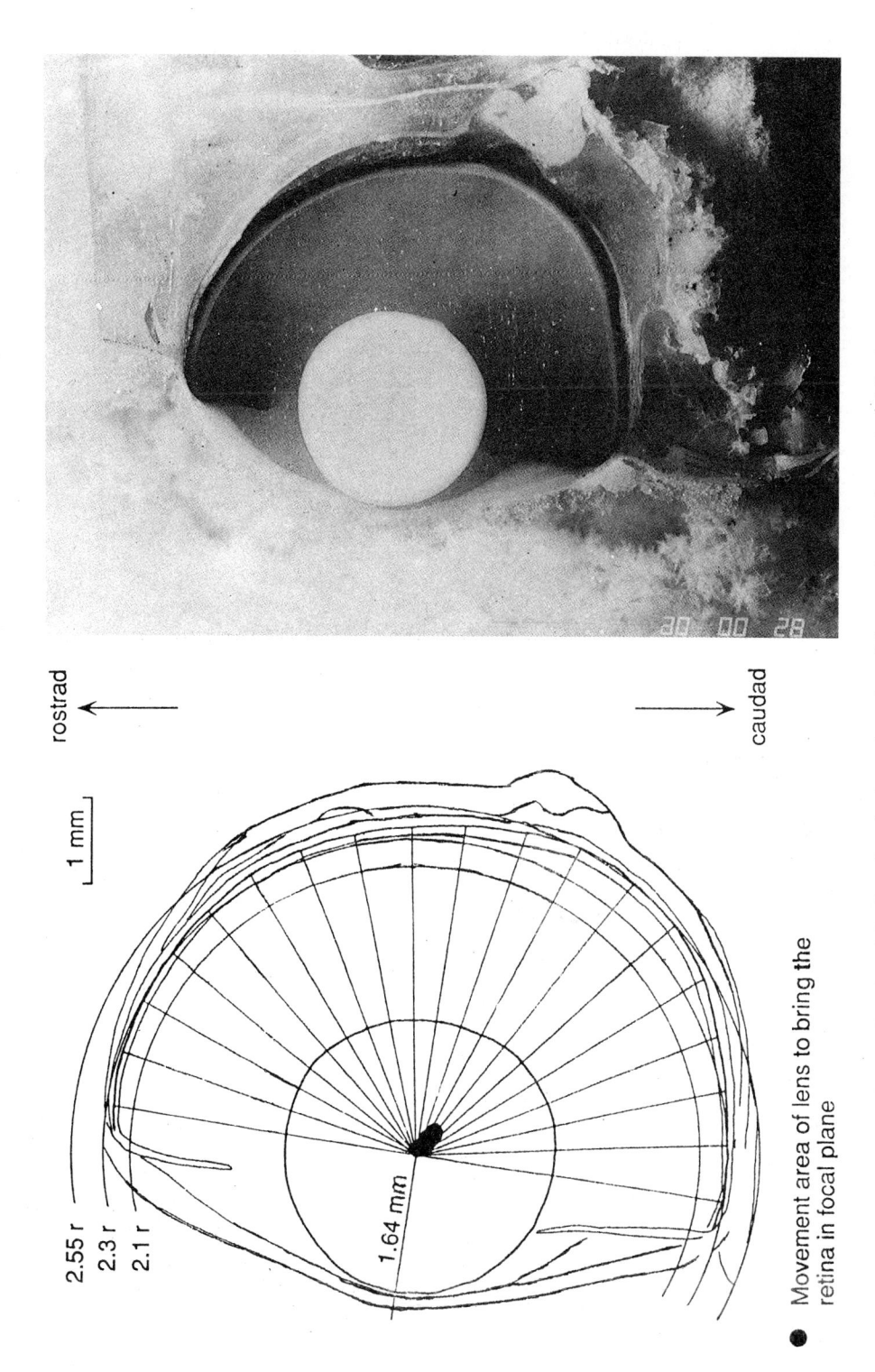

Fig. 1. Position of the lens in a horizontal section of the eye through the center of the lens.

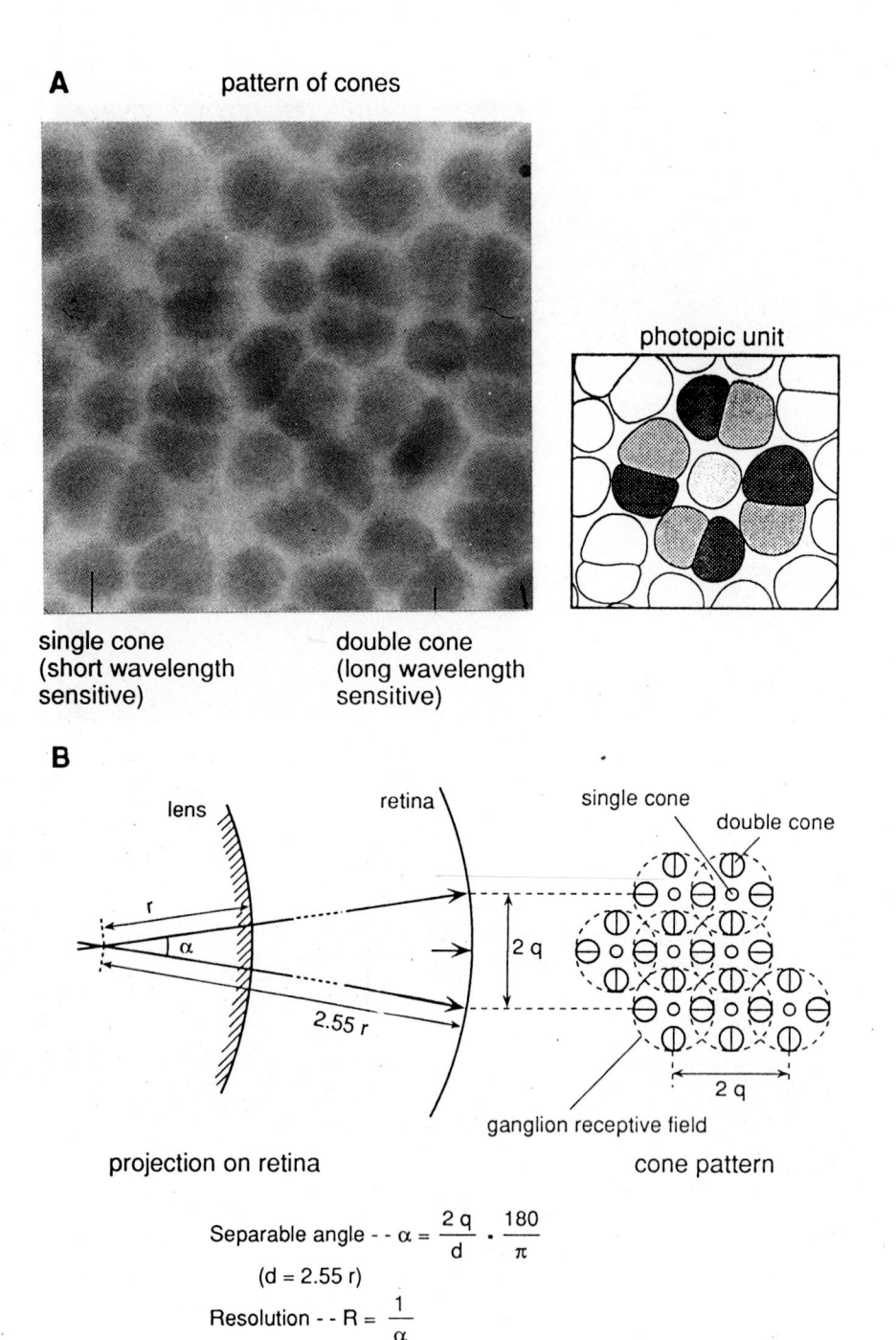

Fig. 2. A square pattern of cones and the supposed photopic unit in this pattern. B. Manner in which separable angle and resolution are calculated from the supposed ganglion receptive field (after Van der Meer and Anker, 1984).

$$\text{Separable angle} - \alpha = \frac{2q}{d} \cdot \frac{180}{\pi}$$

$$(d = 2.55\, r)$$

$$\text{Resolution} - R = \frac{1}{\alpha}$$

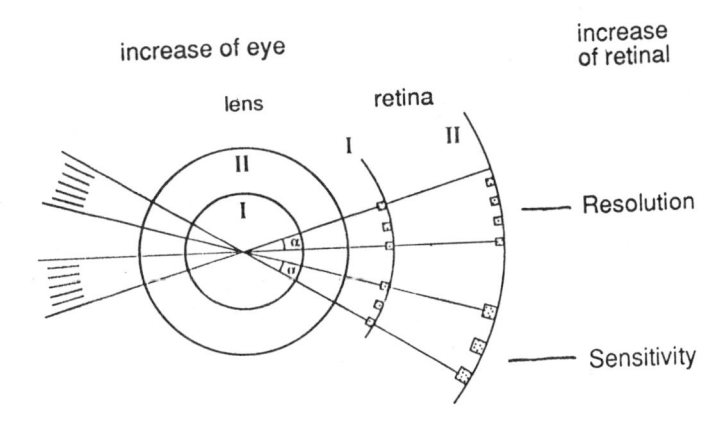

Fig. 3. Possible enhancement of retinal resolution or sensitivity (or both) when the size of the eye increases (after Van der Meer and Anker, 1986).

a trade-off relation between resolution and sensitivity). The actual solution will probably be a compromise of which the degree is related to the niche of the animal. In general, we can expect fishes feeding on large prey and detecting prey from a long distance to have large eyes with big photoreceptors, whereas fishes feeding on nearby small items will have high densities of small receptors, particularly in the caudal part of the retina. There are, however, many exceptions to these expectations. Apparently there are more ecological demands. Also, it is possible that in the retina different areas serve different functions and consequently show differences in structure. Moreover, light intensity is a very important factor. Exceptions may also be due to the limited capacity and space, so that the demands cannot be met.

The following examples show some of these expectations. The fact that small fry of most species grow up in shallow coastal areas can be a consequence of the limited capacity of their small eyes for vision in darker habitats. The actual increase of resolution during ontogeny, as experimentally determined by Van der Meer (1991), appeared to be related to the size of photopic units, which were twice as small as calculated (Fig. 2). An experimentally determined increase in sensitivity was found to be related to increase in receptor volume and probably to increase in photosensitive pigment in the receptor outer segment (Van der Meer, 1991).

A predicted general increase in resolution and sensitivity related to eye size was found during ontogeny (Van der Meer, 1991) and in adults of eleven cichlid species differing in eye size (Van der Meer and Anker, 1984). Also the predicted relatively higher resolution of the temporal area of the retina was confirmed. Much variation occurred depending on the behaviorally and environmentally required balance between resolution and sensitivity.

To make the model more operational to testing, it has to be refined for specific cases for which the ecological situation is well known. It can be refined by adding the property that the single cones contain short wavelength sensitive pigment and the double cones contain long wavelength sensitivity pigments (Van der Meer and Bowmaker, 1955). If the wavelength is an important demand, it is predictable that in darker habitats those receptors should be increased in size which are most sensitive to the available wavelength and the others should be decreased to keep sufficient resolution. Ultimately the ratio in numbers between both receptor types may change.

In the turbid water of Lake Victoria, long wavelengths penetrate deeper into the water layers than shorter ones (de Beer, 1989). From these facts it can be predicated that in cichlids living in deeper water layers, single cones should decrease in size, or even be absent to accommodate larger double cones. From a model study (Van der Meer, 1992), it appears that a change from a square pattern with larger single cones to a row pattern with small or without single cones becomes necessary to maintain an optimal packing.

These predictions have been tested by breeding fish under several artificial light conditions (Van der Meer, 1993) and by comparing fishes with more or less the same feeding behavior living in optically different surroundings (Van der Meer, 1991). Indeed, in fry of *Haplochromis sauvagei* deprived completely of short wavelength light, a rowlike pattern with enlarged double cones and smaller single cones develops, whereas square patterns of receptors are observed in wild specimens and in specimens raised in the laboratory under normal light conditions. Such a change in pattern can also be expected when growing fry migrate to deeper water. However, retinal development of different species (*Haplochromis pyrrhocephalus* and *H. argens*) raised under similar laboratory conditions showed a different retinal configuration from the start. In general, species of deeper habitats in Lake Victoria have row patterns, whereas most species living in shallow waters have a square pattern with distinct single cones (Fig. 4).

Goldschmidt et al. (1990), made an interesting comparison of four related *Haplochromis* species sampled from the same water column (14 m deep) in the Mwanza Gulf of Lake Victoria (Fig. 5). One species, *H. argens*, was a surface dweller to a depth of 4 m during day and night. *H. heusinkveldi* was found distributed over the whole column during

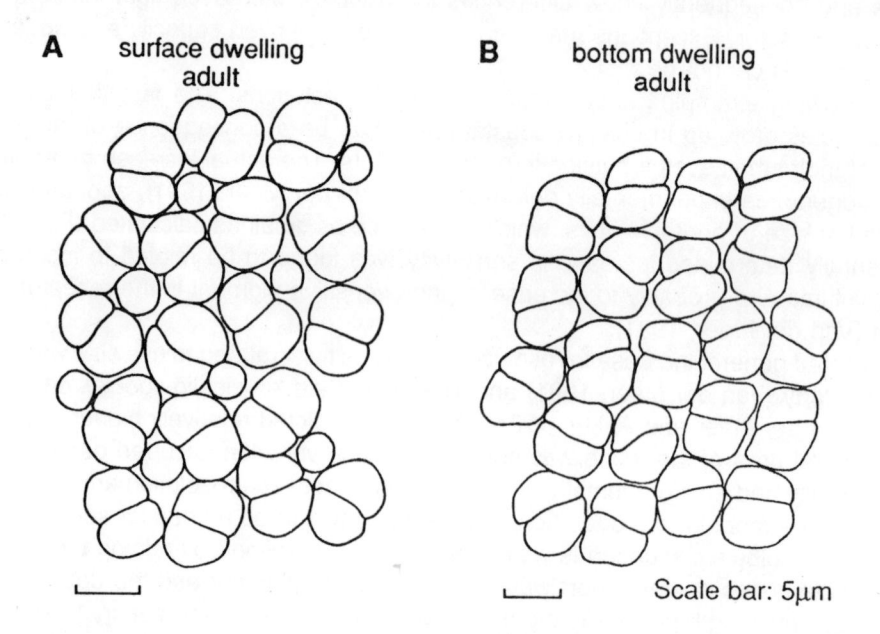

Fig. 4. A. Square pattern of double cones surrounding single cones in *H. argens*. B. Row pattern of double cones in *H. reginus*.

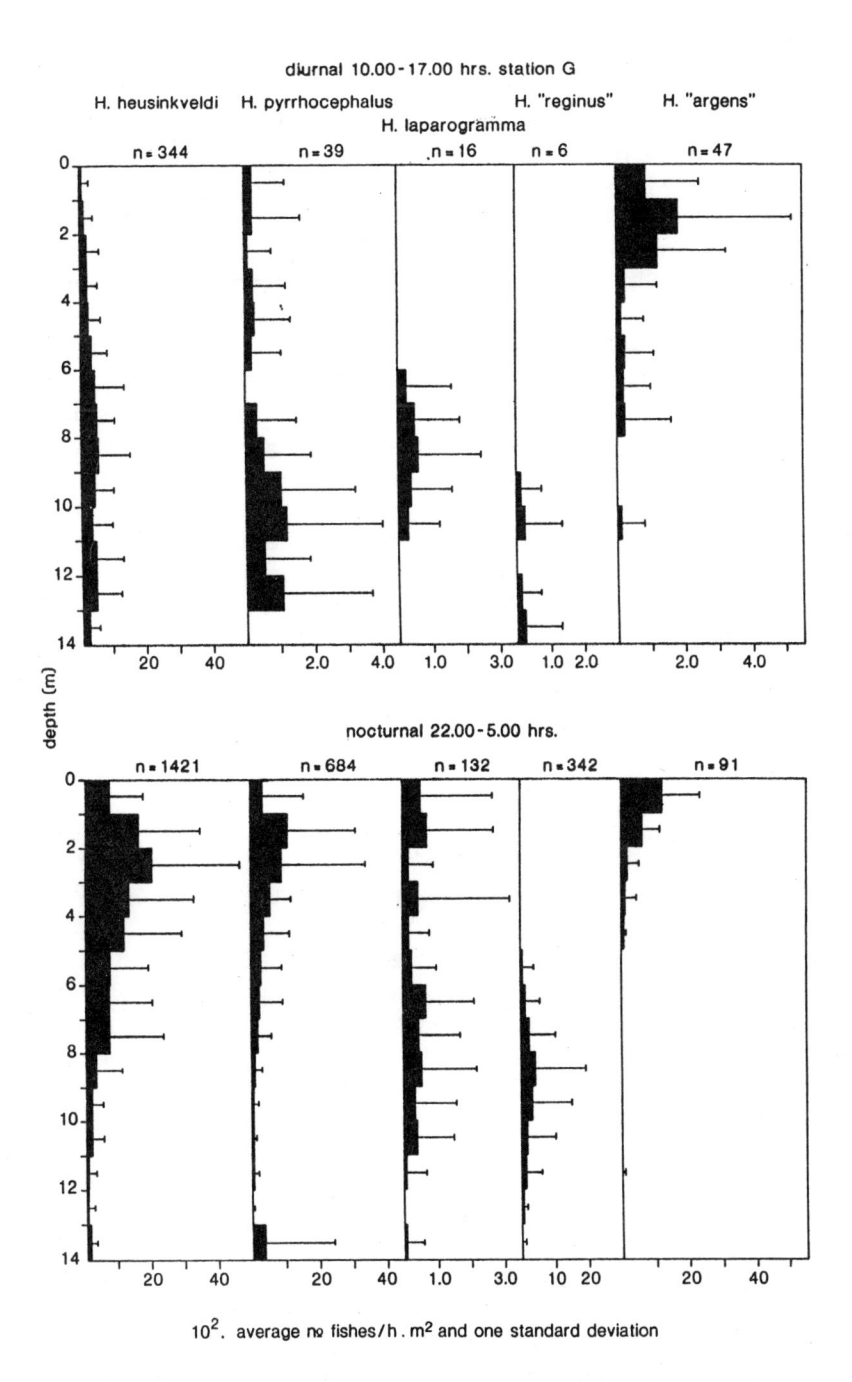

Fig. 5. Diurnal and nocturnal vertical distributions of five zooplanktivores. Average number/h.m² of net (and one standard deviation) (abscissa) for every meter in depth (ordinate) plotted. Averages based on 33 diurnal and 43 nocturnal observations, from 1981-1984, at station G on the Mwanza Gulf transect. n = number of individuals (after Goldschmidt et al., 1990).

the day and more concentrated in the upper 4 m during the night. *H. pyrrhocephalus* was found concentrated in the lower part of the column during the day and also moved to the upper layer during the night. *H. reginus* was a bottom dweller during day and night and migrated from deeper locations during its breeding period. The predicted change from a square pattern to a row is confirmed by comparing the retinas of the four species. *H. argens* has a square pattern with distinct single cones, *H. reginus* a row pattern without single cones (Fig. 4). The retinas of *H. heusinkveldi* and *H. pyrrhocephalus* have an intermediate configuration. The expected shift of balance between resolution and sensitivity in favor of sensitivity related to depth and consequently to decreased luminance of longer wavelength was not confirmed. Mean cone densities in specimens with equal eye size of these four species did not decrease with depth and no distinct decrease in mean cone size was found (Van der Meer, 1991). Only *H. pyrrhocephalus* has distinctly larger cones and a slighter resolution if mean values for whole retinas are compared. During the ontogeny of the four species cone density per visual angle as well as cone size increases with eye size. The intraretinal variation in the four species is much more difficult to interpret (Fig. 7).

Hence we are compelled to look for other ecological factors, e.g. food. From the investigation of Goldschmidt et al. (1990) of the four species, and the food distribution over the water column in the period 1981-1984, it appears that (Fig. 6):

a. During the day all four species fed on zooplankton, which was distributed over the whole column with, in the dry season, peak values in the top layers during both day and night. Besides zooplankton, *H. reginus* fed also on detritus and phytoplankton, mainly diatoms (*Melosira*), and *H. heusinkveldi* also ate bluegreen algae. Except for *H. reginus*, insect larvae were found only occasionally in the stomach of the other species; and

b. During the night *H. argens*, *H. pyrrhocephalus*, and *H. heusinkveldi* took *Chaoborus* larvae besides zooplankton. These larvae spent the daytime in the mud and migrated during the night to the top layer of about six meters. *H. reginus* shifted from zooplankton by day to phytoplankton (*Melosira*) at night.

Chaoborus larvae and zooplankton were present in the habitat of all four species. Only *H. reginus* did not eat *Chaoborus*. During *Chaoborus* abundance *H. argens*, *H. pyrrhocephalus*, and *H. heusinkveldi* selected IV-instar larvae in the top layer at night. At low density of this prey only *H. argens* showed this selectivity; the other species consumed smaller instars also (Goldschmidt, 1989). If this selection of food is performed by vision, then light conditions in the water would necessarily play a role also.

Light measurements by day in the spring of 1987 (de Beer, 1989) revealed that at a depth of about 2.5 m the light is about 5% of the intensity at the surface. Secchi disk readings in 1981 showed visibility at a depth of 2.5 m (Van Oijen et al., 1981) against 1.1 m in 1987 by de Beer. Therefore, the visibility of colors may have been twice as good before the Nile perch boom during the investigation of Goldschmidt et al. compared to those measured by de Beer. During the intermediate period dramatic changes related to the Nile perch boom occurred in Lake Victoria.(Witte et al., 1992). We can estimate that in the water column during the day red-orange light might have reached the bottom at 14 m and blue light at a depth of about 10 m. Upwelling and sidewelling light in the top layer must have been bluish-green and must have become more brownish and fainter in the deeper layers. Downwelling light would have been brighter than the others at any depth.

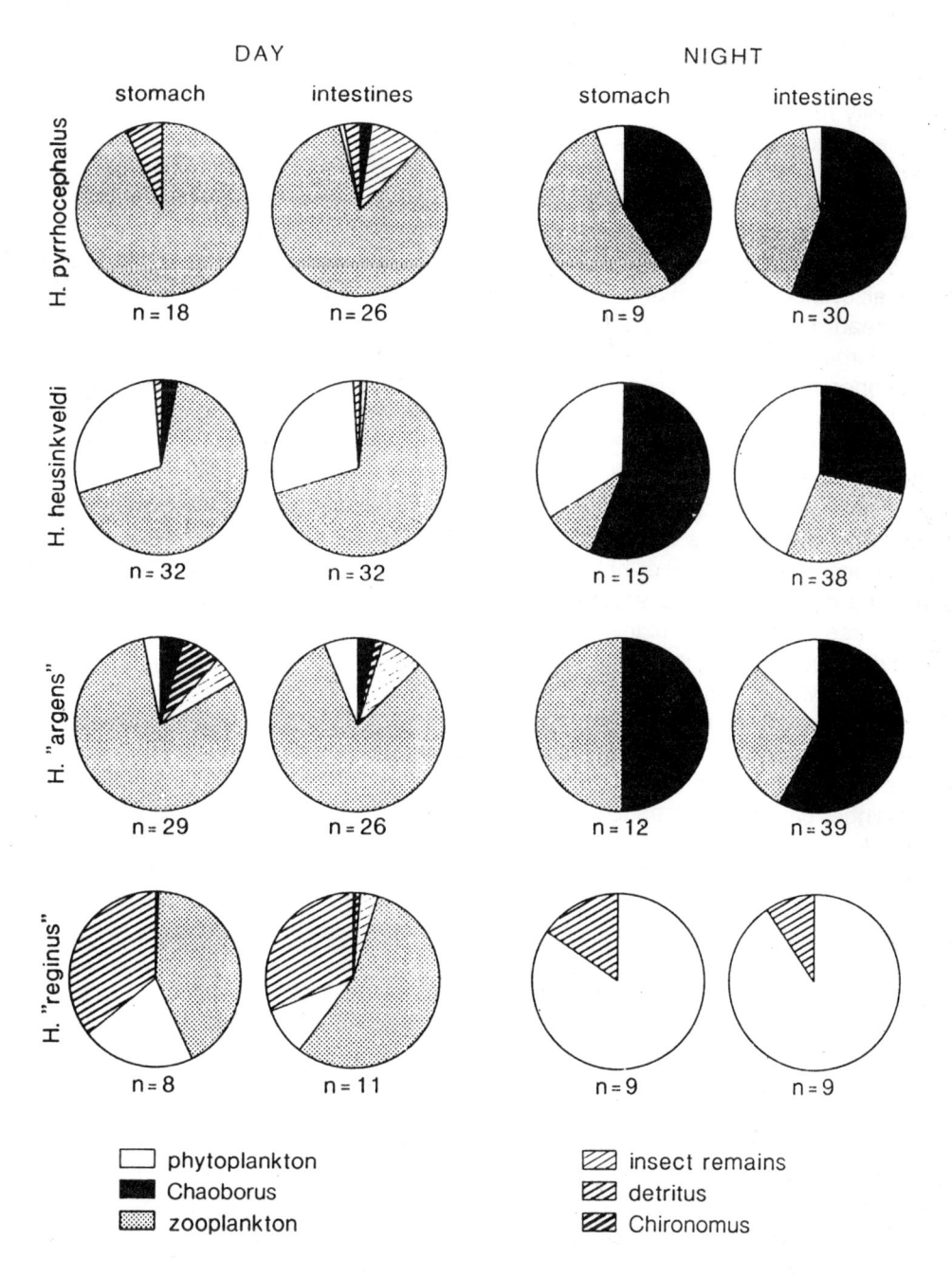

Fig. 6. Diet composition (mean volume percentage) by day and night of four coexisting haplochromine species at station G on the Mwanza Gulf transect throughout the year. Data on stomach and intestinal contents are presented separately. Empty stomachs and/or intestines were discarded. Prey categories; phytoplankton, zooplankton, *Chaoborus* larvae (instar 3 and 4, only rarely instar 2) and pupae and insect remains (which usually were digested too far to identify). Note that food categories can be subdivided: e.g. phytoplankton in *H. heusinkveldi* comprises mainly bluegreen algae and in *H. reginus* mainly diatoms (after Goldschmidt et al., 1990).

Especially in a darker environment, distinguishing food will mainly depend on visibility by contrast rather than by color differences, although wavelength has, of course, an effect on visibility related to photosensitive pigments present in the retina.

Speculation about the relation between the ecological and behavioral evidence mentioned above and collection of more details concerning structural differences, might reveal which aspects require much closer investigation.

All four species have a higher cone density in the temporal region compared to other parts of the retina. According to the model, this less sensitive part with a higher resolution allows all of them to detect small items of zooplankton and *Chaoborus* larvae just in front of the head. Even *H. reginus* ate zooplankton during the day although its double cones are not larger in this part of the retina compared to those in the other species. Therefore, these cones seem to be sensitive enough for prey detection at the very low light intensity near the bottom provided that the rod system or another sensory system is not involved. As in most teleosts, the rods can be put into the circuit by means of the retinomotor response, which screens off rods from light during the day and exposes them during the night. There seems to be an endogenous control of this circadian rhythm but the amount of light may overrule this control (Wagner, 1990). Because there is no evidence of what actually occurs, we suppose that during the day the cones of the photopic system, and during the night rods of the scotopic system are involved. It is surprising that *H. argens* as a surface dweller has a higher rod density and a higher rod convergence (ratio number of rods/number of ganglion cells) than *H. reginus* (Van der Meer, 1991). This might be the reason why *H. reginus* did not eat zooplankton and possibly missed the passing *Chaoborus* larvae during the night. If so, this species has to use another sensory system to secure food. This system may be taste. Hoogenboezem (pers. comm.), comparing trophic groups, found in cichlids foraging on tiny food items in muddy environments, a much higher density of tastebuds all over the epithelium of the buccal cavity than in other groups.

H. pyrrhocephalus and *H. heusinkveldi*, with rod densities between those of *H. argens* and *H. reginus*, may have to migrate upward during the night to get zooplankton and *Chaoborus* larvae. Rod convergence is significantly lower in *H. reginus*. This suggests surprisingly a higher resolution, but also a less sensitive system. However, we have taken into account that rod convergence may depend on the level of illuminance. The last aspect means that in decreasing light conditions more receptor units send their information to one ganglion cell and consequently the visual field of a ganglion cell increases, resulting in an increase in sensitivity and a decrease in resolution.

Most interesting are the differences between *H. pyrrhocephalus* and *H. heusinkveldi*. Both species fed on zooplankton in the whole column during the day and ate *Chaoborus* larvae and zooplankton when migrating to the top layer during the night. It may be that *H. pyrrhocephalus* with its more sensitive cone system detected planktonic items better when in the lower half of the column during the day. There the specimen concentration of *H. pyrrhocephalus* might be higher (Fig. 5), but the zooplankton concentration was higher in the top six meters (Goldschmidt et al., 1990). During the night both species, having similar rod density and rod convergence, fed on similar prey, although *H. heusinkveldi* ate more bluegreen algae. Thus the two species were not distinctly segregated in time and space, and the significant different photopic systems can hardly explain the slightly different food preferences.

G.C. Anker

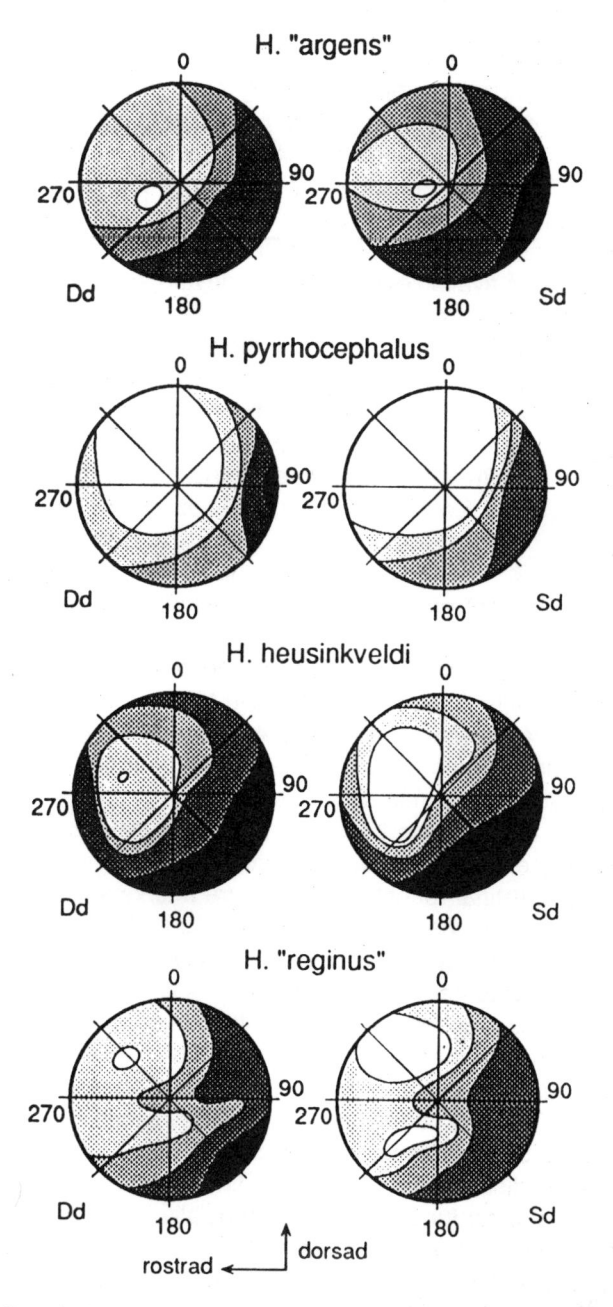

Fig. 7. Intraretinal distribution of the density (Dd) and size (Sd) of the double cones in four representatives of zooplanktivorous cichlids from station G with equal eye size (radius lens = 1.40 mm). From light to dark increase in density (Dd) and decrease in size (Sd) (after Van der Meer, Ch. 2, 1991).

Not only feeding but also other demands may be important, e.g. recognition of partners in schooling and sexual behavior, and hiding by protective coloration. To be inconspicuous in the top layer of the column an object has to be bluish-green at its upper side and silvery at its lower side in a countershading fashion. In deeper layers, brownish-green at the upper side will be inconspicuous and a higher reddish-brown or even white may be necessary for reflection of enough upwelling light to mimic downwelling light. Females appeared to be rather inconspicuous which would not set a demand on the eyes. However, bright coloration of the body with black spots or bars of sexually active females might enhance the contrast with background light. This plays a role in partner recognition and would be related to sensitivity to the wavelengths present. Only *H. argens* might profit from the broad spectrum of available light. Its well-developed single cones with sensitive pigment of 455 nm and double cones of which the partners have 533 nm and 567 nm sensitive pigments respectively, could probably mean color vision. Next to these peaks of absorbance spectra, Van der Meer and Bowmaker (1995) found in *H. pyrrhocephalus* 462 nm for single cones and 539 nm and 594 nm for the partners of a double cone. This shows also a shift to longer wavelengths of the photopic sensitive pigments that may be related to an increased sensitivity for long wavelengths, penetrating deeper into the water column. Absorbance spectra for the other two species are not known. Only *H. argens* has distinct single cones. They are absent in *H. reginus* and scarce in the other species. The role of single cones in a pattern is not well understood but since they are generally present in fishes living in light environments it is suggested that they are important for color vision.

The inductive approach in the investigation of the four species is far from satisfactory but illustrates nevertheless how many aspects have to be incorporated to develop deductive models which can predict size, shape, and structure of fish eyes. Without quantitative data on the visual capacity which features as a boundary condition in the model, related to structural configurations of retinas, it is impossible to predict the structural composition of retinas and how the increased opportunities of larger eyes is exploited even if all behavioral and environmental aspects are quantitatively known.

If it is supposed that retinal receptor patterns are mainly adapted to vision for finding food during day and (or) night, this adaptation also includes the proportion between cones and rods in the available space of the retina. Improving simultaneous resolution and sensitivity implies an enlargement of the eye. With optimal packing this happens when one of these functions has reached its functional limit in the balance between them if the other cannot be diminished further. If long-distance vision implies less image-forming light and requires a more sensitive system, bigger eyes would be necessary to accommodate this, e.g. larger cones. This may be the case in predatory fishes and hunters, mostly larger species among cichlids. Small fishes often detect small items that are picked one or a few at a time, which must be distinguished from other fine granular structures. Then resolution may have to be at its highest functional limit, especially in the caudal part of the retina. Small fishes usually have relatively big eyes.

There is much information about eye sizes; however, there is very little information about the quantitative demands. Therefore, modelling these demands to predict the size is not possible at this moment.

For example, varying light intensities in the habitat of a species may be a demand for the limits between which the functional receptor convergence has to change to vary sensitivity and resolution. This might not only be related to the structural receptor convergence, but also to the available bipolar and amacrine cells in the internuclear layer, which are involved in signal transfer to the ganglion cells. Although hypotheses have been formulated, this mechanism is not well understood.

Plasticity of the retinal cone configuration as a response to changed environmental condition was shown in *H. sauvagei* by Van der Meer (1993). There are strong indications that such a retinal plasticity exists in *H. pyrrhocephalus* and *H. argens*. This would also hold for the other neurons in the retina. However, the limits of potential plasticity differ among species, which poses a difficulty in deduction.

MORPHOLOGICAL CONNECTION OF THE EYE WITH OTHER FUNCTIONAL COMPONENTS

The case study of the eye has taught us that according to optical and movement demands eye shape cannot be affected very much and thus will not be very plastic in ontogeny and phylogeny (eye shape is dominant in the architecture of the head). The size of the eye differs among species and is only partly related to the size of the animal. Size is constrained between the demand of necessary vision, i.e., resolution and sensitivity, and the available space related to the size of the animal. The result is that we observe relatively big eyes in small animals and only for specific functions in big animals. The position of the eye varies very little, partly due to strong visual demands, partly to the interconnection with other functional components (Van der Klaauw, 1950; Barel, 1984; Dullemeijer, 1974). These circumstances leave very little room for plasticity and late ontogenetical changes, once the eye is formed in early ontogeny (Otten, 1981). It seems that *mutatis mutandis* the same holds in phylogeny.

The only property which allows a great deal of flexibility is the structure of the retina, a flexibility which occurs within the limits set by the size of the eye. In this property we find the potential adaptation to a great variety of ecological circumstances, as shown in the section above.

From an architectonic point of view the connection of the eye with the brain gives little reason for special discussion here. It follows the normal pattern. Of the special aspects of this connection, which would distinguish various fishes (and not at all cichlids), nothing is known. Even about the internal connections in the layers inside the retina there is no information on particular constructions that could be connected to specific adaptations on the species level except for the few data mentioned earlier. This holds also for the vascular system. Blood from the first gill arch passes through a well-developed pseudobranch in the roof of the buccal cavity and a rete chorioidea inside the eyeball. The rete regulates the oxygen pressure in the capillaries of the tela chorioidea. This pressure is higher than that in the blood of the efferent branchial vessels (Wittenberg and Wittenberg, 1962). There are no capillaries inside the retina or at the vitreous side of it. Also the eye muscles do not warrant description and discussion. They do not differ in the various cichlids. In general, the major effect of the eye, in particular its shape and position, on skeletal elements of teleosts is well known and very apparent. They are not discussed here, as they are well covered in various other publications (Van der Klaauw, 1948–1952; Dutta, 1968; Sarkar, 1960; Dullemeijer, 1974).

The foregoing induces the hypothesis that in cichlids only eye size will be affected by surrounding structures of functional components and limit the possible adaptation of the retina (resolution versus sensitivity). At first glance, the position of structural elements in a fish head gives the impression of dominance in the shape of the eye (Fig. 8). Therefore, increase in eye size would affect the structure of surrounding functional components and their function. Barel (1984) developed a model for the orbital part of the head in which three major factors were varied (Fig. 9). These factors were head profile, eye size, and the size of the musculus adductor mandibulae (mAM) associated with locomotion (streamlining), vision, and biting force respectively. This model leads to a number of combinations which illustrate a relative increase or decrease in functions. So, from a supposed change of a function, the possible redistributions of space in this part of the head can be traced and the concomitant functional changes of the other functional components formulated. Also note that in Figure 9 eye shape is varied (elliptic eyes), which implies differences inside the eye, especially of the accommodation mechanism. In such models, functional demands lead to constructional demands which predict configurations reflected by different trophic groups with different functional capacities (Fig. 10).

In general, there are small positional and shape features which modulate or minimize strong antagonistic effects. First the eye can be elliptical in a lateral view and yet keep its essential function, which is mainly related to medial depth. Then such an eye can be tilted and slightly shifted rostrally and caudally. Second, a spherical eye can bulge out

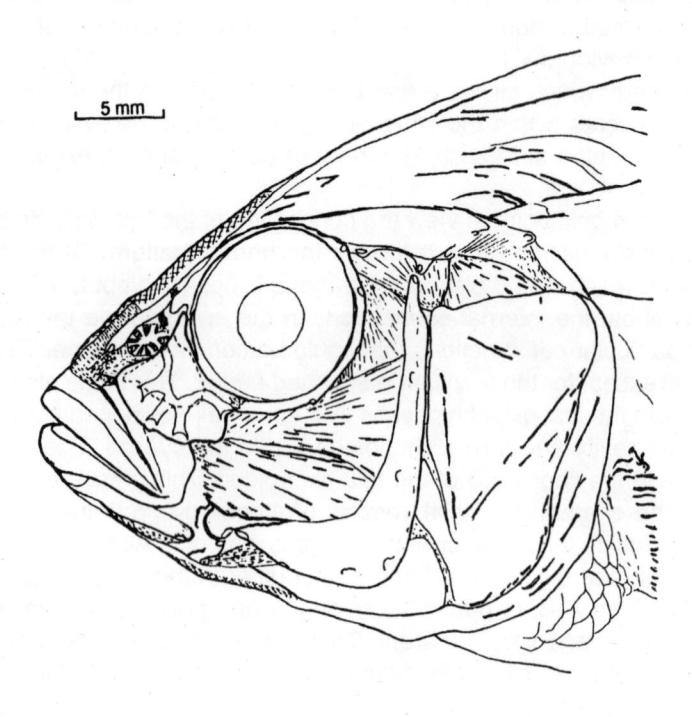

Fig. 8. Lateral view of the head of *H. elegans* (skin removed).

G.C. Anker

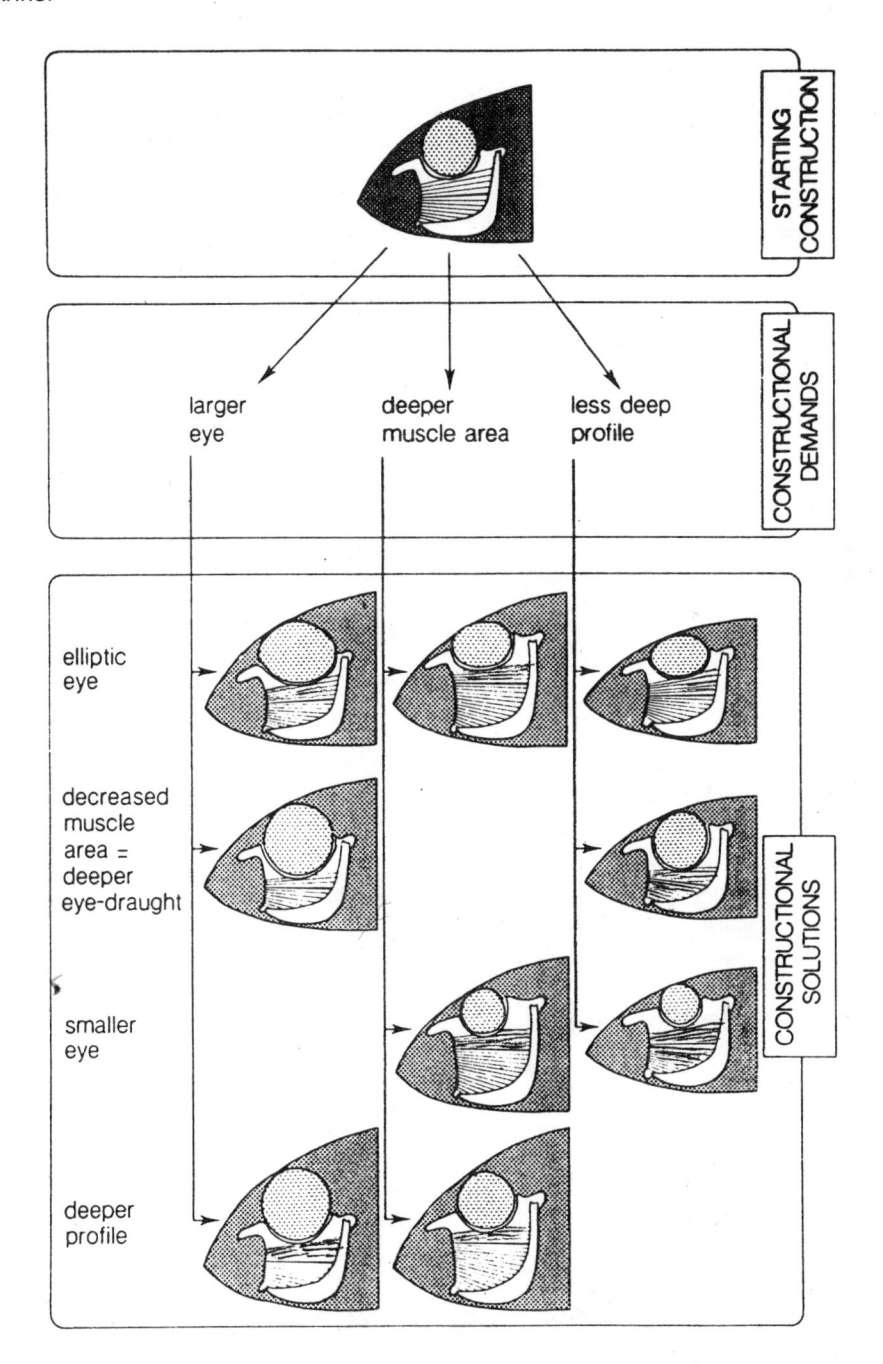

Fig. 9. Schematic representation of geometric transformations in response to three constructional demands in the constructional component consisting of head profile, m. adductor mandibulae, and eye (after Barel, 1984).

somewhat, affecting streamlining. Third, the outer shape can give way somewhat, making more space for all elements at the cost of fast swimming. Of course, the size of the

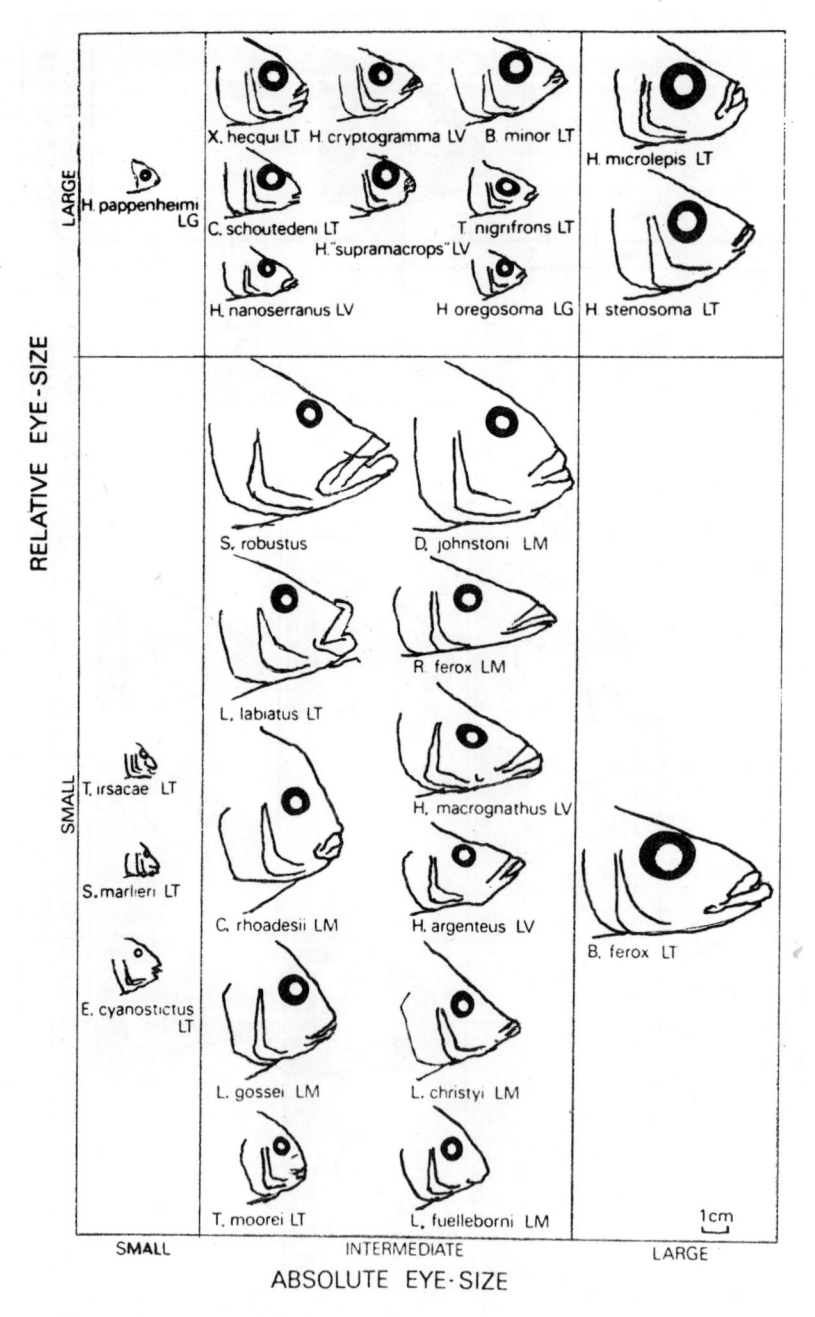

Fig. 10. Illustration of the diversity in absolute and relative eye size among cichlids. Relatively small and large eyes can be found in every size class. All figures are from adult specimens and drawn at the same magnification.
Legend: B—*Bathybates*; C—*Cyathopharynx* or *Chilotilapia* (*rhodesi*); D—*Decimodus*; E—*Erethmodes*; H—*Haplochromis* or *Haplotaxodon* (*microlepis*) or *Hemibates* (*stenosoma*); L—*Lethrinops* or *Labeotropheus* (*fuelleborni*) or *Lobochilotes* (*elabiatus*); R—*Rhamphochromis*; S—*Serranochromis* or *Spathodus* (*marlieri*); T—*Tanganicodus* or *Tropheus* (*moorii*) or *Trematocara* (*nigrifrons*); X—*Xenochromis* (=*Perissodus*).

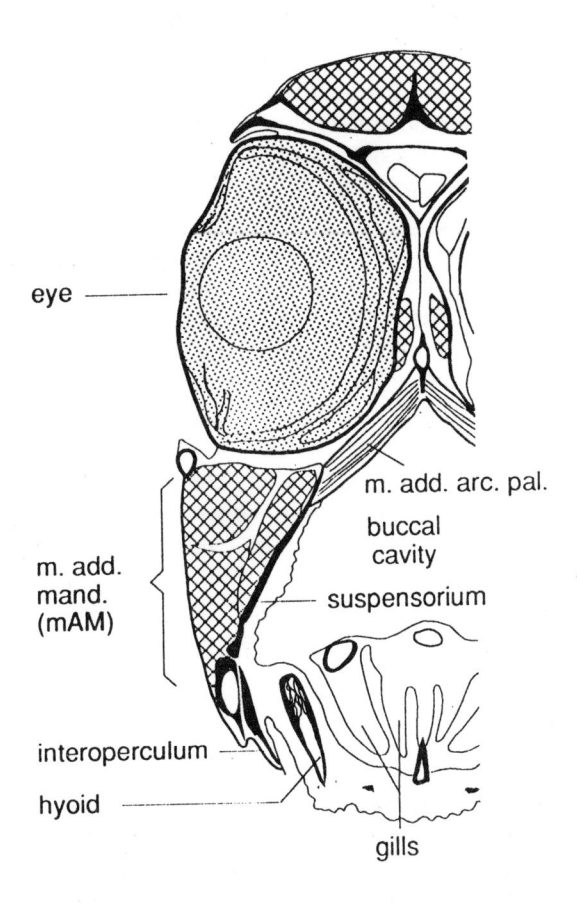

Fig. 11. Transverse section of the head through the center of the eye (after Barel, 1993).

adductor muscle is related to the feeding mechanism but the muscle in turn affects the skeletal elements and consequently other constructional components, such as the branchial basket and the oral cavity (Fig. 11).

Shape dominance of a spherical eye does indeed exist but is not absolute and has a distinct limitation. The major reason for this limitation is the functional significance of the functional components integrated and cooperating within a limited space (Barel, 1993). Space also shows a slight variation. This entire, rather complex system implies that the animal could occupy another habitat, or fulfill another niche. How the transformation takes place is the main topic of modern transformation morphology (Dullemeijer, 1974; Zweers, 1991; Barel, 1993). We can resolve this problem into three operational questions: 1. What must primarily be transformed to enable an animal to live in another way? 2. What are the consequences of the transformation of one or a few members of the system? 3. What are the limitations of the transformation due to the integration? As a corollary of the answers we can expect to find a number of suddenly appearing, potentially new opportunities for the animal (Zweers, 1991) and a number of trade-off features (Barel et al., 1989).

Suppose we have a fish feeding by means of sucking in small plankton. This fish will probably have moderately sized eyes and adductor muscles, small teeth, a streamlined body shape, and an appropriate buccal cavity for sucking, thus a "common" fish. Suppose now that we want to transform this fish into an effective biter of somewhat larger pieces of food from the substrate by means of the oral jaw apparatus. The model for such a fish would likely contain larger adductor muscles and for the needed stronger biting force, bigger jaws with a more robust dentition. The eye will not differ much from the "normal" eye in which only the retinal structure may show independent changes. A larger adductor muscle could affect the streamline profile if fast swimming is not necessary. Otherwise an inward extension of the adductor muscle, if allowed by the sucking capacity, is an option to increase force. However, in small fishes that bite small items from rocks and plants, there is little or no increase in the muscles and hardly any reinforcement of skeletal parts and dentition. This may be due to a lower mechanical demand for biting than expected, and as Otten (1983) showed, increase in biting force is obtained by changing length proportions of bony bars involved in the biting mechanism. He showed that on changing the positions of 37 anatomical points of connections in a three-dimensional model of the jaw apparatus of a "normal" fish, only seven changes resulted in a considerable increase in biting force. Changes in position of the other connections were neutral to biting force. From the seven so-called "hot" points changing the biting force, six were found at their predicted positions in a biter, *H. nigricans*.

In a model of the pharyngeal jaws, Galis (1992) demonstrated that only a change of the working lines of a muscle enhanced the output force of this biting apparatus. Such changes of a function by small changes in proportions and (or) positions of the bars are very limited. Further increase requires more space to enlarge the involved muscles and reinforce the bars. Then the problem arises as to how the needed enlargement could be accommodated in a restricted space, or otherwise what would happen to the shape and volume of the space. These kinds of questions are discussed by Barel (1993) to whom we refer for the theoretical model and methodology and from whom we borrow the following example.

Biting capacities are improved by increasing the cross section of the musculus adductor mandibulae (mAM) (Fig. 11). Indeed the cross sections of mAM in biters are generally greater. The shape varies from triangular in suckers to squarish in biters. It is thus assumed that increasing the cross section is a structural prerequisite for adding the biting function to the oral jaw apparatus of a sucker. The enlargement can be accommodated in a pattern of a sucker by increasing width and depth of the mAM.

a. A larger width can be achieved by an inward and/or outward extension of the muscle. An *inward* extension will diminish the area between the medial face of the left and right mAM. Dissections of many cichlid species suggest that such an inward extension is indeed used. In biters compared to suckers, the m. adductor arcus palatini is shorter, and in specimens with maximally adduced suspensoria the rostral section of the branchial basket takes a lower position (as if forced away from the intruding medial face of the suspensorial muscle area). The shifts of the branchial basket and hyoid have several structural and functional consequences for the gills (Barel, 1983) and the expansion apparatus involved in sucking (Barel and de Visser, in press). Most lacustrine cichlids are mouth brooders and diminishing their buccal cavities may influence per batch fecundity, an important parameter of a species life history strategy. Outward extension of the muscle affects the head profile and several hydrodynamic parameters for swimming propulsion determined by shape and size of the head. An increase in head width of biters and a significant positive correlation between mAM width and head width is demonstrated by Barel (1983) and Barel and de Visser (in press). Laterad extension of the mAM has several less conspicuous consequences also, e.g. another negative effect on the efficiency of the suction apparatus (Barel and de Visser, in press), which is also affected by another requirement for biting, viz. diminished protrusion.

b. A larger depth of the mAM may have any of the following three structural effects (see Figs. 9, 12, and Barel, 1984): 1) decreasing eye size, 2) dorsal shift of the eye, thereby affecting the dorsal head profile, and 3) ventrad shift of the ventral head profile. The hydrodynamic consequences of affecting the head profile would be similar to those mentioned above. Dorsad extension of the mAM depth could be achieved by diminishing eye size or shifting the eye dorsad, thus not affecting the suspensorial outline (Fig. 12); the required space is obtained by diminishing the suspensorial orbita (Barel, 1984). Ventral extension of the mAM depth is accompanied by ventral shift of the suspensorial outline, which implies an increase in rostral and caudal outline of the suspensorium, because the trapezoid shape of the suspensorium with the opercular-suspensorial articulation dorsally on its caudal border and the quadrato-mandibular articulation in its ventrorostral corner seem to be a part of the basic anatomical design of cichlids. Thus, ventral extension of the mAM means a ventral shift of the interoperculum and lengthening of the caudal bar of the opercular four-bar system (Fig. 12, SO) for opening of the mouth, which possibly affects the kinematic transmission coefficient (Anker, 1986). So ventral extension of the mAM affects another functional component. A rostrad-caudad extension of the muscle area does not affect the transmission coefficient (Dullemeijer and Barel, 1977) but neither does it contribute to an increase in the transverse section related to force.

Searching for hot points and neutral ones is useful for predicting and analyzing transformation possibilities and for understanding the differences in the enormous variety of species once found in Lake Victoria. When a functionally required transformation of a structure should be achieved with little effect on other structures or with little cost of material, only those features should change to which the function to be changed is sensitive, and, if possible, only those features in other functional components should be changed that are neutral to the function. Such features provide the organism with structural plasticity (Barel, in press).

When a mathematical relation between a biological function (F) and the structural features (a, b, c) executing this function is known $(F = f(a, b, c))$, an advisable approach would be to calculate the sensitivity of each function to changes of structural features separately $(dF'/da, dF''/db, dF'''/dc)$ (Barel, 1993), since the reverse to predict structural changes from a change of a biological function $(d(a, b, c)/dF)$ appears too difficult although the ultimate goal in functional morphology. The sensitivity of a function to one structural change (dF/dx) will be high for changes of hot points and low for neutral ones. Such values will also depend on more functional components because they are constructionally mutually constraining. Moreover, structural changes ultimately have to be evaluated for their effects on adaptation and fitness and may then be called "hot" structures to fitness. Whereas various functional components differ qualitatively, it is hard to estimate

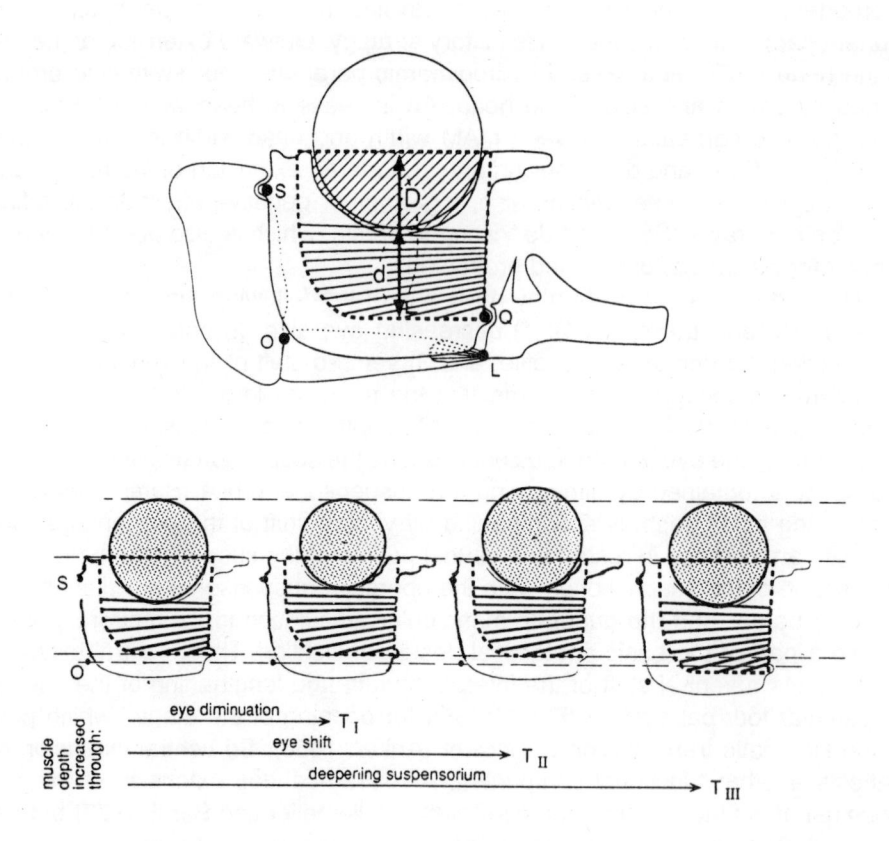

Fig. 12. Schematic representations of eye and m. adductor mandibulae in the outline of the suspensorium with the opercular four bar system indicated (SOLQ) and of effects of increased muscle depth (after Barel, 1993).

their total and integrated effect for adaptation and inclusive fitness. In spite of the enormous variety of species, little variability is found inside populations of species in the same habitat over a limited period. Certain "forms" stay rather stable. Three reasons may be advanced. First, the ecological factors did not change and the functional demands which proceed from them are fulfilled in balance. Second, the changes in functional demands to which the functional components cannot respond properly in one generation because of constructural or genetic constraints. Third, the absence of potential plasticity of tissues. On the other hand, distinct evidence exists that structural changes can be induced experimentally during one generation, following changes in functional demands, e.g. change in the ascending process of the premaxilla in piscivorous cichlids (Witte, 1984), in retinal receptor pattern (Van der Meer, 1993), in shape and dentition of pharyngeal jaws in molluscivorous pharyngeal crushers (Greenwood, 1965; Hoogerhoud, 1986). Such a phenotypic plasticity seems to be genetically limited. Plasticity can also be derived from differences found in populations of such species living in different habitats of Lake Victoria (Hoogerhoud, 1984). Plasticity, phenotypic and/or genetic, is obvious from the appearance of dwarfs of several species during the Nile perch boom (Goldschmidt and Witte, 1990; Witte et al., 1990). The decrease in size with consequences for the proportion of the composing parts and even for the relative fecundity (Wanink, 1991), may be related to high predation but how it affects this decrease is not well understood.

The strength of deductive explanation in functional morphology and in morphological transformations has been shown. It is of great help in understanding ontogeny and phylogeny. However, it is also clear that we were forced to inductive reasoning due to lack of information concerning functional demands and sensitivity of function to changes in structural features. Both methods were used by us in iterative sequences. We followed the procedure of inductive collection and selection of data, of deductive as well as inductive explanation to establish the relation between form and function, and estimation of the effects of transformations for adaptation. The aim was to investigate the value of such adaptations for the inclusive fitness of the organism.

References

Anker, G.G. 1986. The morphology of joints and ligaments in the head of a generalized *Haplochromis* species: *H. elegans* Trewavas 1933 (Teleostei, Cichlidae). I. The infraorbital apparatus and the suspensorial apparatus. *Neth. J. Zool*. 36:498-530.

Barel, C.D.N. 1983. Towards a constructional morphology of cichlid fishes. *Neth. J. Zool*. 33:357-424.

Barel, C.D.N. 1984. Form-relations in the context of constructional morphology: the eye and suspensorium of lacustrine Cichlidae (Pisces, Teleostei). With a discussion on the implications of phylogenetic and allometric form interpretations. *Neth. J. Zool*. 34:439-502.

Barel, C.D.N. 1993. Concepts of an architectonic approach to transformation morphology. *Acta Biotheoretica* 41:345-381.

Barel, C.D.N., G.C. Anker, F. Witte, R.J.C. Hoogerhoud and T. Goldschmidt. 1989. Constructional constraint and its ecomorphological implications. *Acta Morphol. Neerl.-Scand*. 27:83-109.

Barel, C.D.N. and J. de Visser. Architectonic constraints on the hyoid's optimal starting position for suction feeding of fish. *J. Morphol*. (In press).

de Beer, M. 1989. Light measurements in Lake Victoria, Tanzania. *Ann. Mus. Roy. Afr. Centr. Sc. Zool*. 257:57-60.

Dullemeijer, P. 1974. *Concepts and Approaches in Animal Morphology*. Van Gorcum, Assen.

Dullemeijer, P. and C.D.N. Barel. 1977. Functional morphology and evolution. *In* M.K. Hecht, P.C. Goody, and B.M. Hecht (eds.), *Major Patterns in Vertebrate Evolution*, series A, vol. 14:83-117. NATO Advanced Study Institute. Plenum Press, New York-London.

Dutta, H.M. 1968. *Functional morphology of the head of Anabas testudineus (Bloch)*. Thesis, Leiden, 1-146.

Galis, F. 1992. A model for biting in the pharyngeal jaw of a cichlid fish: *Haplochromis piceatus*. *J. Theor. Biol*. 155:343-368.

Goldschmidt, T. 1989. Reproductive strategies, subtrophic niche differentiation and the role of competition for food in haplochromide cichlids (Pisces) from Lake Victoria, Tanzania. *Ann. Mus. Roy. Afr. Centr. Sc. Zool*. 257:119-132.

Goldschmidt, T. and F. Witte. 1990. Reproductive strategies of zooplanktivorous haplochromine cichlids (Pisces) from Lake Victoria before the Nile perch boom. *Oikos* 58:356-386.

Goldschmidt, T., F. Witte and J. de Visser. 1990. Ecological segregation in zooplanktivorous haplochromine species (Pisces: Cichlidae) from Lake Victoria. *Oikos* 58:343-335.

Greenwood, P.H. 1965. Environmental effects on the pharyngeal mill of a cichlid fish, *Astatoreochromis alluaudi*, and their taxonomic implications. *Proc. Linn. Soc. Lond*. 176:1-10.

Hoogerhoud, R.J.C. 1984. A taxonomic reconsideration of the haplochromine genera *Gaurochromis* Greenwood, 1980 and *Labrochromis* Regan, 1920 (Pisces, Cichlidae). *Neth. J. Zool*. 34:539-565.

Hoogerhoud, R.J.C. 1986. Taxonomic and ecological aspects of morphological plasticity in molluscivorous haplochromines. *Ann. Mus. Roy. Afr. Centr. Sc. Zool*. 251:131-134.

Otten, E. 1981. Vision during growth of a generalized *Haplochromis* species, *H. elegans* Trewavas 1933 (Pisces, Cichlidae). *Neth. J. Zool*. 31:650-700.

Otten, E. 1983. The jaw mechanism during growth of a generalized *Haplochromis* species: *H. elegans* Trewavas 1933 (Pisces, Cichlidae). *Neth. J. Zool.* 33:31–48.

Sarkar, S. 1960. *A study of functional morphology of the head of an Indian puffer fish, Spaeroides oblongus (Bloch)*. Thesis, Leiden, 1–119.

Van der Klaauw, C.J. 1950. Architectuur van de schedel. *Verslagen Kon. Ned. Akad. Wet. Amst.* 59(3):1–5.

Van der Klaauw, C.J. 1948-1952. Size and position of functional components of the skull. A contribution to the knowledge of the architecture of the skull, based on data in the literature. *Arch. Neerl. Zool.* 9:1–558.

Van der Meer, H.J. 1991. *Ecomorphology of photoreception in haplochromine cichlid fishes*. Thesis, Leiden, 1–13.

Van der Meer, H.J. 1992. Constructional morphology of photoreceptor patterns in percomorph fish. *Acta Biotheoretica* 40:50–85.

Van der Meer, H.J. 1993. Light-induced modulation of retinal development in the cichlid fish *Haplochromis sauvagei* (Pfeffer, 1896). *Zool. Linn. Soc.* 108:271–285.

Van der Meer, H.J. and G.C. Anker. 1984. Retinal resolving power and sensitivity of the photopic system in seven haplochromine species (Teleostei, Cichlidae). *Neth. J. Zool.* 34:197–207.

Van der Meer, H.J. and G.C. Anker. 1986. The influence of light deprivation on the development of the eye and retina in the cichlid fish *Sarotherodon mossambicus* (Teleostei). *Neth. J. Zool.* 36:480–498.

Van der Meer, H.J. and J.K. Bowmaker. 1995. Interspecific variation of photoreceptors in four co-existing Haplochromine Cichlid Fishes. *Brain Bchav. Evol.* 45:232–240.

Van Oijen, M.J.P., F. Witte, and E.L.M. Witte-Maas. 1981. An introduction to ecological and taxonomic investigations on the haplochromine cichlids from the Mwanza Gulf of Lake Victoria. *Neth. J. Zool.* 31:149–174.

Wagner, H.J. 1990. Retinal structures in fishes. *In* R.H. Douglas and M.B.A. Djamgos (eds.), *The Visual System of Fish*, pp. 109-157. Chapman & Hall, London.

Wanink, J.J. 1991. Survival in a perturbated environment: the effects of Nile perch introduction on zooplanktivorous fish community in Lake Victoria. *In* O. Ravera (ed.), *Terrestrial and Aquatic Ecosystems: Perturbation and Recovery*, pp. 271–275. Ellis Horwood Ltd., Chichester, England.

Witte, F. 1984. Consistency and functional significance of morphological differences between wild-caught and domestic *Haplochromis squamipinnis* (Pisces, Cichlidae). *Neth. J. Zool.* 34:596–612.

Witte, F., C.D.N. Barel and R.J.C. Hoogerhoud. 1990. Phenotypic plasticity of anatomical structures and its ecomorphological significance. *Neth. J. Zool.* 40:278–298.

Witte, F., T. Goldschmidt, P.C. Goudswaard, W. Ligtvoet, M.J.P. van Oijen, and J.H. Wanink. 1992. Species extinction and concomitant ecological changes in Lake Victoria. *Neth. J. Zool.* 42:214–232.

Wittenberg, J.B. and B.A. Wittenberg. 1962. Active secretion of oxygen into the eye of fish. *Nature* 194:106–107.

Zweers, G. 1991. Transformation of the avian feeding mechanisms: a deductive method. *Acta. Biotheoretica* 39:15–36.

2

Catfish Morphology: A Reappraisal

C.B.L. Srivastava

In the last decade catfish morphology has yielded some features which call for a reappraisal of their character-states, particularly from an evolutionary point of view. These features have been either wholly unseen or seen but little understood. Indeed, catfish may be endowed with many more structures yet to be understood. Believing that "no structure is without function," understanding the new structures is rewarding for better evaluation of not only the biology, but also of the evolutionary relationship of the catfish group—a subject of hot debate in recent years. The importance of "soft anatomy" has been greatly enhanced in the cladistic approach to phyletic schemes of classification of extant fishes. The cladistic approach is particularly important for the teleostean forms for which fossil records are scant. For fish, as for higher vertebrates, old views on phylogeny have been replaced by new ones, because of cladistic-based systematics. At the end of the twentieth century we stand on the threshold of a new vista in the study of fish morphology, physiology and behavior. Under the cladistic concept of fish classification, the character-states no longer need be given the same weight for consideration of phylogenetic relationships. The change in outlook is significant. It requires reevaluation of what is already known, reinterpreting or redefining it, apart from investigations into aspects of fish morphology, physiology, and behavior that still elude understanding. This paper seeks to make a contribution in this direction, which may serve as a model for other fish groups.

The commonly considered autapomorphies of the catfish group (see Day, 1889; Nikolskii, 1954; Bertin and Arambourg, 1958; Greenwood et al., 1966; Migdalski and Fichter, 1976; Nelson, 1984; Datta Munshi and Srivastava, 1988) may be enumerated as follows:

1. Eyes small and ganoid type.
2. Maxilla rudimentary and supporting a barbel.
3. Skin naked, with scales absent or reduced, or with scutes.
4. Mouth nonprotrusible and broad.
5. Jaws with teeth.
6. Dermal bones sculptured.
7. Pineal foramen present.

Dedicated to the memory of Dr. Thomas Szabo who died 28 November 1993.

8. Pectoral girdle massive and attached to skull.
9. Palatine, pterygoid, prevomer, and vomer toothed.
10. Spines on dorsal and pectoral fins.
11. Branchiostegals numbering 4 to 17.
12. Parietals, symplectic, subopercular, epipleurals and epineurals absent (parietals fused to supraoccipitals).
13. Intermuscular bones absent.
14. Posttemporal fused to supracleithrum.
15. Preopercular and interopercular small.
16. Mesopterygoid reduced.
17. Single dorsal fin.
18. Adipose fin present.
19. Pseudobranch absent.
20. Sensory barbels (1 to 4 pairs) important in food detection.
21. Airbladder subdivided.

The character-states which the catfish group shares (synapomorphies) with other bony fish groups are given below:

A. Characters shared with other ostariophysian forms.
   1. Otophysic connection through Weberian ossicles.
   2. Well-toothed predatory and vegetarian types.
   3. Predominantly freshwater.
B. Characters shared with osteoglossomorphan forms
   1. Pterygoid toothed.
   2. Subopercular reduced.
   3. Somatic electric organs.
   4. Insectivorous or piscivorous.
C. Characters shared with elopomorphan and clupeomorphan forms
   1. Opercular series reduced.
   2. Scales caduceous.
   3. Pterygoid teeth.
   4. Otophysic connection.
D. Characters shared with salmonids
   1. Exclusion of the maxilla from the gape.
   2. Branchiostegals few or numerous.
   3. Adipose fin present.
E. Characters shared with holosteans
   1. Air-breathing habit.
   2. Jaws adapted for catching fish.
   3. Crushing teeth.
F. Characters shared with chondrosteans
   1. Pineal foramen.
   2. Pectoral girdle united with skull.
   3. Thick sculptured skull dermal bones.

Teleostei are considered advanced bony fishes, very competitive and successful. They comprise the bulk (96%) of all living fishes, estimated to be around 20,000 species. Among them, however, some groups are considered primitive, close to or identical with Eocene,

Paleocene or Cretaceous forms, viz., elopids, salmonids, clupeids and osteoglossids (see Greenwood et al., 1966); the elopids and the salmonids have been thought to serve as the morphotype for the teleostean radiation. Although these primitive forms retain features of holostean grade (pholidophorids), they are a mosaic of primitive and derived (specialized) characters.

Ostariophysian forms have been considered by some ichthyologists (Weitzman, 1964) to be far older than previously believed. Ostariophysian lineages can be traced back to the Mesozoic (Paleocene) of all continents and are very close to salmonids among living forms (Greenwood et al., 1966). However, due to the presence of one specialized character (Weberian ossicles) they were cut off from the main course of early teleostean evolution.

Extant catfishes, widely distributed (N. America, Europe, Japan, Asia and Australia), have been considered primitive among ostariophysian forms by some ichthyologists, recalling similarities with even lower actinopterygians (Bertin and Arambourg, 1958). While ostariophysian forms comprise 23% of all known living fish (72% of all living freshwater fishes), the catfish group makes up 50% of the living ostariophysian species, with 2,211 species belonging to 400 genera of 31 families (Nelson, 1984). These are mostly freshwater. Views differ on the phylogenetic relationships of catfishes. Nikolskii (1954), Berg (1955) and Verigina (1969) believed that catfishes are a sister group of characinoids. Alexander (1965) held them as descendants of primitive characinoids due to their acquiring a dorsoventrally flattened body, for a secondary bottom-living mode of life a night-adapted sense organ and defensive spines. Wu et al. (1981), on the other hand, stated that characiforms are more primitive than cypriniforms, whereas Fink and Fink (1981) considered the latter more primitive. On the basis of hard tissues, Orvig (1957) and Bertin and Arambourg (1958) considered catfishes more primitive than not only ostariophysians, but all other teleosts. The recent reappraisal of soft anatomy (see later) has favored the very primitive nature of catfishes as a group, necessitating their dismemberment from Ostariophysi and assignment of the status of a separate taxon (see Srivastava et al., 1990; Srivastava, 1993).

For proper evaluation of the character-states of living catfishes, it is important to ascertain criteria for distinguishing between primitive and derived characters based on cladistic principles. The following criteria (see Pough et al., 1989) may prove useful:

1. Morphological, physiological and behavioral characteristics should be weighed not equally but relative to the phylogenetic lineage of the concerned group because some may be found more significant than others.
2. A taxonomic group may be named solely on the basis of monophyletic lineage, determined by autapomorphy which must not suffer from convergent character-states or paraphyly.
3. Primitive (ancestral, plesiomorphic) characters and derived (specialized, advanced, apomorphic) characters may coexist in a primitive extant taxon, which may not be thought of as a mere pathetic remnant of antiquity but a competing and successful group in present times.
4. Primitive characters are verifiable features of modern living forms, identifiable not only in terms of morphology, but also in physiological and behavioral aspects of biology. Such features may be recognized as those which resemble the known features of the probable ancestor of the group under consideration. Retention of ancestral features

through the ages may represent either their neutral selective value or their definite advantage or both on the belief that primitive character-state is simply older not necessarily inferior or less adaptive. Hence a primitive character could be either generalized or specialized, like a derived character.
5. In contrast to primitive characters, derived characters are temporally of later appearance and advanced, of great selective value yet usually noncompetitive. They replace the primitive character and are transformed for a more efficient performance of the same function or changed function(s) with regard to the function of the related primitive character. Derived characters could also be generalized or specialized in the functional context.
6. Transformation of a primitive character of an ancestral group into derived character(s) along several lineages descendant from it may be so diverse that the derived characters of the descendants are no longer similar in structure or function or both, rendering their identification difficult. However, they may retain similarity of embryonic origin and development, making their identification possible failing a fossil record. Such derived characters are homologous among themselves and with the ancestral character in an evolving group of forms.

To create a taxon and subdivide it into its sister groups on the basis of their phylogenetic relationships of evolutionary divergence, i.e., their relative closeness to the ancestral group, the cladistic approach uses the following criteria:
1. A taxon may have one or more lineages. If more, the various lineages are arranged into sister groups according to the degree of their relatedness to the ancestral group. Shared derived characters, called synapomorphies, unite sister groups, whereas, unique characters called autapomorphies make them distinct.
2. The ancestral condition of character-states, called the morphotype, may be present in the descendant sister groups. Shared ancestral characters are called symplesiomorphies and it may be helpful in highlighting synapomorphies of sister groups but will not reflect the degree of their relatedness.
3. Sequencing taxa within the fish group is based on their phylogenetic lineages from the primitive to the advanced. This sequencing is represented in a cladogram by branchings (dichotomies), giving names for synapomorphies that determine a branch point. The taxon which is the earliest ascendant from the branch point is represented by the left branch and so recognized on the basis of certain autapomorphies. However, it is related through synapomorphies that define the branch point and hence also stands as a primitive sister group (closest relative) to the rest of the sister groups represented by the right branch.
4. There is no scope for taxa created on the basis of what may be called "structural grade of evolution" as this leads only to polyphyletic lineages or paraphyletic lineages—a common weak point of most of the classification schemes of the past.

Turning now to the current views on the status of the catfish group, a number of "new" character-states have been added to the previously held autapomorphies enumerated above. These include:
1. Third eye equivalent pineal window associated with the pineal foramen.
2. Carotid labyrinth.
3. Electroreceptors.
4. Pseudobranchial neurosecretory cells.

PINEAL WINDOW AND CAROTID LABYRINTH

A pineal foramen is reported in ostracoderms, placoderms, chondrichthyans, paleoniscids and sarcopterygians but not in polypterids and other actinopterygians except the catfish group. This feature immediately brings catfishes close to sarcopterygians and paleoniscids rather than extant actinopterygians among the Osteichthyes. However, the homology of bones of the skull roof which lie on either side of the pineal foramen is an unsettled issue. The foramen is present in a peculiar plaque pineal in placoderms (Stensio, 1958). Frontal bones border it in paleoniscids. However, in rhipidistians (*Osteolepis, Eusthenopteron*) and dipnoans (*Dipnorhynchus*) the interpretation of related bones is not clear with regard to the condition in tetrapods (*Ichthyostega*) (see Panchen and Smithson, 1987; Pough et al., 1989). Some evidence has been presented (Srivastava and Srivastava, 1991; Srivastava, 1992) to show that in catfishes the pineal foramen is not merely a vestige but a functional one, not a derived character but retains a primitive condition insofar as it is associated with a primitive disposition of the dorsal pineal eminence, indicating, in some way, retention of a pineal (third) eye of the ancestral gnathostomes (see Smith, 1960). Alternatively, a phylogenetic lineage of catfish ascendant at a later stage in actinopterygian evolution may be considered if the pineal foramen turns out to be a secondarily acquired character and not the retention of a primitive one.

Catfishes are singular among extant fishes in lacking a pseudobranch and in possessing a cavernous carotid labyrinth, the morphology of which is comparable to that of amphibians (see Srivastava and Singh, 1980; Singh, 1982; Srivastava and Pandey, 1984a, b; Pandey, 1985; Tripathi, 1985; Srivastava et al., 1988). There is great controversy over the homology of this unique character of catfishes. The hypothesis of the transformation of the pseudobranch into the carotid labyrinth supported on evidence from topography and microvascular architecture (Srivastava et al., 1988) is not found tenable on grounds of lack of pillar cells. The catfish carotid labyrinth is found to be a plexus of small-diameter channels lined with endothelium, considered only arterioles (Hughes, 1984; Munshi and Hughes, 1987; Olson et al., 1990) or only capillaries (Singh, 1982; Pandey, 1985), or both (Prakash, 1993). These were either held to represent the secondary vessels of Vogel (see Olson et al., 1990) or shunt vessels (see Srivastava et al., 1994). A recent reinvestigation has revealed the presence of smooth muscle cells and sphincters (Prakash, 1993). A recently proposed hypothesis (Prakash, 1993; Srivastava, 1993 unpublished) envisages derivation of the carotid labyrinth not from the pseudobranch but from its precursor, the mandibular gill. According to this hypothesis the microvascular architecture of the carotid labyrinth seems to have evolved by elaboration of the corpus cavernosum (see Laurent, 1984) and/or marginal channels (Hughes and Datta Munshi, 1979) of the arterio-arterial pathway, and the pseudobranch by retention of the pillar cell sheet flow system. It is likely that the piscine and amphibian carotid labyrinth may have derived from the same structure though independently. This would bring the catfish close to primitive osteichthyans. However, a prominent corpus cavernosum is found in chondrostean gills, but they possess a pseudobranch like holosteans and teleosteans. The corpus cavernosum is also found in extant chondrichthyans, indicating its presence in early gnathostomes. The dipnoan gill (see Laurent, 1984) probably shows a different though parallel morphological elaboration of the corpus cavernosum into a plexus of arterioles forming an arborescent gill. Thus, while the character-state of the carotid labyrinth is still uncertain, it does point to a more primitive teleostean lineage than hitherto thought, if one holds in replacement of the pseudobranch by the carotid labyrinth in the catfish group. It is also possible, alternatively, that the catfish carotid labyrinth is derived specialized character, occurring due to convergence.

ELECTRORECEPTORS

Apart from the catfish group, a few more teleosts possess electroreceptors (see Pough et al., 1989; Jorgensen, 1989; Bodznick, 1989). It is also well known that practically all groups of lower fishes (except Holostei) and amphibians (except Anura) are endowed with electroreceptors. However, whereas the electroreceptors of nonteleostean forms are considered to be homologous, those of teleosts are not on the grounds of recent morphological findings. The former are characterized with a kinocilium and a cathodal stimulus polarity, and the latter with microvilli and an anodal stimulus polarity. Given these facts, the electroreceptors of catfish, especially the ampullary ones, along with those of other teleosts (notopterids, mormyrids, gymnarchids and gymnotids) (Jorgensen, 1989; Bodznick, 1989) would not represent retention of a primitive condition but several independent acquisitions. Since electroreception is thought to have been acquired in ancestral chordates as well, electroreceptors may represent ancient characteristics (see Pough et al., 1989) from which have evolved the electroreceptors of all the nonteleostean lineages before the neopterygians (Holostei lack electroreceptors) but not those of the teleosts. Based on this presumption the catfish lineage may not be very primitive.

However, the view that nonteleost electroreceptors are homologous with each other but not with those of teleosts presupposes that the former electroreceptors may be as old or even older than the lateral-line neuromasts. There is much disagreement as to the evolutionary relationship between electroreceptors and lateral-line mechanoreceptors (neuromasts). The origin of electroreceptors is not yet settled (see Bodznick, 1989). Three possibilities are suggested:
1. Neuromasts evolved from pre-existing electroreceptors.
2. Electroreceptors evolved from pre-existing neuromasts.
3. Both evolved independently from a more primitive receptor type of the protochordates.

So, the question of origin of electroreceptors is still open and a difficult problem (see Bullock, 1989) arises. Which features of the two are primitive and which are derived? Or, which features are advanced and which secondarily simplified?

The ampullary (tonic, low frequency) electroreceptors of catfishes (Friedrich-Freksa, 1930; Sato, 1956; Lissmann and Machin, 1963; Mullinger, 1964; Wachtel and Szamier, 1969; Szabo, 1972; Bauer and Denizot, 1972; Szabo, 1974; Srivastava et al., 1978; Seal and Srivastava, 1979; Das, 1980; Srivastava and Seal, 1981; Seal, 1984; Srivastava, 1984), however, turn out to be a different kind in themselves, different from those of other teleosts in that their microvilli may be modified stereocilia (Szabo, 1972) and their ontogenetic development shows similarity of morphogenesis with that of the neuromast (Srivastava and Seal, 1981). Such a finding calls for a fresh look at the problem of homology of catfish electroreceptors. It is likely that the electroreceptors of both nonteleosts and catfish evolved from neuromasts; in the case of the former by loss of stereocilia and in the case of the latter by loss of kinocilia, unlike those of noncatfish teleosts which may be regarded as a secondary acquisition. According to this view, the catfish electroreceptors may also be as primitive as those of nonteleosts and homologous with them.

PSEUDOBRANCHIAL NEUROSECRETORY SYSTEM

A third system of neurosecretion, called the pseudobranchial neurosecretory system, has been found to occur in catfishes (Srivastava et al., 1981; Gopesh, 1983; Srivastava and Gopesh, 1987; Pandey, 1987; Devi, 1987). This system is characterized by neurosecretory cells of the paraneuron type. A ubiquitous character for the catfish group, it is also shared by both lower and advanced teleosts such as Osteoglossomorpha, Atheriniformes, and Perciformes. Nothing is known about the presence or absence of the pseudobranchial neurosecretory cells in lower actinopterygians. Nevertheless, their association with the carotid labyrinth (being situated in close proximity) has been reported in catfish (Singh, 1982; Gopesh, 1983; Prakash, 1993). However, it is premature to assess whether this character-state is primitive or derived and advanced.

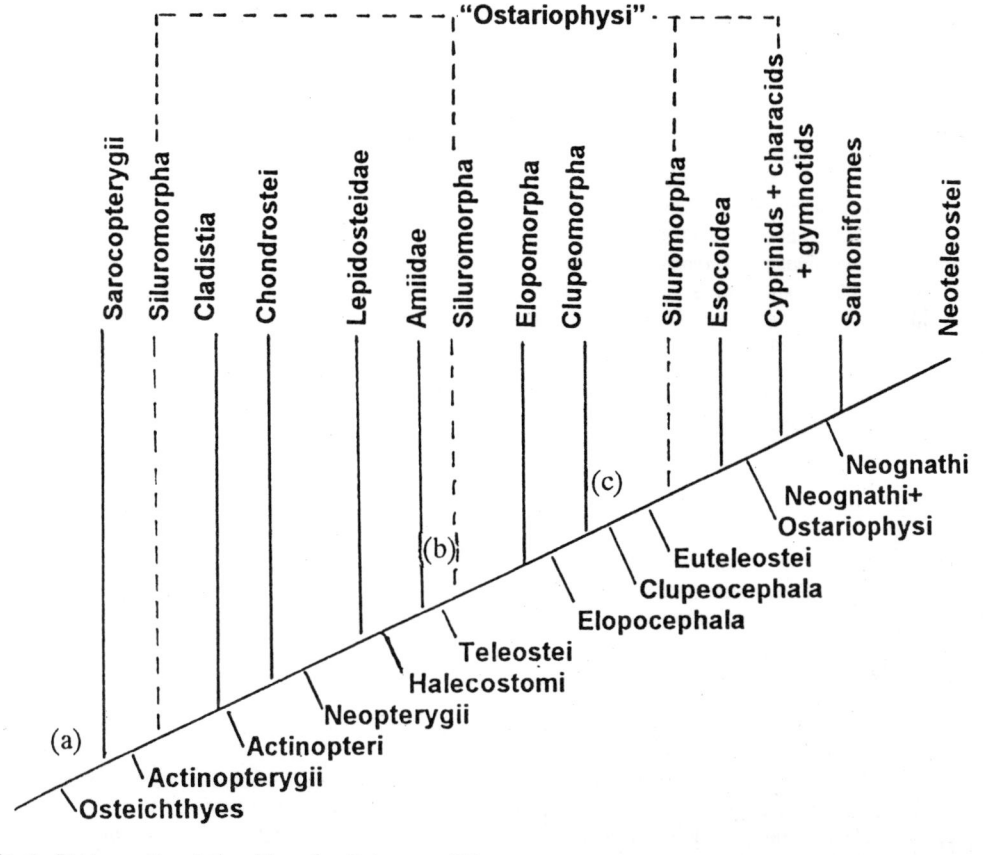

Fig. 1. Phylogenetic relationships of catfish group "Siluromorpha" with other extant bony fish groups. Solid lines show best corroborated lineages. Broken lines show possible alternative lineages proposed herein. Siluromorpha may also include gymnotids apart from catfishes to form a monophyletic group following Fink and Fink (1981).

Thus, from the foregoing account of autapomorphies of the catfish group it is clear that they may no longer be considered a sister group to cyprinids and other ostariophysians. The newly discovered characters of catfish probably present a mixture of derived and primitive characters as determined by the phylogenetic distribution of these character-states among the extant lower groups of fishes. The recent additions to catfish autapomorphies do suggest an independent evolution of the catfish group as a phylogenetic lineage that originated earlier than hitherto thought. In the light of the above discussion, three possible alternative relationships for the catfish group are suggested in Fig. 1a, b and c. This summarizes the current thinking on the phylogenetic relationships of the extant bony fishes (see Pough et al., 1989). The first indicates their closeness to lower living Actinopterygii (chondrosteans). The second indicates closeness to lower living Teleostei. The third indicates closeness to lower living Euteleostei. The phylogenetic lineage at a, b or c has been named Siluromorpha. The quotation marks around "Ostariophysi" indicate that this is a paraphyletic grouping.

In summary, the phylogenetic and systematic status of the catfish group is subject to revision, yet recent evidence suggests that they represent a more primitive status than hitherto held. Future researchers of the twenty-first century will illuminate this and similar issues unresolved in fish morphology.

References

Alexander, R.McN. 1965. Structure and function in the catfish. *J. Zool.* 148:88–152.

Bauer, P.R. and J-P. Denizot. 1972. Sur la présence et la répartition des organes ampullaires chez *Plotosus anguillaris*. *Arch. Anat. Micr. Morph. Exp.* 61:85–90.

Berg, L.S. 1955. Systema Ryb. *Trudy* 21:1–286.

Bertin, L. and C. Arambourg. 1958. Super-ordre des téléosteens. In P.P. Grassé (ed.), *Traité de Zoologie, Anatomie, Systematique, Biologie*, 13/III. Masson et Cie., Paris.

Bodznick, D. 1989. Comparisons between electrosensory and mechanosensory lateral line systems. In S. Coombs, P. Gorner and H. Munz (eds.), *The Mechanosensory Lateral Line*. Springer-Verlag, New York.

Bullock, T.H. 1989. Lateral line research: Prospects and opportunities. In S. Coombs, P. Gorner and H. Munz (eds.), *The Mechanosensory Lateral Line*. Springer-Verlag, New York.

Das, P.K. 1980. *Degeneration and regeneration studies on the lateral line organs of certain teleosts*. D.Phil. thesis, University of Allahabad, India.

Datta Munshi, J.S. and M.P. Srivastava. 1988. *Natural History of Fishes and Systematics of Freshwater Fishes of India*. Narendra Publishing House, New Delhi.

Day, F. 1889. *The Fauna of British India. Including Ceylon and Burma. Fishes*. William Dawson & Sons Ltd., London.

Devi, U. 1987. *Studies on the pseudobranchial neurosecretory system of certain teleosts with particular reference to its function*. D.Phil. thesis, University of Allahabad, India.

Fink, S.V. and W.L. Fink. 1981. Interrelationships of the ostariophysian fishes (Teleostei). *J. Linn. Soc. (Zool.)* 72:297–353.

Friedrich-Freksa, H. 1930. Lorenzinische ampullen bei dem Siluroiden *Plotosus angullaris* Bloch. *Zool. Anz.* 87:49–66.

Gopesh, A. 1983. *Studies on peculiar pseudobranchial neurosecretory cells in certain teleostean fishes*. D.Phil. thesis, University of Allahabad, India.

Greenwood, P.H., D.E. Rosen, S.N. Weitzman, and G.S. Myers. 1966. Phyletic studies of teleostean fishes with a provisional classification of living forms. *Bull. Brit. Mus. Nat. Hist. (Zool.)* 41 : 213–234.

Hughes, G.M. 1984. General anatomy of the gills. In W.S. Hoar and D.J. Randall (eds.), *Fish Physiology* X/A. Academic Press, New York.

Hughes, G.M. and J.S. Datta Munshi. 1979. Fine structure of the gills of some Indian air-breathing fishes. *J. Morph.* 160:169–194.

Jorgensen, J.M. 1989. Evolution of Octavolateralis sensory cells. *In* S. Coombs, P. Gorner, and H. Munz (eds.), *The Mechanosensory Lateral Line*. Springer-Verlag, New York.
Laurent, P. 1984. Gill internal morphology. *In* W.S. Hoar and D.J. Randall (eds.), *Fish Physiology*, X/A. Academic Press, New York.
Lissmann, H.W. and K.E. Machin. 1963. Electric receptors in a nonelectric fish (*Clarias*). *Nature* 199:88-89.
Migdalski, E.C. and G.S. Fichter. 1976. *The Fresh and Salt Water Fish of the World*. Manderin Publishers Ltd., Hong Kong.
Mullinger, A.M. 1964. The fine structure of the ampullary electric receptors in *Amiurus*. *Proc. Roy. Soc. B.* 160:345-359.
Munshi, J.S.D. and G.M. Hughes. 1987. Microvascular organization of the pseudobranch and carotid labyrinth of two species of air-breathing teleostean fishes. *J. Fish Biol.* 31:705-714.
Nelson, J.S. 1984. *Fishes of the World*. John Wiley & Sons, New York.
Nikol'skii, G.V. 1954. *Special Ichthyology*. Israel Program for Scientific Translations Ltd. (1963), Jerusalem.
Olson, K.R., J.S.D. Munshi, T.K. Ghosh, and J. Ojha. 1990. Vascular organization of the head and respiratory organs of the air-breathing catfish, *Heteropneustes fossilis*. *J. Morph.* 203:165-179.
Orvig, T. 1957. Paleohistological notes. I. On the structure of the bone tissue in the scales of certain paleonisciformes. *Arkiv. Zool.* 10:481-490.
Panchen, A.L. and T.R. Smithson. 1987. Character diagnosis, fossilis and the origin of tetrapods. *Biol. Rev.* 62:341-438.
Pandey, A. 1987. *Studies on the surfacing activity of certain catfishes with particular reference to the role of pseudobranchial neurosecretory system*. D.Phil. thesis, University of Allahabad, India.
Pandey, K.N. 1985. *Studies on the structure of carotid labyrinth in certain catfishes*. D.Phil. thesis, University of Allahabad, India.
Pough, F.H., J.B. Heiser, and W.N. McFarland. 1989. *Vertebrate Life*. MacMillan Publishing Co., New York.
Prakash, S. 1993. *Studies on carotid labyrinth of some bony fishes*. D.Phil. thesis, University of Allahabad, India.
Sato, M. 1956. Studies on the pit-organs of fishes: I. The distribution, histological structure and development of small pit-organs. *Ann. Zool. Jap.* 29:207-212.
Seal, M. 1984. *Studies on the structure and development of the so-called "pit-organ" of certain fishes*. Ph.D. thesis, University of Allahabad, India.
Seal, M. and C.B.L. Srivastava. 1979. On the occurrence of ampullary (electroreceptors) organs in another Indian catfish, *Rita rita* (Ham.). *Nat. Acad. Sci., Letters* 2:41-42.
Singh, M. 1982. *Studies on the structure and development of the pseudobranchs in certain fishes, with particular reference to their innervation*. D.Phil. thesis, University of Allahabad, India.
Smith, H.M. 1960. *Evolution of Chordate Structure*. Holt, Rinehart & Winston Inc., New York.
Srivastava, C.B.L. 1984. Electroreceptive behaviour response of Indian catfish *Clarias batrachus* (Linn.) to experimental test objects. *Proc. Nat. Acad. Sci. India* B 54:267-271.
Srivastava, C.B.L. 1993. Evolutionary relationships of major groups of fishes with particular reference to Indian siluroids. (submitted)
Srivastava, C.B.L. and M. Singh. 1980. Occurrence of carotid labyrinth in the catfish group of teleost fishes. *Experientia* 36:651-652.
Srivastava, C.B.L. and M. Seal. 1981. Electroreceptors in Indian catfish, teleost. *In* T. Szabó and G. Czeh (eds.) *Sensory Physiology of Aquatic Lower Vertebrates*. Pergamon Press, Oxford.
Srivastava, C.B.L. and K.N. Pandey. 1984a. Carotid labyrinth of a marine catfish, *Tachysurus thalassinus* (Rupp). *Proc. Nat. Acad. Sci. India* B 54:6-8.
Srivastava, C.B.L. and K.N. Pandey. 1984b. Carotid labyrinth of catfishes. *Proc. Nat. Acad. Sci. India* B 54:204-212.
Srivastava, C.B.L., S. Tripathi, K.N. Pandey, and M. Singh. 1988. Fish carotid labyrinth—a transformed pseudobranch. *Arch. Biol.* (Bruxelles) 99:33-50.
Srivastava, C.B.L. and A. Gopesh. 1987. A third system of neurosecretion in fish—the pseudobranchial system. *In* V.S. Bhatnagar (ed.), *Advances in Cytology and Genetics*, Allahabad Univ. Press, Allahabad, India.
Srivastava, C.B.L., A. Gopesh, and M. Singh. 1981. A new neurosecretory system in fish, located in the gill region. *Experientia* 37:850-851.
Srivastava, C.B.L., S. Tripathi, and K.N. Pandey. 1994. Fish carotid labyrinth. *In* H.R. Singh (ed.), *Advances in Fish Biology*. Hindustan Publication Corporation, New Delhi.

Srivastava, C.B.L., M. Seal, P.K. Das, and A. Gopesh. 1978. Anatomical identification of the presumed electroreceptors of two air-breathing catfishes, *Clarias batrachus* and *Heteropneustes fossilis. Experientia* 34:1345–1346.

Srivastava, C.B.L., A. Gopesh, S. Tripathi, S. Srivastava, K.P. Singh, D. Srivastava, and M. Seal-Prasad. 1990. Autapomorphies for a separate taxon status for catfishes. Diamond Jubilee Session Nat. Acad. Sci., India. p. 53 (abstract).

Srivastava, S. 1992. *Studies on the pineal window of certain catfishes*. D.Phil. thesis, University of Allahabad, India.

Srivastava, S. and C.B.L. Srivastava. 1991. A lens-like specialization for photic input in the pineal window of an Indian catfish, *Heteropneustes fossilis. Experientia* 47:698–700.

Stensio, E. 1958. Les cyclostomes fossiles ou Ostracodermes. In P.P. Grassé (ed.), *Traité de Zoologie, Anatomie, Systematique, Biologie*, 13. Masson et Cie., Paris.

Szabó, T. 1972. Ultrastructural evidence for a mechanoreceptor function of the ampullae of Lorenzini. *J. Microscopie* 14:343–350.

Szabó, T. 1974. Anatomy of the specialized lateral line organs of electroreception. In A. Fessard (ed.), *Electoreceptors and Other Specialized Receptors in Lower Vertebrates*. Springer-Verlag, New York.

Tripathi, S. 1985. *Studies on the carotid labyrinth in certain teleosts*. D.Phil thesis, University of Allahabad, India.

Verigina, I.A. 1969. Ecological and morphological peculiarities of the alimentary system in some carps. In L.P. Poznannin (ed.), *Ichthyology*. Amerind Publishing Co. Ltd., New Delhi.

Wachtel, A.W. and R.B. Szamier. 1969. Special cutaneous organs of fish: IV. Ampullary organs of the nonelectric catfish, *Kryptopterus. J. Morph*. 128:291–308.

Weitzman. 1964. Cited in Greenwood et al., 1966.

Wu, X., Y. Chan, X. Chen, and T. Chen. 1981. A taxonomic system and phylogenetic relationship of the families of the suborder Cyprinoidei (Pisces). *Sci. Sinica* 24:563–572.

3

Scanning Electron Microscopy of the Fish Gill

Kenneth R. Olson

The delicate and elaborate structure of the fish gill has long fascinated biologists as perhaps the most exquisite example of an external respiratory organ. The pioneering work of Krogh, and later Maetz, clearly showed that the gill was also the major osmoregulatory organ in these vertebrates and Smith added yet another dimension, nitrogen excretion, to gill function. Additional work has since demonstrated that acid-base regulation and detoxification are also mediated by branchial tissues (see, Hoar and Randall, 1984a, b for an excellent compilation of reviews on gill anatomy and physiology). It thus became apparent that the gill was a multifunctional organ engaging in a number of seemingly unrelated homeostatic activities whose only common feature was transepithelial exchange between internal and external milieu. Morphological studies have revealed a number of distinct cells in gill tissue and numerous attempts have been made to correlate cell type with function (see below). An interesting, and perhaps not unexpected, finding from these studies was that specific cells are grouped into discrete areas within the gill and thus gill tissues are organized into anatomical, and probably functional units. The most obvious common modality between these anatomical units is the vasculature that perfuses them. Recent studies have shown that the gill circulation forms one of the most complex vascular networks found in any vertebrate and that it may, through vasoactive mechanisms, have considerable impact on gill function. Furthermore, the extensive gill vascular surface area and the in-series position of the gill and systemic circulations provide the gill with a unique opportunity to metabolize plasma-borne molecules and thus adds an additional regulatory role to the branchial tissues.

The intent of this article is to briefly summarize gill anatomy and describe the application of scanning electron microscopy (SEM) to studies of gill form and function. Emphasis is placed on key or unique SEM studies that raise interesting questions to be pursued further, suggest unique methodological applications, or define specific areas that are actively being investigated. It is beyond the scope of this article to review all aspects of gill anatomy and physiology; many of these topics are described in great detail elsewhere (Hughes and Morgan, 1973; Hoar and Randall, 1984a, b; Olson, 1991). The present discussion is also confined to the teleosts (bony fish). It will become evident that we are just beginning to understand the relationship between form and function in the gill and that SEM investigations have played, and will continue to play, a key role in this process.

GILL ANATOMY

Teleost fish typically have four pairs of gill arches. Each arch consists of a bony support skeleton attached anteroventrally to the tongue and lower jaw and posterodorsally to the roof of the buccopharyngeal chamber. Two rows of gill filaments, each row comprising a hemibranch, radiate posteroventrally from the arch. The tips of the filaments of hemibranchs from adjacent arches are closely appositioned and form the gill curtain. This curtain causes water to sieve between the filaments on each hemibranch and prevents the respiratory medium from bypassing the filaments as water is pumped from buccal to opercular chambers.

Figure 1a shows the general structure of the gill and its vasculature. Labeled coordinates in Fig. 1a, and the following nomenclature, are used for purposes of orientation: basal (B), the area of the filament nearest to the arch skeleton; peripheral (P), toward

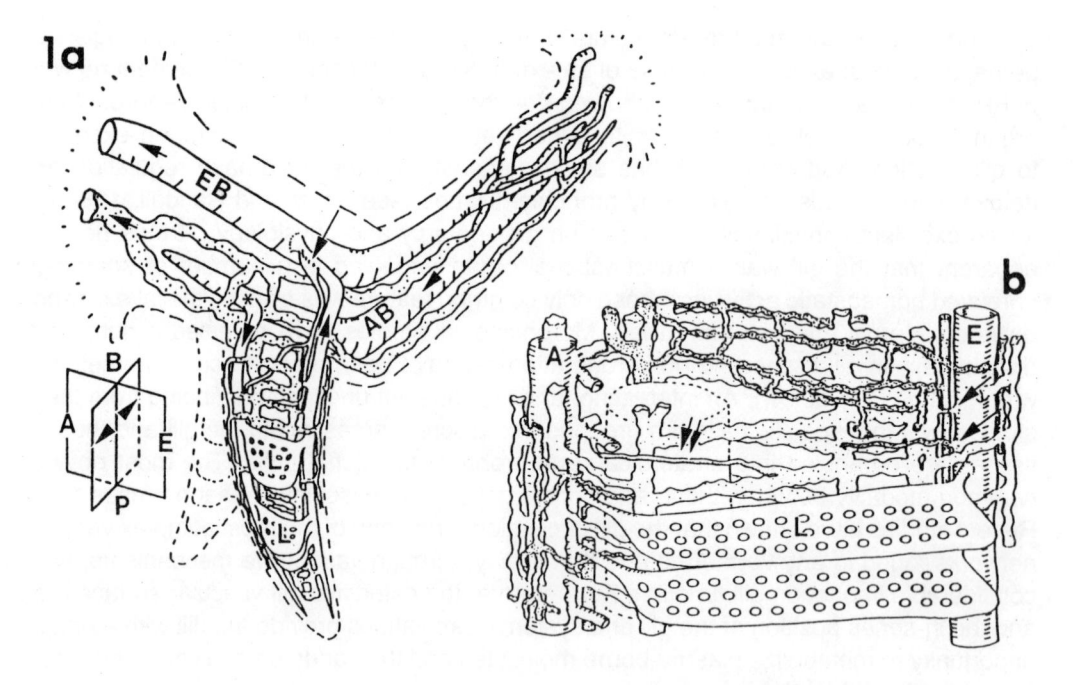

Fig. 1. The gill arch and major vessels. **a:** The arch support skeleton contains the afferent (AB) and efferent (EB) branchial arteries that supply blood to, or drain it from the filaments respectively. Afferent filamental arteries originate from the AB and travel along the filament, giving rise to the lamellar sinus (L). Blood drains from the lamellae into the efferent filamental artery and then into the EB for delivery to systemic circulation. Alternatively, some blood enters the vessels in the body of the filament, usually via efferent filamental or branchial arteries and drains into the gill venous system (stippled vessels). Long arrow, origin of secondary circulation; short arrow, direction of blood flow. A filament from the other hemibranch of the arch is outlined. **Inset:** Orientation of filament; ABEP medial plane; B, base of filament; P, peripheral end of filament; A and E, lines of afferent and efferent filamental arteries respectively; arrowheads indicate lateral to medial plane. **b:** Enlargement of filamental circulation of catfish, *Ictalurus punctatus:* A and E, afferent and efferent filamental arteries respectively; L, lamellar sinusoid. Body of filament has two vasculature networks; light stippling—interlamellar system, heavy stippling—nutrient system. Double arrows—prelamellar AVA; single arrows—postlamellar AVA. See text for details (from Olson, 1991, with permission).

the free tip of the filament, away from the arch skeleton; afferent (A), and efferent (E), areas along the filament associated with either afferent or efferent filamental arteries respectively; medial, the plane A-B-E-P which passes through the afferent and efferent filamental arteries and the core of the filament; lateral, extending away from either side of the medial plane (arrowheads in Fig. 1a coordinates).

The filament is the functional unit of the gill. Diffusive exchange of respiratory gases or other molecules presumably takes place across thin shelflike respiratory lamellae. Lamellae extend laterally from both sides of the filament, perpendicular to the medial plane and perpendicular to the longitudinal axis of the filament. The lateral (outer) margins of lamellae from adjacent filaments in a common hemibranch are closely apposed and often appear to slightly interdigitate. Ventilatory activity moves water across the lamellae in an efferent to afferent direction. Water flow over the lamellae is countercurrent to the direction of blood flow through the lamellae and this maximizes gas exchange efficiency.

At least two general types of epithelium have been described on the gill surface, one on the body of the filament and the other on the lamellae (Laurent and Dunel, 1980; Laurent, 1984). Three cell types—pavement, chloride (ionocytes or mitochondria-rich cells), and mucous—are most prevalent on both epithelia. Other cells such as rodlet, taste bud, and Merkel cells are also observed, but they are less common.

Pavement cells

Simple squamous (pavement) cells are the predominant cell type of both filamental and lamellar epithelia and in some species may be the only cells found on all but the most medial areas of the lamellae (Newstead, 1967; Morgan and Tovell, 1973). The characteristic topographical features of pavement cells, broad pentagonal or hexagonal shape, elaborate system of surface microridges (microplicae), and a double-ridged or single-ridged border between a pavement cell-to-pavement cell or pavement cell-to-chloride cell junction respectively, were described in the initial SEM study of the trout gill (Olson and Fromm, 1973a). Since then, pavement cells have generally received only cursory attention and subsequent reports have, until recently, been largely confirmatory.

Laurent (1984) observed that pavement cell microridges vary in length from 5 to 15 μm and in width from 0.15 to 0.20 μm. He also indicated that they differ in size and shape between species. Variations in microridge organization can also be found in individual species and within a single arch, perhaps indicative of functional subtypes of pavement cells. The combined effects of interspecies and intrabranchial variation and difficulty in completely removing external mucus make comparisons difficult; however, a few tentative generalizations may be drawn. Typically, there is a progressive reduction in microridge length and complexity in epithelia from the arch, filament body, and respiratory lamellae (Fig. 2; see also Hossler et al., 1979a; Dunel-Erb and Laurent, 1980; Karlsson, 1983). Microridges from the nonfilamental arch supporting skeleton pavement cells are often organized into elaborate whorls (Fig. 2a), similar to that found on fish skin epidermis (Bereiter-Hahn et al., 1979; Schwerdtfeger, 1979). By comparison, lamellar pavement cell microridges, when present (see below), are usually shorter and randomly oriented (Fig. 2c). Pavement cells on the body of the filament may have features similar to, or in between, either of the above extremes (Fig. 2b).

The greatest interspecies variability in microridge organization is found in the lamellar pavement cells and in pavement cells located on the filament body between the basal areas of the lamellae (i.e., the interlamellar filamental epithelium). Species with microridged pavement cells are common. Most fish examined in this laboratory (over twenty freshwater and saltwater species) have had conspicuous microridges on lamellar pavement cells (Fig. 2c). Typically, these microridges were somewhat shorter and more randomly organized than those found on either the filament body or over the arch support skeleton, although whorled microridges were occasionally observed as well (Kendall and Dale, 1979; Hughes and Umezawa, 1983). Pavement cells with long microridges often have filamentous strands that bridge adjacent microridges (Kendall and Dale, 1979). These are probably due to glycocalyx or condensed mucus because membranous interconnections have not been observed in ultrathin sections. Perhaps these strands are mucus invested with calmodulin and are involved in regulating epithelial cell permeability (see below).

A few species of fish have microvillous or smooth-surface lamellar pavement cells. Channel catfish (*Ictalurus punctatus*) and black bullhead (*Ictalurus melas*; Olson, unpublished) or walking catfish (*Clarias batrachus*; Munshi and Olson, unpublished) have lamellar pavement cells with short, stubby microridges that are probably more appropriately termed microvilli (Fig. 2d). The interlamellar epithelium also appears to be covered with a microvillous epithelium. A similar epithelium has also been reported on *Trachurus mediterraneus* lamellae (Hughes and Mondolfino, 1983). Lamellar epithelia from pelagic fish such as bluefish (*Pomatomus saltatrix*), Atlantic mackerel (*Scomber scombrus*), or Atlantic bonito (*Sarda sarda*) are almost, or completely devoid of microridges (Fig. 2e; Olson, unpublished). The interlamellar epithelium in all three of these species has two types of pavement cells. The majority of pavement cells have no microridges while a few other pavement cells with short microridges are only occasionally observed (Fig. 3b). In all three of these species, however, pavement cells along the afferent and efferent borders of the filament and those covering the arch support skeleton are heavily invested with long microridges. A similar distribution of ridged (filamental) versus smooth (lamellar) pavement cells has been reported in the striped bass (*Morone saxatilis*; Hossler et al., 1986).

These findings suggest that the variation in topography of lamellar and interlamellar pavement cells may have a phylogenetic or physiological basis. To date, however, the function of the pavement cell surface has not been determined. Microridges have been proposed to trap, hold or distribute mucus, provide structural integrity to the epithelium, or increase the surface area of the apical membrane (Hughes and Wright, 1970; Olson and Fromm, 1973a; Sperry and Wassersug, 1976). It seems logical that microridges would create an unstirred boundary layer at the cell surface which would impede the convective delivery of oxygen to the epithelium. Mucus trapped in the microridges would exacerbate this problem. Interestingly, an abrupt transition between microridged and smooth epithelial cells is also found in the accessory respiratory organs of the air-breathing fishes, *Anabas testudineus, Monopterus cuchia*, and species of *Channa* (Hughes and Munshi, 1986; Munshi et al., 1986, 1989, 1990; Munshi and Hughes, 1991; Olson and Munshi, unpublished). In these fish, microridged cells on nonrespiratory tissues give way to completely smooth epithelial cells that cover the respiratory surface. Mucus may impart a degree of ionic or hydraulic permselectivity to the underlying cells and thereby serve an important role in regulating water and electrolyte transfer across the epithelium (see page 37). Thus the presence or absence of mucus-trapping microridges might be indicative of a trade-off between hydromineral regulation and gas exchange.

Microridges may also enhance mechanical flexibility. Knutton et al. (1976) proposed that microridges on Lettree cells allow rapid cell swelling without concomitant lysis, thereby allowing a change in cell size or shape without the need for membrane protein synthesis. This hypothesis, extrapolated to fish gills, would provide flexibility to the lamella and permit the epithelial cells to stretch or fold (the reverse occurring on the opposite side of the lamella) as the lamella moves in the water current. Lamellar-type microvilli would not be necessary on the ridged filament or arch support skeleton and these epithelia predictably would have more typical epidermal microridges, which is indeed the case.

Microridges on fish epidermal (nongill) cells are attached to subcellular contractile elements and spontaneously change in size and appearance (Bereiter-Hahn, 1971; Bereiter-Hahn et al., 1979). Size and shape of epidermal microridges are also affected by ambient salinity, electrolytes and hormones (Schwerdtfeger and Bereiter-Hahn, 1976; Schwerdtfeger, 1979; Wendelaar Bonga and Meis, 1981). Similar responses could be produced on gill pavement cells, although it seems likely that the variations shown in Figs. 2c–e reflect actual species differences. Clearly, further investigation of other species and the effects of environmental and hormonal manipulations are required.

Little attention, from an experimental perspective, has been given to the double-ridged junction between pavement cells. Adaptation of mullet (*Mugil capito*) or eels (*Anguilla anguilla*), to freshwater or saltwater did not alter the junctions between lamellar pavement cells, implying that paracellular solute and solvent transfer across the lamella is not affected by salinity (Sardet et al., 1979). Further work is needed in fish, especially catfish, in which the cell junctions appear somewhat different (compare Figs. 2c, d).

Chloride cells

Chloride cells (also called ionocytes after their presumed role in electrolyte transport, or mitochondria-rich cells because of the latter's abundance) are found on the arch support epithelium, alamellar portion of the filament, and medial margin of the lamellae. These cells are most frequent on the filament surface between the lamellae (interlamellar filamental epithelium) and around the afferent (trailing edge with respect to water flow) filament border. Chloride cells can be classified into two categories: 1) surface cells, those with apical membranes flush with the pavement epithelium and 2) recessed cells, those with apical membranes invaginated and opening into an "apical pit" partially covered by overlying pavement cells. The former are commonly found in freshwater fish and the latter associated with fish inhabiting a saltwater environment, although there are numerous exceptions and both types may even be found in a single fish (Karlsson, 1983). Several chloride cells usually share a common apical pit in saltwater fish (Sardet et al., 1979). The general anatomical features of these cells have been reviewed by Laurent (1984). Scanning electron microscopy has been useful in examining the topography of chloride cells, following the changes in apical membrane organization during salinity transfer, and quantifying the chloride cell population during various experimental maneuvers.

Surface chloride cells lack microridges and are typically covered with numerous stubby microvilli (Fig. 3a, c; Olson and Fromm, 1973a; Hossler et al., 1979a; Kimura and Kudo, 1979; Karlsson, 1983; Hossler et al., 1986; Franklin and Davison, 1989; Perry and Laurent, 1989; Laurent and Perry, 1990; Munshi and Hughes, 1991). In trout the microvillar structure may be correlated with sodium and chloride transport; microvilli become greatly elongated, or even ridgelike, and increase in density in fish with high ionic fluxes in hypoosmotic media (Laurent and Perry, 1990). Old and degenerating chloride cells develop a smooth apical membrane (Olson and Fromm, 1973a; Franklin and Davison, 1989; Laurent and Perry, 1990). Recessed chloride cells may have a few long (up to 1 µm) filamentous microvilli that extend into the apical pit (Fig. 3f; see also Hossler et al., 1979b; Dunel-Erb and Laurent, 1980). Large chloride cells in the interlamellar epithelium of the mackerel (*Scomber scombrus*) appear to contain a filigree of microvilli-like filaments in their apical pit (Fig. 3b; Olson, unpublished). An additional cell, the accessory cell, has also been identified in the chloride cell complex of saltwater fish. This cell has the appearance of a long, thin chloride cell with a narrow cytoplasmic fold extending into the surface of the apical pit (Sardet et al., 1979; Dunel-Erb and Laurent, 1980; Laurent, 1984). The accessory cell surface fold has neither microvilli nor microridges (Dunel-Erb and Laurent, 1980).

Conventional histology and transmission electron microscopy have provided the basis for most of what is known regarding morphological changes accompanying adaptation of fish to varied environments. SEM methods can provide valuable information on the structural and/or chronological changes in microvilli, pit diameter, and surface interdigitations between chloride and accessory cells. However, there is surprisingly limited morphometric data on gill chloride cell responses to salinity adaptation in spite of the intense interest in the physiological and biochemical aspects of teleost osmoregulation (Laurent and Perry, 1991). Transfer of euryhaline fish between hypo- and hyperosmotic media produces profound changes in the ultrastructure of chloride cells yet few attempts have been made with SEM techniques to document the chronology and extent of these events. Laurent and colleagues (Laurent et al., 1985; Laurent and Hebibi, 1989; Perry and Laurent; 1989; Laurent and Perry, 1990) were the first to quantify the morphological changes associated with adaptation of euryhaline fish to freshwater, saltwater, ionically modified environments, or hormonal manipulation. The authors employed a computer-assisted digitizer in their studies to examine SEM micrographs of controlled areas of the gill filament. Parameters such as the density of chloride and mucous cells as well as the apical surface of these cells were readily quantified and statistically analyzed. In some instances these measurements were also correlated with physiological measurements of ion transport performed previously on the same fish. These methods are clearly the most advantageous application of the SEM and should provide a new direction into structure-function analysis with broad application.

Transfer of euryhaline fish between freshwater and seawater produces reversible changes in chloride cell topography and/or density. Most SEM studies to date have been of a descriptive nature. Transfer of euryhaline fish, such as mullet (*Mugil cephalus*), from freshwater to saltwater causes invagination of the apical membrane, a decrease in the aperture above the cell, due to partial covering by adjacent pavement cells, and reduction or elimination of surface microvilli (Figs. 3c–f; Hossler et al., 1979b; Hossler, 1980). Anatomical changes in the mullet gill appeared within six hours and were virtually completed twenty-four hours after transfer (Hossler, 1980). In a quantitative SEM study Laurent and Hebibi (1989) found that adaptation of rainbow trout to seawater more than doubled the number of chloride cells on the gill filament but, surprisingly, did not affect either the apical surface area or the topographical morphology of the chloride cells.

Dramatic morphological responses of gill chloride cells from fish exposed to ion-poor media have recently been reported using SEM methods. Laurent et al. (1985) demonstrated that as ambient [NaCl] was lowered below 0.05 mmol/l, chloride cell numbers increased and chloride cells appeared on the lamellar surface (Laurent et al., 1985). Perry

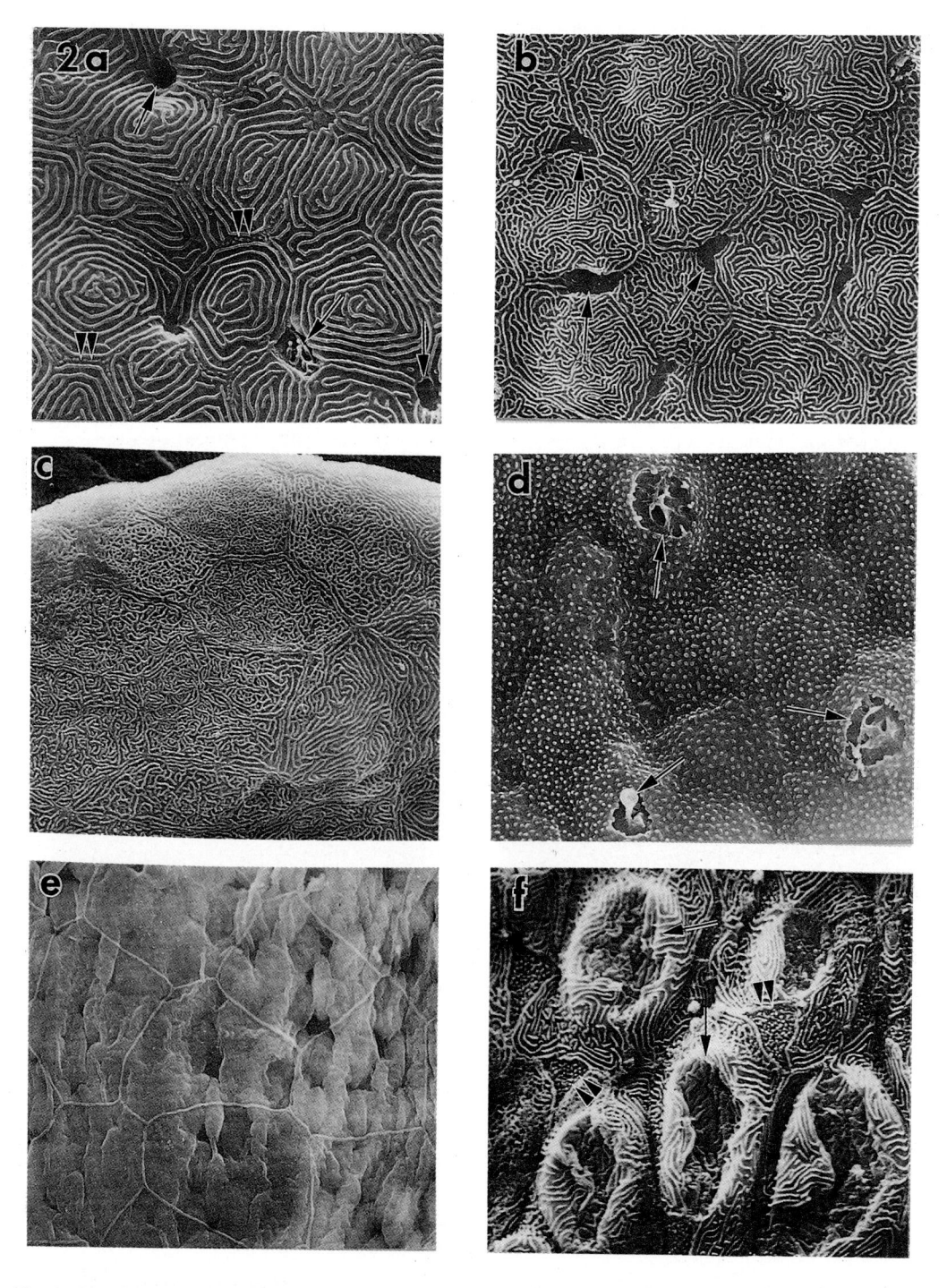

Fig. 2. Variations in gill pavement cell topography: arrows—mucous cells; double arrowheads—chloride cells **a:** Whorled microridges on *Clarias batrachus* arch support (3500 ×). **b:** Random microridges on *Oncorhynchus mykiss*, filament near efferent border (2000 ×). **c:** Microridges on *O. mykiss*, lamella near leading (efferent) border (2000 ×). **d:** Microvilli on lamella of *C. batrachus* (3500 ×). **e:** Smooth lamellar epithelium from *Pomatomus saltatrix* (1600 ×). **f:** Mucus, chloride, and pavement cells on filament epithelium of *Esox lucius* (200 ×).

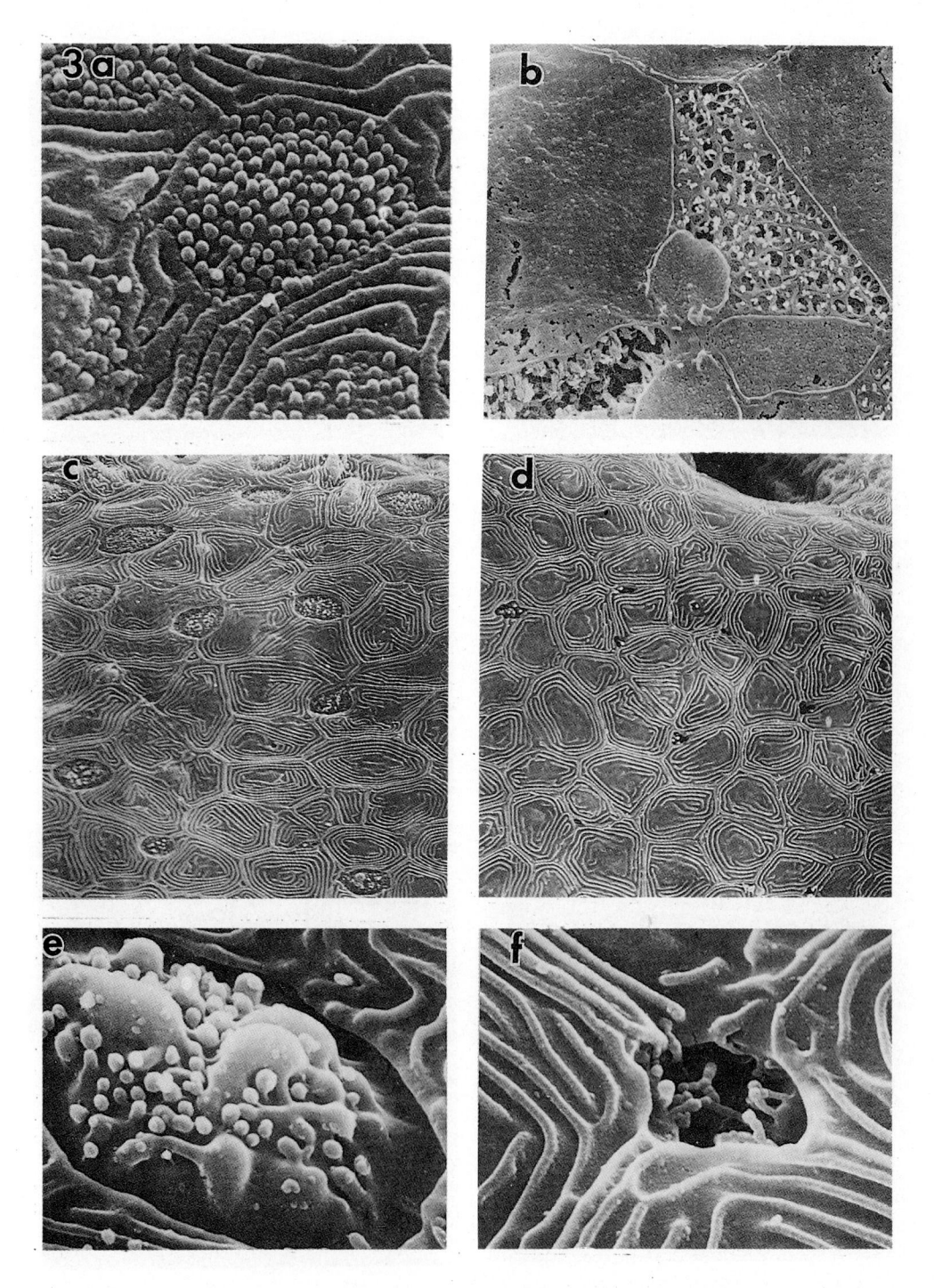

Fig. 3. Chloride cells **a:** Microvillous chloride cell from freshwater *Anabas testudineus* (10,000 ×). **b:** Large microfilamentous chloride cell from saltwater *Scomber scombrus* (4000 ×). **c and e:** Surface chloride cells bearing microvilli on filament of freshwater adapted *Mugil cephalus* (**c**, 2000 ×; **e**, 20,000 ×). **d and f:** Chloride cells with few microvilli recessed in apical pit on filament of 4.5% saltwater adapted *Mugil cephalus* (**c**, 2000 ×; **e**, 20,000 ×) (Figures **c** and **d** courtesy of F.E. Hossler; Figures **e** and **f** from Hossler et al., 1979b, with permission).

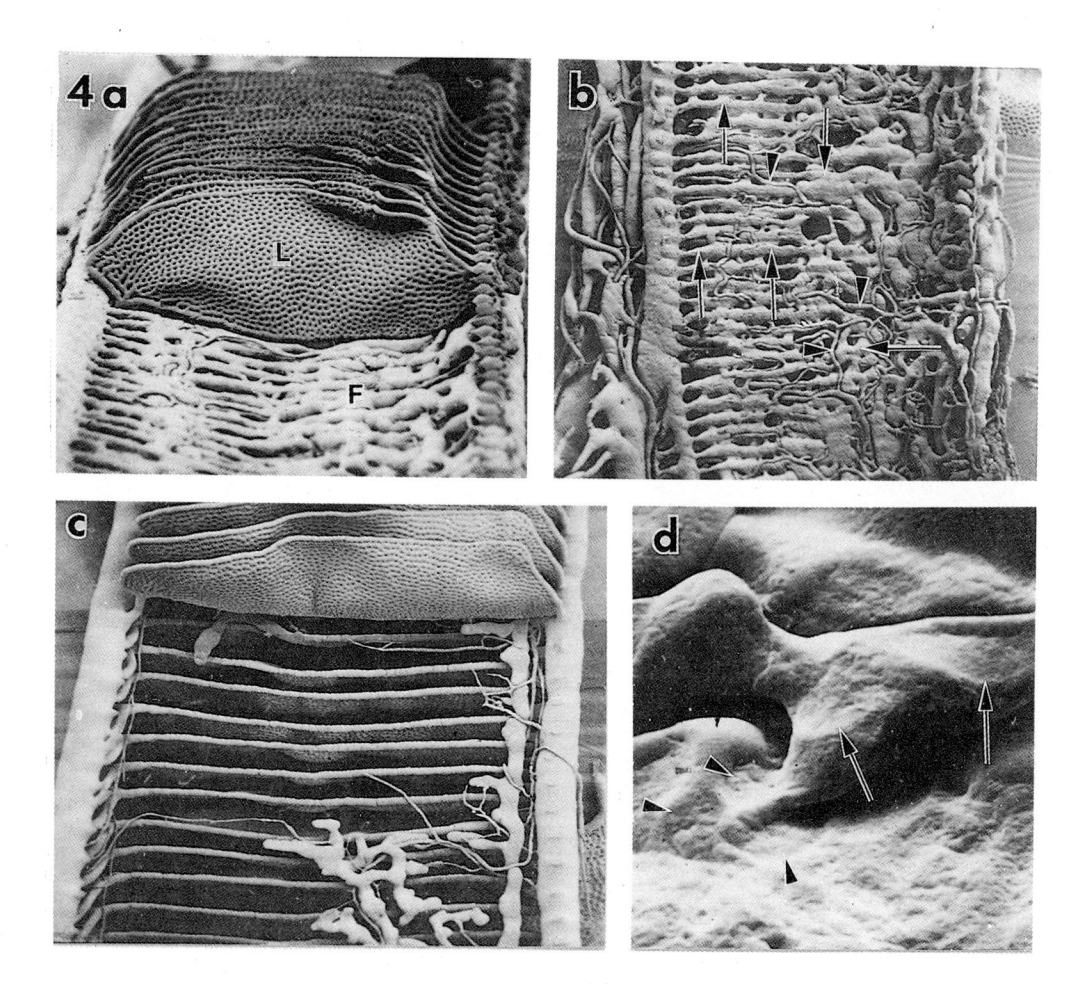

Fig. 4. Corrosion replicas of gill vessels. Afferent border of filament is at left in **a, b,** and **c a:** Filament from *O. mykiss* showing respiratory lamellae (L) and oblique view of filamental vasculature (F). Small holes in lamellae are occupied by pillar cells *in vivo* (100 ×). **b:** Filamental vasculature of *O. mykiss* with lamellae removed. The interlamellar system (larger vessels; arrows) and nutrient system (narrow vessels; arrowheads) are often in parallel and form an extensive network in the filament body (100 ×). **c:** Preperfusion of *Ictalurus punctatus* gill arch with 10^{-7} M epinephrine prevents the resin from entering the intrafilamental and nutrient systems (100 ×). **d:** Arterio-venous anastomosis in *O. mykiss* (2000 ×). Impressions of endothelial cells in the resin can be seen in the small anastomotic vessel (arrows) and at the constricted junction with the interlamellar vessel (arrowheads) (Fig. 4c from Olson, 1980, with permission).

and Laurent (1989) used SEM morphometry to quantify the response of trout adapted to low [Na^+] (0.017 mmol/l) and [Cl^-] (0.014 mmol/l) freshwater and found a 1.5–2.5-fold increase in apical surface area of individual chloride cells while the number of cells did not change appreciably. The apical surface also correlates with [NaCl], microvilli become dense and elongated in low [NaCl] medium, perhaps to enhance the ion absorptive processes (Perry and Laurent, 1989). In contrast, Laurent et al. (1985) reported that the number of smooth-surface chloride cells increased in low [NaCl] environments. This dichotomy may be explained by an increased cell turnover, thereby increasing the population of degenerating ("worn out"; Laurent and Perry, 1990) cells. However, Laurent and Hebibi (1989) also observed an increase in "normal" chloride cell number as well as nearly a fourfold increase in apical surface area in trout adapted to ion-poor media. Interestingly, the number and surface area of chloride cells examined with SEM also increased when trout were placed on a high salt (sodium chloride) diet (Salman and Eddy, 1987). Obviously, additional work is necessary to fully characterize these adaptive processes. Nevertheless, the importance of SEM in physiological studies has been clearly shown.

Other studies with SEM have observed changes in chloride cell topography associated with acid base regulation in freshwater fish or hormonal manipulation. Hyperoxic-induced hypercapnic acidosis increased the apical surface area of individual trout chloride cells (Laurent and Perry, 1991), whereas ambient hypercapnia increased the apparent chloride cell density in the catfish *Ictalurus punctatus* (Cameron and Iwama, 1987). Chloride cells with apical pits, and occasionally chloride cells with bulbous apical evaginations, developed in chloride cells of freshwater fathead minnows (*Pimephales promelas*) exposed to acid (pH 5.0) water (Leino et al., 1983). Daily injection of cortisol for ten days increased chloride cell density three times and individual cell surface area two times; this finding has been offered as evidence of the role of this steroid hormone in the adaptation response (Laurent and Perry, 1990). Similar changes in apical membrane topography were produced by cortisol and low [NaCl] environments (Laurent and Perry, 1990). While additional work is indicated, these studies clearly show the value of SEM methods in examining morphological and physiological correlations.

Mucous cells

Mucous cells are scattered over the arch support epithelium, filament, and lamellae (Figs. 2a, b, d, f). In most fish the relative cell density of mucous cells on the filament is: efferent filamental border > afferent filamental border > interlamellar filamental epithelium > base of lamella >>> outer margin of lamella. There are, however, notable species variations in this pattern (Laurent, 1984).

Mucus probably performs a variety of functions on the gill. Interposition of the mucus coat between the environment and epithelium suggests that the primary function of mucus is insulative in nature. The barrier function of mucus may be: 1) physical, i.e., prevent mechanical abrasion or limit parasitic, bacterial, or viral access; 2) biochemical, through an ability to sequester or precipitate specific molecules, e.g., heavy metals (Olson and Fromm, 1973b; Varanasi and Markey, 1978); 3) immunological through secretion of immunoglobulin (Ourth, 1980; Itami et al., 1988); or 4) permselective through an ability to affect oxygen (Ultsch and Gros, 1979) or electrolyte (Pärt and Lock; 1983; Shephard, 1984; Simonneaux et al., 1987; Handy, 1989) diffusion. Other functions for mucus have also been described. Calmodulin has been found in fish mucus and it has been proposed

(Flik et al., 1984) that ambient calcium binds with mucus and thereby affects epithelial permeability. This hypothesis, it would seem, should be amenable to analysis with SEM. Mucus may also decrease the coefficient of drag for water flow across the gill in a manner similar to its effects on the general body surface (Daniel, 1981). This would minimize the tendency for disruptive turbulent flow and thereby reduce the resistance of the gill sieve.

A number of investigators have reported changes in mucous cell distribution and density as a result of salinity adaptation or hormonal stimulation (see Leatherland and Lam, 1969; Marshall, 1976; Laurent, 1984). Comparatively few studies have used SEM methods for evaluating mucous cell populations or secretory activity (Hossler et al., 1979b; Boyd et al., 1980; Laurent, 1984). To my knowledge, only one study used SEM to quantify mucous cell dynamics (Laurent and Hebibi, 1989). These authors examined mucous cell populations and topography in a concurrent study of the response of chloride cells to freshwater, saltwater, or ion-poor water (see page 37). Their results show little difference between freshwater and seawater fish; however, ion-poor water increased the size and number of mucous cells. Perry and Laurent (1989) also observed an apparent increase in number of filamental mucous cells in trout adapted to low [Na^+] and [Cl^-] freshwater; however, this was not quantified. As pointed out by Laurent and Hebibi (1989), additional studies on mucous cell dynamics using SEM morphometry are clearly warranted.

The number of chloride and mucous cells on the lamellae varies, depending on species, and may be affected by environmental conditions. Laurent and Hebibi (1989) found an increase in mucous cell size, but not number, on the afferent site of filaments of trout adapted to seawater. Mucous cells were distributed across the entire lamellar surface in the air-breathing walking catfish (*Clarias batrachus*; Olson and Munshi, unpublished), perhaps indicative of the harsh, muck environment in which these fish often congregate. It would be interesting to determine whether suspended particulates, which may have an abrasive action on the gill, can stimulate mucous cell proliferation on the lamella.

Pathophysiology and other applications

SEM methods have proven invaluable in a variety of pathophysiological studies in fish. Techniques are available for assessing the degree of damage produced by a specific pathogen, the nature of the pathogen, and the mechanisms and degree of success of homeostatic responses. Several excellent studies employing SEM and other morphological methods to investigate the effects of toxicants on gill structure have recently been published (Mallatt, 1985; Richmonds and Dutta, 1989; Roy et al., 1990; Lauren, 1991; Munshi and Singh, 1992; Dutta et al., 1993, 1994a, 1994b) and the reader is referred to them for further details.

SEM examination of gill anatomy has also provided other information. Coughlan and Gloss (1984) used SEM to follow the early morphological development of the gills of 11- to 180-day-old smallmouth bass (*Micropterus dolomieui*). King and Hossler (1988) used an ingenious combination of fixatives and sonication to microdissect the gills of striped bass (*Morone saxatilis*) in order to examine the basal lamina and other subepithelial tissues. Both these studies (Coughlan and Gloss, 1984; King and Hossler, 1988) demonstrated innovative use of SEM and provide a background for further investigations. Undoubtedly, a variety of other related applications await development.

GILL VESSELS

Early work on gill microcirculation, using standard light microscopy, suggested that there were several circulatory pathways through the gill vasculature. Steen and Kruysse (1964) described several shunt pathways in the eel gill that would allow blood to bypass the respiratory lamellae en route from the ventral aorta to the dorsal aorta. Based on transmission electron microscopic studies, Newstead (1967) and Morgan and Tovell (1973) questioned the existence of these shunts in other species of fish and considerable attention has since been given to resolving this issue.

Perhaps the most significant use of SEM in fish morphology has been in the analysis of corrosion replicas of the complex gill microcirculation. This technique, pioneered by Murakami (1971) for use on mammalian tissues, was initially adapted to a study of gill vasculature by Gannon and colleagues (Gannon et al., 1973). The method is based on perfusion replacement of blood with a low viscosity polymerizable resin, methyl methacrylate. After the resin polymerizes, the tissue is digested and a three-dimensional vascular replica is obtained that is stable under the electron beam. With proper selection of a low-viscosity resin, several capillary beds in series can be successfully filled and resolution of the imprint of individual endothelial cells is usually possible (Fig. 4d). Orientation of the endothelial cells, axial or random, has been used to distinguish between arterial and venous vessels respectively (Olson, 1980, 1981; Olson et al., 1986). Several reviews (Gannon, 1978; Olson, 1985) detail the method for preparing gill corrosion replicas.

Figures 4a–c show typical corrosion replicas of the gill vasculature; Figure 1b provides a schematic diagram for comparison. An afferent branchial artery distributes blood through the arch to the filaments and the afferent filamental artery delivers blood to the filamental circuits. Although there is still controversy over the vascular pathways within the filament, the present discussion assumes at least three perfusion networks in the filament.

The arterio-arterial pathway consists of the afferent filamental artery, afferent lamellar arteriole, lamellar sinus, efferent lamellar arteriole and efferent filamental artery. This system is drained from the arch into the dorsal aorta (systemic arterial circulation) by the efferent branchial artery. In typical water-breathing teleosts there are no shunts that bypass the lamellae en route from the ventral to the dorsal aorta. A variety of ingenious, and often elaborate, branchial, filamental, and lamellar shunts have been described in air-breathing fish by SEM examination of corrosion replicas (Munshi et al., 1986; Olson et al., 1986; Munshi et al., 1990; Olson et al., 1990).

The arterio-venous circulation in most fish arises from the postlamellar vasculature, typically from the efferent filamental or efferent branchial artery, and is found in the body (alamellar portion) of the filament. This vascular system has variously been called the central sinus, central venous sinus, or venolymphatic system. Careful examination of gill corrosion replicas strongly suggests that the arterio-venous circulation may be two separate vascular networks within the filament body (Boland and Olson, 1979; Olson, 1983; Olson et al., 1986; see also Figs. 4b, c). The tentative nomenclature "nutrient" and "interlamellar" pathways is suggested until their function(s) are resolved. These two systems are closely apposed and although they appear to perfuse the same tissue, there are no anastomoses between them in the filament. They are best distinguished by the size of the vessels.

The nutrient system originates from the base of the efferent filamental artery or the efferent branchial artery in the form of numerous tortuous arterioles that reanastomose to form a nutrient artery. The nutrient artery reenters at the base of the filament and travels along the filament in one to three branches, often close to the inner border of the filamental arteries. This system has been described as the secondary circulation by Vogel (Vogel et al., 1974; 1976; Vogel, 1985). Nutrient capillaries are uniformly smooth and round in appearance and have a narrow (6–15 µm) diameter (Figs. 4b, c). They are drained from the filament by an extensive venous network. Both capillaries and veins are closely associated with the interlamellar system.

The interlamellar system appears to be present in all fish, yet its relationship with other gill vessels, much less its function, remains an enigma. The interlamellar system is characterized by large-bore, irregular vessels that run the length of the filament and traverse the filament beneath the interlamellar filamental epithelium (Fig. 4b). It is a highly organized network whose volume is second only to the arterio-arterial system. This system appears to be highly distensible and its appearance is greatly affected by the pressure used in preparation of the corrosion replica (Olson, 1983). Inlet and outlet vessels for the interlamellar system have been identified in only a few fish. In the catfish *Ictalurus punctatus* the interlamellar system is perfused from narrow vessels arising from the afferent lamellar arterioles (Boland and Olson, 1979). In the air-breathing walking catfish *Clarias batrachus* the interlamellar system is perfused from narrow-bore vessels that originate on the medial wall of the efferent filamental artery (Olson and Munshi, unpublished).

A number of important questions regarding the gill microcirculation, and amenable to resolution with corrosion replication techniques, remain to be answered. So little is known about the nonrespiratory pathways (interlamellar and nutrient) that these networks should be given special attention. Areas for future study include the following:

1. *Origin and relationship of interlamellar and nutrient pathways in the filament.* What are the feeder vessels and how are these systems drained? How distinct are these two systems in different species? Are there any environmental effects?
2. *Gill vascular morphometrics.* Careful analysis of corrosion replicas should provide quantitative estimates of individual vessel dimensions, including resistances. These parameters can also be measured under different conditions such as perfusion pressure (compliance), vasostimulation (see below), or factors that affect angiogenesis.
3. *Regulation of perfusion through the gill microcirculation.* It has already been demonstrated that preperfusion of gills with vasoactive hormones alters the resin filling pattern, thereby providing a visual image of the sites of perfusion regulation (Fig. 4c; Olson, 1980). This method has a variety of applications for examination of hormonal effects on perfusion distribution and the chronological sequence of network perfusion. The parallel nature of these vessels suggests the potential for counter- or concurrent perfusion; is this possible?
4. *Specialized vascular adaptations.* Air-breathing fish have been especially interesting in this regard, not only for shunt pathways, but also for a variety of modifications in their capillary organization (Olson et al., 1986; Munshi et al., 1990). Other fish with unique environmental or phylogenetic history are also of interest.

In addition to analysis of corrosion replicas, direct SEM examination of blood vessels can provide additional insight into vascular function. For example, Vogel and coworkers combined scanning and transmission electron microscopy to show that the entrance to arterio-venous anastomoses branching off from the efferent filamental artery of *Tilapia* was guarded by cells with long microvilli. These microvilli appear to restrict red cells from entering the AVA. Other endothelial specializations have been described in gills and accessory respiratory organs of air-breathing fish (Olson et al., 1986; Munshi et al., 1990; Olson, 1991). Virtually nothing is known regarding the physiological functions of many of these specializations and this should be an interesting field for future work.

X-RAY ANALYSIS

X-ray microanalysis has been used only sparingly in fish studies, although the potential of this technique in areas such as osmoregulation and toxicology is considerable. With this method x-rays characteristic of specific elements are measured as the electron beam sweeps across the tissue. Thus an "image" of the elemental profile of tissues is obtained and can be compared with an actual image of a tissue obtained from the back-scattered electrons. In the first such study, Olson and Fromm (1973b) used an energy-dispersive electron microprobe to examine mercury uptake and binding by the gill. Since then, several investigators have used the more modern energy dispersive x-ray microanalytical systems to examine ion or heavy metal concentrations and distribution in gill tissue (Lehtinen and Klingstedt, 1983; Potts and Oates, 1983; Ishihara and Mugiya, 1987). Although there are still difficulties with this method, the results are encouraging and this technique has tremendous potential in the future.

CONCLUSION

The gill is the primary corridor for molecular exchange between the internal milieu of a fish and its environment. Gills perform myriad seemingly unrelated homeostatic functions that include respiration, osmoregulation, acid-base balance and nitrogen excretion. These varied activities are accomplished in discrete areas within the gill and it is now evident that the gill vasculature is as complex as the functions it subserves. Recent work has also shown that gills have an additional, internal, regulatory function through their ability to modify plasma hormones and thus "condition" arterial blood prior to systemic perfusion. The challenge at hand for anatomists and physiologists alike is to correlate form and function in this sophisticated vasculature. The application of scanning electron microscopy (SEM) to studies of gill anatomy and physiology was examined in this article. The advantages of SEM methods include: ease of sample preparation, three-dimensional perspective, high resolution, and the ability to survey large areas of tissue. X-ray analysis, in conjunction with SEM, offers a quantitative elemental analysis of gill tissue and has considerable potential in physiological as well as toxicological studies. Collectively, these studies have had, and will continue to have, an impact upon many areas of fish research.

Acknowledgments

The author acknowledges collaborators Drs. F.E. Hossler, J.S. Datta Munshi, present and past students, Mr. D. Conklin and Drs. E. Boland and P. Holbert, who have been instrumental in both published and unpublished studies on gill anatomy. The secretarial assistance of Mrs. C. Gordon is also greatly appreciated. This work has been supported by National Science Foundation Grants PCM 76-16840, PCM 79-23703, PCM 84-04897, DCB 86-16028, DCB 90-04245, DCB 91-05247, INT 83-00721, INT 86-02965, INT 86-18881, and INT 90-0944.

References

Bereiter-Hahn, J. 1971. Light and electron microscopic studies on function of tonofilaments in teleost epidermis. *Cytobiologie* 4:73-102.

Bereiter-Hahn, J., M. Osborn, K. Weber, and M. Voth. 1979. Filament organization and formation of microridges at the surface of fish epidermis. *J. Ultrastruct. Res.* 69:316-330.

Boland, E.J. and K.R. Olson. 1979. Vascular organization of the catfish gill filament. *Cell Tissue Res*. 198:487-500.

Boyd, R.B., A.L. DeVries, J.T. Eastman, and G.G. Pietra. 1980. The secondary lamellae of the gills of cold water (high latitude) teleosts. A comparative light and electron microscopic study. *Cell Tissue Res*. 213:361-367.

Cameron, J.N. and G.K. Iwama. 1987. Compensation of progressive hypercapnia in channel catfish and blue crabs. *J. Exp. Biol*. 133:183-197.

Coughlan, D.J. and S.P. Gloss. 1984. Early morphological development of gills in smallmouth bass (*Micropterus dolomieui*). *Can. J. Zool*. 62:951-958.

Daniel, T.L. 1981. Fish mucus: *In situ* measurements of polymer drag reduction. *Biol. Bull*. 160:376-382.

Dunel-Erb, S. and P. Laurent. 1980. Ultrastructure of marine teleost gill epithelia: SEM and TEM study of the chloride cell apical membrane. *J. Morphol*. 165:175-186.

Dutta, H.M., T. Zeno, and C. Richmonds. 1993. Effects of diazinon on the gills of bluegill sunfish *Lepomis macrochirus*. *J. Environ. Pathol. Oncol*. 12(4):219-227.

Dutta, H.M., J.S.D. Munshi, P.K. Roy, N.K. Singh, and L. Motz. 1994a. Effect of diazinon on gill of bluegill sunfish, *Lepomis macrochirus*: Observation with scanning electron microscope. (submitted).

Dutta, H.M., J.S.D. Munshi, P.K. Roy, and J. Killius. 1994b. Malathion induced changes in the respiratory lamellae of a catfish, *Heteropneustes fossilis*, with particular reference to chloride cells. An electron microscopic study. (submitted).

Flik, G., J.H. Van Rijs, and S.E. Wendelaar Bonga. 1984. Evidence for the presence of calmodulin in fish mucus. *Eur. J. Biochem*. 138:651-654.

Franklin, C.E. and W. Davison. 1989. S.E.M. observations of morphologically different chloride cells in freshwater-adapted sockeye salmon, *Oncorhynchus nerka*. *J. Fish Biol*. 34:803-804.

Gannon, B.J. 1978. Vascular casting. *In* M.A. Hayat (ed.), *Principles and Techniques of Scanning Electron Microscopy*, Vol. 6, pp. 170-193. Van Nostrand Reinhold, New York.

Gannon, B.J., G. Campbell, and D.J. Randall. 1973. Scanning electron microscopy of vascular casts for the study of vessel connections in a complex vascular bed—the trout gill. *31st Ann. Proc. Electron Microscopy Soc. Am*. 31:442-443.

Handy, R.D. 1989. The ionic composition of rainbow trout body mucus. *Comp. Biochem. Physiol*. [A] 93A:571-575.

Hoar, W.S. and D.J. Randall. 1984a. *Fish Physiology*, vol. X, pt. A. Academic Press Inc., New York, 456 pp.

Hoar, W.S. and D.J. Randall. 1984b. *Fish Physiology*, vol. X, pt. B. Academic Press Inc., New York, 489 pp.

Hossler, F.E. 1980. Gill arch of the mullet, *Mugil cephalus*. III. Rate of response to salinity change. *Am. J. Physiol. Regul. Integr. Comp. Physiol*. 238:R160-R164.

Hossler, F.E., J.R. Ruby, and T.D. McIlwain. 1979a. The gill arch of the mullet, *Mugil cephalus*. I. Surface ultrastructure. *J. Exp. Zool*. 208:379-398.

Hossler, F.E., J.R. Ruby and T.D. McIlwain. 1979b. The gill arch of the mullet, *Mugil cephalus*. II. Modification in surface ultrastructure and Na, K-ATPase content during adaptation to various salinities. *J. Exp. Zool*. 208:399-406.

Hossler, F.E., J.H. Harpole, Jr., and J.A. King. 1986. The gill arch of the striped bass, Morone saxatilis. I. Surface ultrastructure. J. Submicrosc. Cytol. 18:519-528.

Hughes, G.M. and D.E. Wright. 1970. A comparative study of the ultrastructure of the water-blood pathway in the secondary lamellae of teleost and elasmobranch fishes—benthic forms. Z. Zellforsch 104:478-493.

Hughes, G.M. and M. Morgan. 1973. The structure of fish gills in relation to their respiratory function. Biol. Rev. 48:419-475.

Hughes, G.M. and R.M. Mondolfino. 1983. Scanning electron microscopy of the gills of Trachurus mediterraneus. Experientia 39:518-519.

Hughes, G.M. and S.I. Umezawa. 1983. Gill structure of the yellowtail and frogfish. Jap. J. Ichthyol. 30:176-183.

Hughes, G.M. and J.S.D. Munshi. 1986. Scanning electron microscopy of the accessory respiratory organs of the snake-headed fish, Channa striata (Bloch) (Channidae, Channiformes). J. Zool. Lond. 209:305-317.

Ishihara, A. and Y. Mugiya. 1987. Ultrastructural evidence of calcium uptake by chloride cells in the gills of goldfish, Carassius auratus. J. Exp. Zool. 242:121-129.

Itami, T., Y. Takahashi, K. Okamoto and K. Kubono. 1988. Purification and characterization of immunoglobulin in skin mucus and serum of Ayu. Nippon Suisan Gakkaishi. 54:1611-1617.

Karlsson, L. 1983. Gill morphology in the zebrafish, Brachydanio rerio (Hamilton-Buchanan). J. Fish. Biol. 23:511-524.

Kendall, M.W. and J.E. Dale. 1979. Scanning and transmission electron microscopic observations of rainbow trout (Salmo gairdneri) gill. J. Fish. Res. Board Can. 36:1072-1079.

Kimura, N. and S. Kudo. 1979. The fine structure of gill filaments in the fingerlings of rainbow trout Salmo gairdneri. Jap. J. Ichthyol. 26:289-301.

King, J.A.C. and F.E. Hossler. 1988. The gill arch of the striped bass, Morone saxatilis. III. Morphology of the basal lamina as revealed by various ultrasonic microdissection procedures. J. Submicrosc. Cytol. Pathol. 20:371-377.

Knutton, S., D. Jackson, J.M. Graham, K.J. Micklem, and C.A. Pasternak. 1976. Microvilli and cell swelling. Nature 262:52-54.

Lauren, D.J. 1991. The fish gill: A sensitive target for waterborne pollutants. In M.A. Mayes and M.G. Barron (eds.), Aquatic Toxicology and Risk Assessment, Vol. 14, ASTM STP 1124, pp. 223-244. Amer. Soc. Testing Materials, Philadelphia.

Laurent, P. 1984. Gill internal morphology. In W.S. Hoar and D.J. Randall (eds.), Fish Physiology, vol. A, pp. 73-183. Academic Press, New York.

Laurent, P. and S. Dunel. 1980. Morphology of gill epithelia in fish. Am. J. Physiol. 238:R147-R159.

Laurent, P. and N. Hebibi. 1989. Gill morphometry and fish osmoregulation. Can. J. Zool. 67:3055-3063.

Laurent, P. and S.F. Perry. 1990. Effects of cortisol on gill chloride cell morphology and ionic uptake in the freshwater trout, Salmo gairdneri. Cell Tissue Res. 259:429-442.

Laurent, P. and S.F. Perry. 1991. Environmental effects on fish gill morphology. Physiol. Zool. 64:4-25.

Laurent, P., H. Hobe, and S. Dunel-Erb. 1985. The role of environmental sodium chloride relative to calcium in gill morphology of freshwater salmonid fish. Cell Tissue Res. 240:675-692.

Leatherland, J.F. and T.J. Lam. 1969. Effect of prolactin on the density of mucous cells on the gill filaments of the marine form (trachurus) of the threespine stickleback, Gasterosteus aculeatus L. Can. J. Zool. 47:787-792.

Lehtinen, K.-J. and G. Klingstedt. 1983. X-ray microanalysis in the scanning electron microscope on fish gills affected by acidic, heavy metal containing industrial effluents. Aquat. Toxicol. 3:93-102.

Leino, R.L., J.G. Anderson, and J.H. McCormick. 1983. Development of apical pits in chloride cells of Pimephales promelas after chronic exposure to acid water. Proc. 41st Ann. Meeting Electron Microscopy Soc. Am. 41:462-463 (abstr.).

Mallatt, J. 1985. Fish gill structural changes induced by toxicants and other irritants: A statistical review. Can. J. Fish. Aquat. Sci. 42:630-648.

Marshall, W.S. 1976. Effects of hypophysectomy and ovine prolactin on the epithelial mucus-secreting cells of the Pacific staghorn sculpin, Leptocottus armatus (Teleostei: Cottidae). Can. J. Zool. 54:1604-1609.

Morgan, M. and P.W.A. Tovell. 1973. The structure of the gill of the trout Salmo gairdneri (Richardson). Z. Zellforsch. 142:147-162.

Munshi, J.S.D. and G.M. Hughes. 1991. Structure of the respiratory islets of accessory respiratory organs and their relationship with the gills in the climbing perch, Anabas testudineus (Teleostei, Perciformes) J. Morphol. 209:241-256.

Munshi, J.S.D. and A. Singh. 1992. Scanning electron microscopic evaluation of low pH on gills of Channa punctata (Bloch.). J. Fish Biol. 41:83-89.

Munshi, J.S.D., K.R. Olson, T.K. Ghosh and J. Ojha 1990. Vasculature of the head and respiratory organs in an obligate air-breathing fish, the swamp eel *Monopterus* (=*Amphipnous*) *cuchia*. *J. Morphol.* 203:181–201.

Munshi, J.S.D., K.R. Olson, J. Ojha, and T.K. Ghosh. 1986. Morphology and vascular anatomy of the accessory respiratory organs of air-breathing climbing perch, *Anabas testudineus* (Bloch). *Am. J. Anat.* 176:321–331.

Munshi, J.S.D., G.M. Hughes, P. Gehr and E.R. Weibel. 1989. Structure of the air-breathing organs of a swamp mud eel, *Monopterus cuchia*. *Jap. J. Ichthyol.* 35:453–465.

Murakami, T. 1971. Application of the scanning electron microscope to the study of the fine distribution of the blood vessels. *Arch. Histol. Jap.* 32:445–454.

Newstead, J.D. 1967. Fine structure of the respiratory lamellae of teleostean gills. *Z. Zellforsch* 79:396–428.

Olson, K.R. 1980. Application of corrosion casting procedures in identification of perfusion distribution in a complex microvasculature. In O. Johari (ed.), *Scanning Electron Microscopy*, vol. 3, pp. 357–364. SEM, Inc., Chicago.

Olson, K.R. 1981. Morphology and vascular anatomy of the gills of a primitive air-breathing fish, the bowfin (*Amia calva*). *Cell. Tissue Res.* 218:499–517.

Olson, K.R. 1983. Effects of perfusion pressure on the morphology of the central sinus in the trout gill filament. *Cell Tissue Res.* 232:319–325.

Olson, K.R. 1985. Preparation of fish tissues for electron microscopy. *J. Electron Microsc. Technique* 2:217–228.

Olson, K.R. 1991. Vasculature of the fish gill: anatomical correlates of physiological function. *J. Electron Microsc. Technique* 19:389–405.

Olson, K.R. and P.O. Fromm. 1973a. A scanning electron microscopic study of secondary lamellae and chloride cells of rainbow trout (*Salmo gairdneri*). *Z. Zellforsch*, 143:439–449.

Olson, K.R. and P.O. Fromm. 1973b. Mercury uptake and ion distribution in gills of rainbow trout (*Salmo gairdneri*): Tissue scans with an electron microprobe. *J. Fish. Res. Board Can.* 30:1575–1578.

Olson, K.R., J.S.D. Munshi, T.K. Ghosh and J. Ojha 1990. Vascular organization of the head and respiratory organs of the air-breathing catfish, *Heteropneustes fossilis*. *J. Morphol.* 203:165–179.

Olson, K.R., J.S.D. Munshi, T.K. Ghosh, and J. Ojha. 1986. Gill microcirculation of the air-breathing climbing perch, *Anabas testudineus* (Bloch): Relationships with the accessory respiratory organs and systemic circulation. *Am. J. Anat.* 176:305–320.

Ourth, D.D. 1980. Secretory IGM, lysozyme and lymphocytes in the skin mucus of the channel catfish, *Ictalurus punctatus*. *Dev. Comp. Immunol.* 4:65–74.

Pärt, P. and R.A.C. Lock. 1983. Diffusion of calcium, cadmium and mercury in a mucous solution from rainbow trout. *Comp. Biochem. Physiol.* [C] 769:259–263.

Perry, S.F. and P. Laurent. 1989. Adaptational responses of rainbow trout to lowered external NaCl concentration: Contribution of the branchial chloride cell. *J. Exp. Biol.* 147:147–168.

Potts, W.T.W. and K. Oates. 1983. The ionic concentrations in the mitochondria-rich or chloride cell of *Fundulus heteroclitus*. *J. Exp. Zool.* 227:349–359.

Richmonds, C. and H.M. Dutta. 1989. Histopathological changes induced by malathion in the gills of bluegill *Lepomis macrochirus*. *Bull. Environ. Contam. Toxicol.* 43:123–130.

Roy, P.K., J.D. Munshi, and H.M. Dutta. 1990. Effect of saponin extracts on morphohistology and respiratory physiology of an air-breathing fish *Heteropneustes fossilis* (Bloch). *Freshwater Biol.* 2(2):135–145.

Salman, N.A. and F.B. Eddy. 1987. Response of chloride cell numbers and gill Na^+/K^+ ATPase activity of freshwater rainbow trout (*Salmo gairdneri* Richardson) to salt feeding. *Aquaculture* 61:41–48.

Sardet, C., M. Pisam and J. Maetz. 1979. The surface epithelium of teleostean fish gills. Cellular and junctional adaptations of the chloride cell in relation to salt adaptation. *J. Cell Biol.* 80:96–117.

Schwerdtfeger, W.K. 1979. Morphometrical studies of the ultrastructure of the epidermis of the guppy, *Poecilia reticulata* Peters, following adaptation to seawater and treatment with prolactin. *Gen. Comp. Endocrinol.* 38:476–483.

Schwerdtfeger, W.K. and J. Bereiter-Hahn. 1976. Hormone-induced alterations in the epidermal surface of teleosts. *Beitr. Elektronenmikroskop. Direktabb. Oberfl.* 9:515–523.

Shephard, K.L. 1984. The influence of mucus on the diffusion of chloride ions across the oesophagus of the minnow (*Phoxinus phoxinus* (L.)). *J. Physiol.* 346:449–460.

Simonneaux, V., W. Humbert, and R. Kirsch. 1987. Mucus and intestinal ion exchanges in the sea-water adapted eel, *Anguilla anguilla* L. *J. Comp. Physiol.* [B] 157:295–306.

Sperry, D.G. and R.J. Wassersug. 1976. A proposed function for microridges on epithelial cells. *Anat. Rec.* 185:253–258.

Steen, J.G. and A. Kruysse. 1964. The respiratory function of the teleostean gill. *Comp. Biochem. Physiol.* [A] 12:127–142.

Ultsch, G.R. and G. Gros. 1979. Mucus, a diffusion barrier to oxygen: possible role in O_2 uptake at low pH in carp (*Cyprinus carpio*) gills. *Comp. Biochem. Physiol.* [A] 62:685–689.

Varanasi, U. and D. Markey. 1978. Uptake and release of lead and cadmium in skin and mucus of coho salmon (*Oncorhynchus kisutch*). *Comp. Biochem. Physiol.* [C] 60:187–191.

Vogel, W.O. 1985. Systemic vascular anastomoses, primary and secondary vessels in fish, and the phylogeny of lymphatics. *Alfred Benzon Symp.* 21:143–159.

Vogel, W., V. Vogel, and W. Schlote. 1974. Ultrastructural study of arterio-venous anastomoses in gill filaments of *Tilapia mossambica*. *Cell Tissue Res.* 155:491–512.

Vogel, W., V. Vogel, and M. Pfautch. 1976. Arterio-venous anastomoses in rainbow trout gill filaments. *Cell Tissue Res.* 167:373–385.

Wendelaar Bonga, S.E. and S. Meis. 1981. Effects of external osmolality, calcium and prolactin on growth and differentiation of the epidermal cells of the cichlid teleost *Sarotherodon mossambicus*. *Cell Tissue Res.* 221:109–123.

4

Vascular Organization of Lungfish, a Landmark in Ontogeny and Phylogeny of Air-breathers

Pierre Laurent

Dipnoan lungfish have evoked considerable interest over the last two centuries. These fish belong to a subclass of osteichthyes which flourished during the early Paleozoic era. This subclass comprises only three surviving genera: *Neoceratodus* (Australia), *Protopterus* (Africa), and *Lepidosiren* (South America).

Lungfish have survived because they have developed, in addition to their piscine gill respiratory system, an esophageal diverticulum which, as can be observed in living species, forms a pair of long hollow sacs along the dorsal portion of the abdominal cavity from the pectoral to the pelvic girdle (Fig. 1). Cephalad, the two sacs merge to form a common chamber connected to the glottic sphincter of the pharynx by a short pneumatic duct. Trabeculae extend from the inner wall of the lung into the central cavity and delineate a spongy arrangement of incompletely closed alveoli. In addition to bundles of smooth muscle, trabeculae contain the blood vessels which ultimately supply the capillary network of alveoli (Fig. 2).

In contrast to *Neoceratodus*, a facultative air-breather relying mainly on gills for respiratory gas exchanges, *Protopterus* and *Lepidosiren* are obligatory air-breathers. Due to the thickness of the gill epithelia and the filament organization (Figs. 3, 4) the gill of these lungfish is poorly designed for oxygen exchange. During the part of their life spent in a lake or river, these lungfish use their gill to eliminate carbon dioxide, relying on their lung for about 90% of their oxygen supply (Johansen, 1970). A preferential gill route for eliminating carbon dioxide is due to the high solubility of this gas in water, compared to oxygen, whereas the preferential lung route of oxygen is due to the high O_2 capacitance of air. The respiratory pattern of lungfish consists of a short period of air-breathing recurrently interspersing branchial respiration. During emersion, lungfish expire and then reinflate the lungs by a series of buccal pump maneuvers (MacMahon, 1969). The lung oxygen content gradually declines during the subsequent period of submersion (MacMahon, 1970), which lasts for 10 to 30 minutes.

In addition, lungfish have the capability of withstanding long periods of drought by encystment in a cocoon buried in dry mud. This torpid phase is called aestivation (Swan,

1974). At this time, lungfish rely, for both O_2 and CO_2 exchanges, exclusively on fresh air supplied to the lung through a "snorkel" which communicates from inside the cocoon with the surface of the soil. Lungfish are able to survive in this condition for several months or even years, in spite of an elevated body concentration of urea and a reduced metabolism (Delaney et al., 1974).

The capability of lungfish of withstanding drastic environmental changes necessitates several relevant morphological adaptations. The circulatory system is designed to operate in conjunction with two gas exchangers and differs sharply from the circulatory pattern of a fish which breathes under water exclusively. Moreover, in contrast with tetrapods which adopt a definitive and irreversible pattern of aerial respiration, lungfish keep the capability for life of shifting the mode of respiration and, consequently, the corresponding mode of circulation from one pattern to another.

The present review attempts to summarize past and recent morphological findings which have thrown some light on the strategy of the African and South American lungfish to reversibly accommodate two completely different ways of life: aquatic and aerial. Unfortunately, due to lack of access to Australian lungfish specimens, our study is presently restricted to these two types. In addition to a better understanding of the processes of adaptation to a terrestrial life, another fascinating aspect of the lungfish strategy is that it helps in understanding what happens at the very beginning of the postfetal life in a young mammal. It might also serve as a model for studying the physiology of the ductus arteriosus, the pathology of which is still a major problem in prenatal medicine.

CIRCULATION IN LUNGFISH

The general organization of the visceral arches in Protoperidae is shown in Fig. 5. This description based on the works of Parker (1892) and Chardon (1961), was more recently completed (Laurent et al., 1978). Considering all the data together, it became possible to draw a comprehensive view of the blood circuitry system, and then to propose a new functional concept (Laurent et al., 1978, 1979; Laurent, 1981) in African and South American lungfish.

There are six pairs of arches. Arch I, bearing one row of gill filaments, corresponds to the hyoidean hemibranch. Arches II and III, devoid of filaments, represent direct thoroughfares between the heart and the dorsal aorta. Arches IV to VI bear gill filaments and are functionally interposed between the heart and the pulmonary artery. A crucial feature of this circulatory system is the persistence, throughout life of an embryonic channel, the *ductus arteriosus* (DA). It allows the blood exiting the gills to supply the lung or to flow into the dorsal aorta when it is closed or open respectively. Two large veins return the blood to the heart: the pulmonary vein and the vena cava. Another important feature of the lungfish circulatory system is the (incomplete) partition of the heart. By the virtue of a septum dividing the atria and partially the ventricle, and of a spiral valve inside the conus, the blood coming from the vena cava enters the three posterior arches (Arches IV to A VI). Conversely, the blood leaving the lung, via the pulmonary vein, enters the anterior arches (Arches I to III). After traversing Arch I, the blood goes to the brain. Arches II and III supply the systemic circulation, via the dorsal aorta. Interestingly, the hematosed blood supplying the brain perfuses the gill filaments on arch I, in contrast to the blood supplying

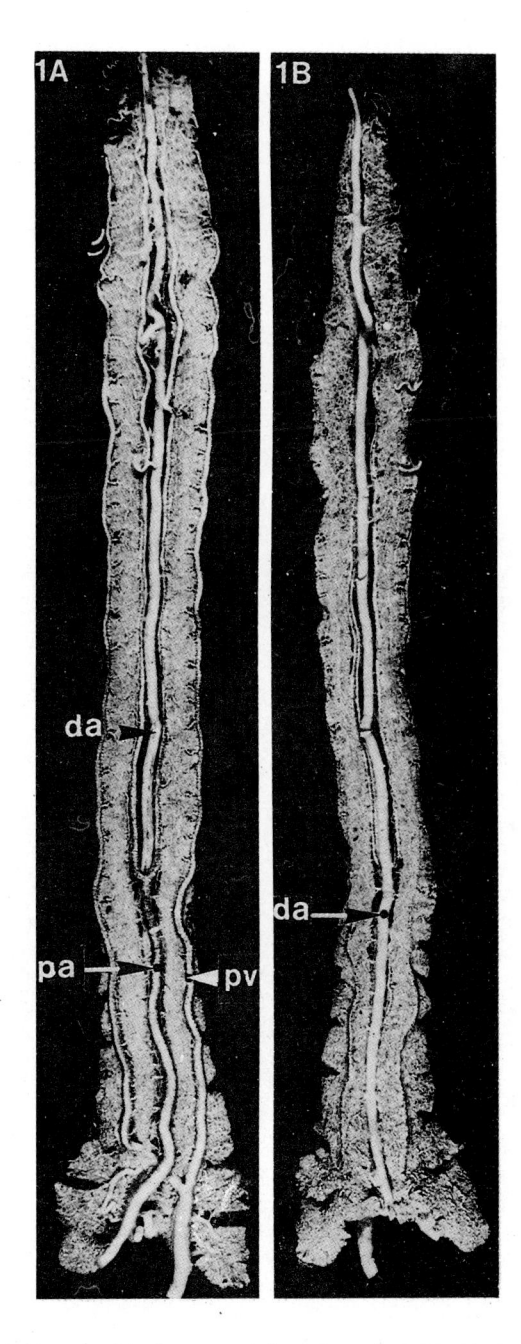

Fig. 1. Microfil cast of *Protopterus aethiopicus* vasculature. A: ventral side; B: dorsal side. In Lepidosirenidae (*Protopterus* and *Lepidosiren* gen.) the parenchyma of the left and right lung is completely individualized and separated by the dorsal aorta (da). Each lung is supplied by tributaries of both right and left pulmonary arteries. Only the left pulmonary artery is visible on the ventral side (pa); the right pulmonary artery approaching the lungs from the dorsal side is not visible here. The pulmonary vein, unique at its entrance in the heart, is formed from two branches running respectively along the external side of the left and right lung (pv). Full scale photo of a specimen weighing 0.400 kg (for more details concerning this organization, see Chardon, 1961).

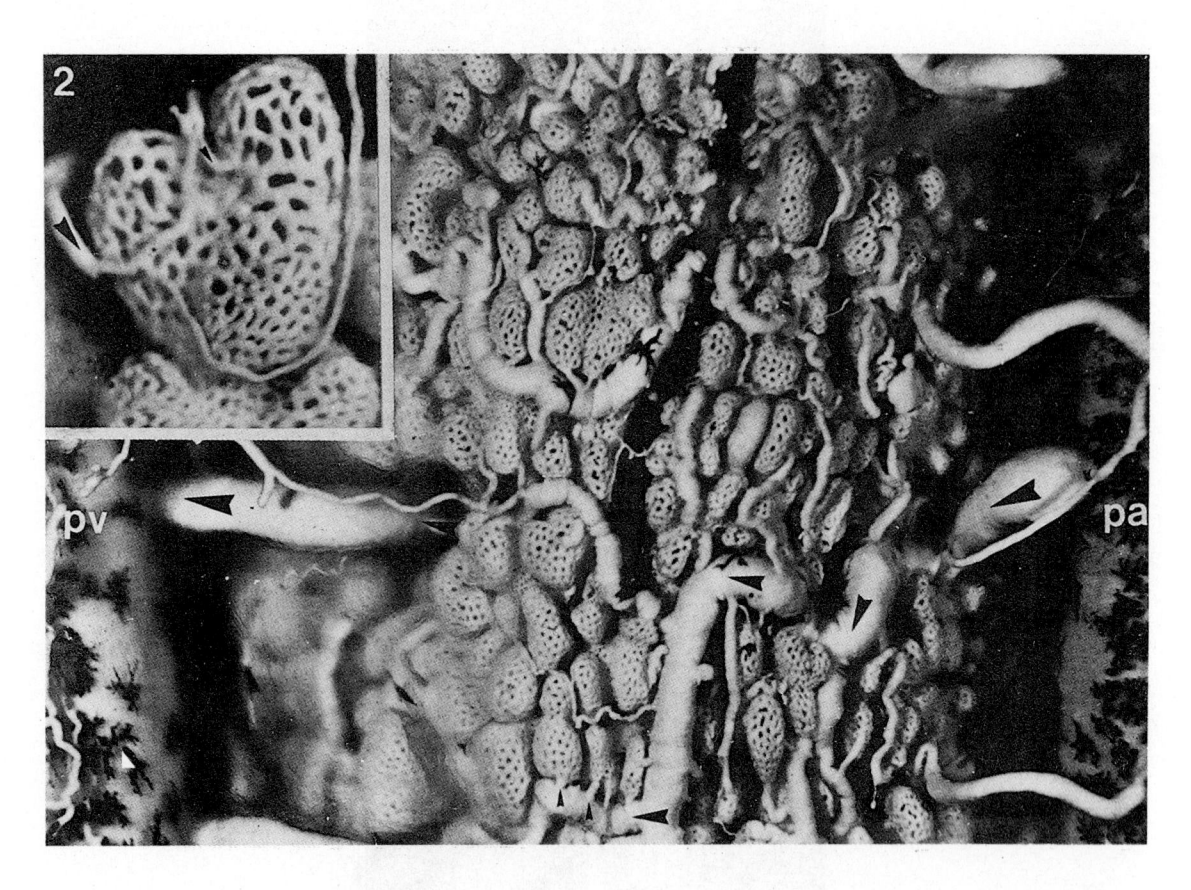

Fig. 2. Microfil injection of the lung of *Protopterus aethiopicus*. Detailed view of distribution in the pulmonary artery supplying the alveolar capillaries. Alveoli seen from their outer side. They form minute baskets, open on the inner side. The efferent, oxygenated blood is collected by veinlets running on the inner side of the lung alveoli. These veinlets are drained into the pulmonary vein (pv). Inset: closeup of an alveolus. Magnification 20 ×, 100 ×.

Fig. 3. Three posterior gill-bearing arches IV–VI viewed from outer left side. Note the dichotomic organization of filaments. aba: afferent branchial arteries. Magnification 5 × (from Laurent et al, 1978).

Fig. 4. Three posterior gill-bearing arches IV–VI viewed from inner (pharyngeal side), showing efferent branchial arteries connected to the pulmonary artery. Magnification 5 × (modified from Laurent et al, 1978).

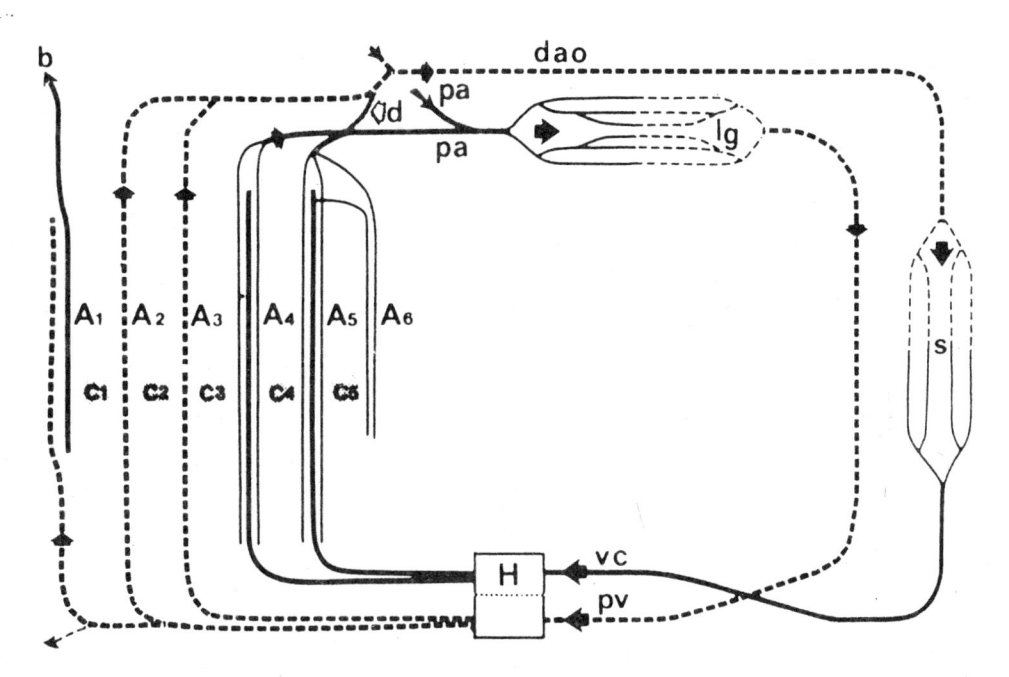

Fig. 5. Schematic diagram of the vascular arrangement of the branchial arches and their relationship with the general circulation in *Protopterus*. Arches shown only from the right side.
Abbreviations: A1 to A6—arches I-VI; b—brain; C1-C5—branchial slits; D—ductus arteriosus; dao—dorsal aorta; H—heart; lg—lung; pa—pulmonary artery; pv—pulmonary vein; vc—vena cava; s—systemic circulation (from Laurent et al.,1978).

the rest of the body, which goes up the filamentless arches II and III. This point will be discussed later.

Thus, persistence of the *ductus arteriosus* and partitioned heart is a trend toward the organization of tetrapods. But coexistence of gills with lung is not possible unless additional structures to control the blood flow exist around that area. These structures are of two kinds. First, anatomical gill shunts allow the blood to bypass the gill exchange surface. In air-breathing teleosts the large bore basal and marginal vessels of the secondary lamellae deviate the blood flow away from the capillary network in order to check loss of oxygen from blood to the hypoxic water of swampy environment (Olson et al. 1994). Second, the sphincter-like segments on the pulmonary artery (pulmonary artery vasomotor segment: PAVS) prevent the blood entering the lung and forces it to enter the dorsal aorta. These structures are described in detail below.

Structure and control of the ductus arteriosus

Vascular casts (using a silicone rubber) show that the location of the ductus is quite similar in *Protopterus aethiopicus* and in *Lepidosiren paradoxa*. Figure 6A shows the ductus in *Lepidosiren* as a short arterial segment connecting the anterior branches of the dorsal aorta with the efferent branchial arteries, which also give rise to the pulmonary arteries. This cast shows that the ductus has a large diameter adapted to a large blood flow. In *Protopterus* (Fig. 6B) the ductus is longer and curves around the branch of the dorsal aorta before entering laterally. The angle formed by the ductus and the efferent arteries of the anterior arches (Fig. 8) makes it unlikely for blood to flow from the anterior arches into the pulmonary arteries, contrary to earlier reports (Szidon et al., 1969). In addition, it is worth mentioning that the efferent branchial arteries of the posterior arches bifurcate in such a way as to allow the blood to flow as easily into the pulmonary artery as into the ductus. A network of vasa-vasorum overlaps the ductus. These vessels originate from the efferent branchial arteries or their tributaries. The ductus wall is continuous with the anterior branches of the dorsal aorta dorsaliy and the pulmonary artery ventrally. As shown on Fig. 9B, the media of the ductus wall is distinguishable by a marked thickening, the part proximal to the pulmonary artery being more muscular. Another interesting observation is the way the ductus joins the dorsal aorta. A valve-like arrangement suggests that the ductus will close down when the hemodynamic pressure is higher in the anterior than in the posterior arches.

The ductus fixed for cytological study often displays an infolded endothelium. This is due to the contraction of the smooth muscle fibers (Fig. 7A, B). As shown in Fig. 10A, the ductus endothelium is characterized by the presence of numerous caveolae, a dense network of presumably actin filaments and rod-shaped Pallade granules. Interendothelial cell junctions are tight. Endothelial cells send cytoplasmic processes across the basal lamina to form gap junctions (nexus) with the smooth muscles of the first media layer.

The most characteristic feature of the ductus is the thickness and organization of the muscular media. This organization is similar in *Protopterus* and *Lepidosiren*. Depending on the orientation of ductus cross sections, up to 30 layers are commonly present. At first glance the disposition of the muscle fibers is helicoid, similar to the ductus of the human fetus (von Hayek, 1935). A detailed observation of differently oriented sections shows that the innermost and outermost layers are disposed more longitudinally and the intermediate layers more circularly. This organization indicates that when muscle fibers contract, the inner and outer layers tend to shorten the ductus and the intermediate layers tend to close the lumen. Observation with the electron microscope of the ductus media shows a tight organization of smooth muscle fibers separated from each other by narrow strips of collagen and a few elastic fibers (Fig. 11). These characteristic features form a resistant (muscular) vessel (Fig. 9). Muscle cells are 40–50 micrometers in length and 5–6 micrometers in width. They contact their adjoining muscle fibers via numerous nexuses. Myofibril tracts anchor on the plasmalemma via dense bodies. In cross section the thick filaments are regularly spaced with numerous thin filaments. Another type of filament, presumably intermediate, is seen variously oriented in the part of the cell rich in organelles (mitochondria, rough endoplasmic reticulum).

The ductus is surrounded by adventitia of connective tissue. In this area there are numerous small arteries and veins (Fig. 9B). Bundles of nerve fibers and different types of cells (fibroblasts, neurones and granular vesicles containing cells) are located close to the external muscle layer. In other words, the ductus is richly vascularized and innervated.

The distribution of nerves and neurones has been studied in detail in lungfish in connection with the problem raised by the nervous control of the ductus in mammals (Laurent, 1981). The nerve bundles which reach the ductus comprise myelinated (2 to 5 micrometers) and nonmyelinated fibers (Fig. 10B, 11A). These bundles of nerve fibers emanate from the nervous intestinalis, a branch of the vagus nerve. Ganglionic cells encapsulated by Schwann cells are distributed in clusters within the adventitia. Terminal nerve fibers are seen beneath the capsule in synaptic relation with short ganglionic dendritic processes. Endings of the ganglionic cell axons are distributed around the ductus muscle fibers forming large varicosities that often lie in contact with the smooth muscle cell membrane (Fig. 10B, 11A). These varicosities, several microns wide, contain agranular vesicles (40 nanometers) which are tightly clustered in large patches. The location of patches on the axonal membrane suggests a process of outward release. More information on granular vesicle cells in air-breathing fish is available in a recent review by Zaccone et al. 1995.

A second type of innervation consists of cells containing granular vesicles. They lie close to blood vessels or to smooth muscle fibers. When they are located close to small vessels or capillaries, as in the case of Fig. 12, it is possible for their content of granular vesicles to be released into the lumen. Another type of relationship consists of granulated cell processes approaching the muscle cells by less than 0.5 micrometers, as shown on Fig. 10C. The structure of granulated cells is identical no matter where they are located, be it close to capillaries or to muscle fibers; they are elongated cells and their pericaryon (15 × 20 micrometers) is occupied almost exclusively by the nucleus. Most of the cytoplasm consists of long varicose processes which form foot-shaped endings close to muscle fibers (Fig. 11B). Granular vesicles range from 70 to 100 nanometers and are more concentrated within the processes. Large nerve endings containing agranular small vesicles are seen contacting granulated cells (Fig. 12), on which they form characteristic cholinergic synapses. Several synapses are present on the same granulated cell. Nerve fibers that synapse with granulated cells are presumably vagal.

Comparison of ductus in lungfish and mammals

From an embryological point of view, the lungfish ductus has the same origin as in the case of mammals. In both cases it is a derivative of the sixth branchial arch artery. Both in mammals and lungfish the ductus media is poor in elastic fibers, a feature contrasting with the aorta or the pulmonary artery which are almost pure capacitance vessels and rich in elastin. In mammals the ductus closes within 10 hours after birth due to intense contraction of the smooth muscle, impairing blood perfusion of the placenta and directing the right ventricular outflow to the pulmonary circulation. This closure, when achieved, is permanent in mammals. It is reversible in lungfish. However, it has been suggested that during fetal life transient or partial constrictions may occur which keep the lung slightly irrigated (Brinkman et al., 1969).

Similitudes of lungfish and mammalian ductus also concern the distribution of nerves and vasa vasorum. As in the fetal lamb (Allan, 1961), the ductus in lungfish is supplied with arterialized blood by a rich adventitial network of dorsal aorta tributaries. This density of vasculature suggests that in lungfish, as in mammals, contraction of the ductus is controlled by blood oxygenation. It has been proposed that high blood O_2 content (or partial pressure) after birth elicits closure of the ductus (Fay, 1971; Fay et al., 1977; MacMurphy et al., 1972; Oberhansli-Weiss, et al., 1972). Ductus O_2 sensitivity has been demonstrated in lungfish (Fishman et al., 1985). This sensitivity plays an important role in the succession of events during transition from emersion to submersion (Laurent, 1981; Fishman et al., 1985).

The existence of a rich innervation in close relationship with the ductus smooth muscle fibers suggests that in lungfish, as in mammals (Heymann and Rudolph, 1975), in addition to the oxygen factor, the nervous system plays an important role in ductus control. In lungfish the number of cholinergic nerve endings is extremely large, which contrasts with the relatively small vasodilation caused by experimental perfusion of the ductus with acetylcholine (Fishman et al., 1985).

In mammals the ductus is innervated by sympathetic nerve fibers, in addition to parasympathetic ones (Boyd, 1941; Aranson et al., 1970; Ikeda, 1970; Silva and Ikeda, 1971). A rich plexus of varicose fibers, revealed by fluorescence studies, is seen within the most external layer of the muscular layer (Boreus et al., 1969; Ikeda, 1970). As a conclusion of similar studies in lungfish (Laurent et al., 1979; Laurent, 1981; Fishman et al., 1985) and because of present knowledge about the lungfish autonomic system (Abrahamsson et al., 1979a), the existence of an adrenergic plexus is unlikely in lungfish. Nevertheless, perfusion experiments have clearly demonstrated the powerful dilative action of dopamine and the constrictor effect of norepinephrine on the lungfish ductus (Laurent, 1981; Fishman et al., 1985). In the absence of a sympathetic network it has been assumed that cells containing granular vesicles (presumably SIF cells) might take over some of the sympathetic functions (Abrahamsson et al., 1979b). Processes of these cells which approach the ductus myocytes by less than 100 Angströms, might release active substances. It is worth mentioning that numerous gap junctions (nexus) between ductus muscle fibers and between endothelial cells and muscle fibers make the ductus a syncytium-like structure, a situation shared with the mammalian ductus.

Pulmonary artery vasomotor segments

These vasomotor segments have been identified in dipnoan fish (Laurent et al., 1978) and are considered crucial components of adaptation to bimodal breathing (Laurent et al., 1979). These structures are similar in localization to those present in amphibians and chelonians (*Rana*: De Saint-Aubain and Wingstrand, 1979; *Chrysemis scripta*: Milsom et al., 1977; Burggren, 1977; *Testudo graeca*: Burggren, 1977). In amphibians and chelonians the pulmonary artery vasomotor segments are situated close to the vestigial ductus (*ligamentum*). In lungfish they consist of thickening of the media of the pulmonary artery, which extends from the *ductus* caudally over one-third of the artery (Fig. 6B). In this region the media is reinforced by a large number of muscle layers (Fig. 13). The pulmonary branch of the vagus, which runs parallel to the artery, sends several bundles within the adventitia which give rise to a rich innervation distributed to the most external layer of the media (Fig. 14A, 14B). Nerve profiles containing nongranular vesicles approach the smooth muscle cells of the first external layer (Fig. 15B). Granular vesicles containing cells are also present in the pulmonary artery adventitia, closely associated with smooth muscle cells (Fig. 15A). It is interesting to mention that these granular vesicles containing cells receive a reciprocal innervation (Fig. 16). This sophisticated structure is commonly observed in tetrapod SIF cells (small intensively fluorescent cells) (Taxi, 1979). Such an organization suggests that some information is sent from the cell to the nerve and that the nerve controls the cell. Although the pulmonary vasomotor segment possesses dual innervation, as in the ductus, perfusion studies failed to demonstrate any effect of epinephrine, norepinephrine or dopamine (Fishman et al., 1985). However, perfusion of the pulmonary artery vasomotor segment with acetylcholine provoked marked constriction, contrary to the ductus which dilated. Another important peculiarity of the pulmonary artery vasomotor segments is insensitivity to oxygen.

Gill shunts

The existence of anatomical gill shunts is a peculiarity of bimodal breathers (Laurent, 1985). They have never been observed exclusively in water-breathing fish (Laurent, 1984). Several types of gill shunts are present in Dipnoi. One type consists of large arteries of the gill-less arches II and III (Parker, 1892). These ventrodorsal shunts play a fundamental role in lungfish circulation (see p. 47 and Fig. 5). More recently, a second type was described in Dipnoi (Laurent et al., 1978), which allows the blood to bypass the gill lamellae circulation in the posterior arches IV to VI. These shunts consist of large vessels with thick walls connecting the afferent and efferent gill arteries at the base of the gill filaments (Fig. 17). This organization is similar to that of the external gills of amphibians (Maurer, 1888; Fige, 1936) and of *Lepidosiren* (Robertson, 1913). These shunt vessels presumably are the remnant of the primitive vascular loops connecting the ventral and dorsal branchial arch vessels. Indeed, during the development of the gill vasculature the first loop does not give rise to a typical branchial system of capillaries, as is usual in teleosts (Morgan, 1974), but remains in the stage of a single vessel. Later, after the external gills have disappeared, these vessels form large straight connections. There is strong functional evidence for neurohumoral control of these shunts in amphibians (Fige, 1936). In lungfish, as in amphibians, structural data on shunt innervation is lacking. Acetylcholine increases the branchial resistance of lungfish, whereas epinephrine and norepinephrine dilate the branchial vascular bed (Johansen and Reite, 1968).

From morphology to physiology of lungfish circulation

When three structures, ductus arteriosus, pulmonary artery vasomotor segments, and gill shunts, are inserted into the classical scheme of the lungfish vasculature, one readily understands how the blood might preferentially perfuse either the lungs or the gills. Sketch 1 of Fig. 18 outlines the blood pathways of a (lung)fish relying on its gills exclusively. In this case the pulmonary artery vasomotor segments (1) and gill shunts (3) are shut and the ductus arteriosus (2) open. Due to reduction in pulmonary blood flow, a large intracardiac shunt compensates for the small pulmonary vein return. The blood perfuses the gill filaments of arch II (see p. 47 and Fig. 5) and arches IV to VI as well, and reaches the systemic circulation. When the fish relies on the lungs exclusively (sketch 2 of Fig. 18), the pulmonary artery vasomotor segments are shut and the ductus arteriosus and gill shunts are open. At this time a double circulation is set up; the cardiac shunt is minimal and blood perfuses lungs and tissues consecutively.

Thus, just by considering the morphology of the vasculature and of its controlling structures, one can assess the distribution of the blood flow according to the conditions the fish has to face. Some physiological data is available which confirms these circulatory patterns (MacMahon, 1969, 1970; Johansen et al., 1968; Johansen and Reite, 1968; Szidon et al., 1969). Other data concerning the effects of blood oxygenation on the motorization of the ductus arteriosus and the pulmonary artery vasomotor segments is also available (Laurent, 1981; Fishman et al., 1985). Due to experimental difficulties, physiological data alone does not solve the fascinating question of lifelong persistence of aquatic and aerial modes

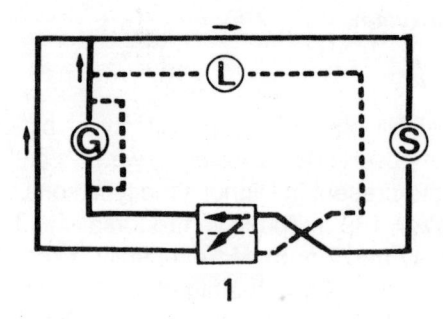

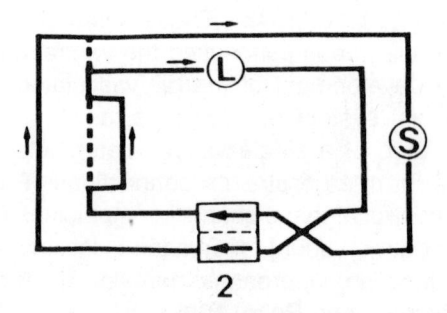

Fig. 18. Schematic representations of lungfish circulation during prolonged submersion (upper scheme) and emersion while air breathing (lower scheme) In [1] gills are well perfused, the ductus is open, and the pulmonary artery is closed; a cardiac right-to-left shunt is present. In [2] gill shunts are open, the ductus is closed, and the pulmonary artery is open. The right-to-left shunt is minimal.

of respiration in parallel. Combined morphological findings now allow us to reconstitute the respiratory cycle in lungfish.

Lungfish presently live in poorly oxygenated water of tropical swamps. They periodically surface to fill their lungs with fresh air. At this time the ductus is closed and the blood is forced to perfuse the lungs through the open pulmonary artery vasomotor segments. The arterial blood becomes well oxygenated and the fish emerge. Now the tissues are well supplied with oxygen for a certain period. While oxygen consumption goes on, arterial blood oxygenation decreases and the ductus progressively opens. Consequently, the branchial arteries of arches IV to VI communicate with the dorsal aorta. The gill shunts close down and the gill filaments, now perfused, release the highly soluble carbon dioxide (MacMahon, 1970). The fish become progressively hypoxic up to a new air-breathing phase. The ductus shuts down again and a new cycle begins. Some of these events are probably under metabolic control (direct effect of oxygen on smooth muscle). The central nervous system plays an additional role due to the opposite effects of some neurotransmitters on the vasomotor structures. During aestivation, the cocooned fish rely on their lungs. Presumably the gills play no role and the blood circulation permanently conforms to the mammalian pattern.

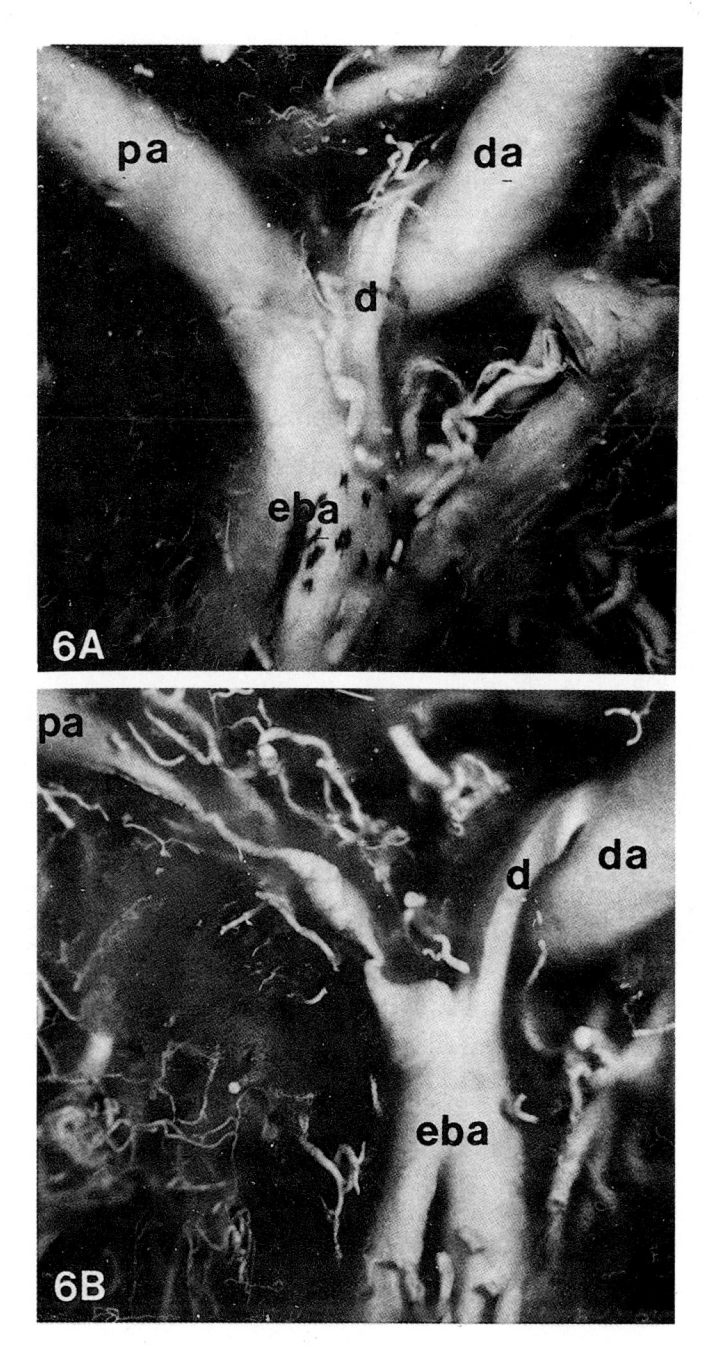

Fig. 6A. Ductus arteriosus in *Protopterus aethiopicus*. Microfil cast of the junction between efferent branchial arteries (eba IV–VI), pulmonary artery (pa), and dorsal aorta (da) via the ductus arteriosus (d).

Fig. 6B. Same kind of preparation after local application of acetylcholine (10^{-6}). Note constriction of the pulmonary artery in contrast to the ductus and dorsal aorta (modified from Laurent et al., 1978). Magnification 20 ×.

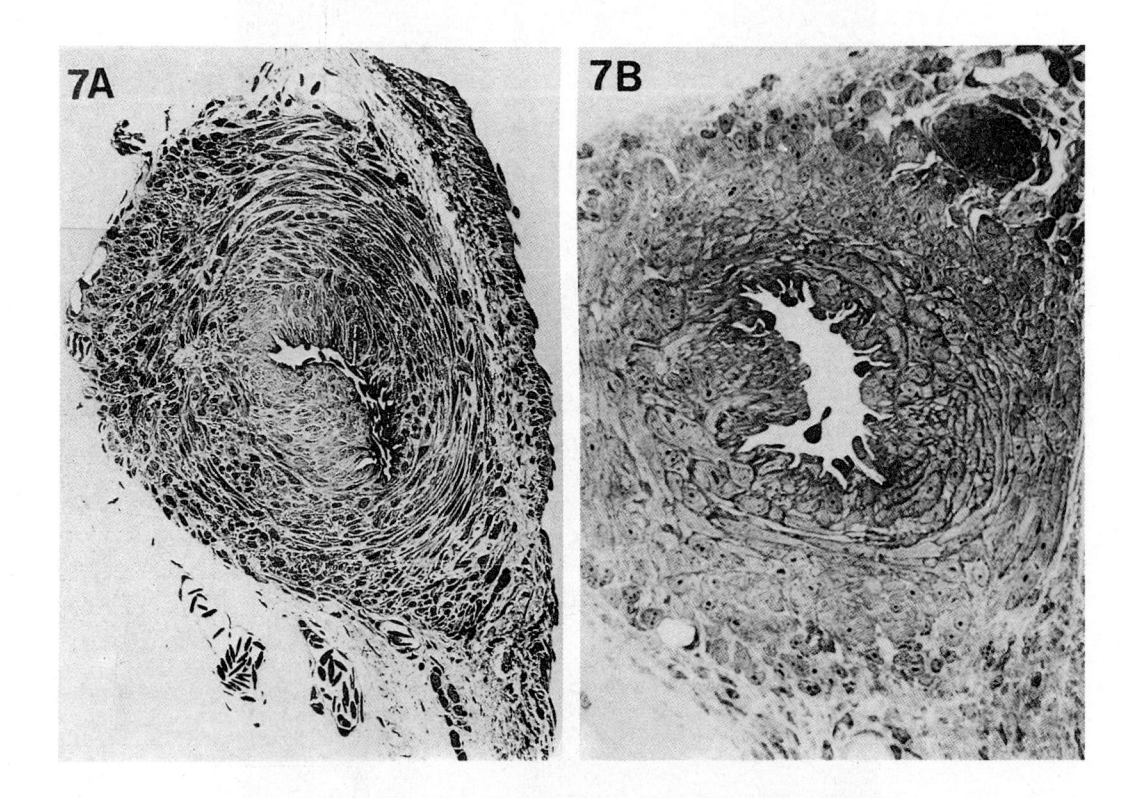

Fig. 7A. Cross section of ductus arteriosus of *Protopterus aethiopicus*. Note thickness of the media and complex arrangement of smooth muscle fibers. Due to contraction of the media, the lumen is narrow. Numerous small vessels are present in the adventitia in addition to nerve cells and fibers (from Fishman, DeLaney and Laurent, 1985).

Fig. 7B. Same type of preparation in *Lepidosiren paradoxa*. The structure is quite similar to that of fig. 7A. Note the infoldings of intima, suggesting a strong contraction of the ductus. Magnification 100 ×.

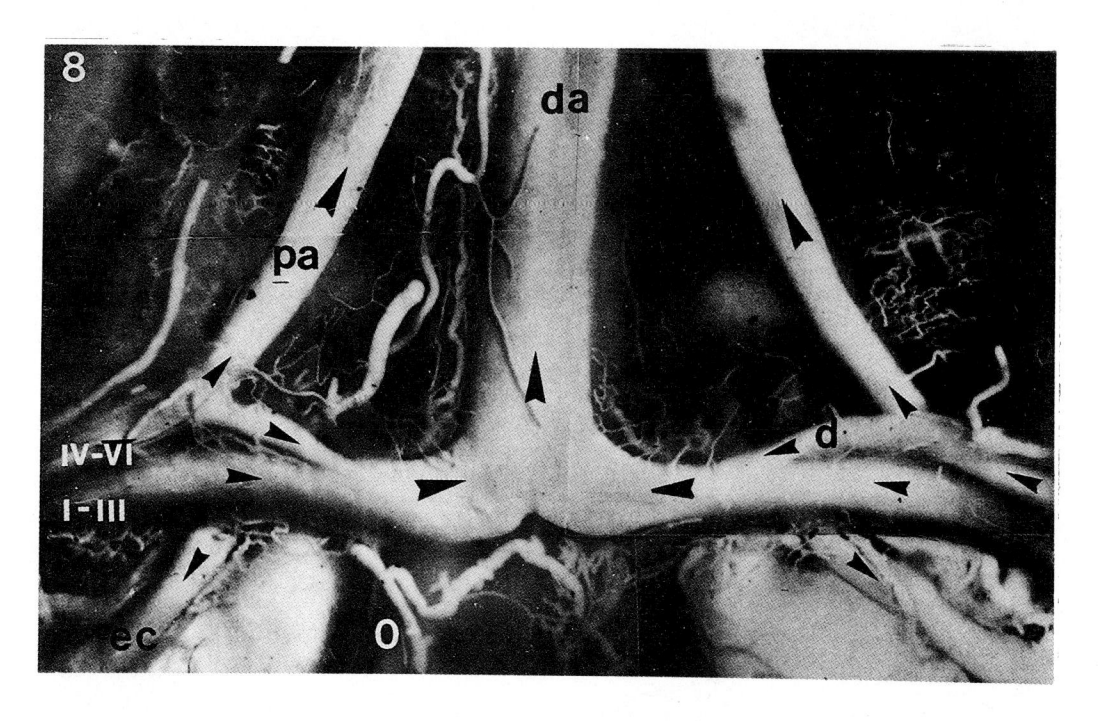

Fig. 8. Organization of the postbranchial arteries in *Lepidosiren paradoxa*. Note division of the dorsal aorta (da) into two aortic roots. Pulmonary arteries appear as extensions of the efferent branchial arteries of posterior arches IV–VI. It is clear from this cast preparation that the blood flow from the efferent branchial arteries of arches I–III cannot reach the pulmonary arteries (pa) via the *ducti arteriosi* (d). ec—external carotid. Magnification 10 ×.

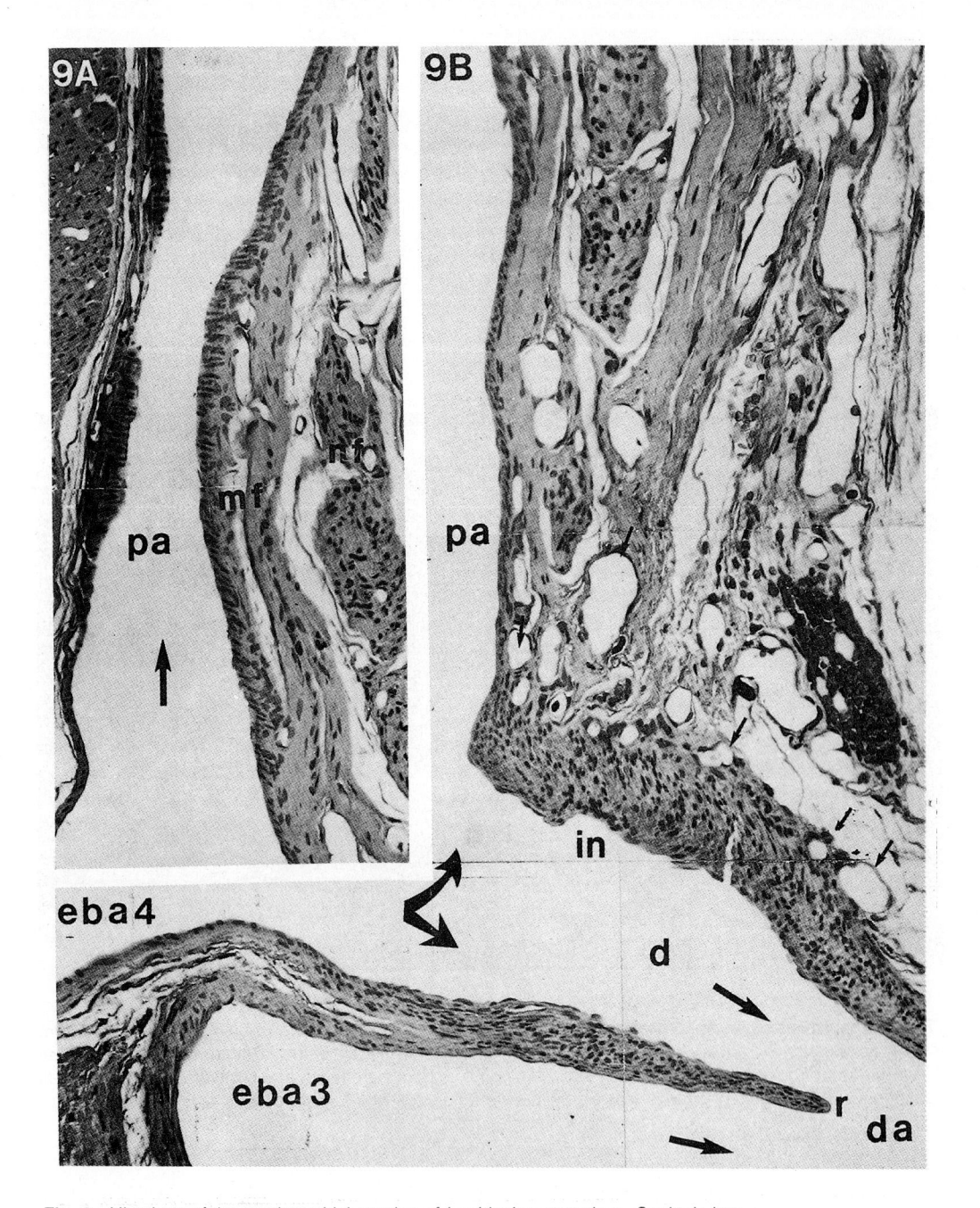

Fig. 9. Histology of the postbranchial arteries of *Lepidosiren paradoxa*. Sagittal view.

Fig. 9A. Longitudinal section of the pulmonary artery. The narrow part represents the pulmonary artery vasomotor segment.

Fig. 9B. Photomontage of several paraffin sections (trioxyhemateine staining method). On the left, lumens of the pulmonary artery (pa) and the efferent branchial artery or arch IV (eba 4) converge. The inner side of the ductus (d) is surrounded by a thick musculature and numerous *vasa vasorum* (small arrows); the opposite wall which separates eba 3 and eba 4 gradually vanishes and finally disappears as these vessels converge into the dorsal aorta (da). Presumably the rim (r) functions as a valve impairing possible reversal of the normal ductal flow (shown by straight arrows). In the case of ductal contraction, the curved arrows show the route to the pulmonary artery. nf—nerve fibers; mf—smooth muscle. Animal weight: 0.5 kg. Magnification 40 ×.

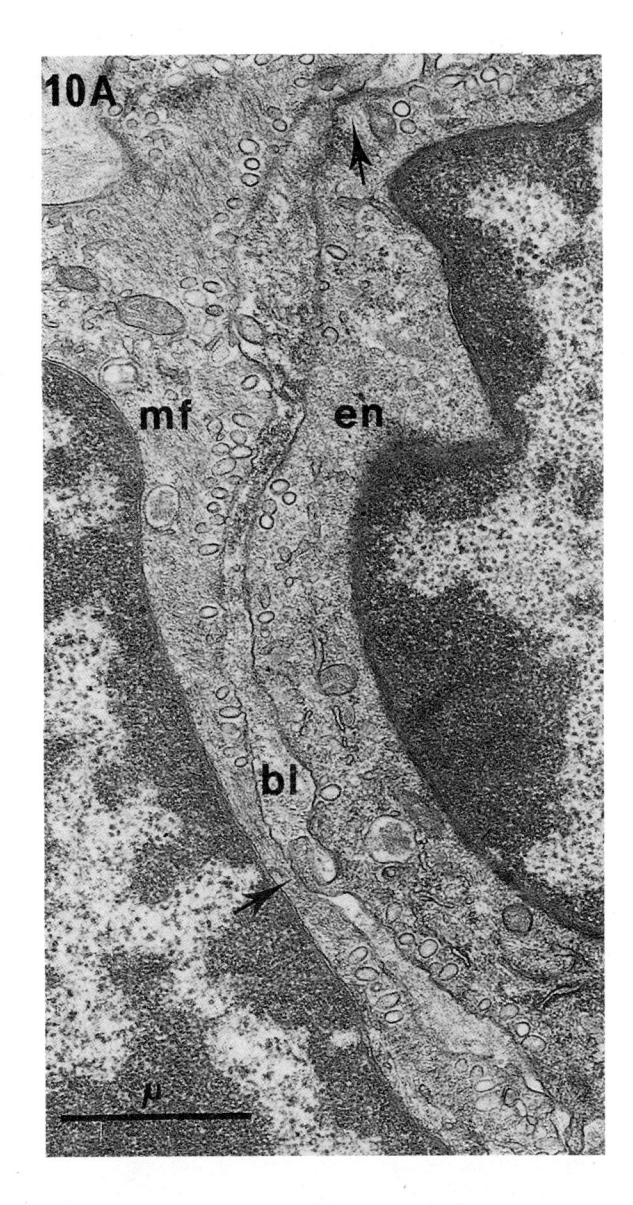

Fig. 10. Ultrastructure of the ductus in *Protopterus*.

Fig. 10A. Relationships between endothelium (en) and smooth muscle cells of the media (mf). Note presence of endothelial processes (arrows) crossing the basal lamina (bl) and forming tight junctions with media muscle fibers.

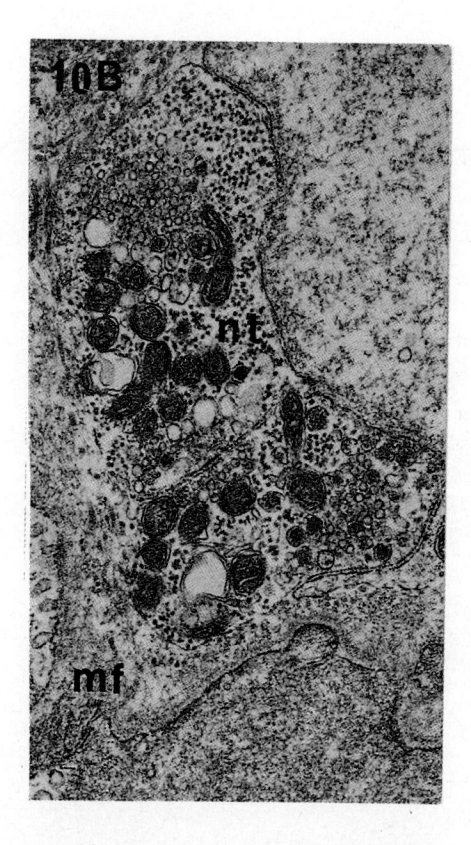

Fig. 10B. Innervation of the ductal smooth muscle by cholinergic nerve terminal (nt). Note presence of cholinergic agranular vesicles (40–50 nm), glycogen granules, and numerous mitochondria.

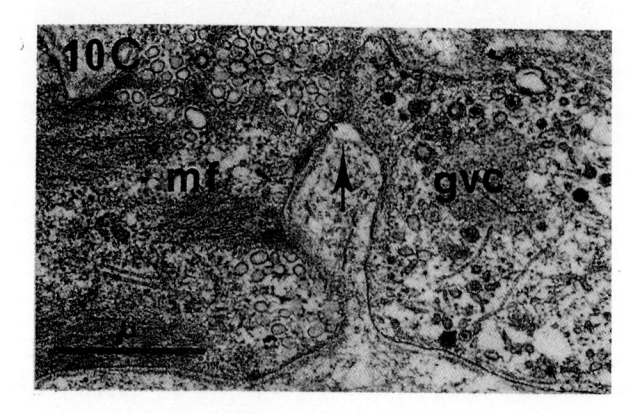

Fig. 10C. Innervation of the ductal smooth muscle by process of granular vesicles cell. Note the close contact between the two structures (arrow).

Fig. 10D. High magnification of junction between neighboring muscle fibers. These junctions, termed nexuses, suggest the ductus functions like a syncytium.

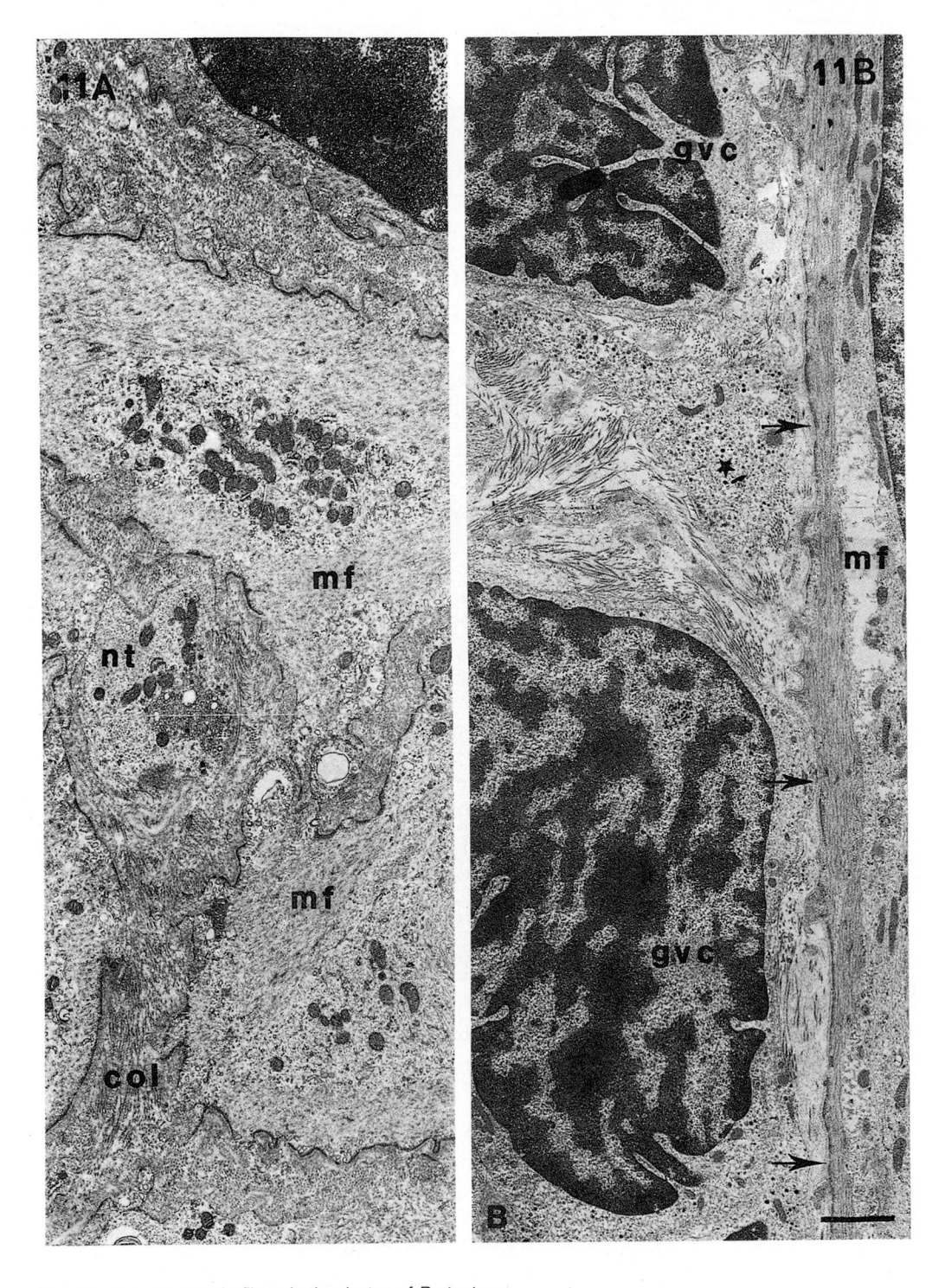

Fig. 11. Smooth muscle fibers in the ductus of *Protopterus*.

Fig. 11A. A nerve terminal (nt) encased between four smooth muscle fibers (mf). Note the densely packed agranular vesicles and collagen (col) between the cells.

Fig. 11B. A second type of control is exerted by granular vesicle cells (gvc) whose processes approach ductal muscle fibers by less than one micrometer (arrows). Bar 1 μm.

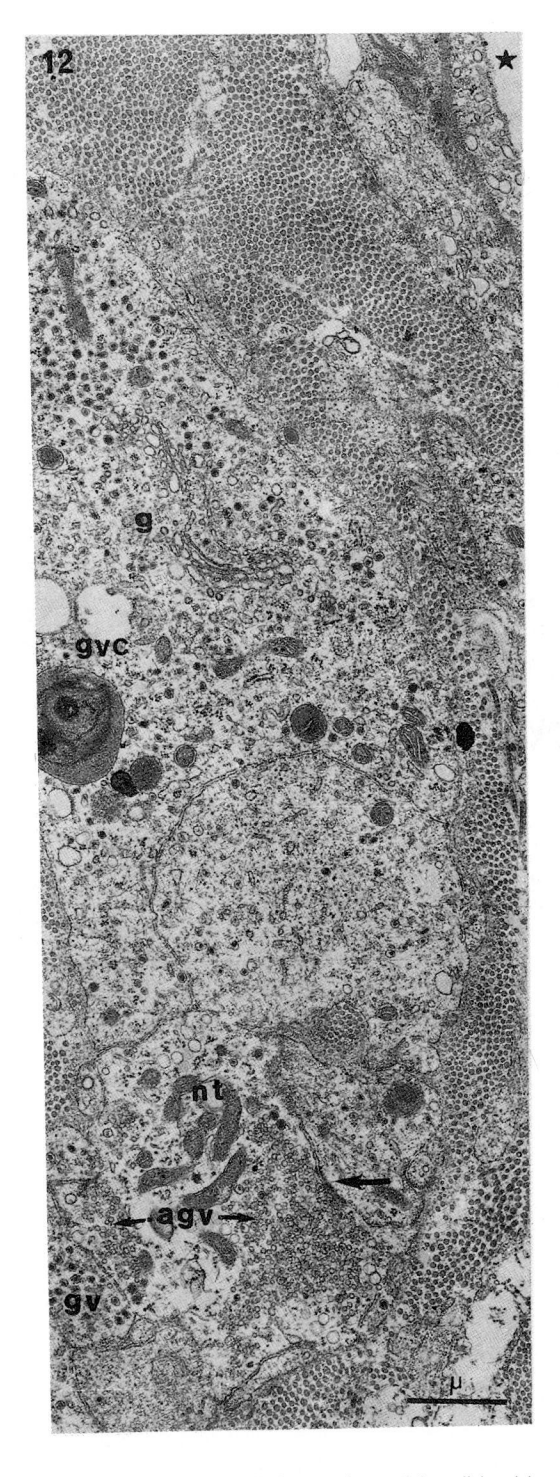

Fig. 12. Structure and control of granular vesicles cell. A granular vesicle cell (gvc) is seen close to an adventitial blood vessel of the ductus (*). Note a cholinergic type nerve ending (nt) encased in gvc processes (identified by granular vesicles, gv) and synapsing with one of them (arrow).

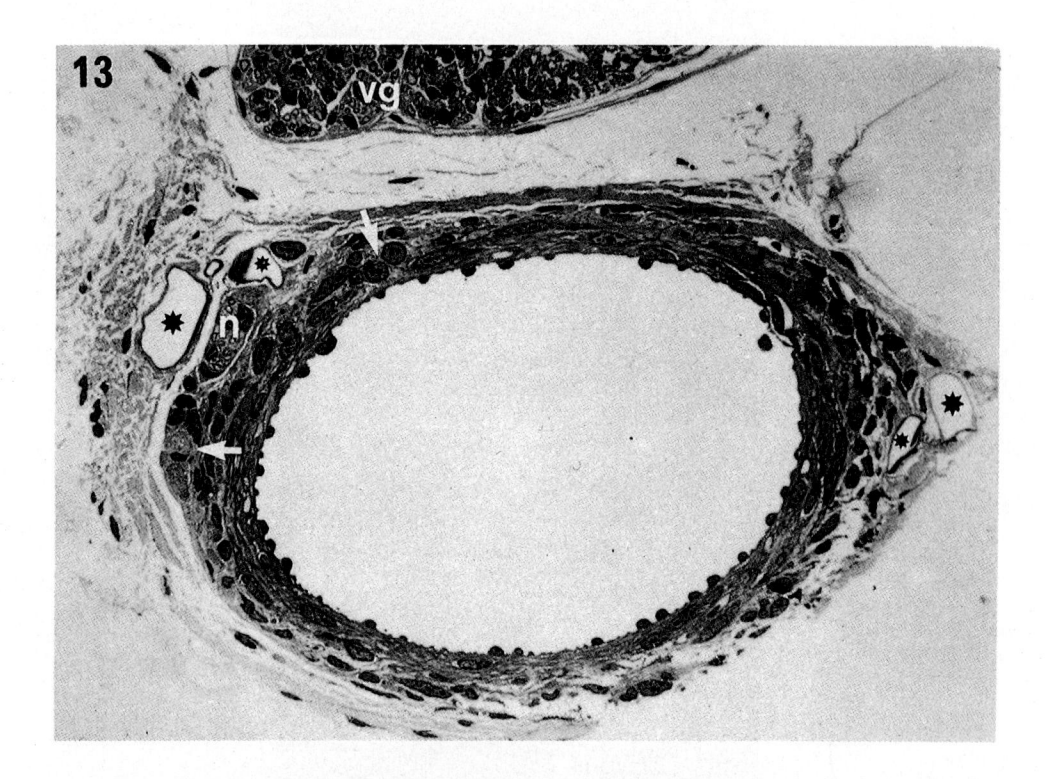

Fig. 13. Histology of pulmonary artery vasomotor segment. The vessel is well dilated due to procaïne perfusion prior to fixation. Note a pulmonary branch of the vagus nerve (vg) running parallel to the pulmonary artery. Small nerve bundles penetrate the adventitia and innervate the segment. In addition, granular vesicle cells (as confirmed with the electron microscope) are present within the thick media. Vasa vasorum are also present in the adventitia (stars). Magnification 50 ×.

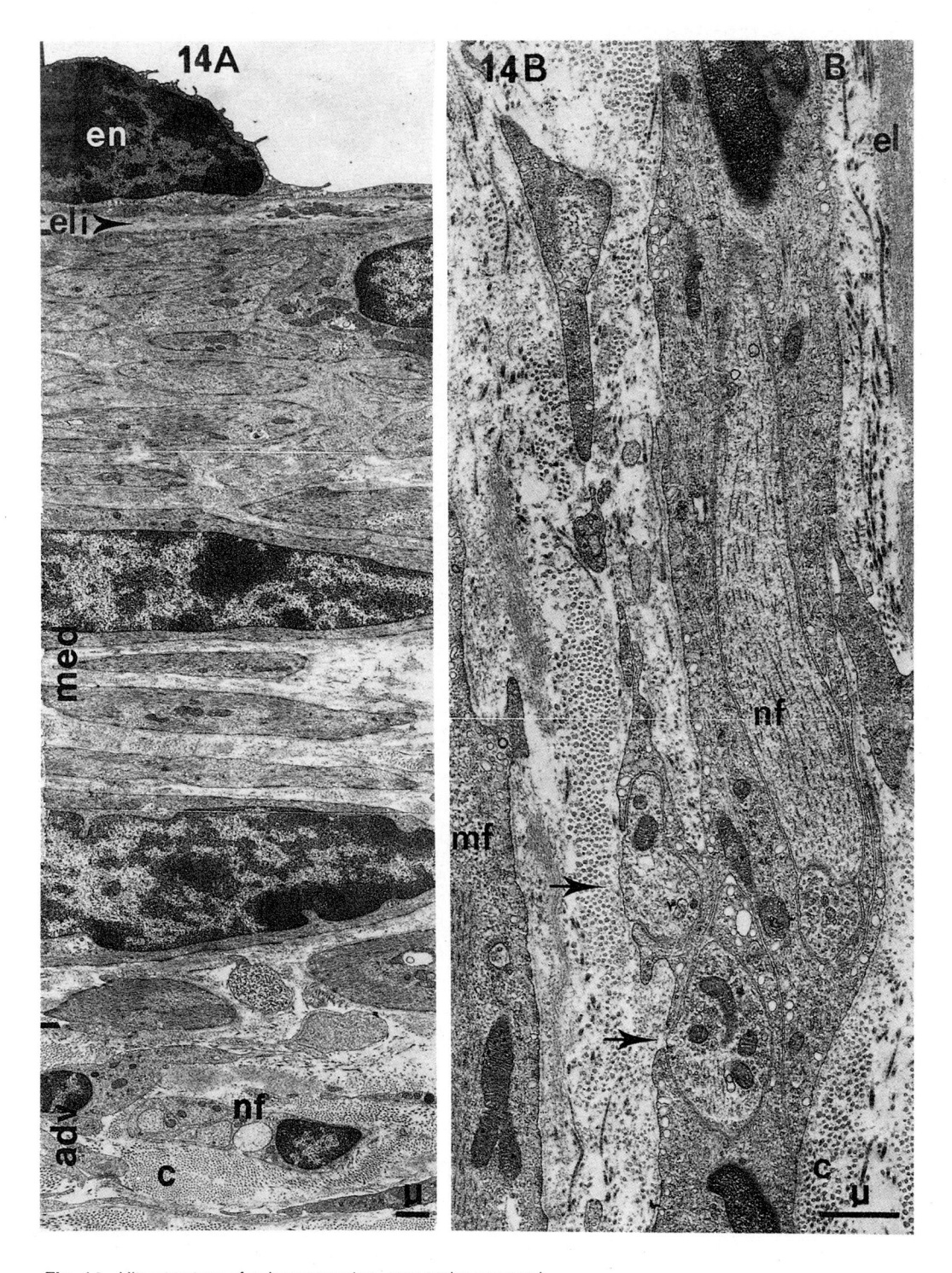

Fig. 14. Ultrastructure of pulmonary artery vasomotor segment.

Fig. 14A. Ultrathin section of the segment wall showing more than 10 smooth muscle layers form the media (med). Note nerve fibers within an adventitia rich in collagen.

Fig. 14B. Distribution of nerve fibers within the media. Normally unsheathed in Schwann cells (nf), fibers of cholinergic type display naked area (arrow) facing muscle fibers (mf).

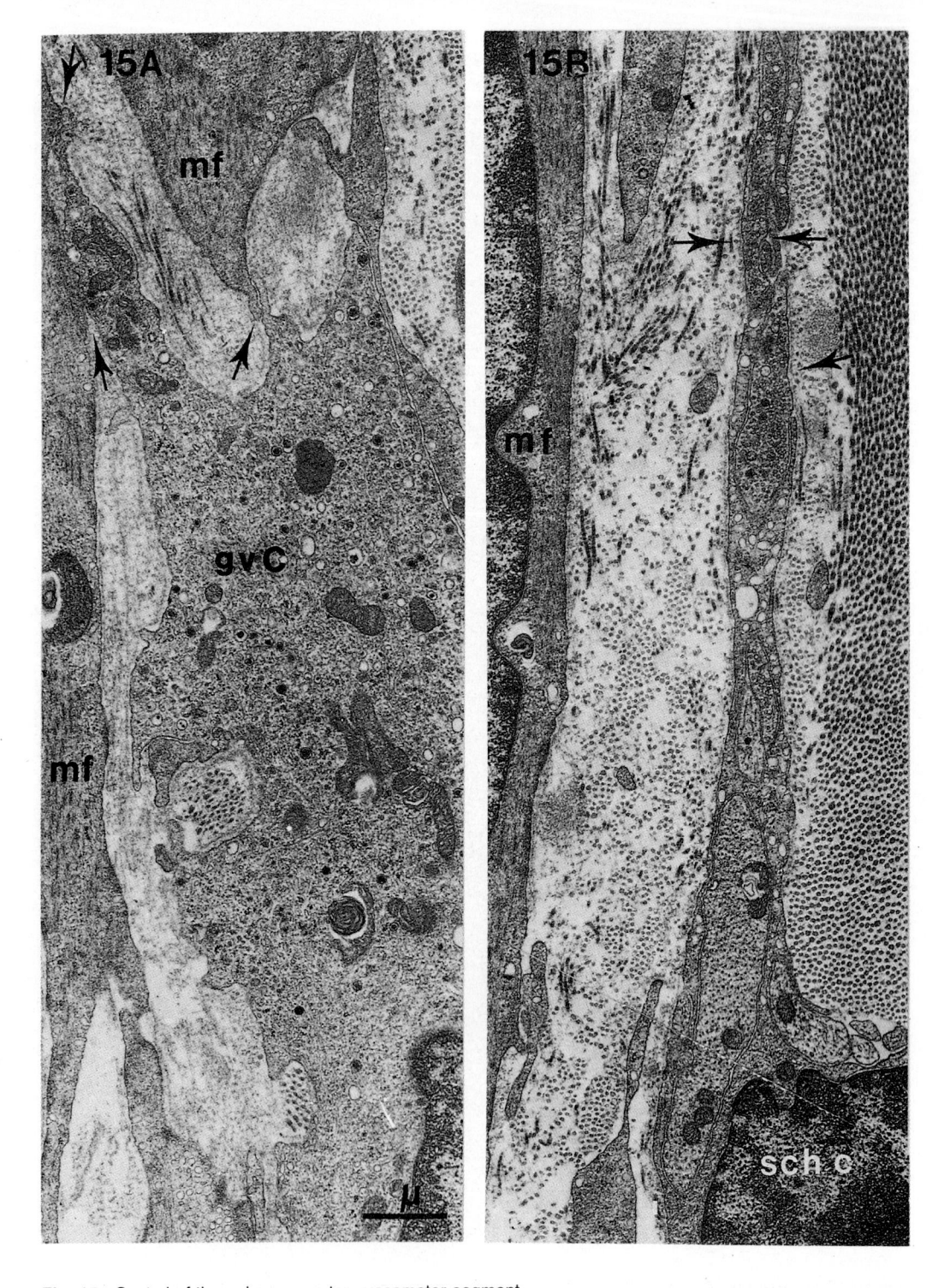

Fig. 15. Control of the pulmonary artery vasomotor segment.

Fig. 15A. A granular vesicle cell (gvc) in close association with several smooth muscle cells (mf). Note the narrow gaps (arrows).

Fig. 15B. Nerve profiles containing agranular vesicles (arrows) close to the first layer of the media. Note abundance of elastin (white matrix), which has been reported scarce in the ductus.

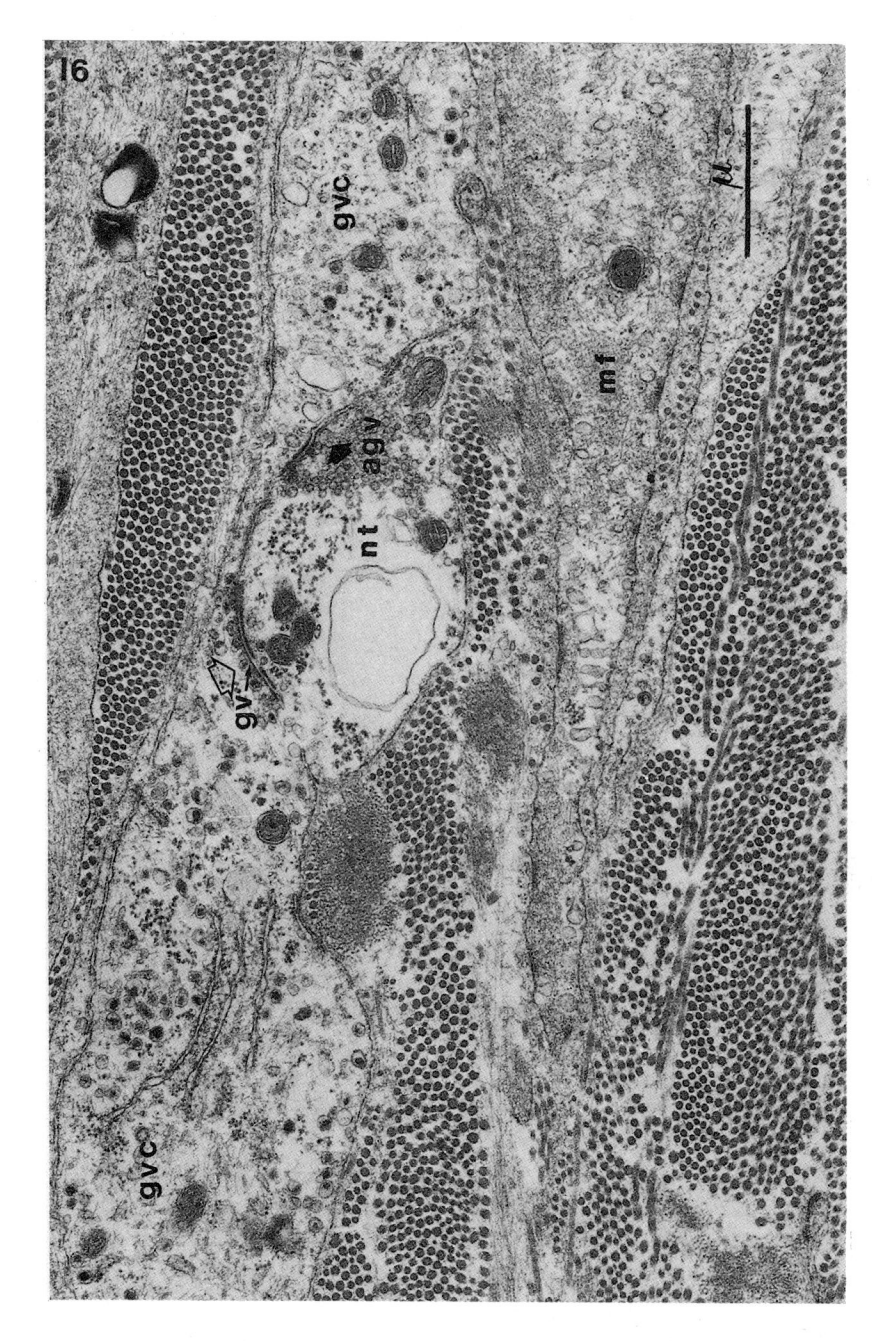

Fig. 16. Pulmonary artery adventitia. A reciprocal synapse between a nerve terminal and a granular vesicles cell. Note the presence of:
—an afferent, centripetal synapse (open arrow) with accumulation of granular vesicles on the postsynaptic membrane;
—an efferent, centrifugal synapse (black arrow) with accumulation of agranular vesicles. Granular vesicle cell surrounded with smooth muscle cells (mf).

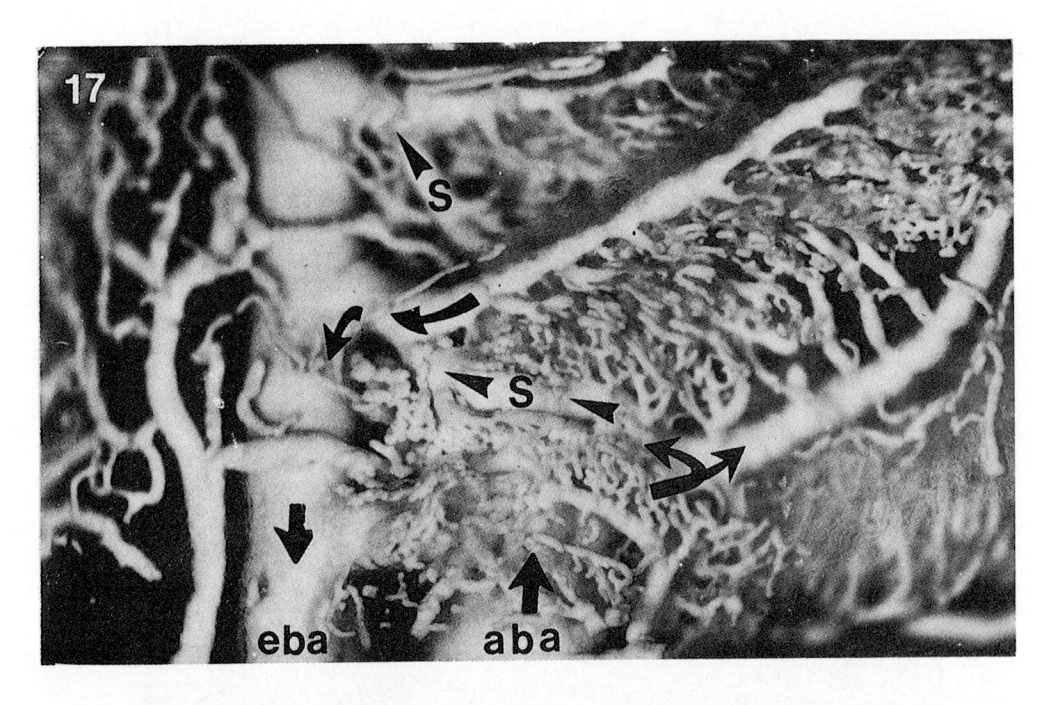

Fig. 17. Vascular cast of the gill filaments in *Protopterus* showing the large shunt vessels (S) connecting the bases of filamental afferent and efferent arteries. Note the direction of the blood flow which might enter the gill capillaries or might pass through the shunt. Magnification 65 × (from Laurent et al. 1978).

SIGNIFICANCE OF THE LUNGFISH MODEL

Ontogeny

According to the present description of the lungfish vasculature, it appears that the dipnoan organization is typical to some extent of that observed in tetrapods during the aquatic stage of their development. The anatomical structures which respectively control perfusion of the lungs and gills are present in amphibians and reptiles before metamorphosis. Some of them disappear after transition to aerial life and completion of the air-breathing pattern. This is the case of the ductus arteriosus. Some structures remain functional, such as the pulmonary artery vasomotor segments, and serve to control the lung circulation in species which may rely on cutaneous exchanges in addition to lungs. In the embryo of amniotic vertebrates (Romer, 1970) the lungs are not functional and the aortic arches, which, of course, do not develop gill filaments, are connected to the dorsal aorta. Gill slits are present as pharyngeal pouches. Gill respiration is replaced by placental exchanges via the dorsal aorta and the umbilical artery. A cardiac right-to-left shunt operates through the foramen ovale which plays the same role as the incomplete partitioning of the ventricle in lungfish. It is not known whether vasomotor segments are present in the mammalian embryo and serve to reduce the blood flow to the lung. Possibly the fetal ductus in mammals is permanently open and does not participate in any form of active regulation of the pulmonary blood flow. As an opposite viewpoint, it has also been suggested that active reflex ductal constrictions occur during fetal life (Brinkman et al., 1969). Thus the way the ductus functions during embryonic life would be, to some extent, similar to that of lungfish. It is suggested that lungfish might serve as a model for studying the physiopathology of circulatory disfunctions which occur in children after birth due to nonclosure of the ductus.

Phylogeny

The relationship between phylogeny and ontogeny has been a matter of debate for more than a century. According to a very popular concept individual development (ontogeny) repeats the history of the race (phylogeny). This "law" has been reexamined many times. Actually, as pointed out by Romer (1970), "it is the fish embryo, not the adult fish, which the mammalian embryo resembles." Unfortunately, data concerning the embryonic development of lungfish is scant and we do not know whether the ancestors of lungfish living during the Paleozoic era had the same capability for relying on atmospheric oxygen as do the actual forms. The atmospheric oxygen partial pressure at the beginning of the Silurian period has been estimated at 14 torr (see Dejours, 1975). Due to the larger air O_2 capacitance and shortage of oxygen in rivers and swamps, a strong pressure was exerted on the animal kingdom to rely on atmospheric respiration. Thus it is likely that the dipnoan ancestors developed bimodal respiration very early, in a way similar to the ancestral crossopterygians evolving to primitive amphibians. A strong argument in favor of this concept is the complete similitude of organization between *Lepidosiren* and *Protopterus*, two genera which evolved on separate continents after the breaking away of Gondwanaland (early Mesozoic). The possibility of resisting seasonal droughts by encystment was probably acquired later. The occurrence of a vasoactive ductus in amphibians is probable, since after metamorphosis a vestigial *ligamentum* remains, indicating that the common ancestors of lungfish and amphibians were equipped with a vasoactive ductus. Thus the ductus arteriosus appears to be a very primitive structure. It is a fascinating fact that this rather complex structure, together with its control apparatus, remains essentially the same as it was 400 million years ago.

CONCLUSION

The morphological approach has been a determinant in revealing the mechanisms by which lungfish are able to rely on bimodal respiration. A precise positioning of the structures controlling blood flow respectively to the lungs and gills enables understanding how adaptation to aerial respiration and life on land was effected during the evolution of vertebrates from fish to mammals.

It is fascinating to realize that the same process, or at least, part of the same process, occurs every time a child is born. Hence it should be kept in mind that lungfish might be a good experimental model for studying some problems of prenatal medicine in man.

Finally, it should be stressed that most of the sophisticated structures which control the basic life processes, such as circulation or respiration, have been present and operating from the very beginning of vertebrate evolution and do not result from a progressive transformation of species. Dipnoan lungfish represent an old step endowed with the full panoply of an air-breathing vertebrate. Some elements of this panoply are still present in newborn humans. Due to lack of information concerning the vasculature of fossilized species, there is no witness to the evolutionary steps that led to its organization in extant lungfish.

References

Abrahamsson, T., S. Holmgren, S. Nilsson, and K. Peterson. 1979a. Adrenergic and cholinergic effects on the heart, the lung, the spleen of the African lungfish, *Protopterus aethiopicus*. *Acta Physiol. Scand.* 107A:141–147.

Abrahamsson, T., S. Holmgren, S. Nilsson, and K. Petterson. 1979b. On the chromaffin system of the African lungfish, *Protopterus aethiopicus*. *Acta Physiol. Scand.* 107B:135–139.

Allan, F.D. 1961. An histological study of the nerves associated with the ductus arteriosus. *Anat. Rec.* 139:531–537.

Aranson, S., G. Gennser, C. Owman, and N.O. Sjoberg. 1970. Innervation and contractile response of the human ductus arteriosus. *Eur. J. Pharmacol.* 11:178–186.

Boreus, I., O. Malmfors, T. MacMurphy, M. Dorothy, and L. Olson. 1969. Demonstration of adrenergic receptor function and innervation in the ductus arteriosus of the human fetus. *Acta Physiol. Lond.* 77:316–321.

Boyd, J.D. 1941. The nerve supply of the mammalian ductus arteriosus. *J. Anat.* 75:457–461.

Brinkman, C.R., C. Ladner, P. Weston, and N.S. Assali. 1969. Baroreceptor functions in the fetal lamb. *Amer. J. Physiol.* 217:1346.

Burggren, W. 1977. The pulmonary circulation of the chelonian reptile, morphology, haemodynamics and pharmacology. *J. Comp. Physiol.* 116:303–323.

Chardon, M. 1961. Contribution à l'étude du système circulatoire lié à la respiration des Protopteridae. *Annales Tervuren Musée Royal. L'Afrique Centrale. Série Sciences Zool.* 103:53–98.

Dejours, P. 1975. *Principles of Comparative Respiratory Physiology*. North Holland Publ., Amsterdam.

Delaney, R.G., S. Lahiri, and A.P. Fishman. 1974. Aestivation of the African lungfish *Protopterus aethiopicus*: cardiovascular and pulmonary function. *J. Exp. Biol.* 61:111–128.

De Saint-Aubain, M.L. and K.G. Wingstrand. 1979. A sphincter in the pulmonary artery of the frog *Rana temporaria* and its influence on blood flow in skin and lungs. *Acta Zool. Stochh.* 60:163–172.

Fay, F.S. 1971. Guinea pig ductus arteriosus. I. Cellular and metabolic basis for oxygen sensitivity. *Amer. J. Physiol.* 221(2):470-479.

Fay, F.S., P. Nair, and W.J. Whalen. 1977. Mechanism of oxygen induced contraction of ductus arteriosus. *In* M. Reivich, R.F. Coburn, S. Lahiri, and B. Chance (eds.), *Tissue Hypoxia and Ischemia,* pp. 123-134. Plenum Press, New York.

Fige, F.H.J. 1936. The differential reaction of the blood vessels of a branchial arch of *Amblystoma tigrinum* (Colorado Axolotl). I. The reaction to adrenaline, oxygen and carbon dioxide. *Physiol. Zool.* 9:79-101.

Fishman, A.P., R.G. DeLaney, and P. Laurent. 1985. Circulatory adaptation to bimodal respiration in the dipnoan lungfish. *J. Appl. Physiol.* 59(2):285-294.

Heyman, M.A. and A.M. Rudolph. 1975. Control of the ductus arteriosus. *Physiol. Rev.* 55:62-78.

Ikeda, M. 1970. Adrenergic innervation of the ductus arteriosus of the fetal lamb. *Experientia* 26:525-526.

Johansen, K. 1970. Air breathing in fishes. *In* W.S. Hoar and D.J. Randall (eds.), *Fish Physiology,* Vol. IV, pp. 361-411. Academic Press, New York.

Johansen, K., C. Lenfant, and D. Hanson. 1968. Cardiovascular dynamics in the lungfishes. *Z. Vergl. Physiol.* 59:157-186.

Johansen, K. and O.B. Reite. 1968. Influence of acetylcholine and biogenic amines on branchial, pulmonary and systemic vascular resistance in the African lungfish, *Protopterus aethiopicus. Acta Physiol., Scand.* 75:465-471.

Laurent, P. 1981. Circulatory adaptation to diving in amphibious fish. *Adv. Physiol. Sci.* 20:305-306.

Laurent, P. 1984. Gill internal morphology. *In* W.S. Hoar and D.J. Randall (eds.), *Fish Physiology,* Vol. 10, pt. A. pp. 73-183. Academic Press, New York.

Laurent, P. 1985. Organization and control of the respiratory vasculature in lower vertebrates: are there anatomical gill shunts. *In* K. Johansen and W.W. Burggren (eds.), *Cardiovascular Shunts,* pp. 57-70. Alfred Benzon Symposium 21, Munksgaard, Copenhagen.

Laurent, P., R.G. DeLaney, and A.P. Fishman. 1978. The vasculature of the gills in the aquatic and aestivating lungfish (*Protopterus aethiopicus*). *J. Morphol.* 156:173-208.

Laurent, P., R.G. DeLaney, and A.P. Fishman. 1979. Circulatory adaptation for bimodal respiration in the Dipnoi lungfish. *The Physiologist.* 22:75 (abstract of FASEB Fall Meeting, New Orleans).

MacMahon, B.R. 1969. A functional analysis of the aquatic and aerial respiratory movements of an African lungfish, *Protopterus aethiopicus*, with reference to evolution of the lung ventilation mechanism in vertebrates. *J. Exptl. Biol.* 51:407-430.

MacMahon, B.R. 1970. The relative efficiency of gaseous exchange across the lungs and gills of an African lungfish, *Protopterus aethiopicus. J. Exptl. Biol.* 52:1-15.

MacMurphy, D.M., M.A. Heymann, A.M. Rudolph, and K.L. Melmon. 1972. Developmental changes in constriction of the ductus arteriosus: responses to oxygen and vasoactive substances in the isolated ductus arteriosus of the fetal lamb. *Pediat. Res.* 6:231-238.

Maurer, F. 1888. Die Kieme und ihre Gefasse bei Urodelen and Anuren. *Jarhb.* 13:23-35.

Milsom, W.K., B.L. Langille, and D.R. Jones. 1977. Vagal control of pulmonary vascular resistance in the turtle *Chrysemys scripta. Can. J. Zool.* 55:359-367.

Morgan, M. 1974. The development of gill arches and gill blood vessels of the rainbow trout *Salmo gairdneri. J. Morphol.* 142:351-363.

Oberhansli-Weiss, I.M., M.A. Heymann, A.M. Rudolph, and K.L. Melmon. 1972. The pattern and mechanism of response to oxygen by the ductus arteriosus and umbilical artery. *Pediat. Res.* 6:693-700.

Olson, K.R., P.K. Roy, T.K. Ghosh and J.S.D. Munshi, 1994. Microcirculation of gills and accessory respiratory organs from the airbreathing snakehead fish, *channa punctata, c. gachua* and *c. marulius. The Anatomical Record* 238:92-107.

Parker, W.N. 1892. On the anatomy and physiology of *Protopterus annectens. Trans. Roy. Irish Acad.* 30:111-230.

Robertson, J.I. 1913. The development of the heart and vascular system of *Lepidosiren paradoxa. Quart. J. Micr. Sc. Lond.* 59:53-132.

Romer, A.S. 1970. *The Vertebrate Body.* Saunders Publ., Philadelphia, 601 pp.

Silva, D.G. and M. Ikeda. 1971. Ultrastructural and acetylcholinesterase studies on the innervation of the ductus arteriosus, pulmonary trunk and aorta of fetal lamb. *J. Ultrastruct. Res.* 34:358-374.

Swan, H. 1974. Aestivation and the aestivating poikilotherm, Kamongo. *In* Thermoregulation and Bioenergetics, pp. 254-280. Amer. Elsevier Publ. Co., New York.

Szidon, J.P., S. Lahiri, M. Lev, and A.P. Fishman. 1969. Heart and circulation of the African lungfish. *Circ. Res.* 25:23–38.

Taxi, J. 1979. The chromaffin and the chromaffin-like cells in the autonomic nervous system. *Int. Rev. Cytol.* 57:283–343.

von Hayek, H. 1935. Der funktionelle bau der nabelarterien und des ductus botalli. *Z. Anat. Entwicklungsgesch* 105:15–24

Zaccone, G., S. Fasulo, and L. Ainis. 1995 Neuroendocrine epithelial cell system in respiratory organs of air-breathing and teleost fishes. *Internat. Rev. Cytology* 157:277–313.

5

Phylogeny, Ontogeny, Structure and Function of Digestive Tract Appendages (Caeca) in Teleost Fish

Amjad M. Hossain and Hiran M. Dutta

Among vertebrates, only teleostean fish species have appendages such as caeca at the gastrointestinal junction (Khanna, 1961; Romer, 1970; Kent, 1983). Although these gastrointestinal appendages are unique features of fish, they are present in only 60% of the known fish species (Khanna and Mehrota, 1971; Kapoor et al., 1975; Lagler et al., 1977; Stroband, 1980). The functional and adaptational significance of these structures in fish is of interest to ichthyologists.

The discovery of caeca dates back to 300 B.C. (Suyehiro, 1942; Rahimullah, 1945, 1947; Al-Hussaini, 1946, 1949) but their exact functional status is still not known. Several functions have been suggested by various researchers for these structures (Dawes, 1929; Barrington, 1957; Kapoor et al., 1975; Wassersug and Johnson, 1976). In earlier years caeca were considered to be pancreas, then accessory food reservoirs (Al-Hussaini, 1946; Dawes, 1929; Kapoor et al., 1975; Wassersug and Johnson, 1976), or breeding pouches for intestinal flora (Andrew, 1959; Kapoor et al., 1975; Reifel and Travill, 1979; Romer, 1970). In most investigations of the gastrointestinal tract (Andrew, 1959; Romer, 1970; Kapoor et al., 1975; Reifel and Travill, 1979), it has been indicated that caeca supplement, the digestive function of the stomach or intestine (Rahimullah, 1945; Mohsin, 1962; Lagler et al., 1977; Bond, 1979; Stroband, 1980; Groman, 1982; Buddington and Diamond, 1987; Kjuravik et al., 1991; Thorarensen et al., 1991) by increasing the surface area for digestion and absorption. However, these different views are still speculative and have yet to be confirmed experimentally. In this paper we have reviewed the information published on fish digestive tract appendages (caeca) in divergent groups of fishes, and have summarized the findings of our own investigations on their phylogeny, ontogeny, structure, and function. Our studies on bluegill sunfish, *Lepomis macrochirus*, indicate that it possesses 6-8 caeca (Hossain and Dutta, 1983, 1986a, 1986b, 1988, 1989, 1991a, 1991b, 1991c, 1992). Fig. 1 depicts the location of the caeca.

PHYLOGENETIC HISTORY OF DIGESTIVE TRACT APPENDAGES

Appendages in various forms exist in the digestive tract of almost all metazoan groups. The phylogenetic history of appendages can be traced back to primitive metazoa, platyhelminthes, and even in the most advanced mammals. Fig. 2 depicts the phylogenetic history of the digestive tract diverticulae shown in a phylogenetic tree of the animal kingdom. The appendages may be located anywhere in the digestive tract, from the tip of the mouth to the tip of the anus (Andrew, 1959; Barnes, 1980; Kent, 1983; Morton, 1967; Romer, 1970). In most cases they develop from the intestine (Hossain and Dutta, 1988). Usually the diverticulae are small structures compared to other parts of the digestive tract. Some invertebrates possess diverticulae which are larger than the parent organ and unusual in shape (Beklemishev, 1969; Barnes, 1980).

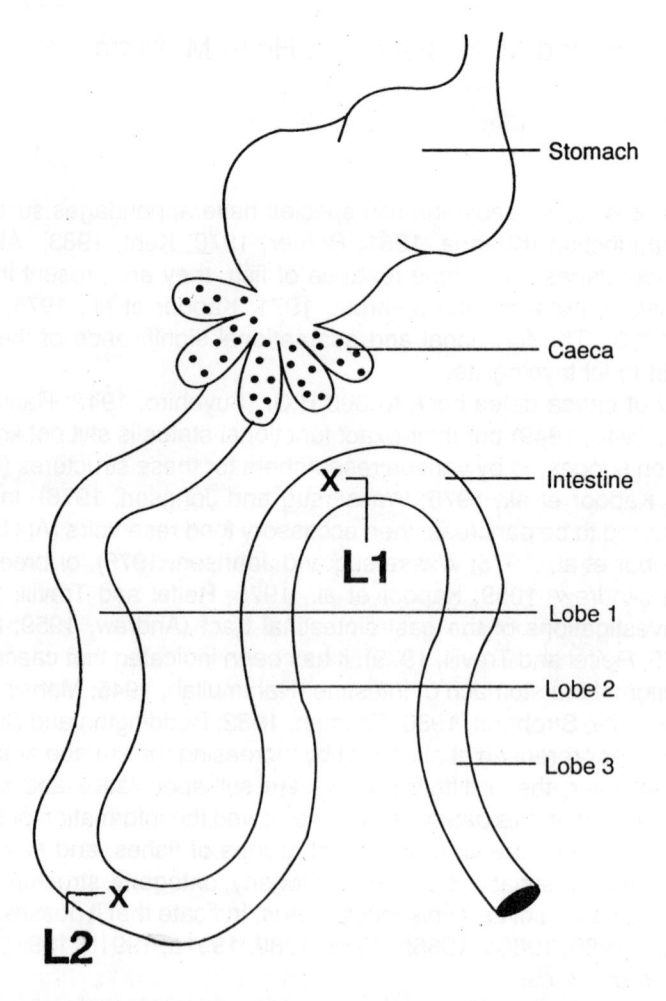

Fig. 1. Diagrammatic representation of the digestive tract compartments of bluegill fish, *Lepomis macrochirus*, showing the caeca at the gastrointestinal junction. The intestine of bluegills is divided into three intestinal lobes (lobe 1, lobe 2 and lobe 3) by two intestinal loops (L1 and L2).

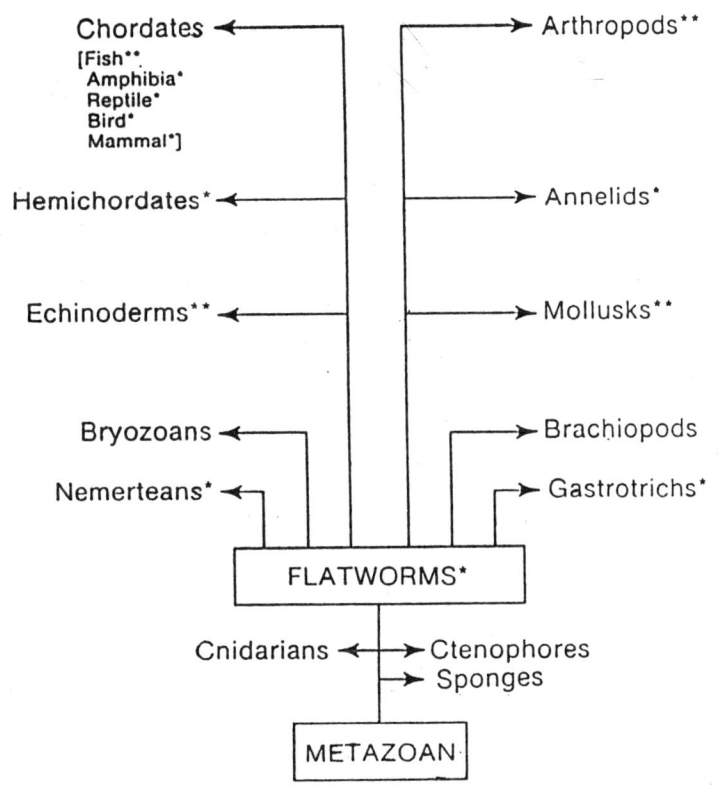

```
Chordates  ◄─────────┐         ┌─────────► Arthropods**
  [Fish**             │         │
   Amphibia*          │         │
   Reptile*           │         │
   Bird*              │         │
   Mammal*]           │         │
Hemichordates* ◄─────┤         ├─────────► Annelids*
                      │         │
Echinoderms** ◄──────┤         ├─────────► Mollusks**
                      │         │
Bryozoans ◄──────────┤         ├─────────► Brachiopods
Nemerteans* ◄────────┤         ├─────────► Gastrotrichs*
                   ┌──┴─────────┴──┐
                   │  FLATWORMS*   │
                   └───────┬───────┘
          Cnidarians ◄─────┼─────► Ctenophores
                           ├─────► Sponges
                       ┌───┴───┐
                       │METAZOAN│
                       └────────┘
```

Fig. 2. Phylogenetic history of the digestive tract diverticulae shown in a phylogenetic tree of the animal kingdom (from Barnes, 1980, modified). * and ** indicate the presence of diverticulae and caeca at the gastrointestinal junction respectively.

In vertebrates, only 60% of the known fish species possess appendages at the gastrointestinal junction similar to some advanced invertebrates and primitive chordates (Fig. 2) (Campbell and Burnstock, 1968; Kapoor et al., 1975; Lagler et al., 1977). The functional and adaptational significance of these appendages in fish is discussed below.

The number of appendages in teleosts is found to be highly variable, ranging from none (absent) to numerous (>1000). The appendage number varies from group to group and species to species. Even within the same species, the number of caeca is not always consistent. A unique pattern of appendage distribution prevails in teleost species which could be considered a reversed "J" pattern (Fig. 3A). This unique "J" pattern of caecal distribution is consistent in the teleosts belonging to one taxonomic or feeding group as shown in Fig. 3B–D. The distribution of caeca in four large diversified taxonomic groups namely clupeiformes, perciformes, cypriniformes, and scorpeniformes can be seen in Fig. 3D. Among these four groups, cypriniformes completely lack caeca while species with and without caeca are found in the other three taxonomic groups. The reversed "J" pattern seen in teleosts in general (Fig. 3A) is also seen in these taxonomic groups (Fig. 3B–D).

Teleost fishes include species with and without a stomach. A survey of both temperate and tropical fish (Suyehiro, 1942; Nikolsky, 1963; Chao, 1973; Scott and Crossman, 1973; Coetzee, 1981) revealed that more than 85% possess a stomach. The teleost stomachs

Fish Morphology

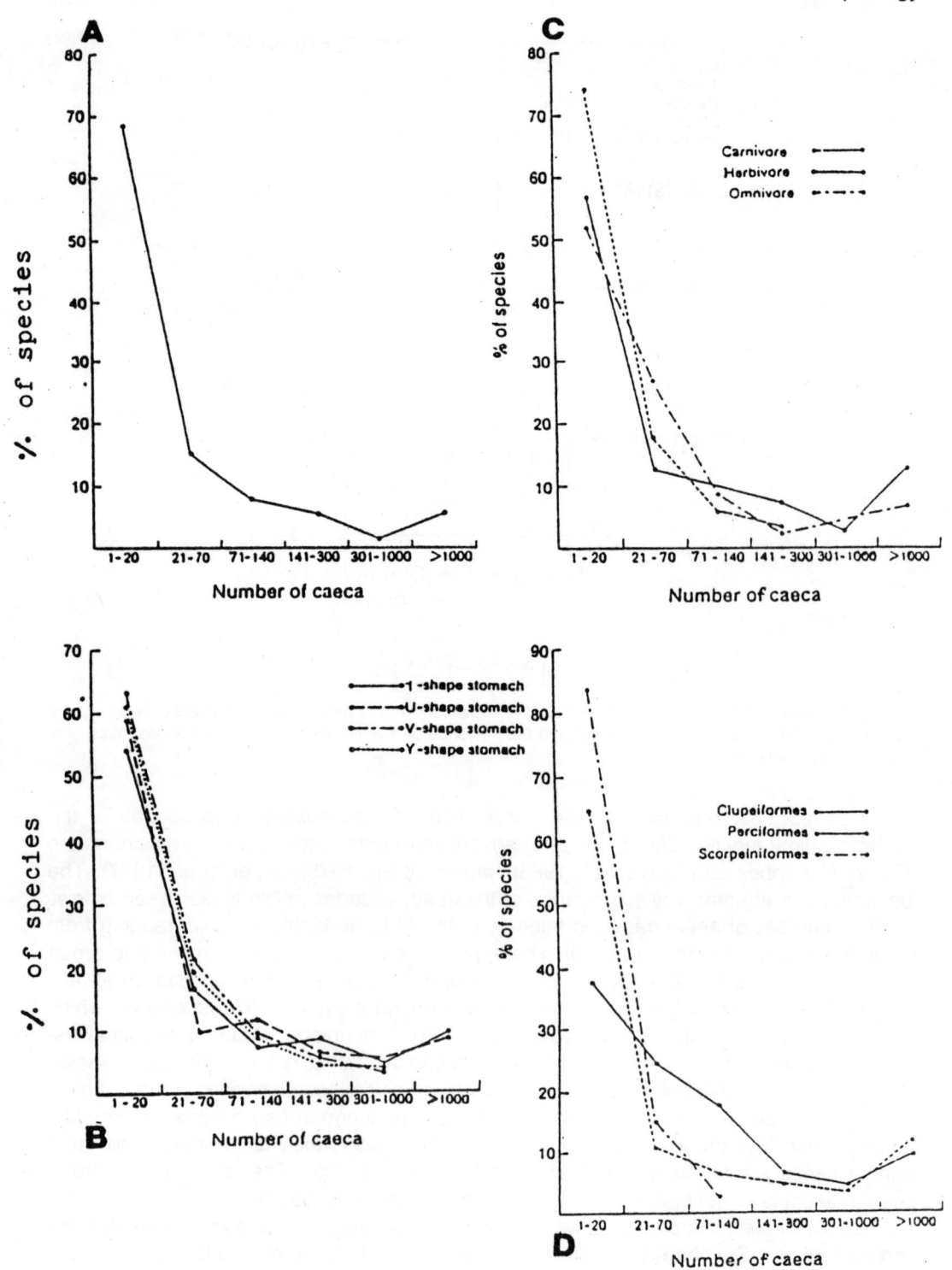

have been classified as I-, 1-, U-, V- or Y-shaped (Suyehiro, 1942; Hale, 1965; Kapoor et al., 1975; Geevarghese, 1983; Hossain and Dutta, 1988). In all stomach groups, except the V-shaped stomach, species with and without caeca are present and the relative percentage for each stomach group is shown in Fig. 4A. All species with a V-shaped stomach possess caeca while the stomachless group possesses no caeca at all. Further, the shape of the stomach exerts no consistent trend of influence on the number of caeca as will be discussed later.

Teleost fishes may be broadly divided into three feeding groups: omnivores, hervibores, and carnivores (Jirge, 1971; Kimball and Helm, 1971; Chao, 1973; Lagler et al., 1977; Marias, 1980; Jobling, 1983; Zihler, 1983; Hossain, 1988). Fishes bearing appendages are not equally distributed in these feeding groups (Fig. 4B). As can be seen in Figure 4B, most of the appendage-bearing fishes are omnivores. Carnivores include more species with caeca than the herbivores. The chances of caeca being present are higher in fish with a relative intestinal length ranging from > 1.0 to < 4.0. However, as will be discussed below, the number of caeca and size of intestine show no correlation.

It may be suggested from the unique "J" pattern of caecal distribution in fish that though the number of caeca varies in a wide range, most teleosts show a tendency toward a smaller number of caeca. The reason for a large number of species having a low number of caeca may be explained by the general limitation of space at the gastrointestinal junction. Our survey revealed that when a large number is present, the caeca spread over the anterior part of the intestine (Suyehiro, 1942; Reifel and Travill, 1978a, b; Hossain, 1988). According to Rahimullah (1945), a large number of species of the family Clupeidae have numerous small caeca arranged in bunches, studded thickly over the duodenum.

MORPHOLOGICAL RELATIONSHIP AMONG STOMACH, CAECA AND INTESTINE

Variation in stomach shape, size of intestine and number of caeca in the fish digestive tract has drawn the attention of many investigators (Mohsin, 1962; Morton, 1967; Labhart and Ziswiler, 1979; Marias, 1980; Ribble and Smith, 1983; Hossain and Dutta, 1988; 1991a, 1991b, 1992). Two hypotheses regarding stomach-caeca-intestine relationship have been tested. First, since the stomach size and shape control the area around the gastrointestinal junction, and since the caeca are the gastrointestinal junctional appendages, it has been suggested that the type of stomach exerts an influence on the number of caeca present. Second, since the function of the caeca is to increase the surface area for the intestine (as experimentally proven in our investigations, see below), there should be some relationship between the number of caeca and size of intestine. The ANOVA test (Table 1) indicated that a variation in number of caeca does exist among the different stomach groups (F ratio 10.06 and F probability < 0.01); the Student-Newman-Keuls test (Table 2) showed that stomach shapes exert no consistent trend of influence on the number of caeca. In stomach groups, only the I-shaped stomach, a small group, was found to differ from the other stomach types. The I-shaped and U-shaped stomach have been considered as one

---

Fig. 3. Pattern of distribution of caeca in teleostei (A), four stomach groups (B), three feeding groups (C), and three taxonomic groups (D).

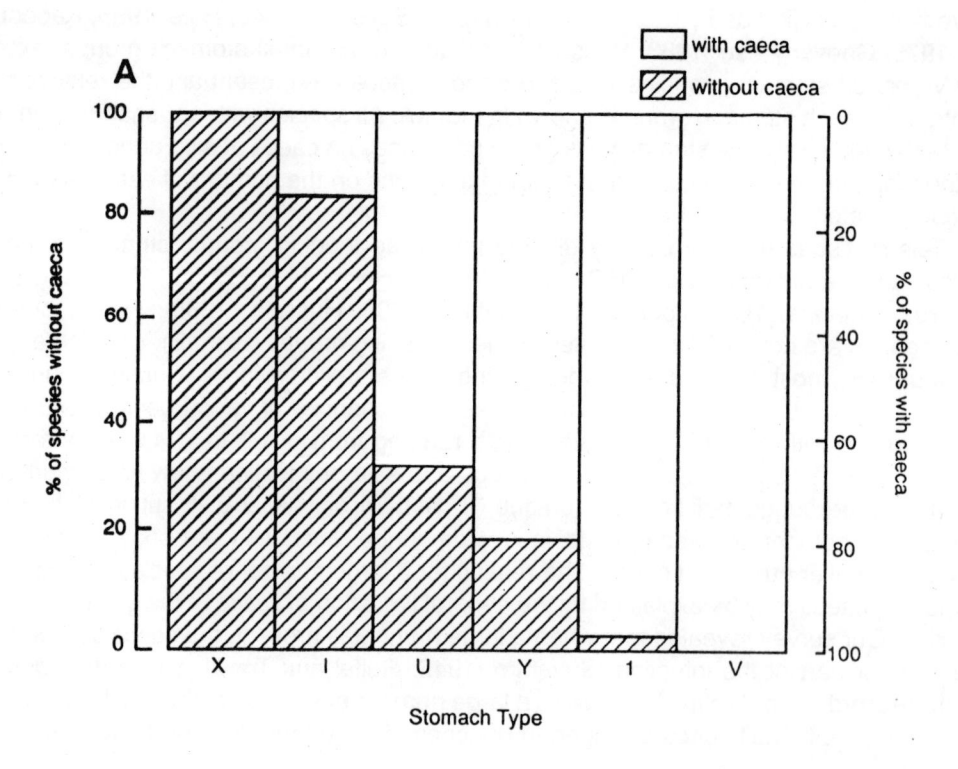

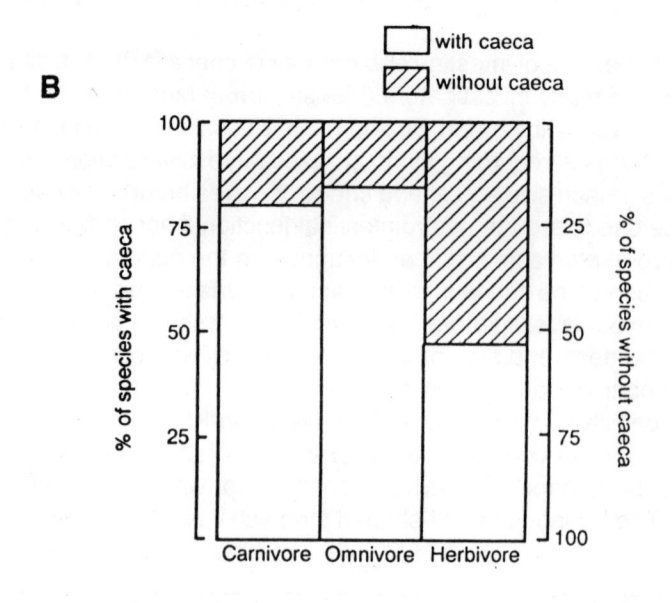

Fig. 4. Percent of species with and without caeca in each stomach group (A) and feeding group (B). X for stomachless fish.

Table 1. Summary of ANOVA test for the number of caeca among various stomach groups and feeding groups. F-probabilities are for the null hypothesis that the mean caeca number among the stomach groups as well as the feeding groups are equal. Probabilities less than 0.05 are considered to reject the null hypothesis

Comparisons	DF	F-ratio	F-prob.
Among 5 stomach groups	4/138	10.0538	< 0.0001
Among 3 feeding groups	2/111	0.6200	0.5398

Source: Hossain, 1988

Table 2. Student-Newman-Keuls (SNK) test of different stomach groups based on the number of caeca. P = distance difference between the two means, *, **, and ns indicate $p < 0.05$, $p < 0.01$ and $p > 0.05$, respectively

P	Comparison (Si vs Sj)	Difference (Xi−Xj)	q-statistics
5	1 vs I	3.52	7.095**
4	1 vs Y	1.42	2.862 ns
	U vs I	2.95	4.185*
3	V vs I	2.79	3.965*
2	Y vs I	2.10	3.541*
	1 U V Y I		

I, 1, U, V and Y represent the shape of the stomach
Source: Hossain, 1988

group by some authors (Barrington, 1957). The feeding groups (carnivore, herbivore, and omnivore) revealed no differences (Tables 1 and 2).

The probable explanation for finding no significant influence of stomach shape on number of caeca is that although the caeca are the gastrointestinal junctional appendages, they are not always confined to the gastrointestinal junction. It has already been pointed out that when caeca are numerous, they spread over the anterior lobe of the intestine (Suyehiro, 1942; Hossain, 1988; Hossain and Dutta, 1992). The second hypothesis, dealing with the number of caeca and size of intestine, has been the subject of controversy in many previous studies (Northcote and Patternson, 1960; Nikolsky, 1963; Pasha, 1964; Morton, 1967; Kapoor et al., 1975; Marias, 1980; Ribble and Smith, 1983). Our study suggests no relationship between the number of caeca and size of intestine. This might be because the number of caeca cannot be used as an index for assessing their functional importance.

A larger number of caeca may not be more functional due to variation in their size and shape. The amount of digestive surface area that a caecum can contribute depends on its overall size. A larger and stouter caecum will obviously produce more surface area than a smaller or tiny one. The number of caeca in two different species may be equal but vary in size; some may be short, some long, some thick, and some thin. Thus two fish species having an identical number of caeca may have significantly different amounts of digestive surface areas provided by their caeca. Even fish with fewer but large caeca may have more surface area than those possessing more but smaller caeca. Further, all the caeca of an individual fish cannot function equally since they differ in size and anatomical position. Our study strongly suggests that the caecal number alone should not be used to assess the functional importance of caeca. It would be more appropriate to presage any relationship between the intestine and the caeca on the basis of biomass of these two compartments.

ERA OF CONFUSION REGARDING TELEOST DIGESTIVE TRACT APPENDAGES

The appendages were frequently regarded as pancreas (Suyehiro, 1942; Rahimullah, 1945, 1947; Kapoor et al., 1975). Twentieth century researchers have not made much effort to establish a correlation between the diffused nature of the fish pancreas and the existence of caeca. Nor have their efforts been directed toward investigating the formation of a compact pancreas and extinction of the gastrointestinal junctional caeca in other vertebrates.

Early as well as recently published reports (Al-Hussaini, 1946, 1949; Barrington, 1957; Mohsin, 1962; Tyler, 1973; Lagler et al., 1977; Bond, 1979; Groman, 1982; Watts and Lawrence, 1986) reveal various names for the teleost gut appendages, viz. pyloric/intestinal appendages, pyloric/intestinal caeca, pyloric/intestinal diverticulae, and pyloric/intestinal outpockets. Aristotle (384–322 B.C.) discovered the caeca in fishes and named them pyloric appendages (Kapoor et al., 1975; Jansson and Olsson, 1960; Suyehiro, 1942) as they were located near the pyloric part of the stomach. It was Jacobshagen (1915) who first observed that the mucosal fold of the caeca is very similar to those of the intestine and proposed that the caeca are intestinal appendages. Subsequent scholars designated these tiny structures arbitrarily as either intestinal caeca or pyloric caeca (Dawes, 1929; Kapoor et al., 1975; Lagler et al., 1977; Stroband, 1980; Groman, 1982). Despite several efforts to eliminate the confusion of nomenclature (Rahimullah, 1945, 1947; Mohsin, 1962), these fish appendages are still designated as either pyloric caeca or intestinal caeca (Suyehiro, 1942; Lagler et al., 1977) of the gastrointestinal tract (Kafuku, 1977; Groman, 1982; Vantue, 1980; Blaxter, 1981; Heinrichs, 1982; Lau and Shafland, 1982; Richard and Applegate, 1982). None of the foregoing authors studied in depth the origin and development of the caeca at the gastrointestinal junction. The controversy concerning caecal nomenclature seems to be due to the absence of studies on caecal development. It is our belief that a study on the development of caeca would provide more convincing evidence for resolving the caecal nomenclatural dispute than simple histological evidence provided by previous investigators (Rahimullah, 1945, 1947; Al-Hussaini, 1946, 1949, 1954; Sastry, 1974). In one of our studies, therefore, we looked at the ontogeny of the G.I. tract appendages, tracing their source of origin and development and their developmental relationship with other parts of the digestive system. The findings of this investigation proved very valuable as they provide more direct evidence to show that the appendages are the property of the intestine, not of the stomach.

ONTOGENY OF FISH CAECA

It can be seen from Fig. 5 that formation of the caeca starts in the mesolarvae and is completed in the metalarvae. During the period of protolarval development the tract remains essentially undifferentiated as a straight nonfunctional tubular canal (Fig. 6). With the onset of exogenous feeding, the digestive tract starts to contract and a constriction forms in the anterior region which divides the gut into stomach and intestine. Until the third day of development, the constricted junction of the stomach and intestine showed no sign of possessing caecal buds (Fig. 6). In four-day-old mesolarvae some of the rapidly proliferating mucosal folds of the proximal bulge of the intestine extended outward and were transformed into caecal buds (Figs. 6–7). Thereafter the number and size of caecal buds increased gradually. So the absence of constriction at this point in the gut might inhibit the formation of any caeca, as evident from this chronological record of development of appendages in the digestive tract of bluegill fish. Why stomachless teleosts do not develop caeca becomes thereby more enigmatic!

Though caecal development appears to be initiated by a constriction between the stomach and intestine, the role of constriction should be considered with caution since not all stomach-bearing fishes have caeca (Suyehiro, 1942; Lagler et al., 1977; Stroband, 1980). The lack of caeca in the gastrointestinal junction in some stomach-bearing fish

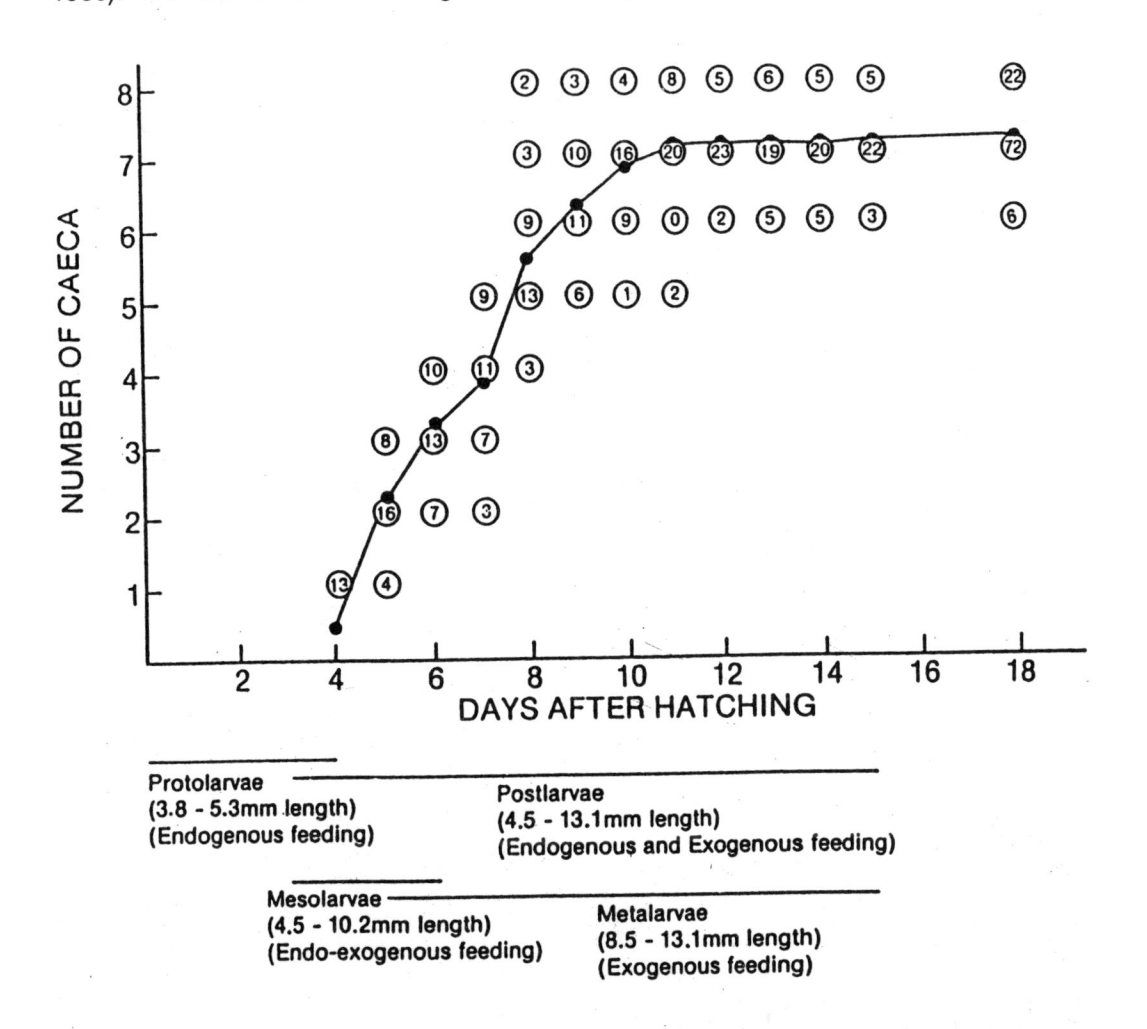

Fig. 5. Number of caeca in three stages of postembryonic development in bluegill larvae. Numbers within circles indicate the number of larvae. *—*—* profile shows the average number of caeca.

(< 20%) suggests that the formation of caeca is not simply related to the existence of a constriction, but to its magnitude and/or other factors.

FUNCTIONAL MORPHOLOGY, GROSS AND MICROSCOPIC ANATOMY OF CAECA

From the very beginning of the twentieth century anatomists have paid a great deal of attention to the structural details of the gastrointestinal tract of fish. Blake (1930), Dawes (1929) and Greene (1912) were among early researchers who made major contributions to microscopic structures of the fish digestive tract with special emphasis on tract appendages. Between 1930 and 1960 more detailed information on the morphology and histology of the fish digestive tract became available, of which the following deserve special attention: Blake (1930), Dawes (1929), Rahimullah (1945), Suyehiro (1942), Berry and Low (1970), Bucke (1971), Clarke and Witcomb (1980), Frange and Grove (1979), Geevarghese (1983), Groman (1982), Khanna and Mehrota (1971), McBee and West (1969), Mohsin (1962), Morton (1967), Sis et al. (1979), Thurmond (1979), and Wassersug and Johnson (1976).

A summary of the above-mentioned microscopic anatomical investigations and of our own (Hossain and Dutta, 1983, 1985, 1986b, 1988, 1991a, 1992) on the teleost digestive tract, and its appendages are given in Tables 3 and 4. The four basic histological layers such as serosa, muscularies, submucosa, and mucosa are present in the region from the esophagus to the tip of the intestine, including the caeca. The two anterior compartments, buccal cavity and pharynx, lack serosa. A comparison of the stomach, caeca, and intestine based on shape, size, goblet cells, glands, muscles, submucosa, and mucosa indicated sharp differences between caeca and stomach, and closer similarities between caeca and intestine. These morphological observations are consistent with our findings from ontogenetic investigations (Figs. 5-7). However, some gross and microscopic dissimilarities still prevail between caeca and intestine, as illustrated in Table 4. The intestine is an open-ended long tube while each caecum is a short tube with a closed end, resembling a test tube. The muscle coats of caeca (circularis and longitudinal) are thinner than that of the intestinal wall. There is regional variation in muscle thickness of the intestinal wall while in the caeca the muscle coat is fairly uniform in thickness except at the distal end. The submucosa of the caeca has boundaries similar to the intestine but less extensive. The goblet cells are more abundant in the intestine than in the caeca. The number of mucus-producing cells per full ocular micrometer are 13 and 22 in the caeca and intestine respectively. Caecal mucosa are more complex compared to the intestine.

The above qualitative histological findings show that despite some microlevel differences, the caeca and intestine are structurally identical; this convinced some researchers (Hossain, 1988; Hossain and Dutta, 1991a, 1991b, 1991c) that the appendages might do the same job that the intestine does. This generalized view about the function of the caeca appears to be an oversimplification of the real fact, which led to neglect of these tiny structures. Studies dealing with fish caeca are very scant in contemporary literature. Whether the structural similarities between caeca and intestine exhibit any functional similarities were reviewed in our several investigations (Hossain, 1988; Hossain and Dutta, 1991a, 1991b, 1992). We found that histologically the gallbladder is to a certain extent similar to the intestine but functionally the two differ. Thus the structural similarity between caeca

Table 3. Histological similarities and differences among different regions of the bluegill (*Lepomis macrochirus*) digestive tract

G.I tract regions	Tissue layers									
	Serosa	Longitudinal muscle		Circular muscle		Sub-mucosa	Muscularis Mucosa	Mucosa		Taste bud
		Striated	Smooth	Striated	Smooth			Squamous	Columnar	
Mouth and buccal cavity	−	−	−	−	−	+	−	+	−	+++
Pharynx	−	+	−	−	−	+	−	+	−	++
Esophagus	+	+	−	+	−	+	−	+	+	+
Cardiac stom.	+	−	+	−	+	+	+	−	+	−
Pyloric stom.	+	−	+	−	+	+	+	−	+	−
Caeca	+	−	+	−	+	+	−	−	+	−
INT1	+	−	+	−	+	+	−	−	+	−
INT2	+	−	+	−	+	+	−	−	+	−
INT3	+	−	+	−	+	+	−	−	+	−

−Absent +Present ++Moderate +++Maximum
Source: Hossain, 1988

Table 4. Gross and microscopic dissimilarities between caeca and intestine in bluegills. All measurements except otherwise mentioned are in ocular micrometer unit (omu) at 400 ×

Parameter	Caeca	Intestine
Size	Short (11.32 ± 3.77 mm)	Long (98.96 ± 11.25 mm)
Shape	Test-tube-like (one end blind) Straight	Hollow tube-like (both ends open) Folded
Circular muscle	Thin (16.59 ± 3.22)	Thick (35.91 ± 5.67)
Longitudinal muscle	Thin (9.32 ± 2.04)	Thick (20.07 ± 3.81)
Pattern of muscle distribution	No regional variation in thickness	Prominent regional variation in thickness
Mucosal folds	More complex, ramified	Less complex
Food passage	Bidirectional (U-turn)	Unidirectional (one way)

Source: Hossain, 1988

and intestine may not necessarily prove their functional similarity. Moreover, because of the blind terminal ends of caeca and their characteristic location at the gastrointestinal junction, the amount of semidigested food which enters from the stomach into the caeca has to come back and pass out through the intestine. If caeca and intestine have the same function, then why is it necessary for semidigested food to pass through the caeca before it enters the intestine? Even if it is accepted that caeca and intestine have the same function, it remains unknown how much of the total amount of food ingested by the fish is digested in the caeca. To focus this issue, we combined quantitative functional morphological and experimental approaches to evaluate the caecal contribution to the digestive tract.

In the functional morphological approach a unit area of caeca is compared with that of intestine. Our computer assisted microscopic analysis of a cross section of caeca and intestine (Table 5) showed that in all aspects (total area, muscle area, lumen, mucosal area etc.) a unit area of intestine is about twice the size that of caeca. The cross section was used to measure some of the structural components, namely the amount of muscle, mucosa, and lumen. The muscle, mucosa, and lumen were considered three important functional components which together play a predominant role in digestion. The total amount of muscle present in the caecal and intestinal wall is considered an indirect estimate of the amount of mechanical propulsive force which may prevail in these two compartments. This in turn indicates the degree of peristaltic movement that may occur in caeca compared to intestine. Similarly, a comparison of the lumen size of caeca and intestine was made to estimate the amount of food which can flow through or be stored in these two compartments. The mucosal area was computed to indirectly infer the degree of absorption. The area of an intestinal cross section is more than twice that of a caecum. The muscle, mucosa, and lumen space varied in a similar way between caeca and intestine. The higher mucosal coefficient in intestine compared to caeca (Table 5) also indicates that more absorption can take place in the main tract than in the diverticulae. These differences suggest that both caeca and intestine may be involved in digestion and absorption, though the digestion and absorption rate in these two tract compartments is not the same.

The structural similarities between caeca and intestine found in a light microscopic study have been supported by electron microscopic study. Our electron microscopic analysis revealed both structural similarities and differences between caeca and intestine at the ultrastructural level (Fig. 8). The same major cell types, columnar epithelial (absorbent cell), goblet cell (secretory cell), and endocrine cells are found in intestinal and caecal mucosa. Although qualitatively the caecal absorbent, secretory, and endocrine cells are similar to those of the intestine, there are quantitative differences in the absorbent cells of these two compartments. The absorbent cells of the intestine are larger and accordingly

Table 5. Summary of quantitative histological comparisons between caeca and intestine. All measurements are in graphic unit (gu) and represent the mean values averaging from 5 specimens

Parameter	Intestine	Caeca	Intestine/Caeca
Total area of cross section (A)	10139	4805	2.11
Total inner area (mucosa and lumen) (B)	6414	3430	1.87
Lumen area (empty space) (C)	2756	1428	1.93
Muscle area (D=A−B)	2668	1442	1.85
Mucosal area (E=B−C)	3593	1985	1.81
Circumference of inner area (F)	411	213	1.93
Length of mucosa (G)	791	478	1.65
Lumen area: inner area (C:B)	0.43	0.42	
Muscle area: inner area (D:B)	0.42	0.42	
Mucosal area: inner area (E:B)	0.56	0.58	
Mucosal area: muscle area (E:D)	1.35	1.38	
Mucosal area: lumen area (E:C)	1.30	1.39	
Mucosal length: mucosal area (G:E)	0.22	0.29	
Mucosal length: circumference of inner area (G:F)	1.92	2.24	
Mucosal circumference/serosal circumference	2.12	1.93	
Mucosal coefficient	1.95 ± 0.69	1.72 ± 0.63	

Source: Hossain, 1988

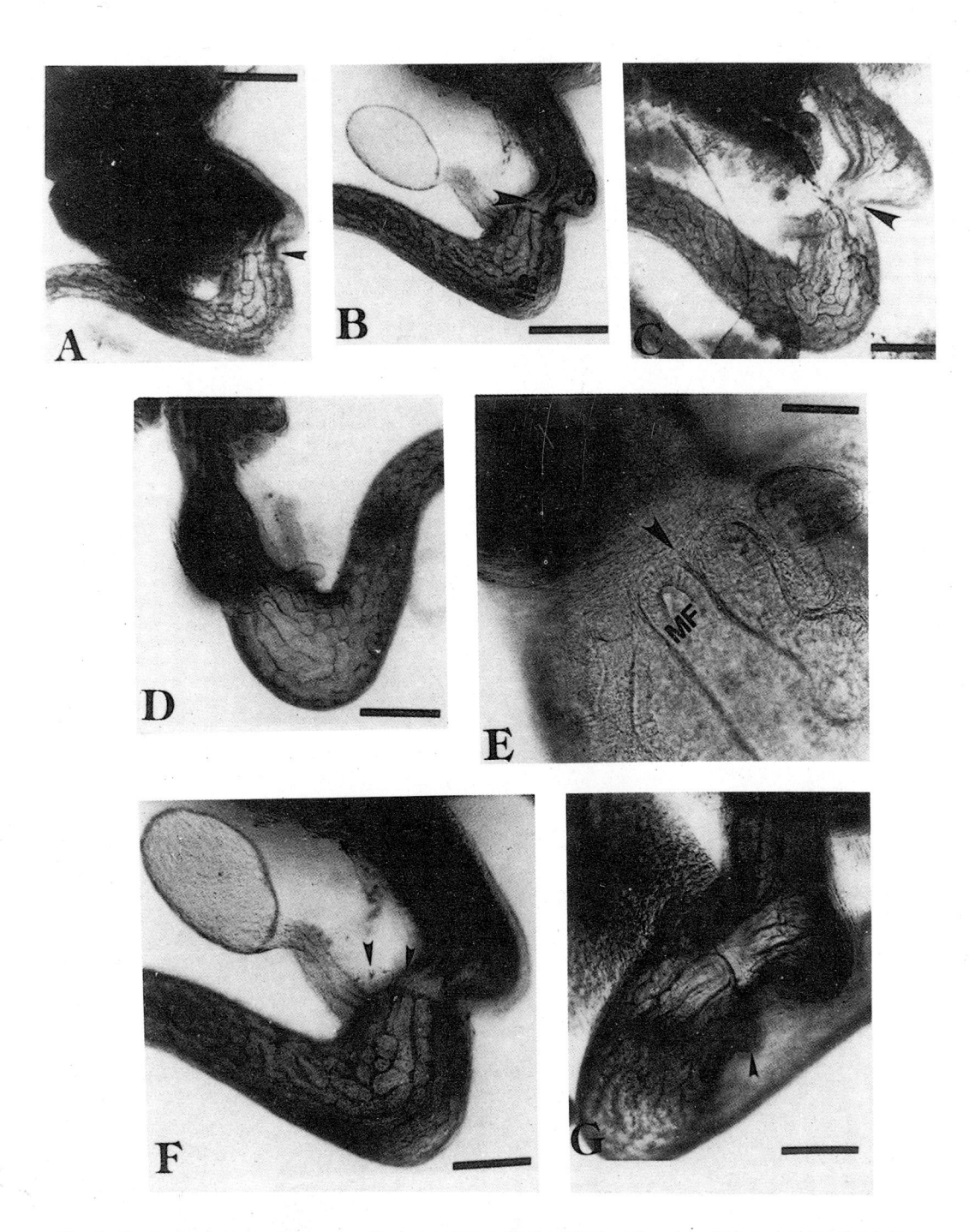

Fig. 6. Postembryonic developmental changes in the gastrointestinal tract and gastrointestinal junction in 3- to 5-day-old bluegills (*Lepomis macrochirus*): A to C—3-day-old larvae at 9:30 am, 2:30 and 11:30 pm respectively; arrows show development of constriction at the gastrointestinal junction. D to E—5-day-old larvae at 2:30 pm show the mucosal folds (MF) obstructed by the gastrointestinal constriction (arrows). F and G—5-day-old larvae at 9:30 am and 9:30 pm respectively, showing the development of terminal caecal buds (arrows). Scale bars for A, B, C = 0.40 . mm; and for D, F, G = 0.25 mm; and E = 0.10 mm.

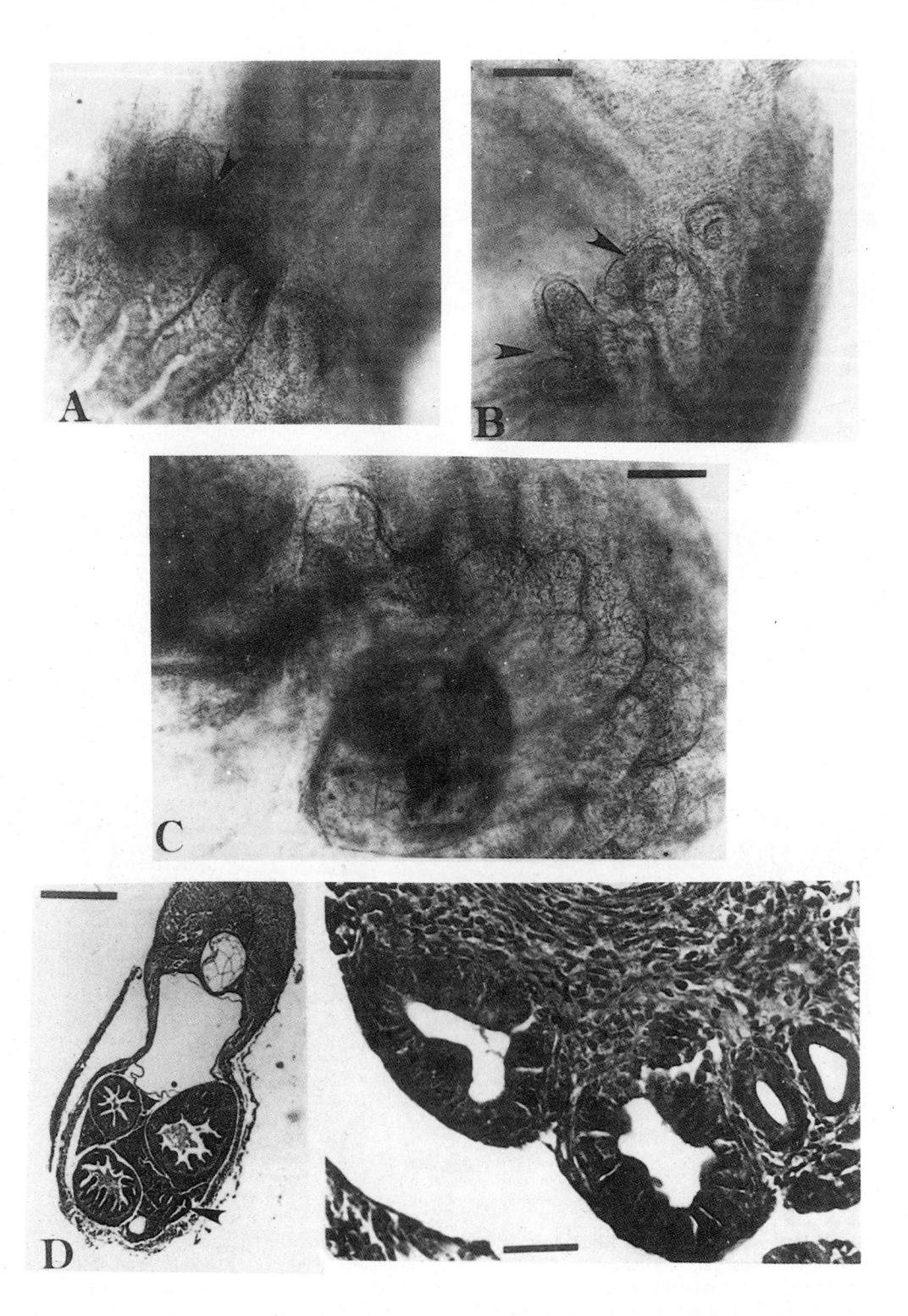

Fig. 7. Postembryonic developmental changes of the gastrointestinal tract and gastrointestinal junction in 6- to 14-day-old bluegills. A and B—6- and 7-day-old larvae respectively, showing splitting of some intestinal mucosal folds (arrows). C—8-day-old larvae, showing appearance of all the caecal buds. D and E—cross sections of 14-day-old larvae, showing nonproliferated mucosa in caeca (arrows). Scale bars for A, B, C = 0.10; for D = 0.25 mm; for E = 0.025 mm.

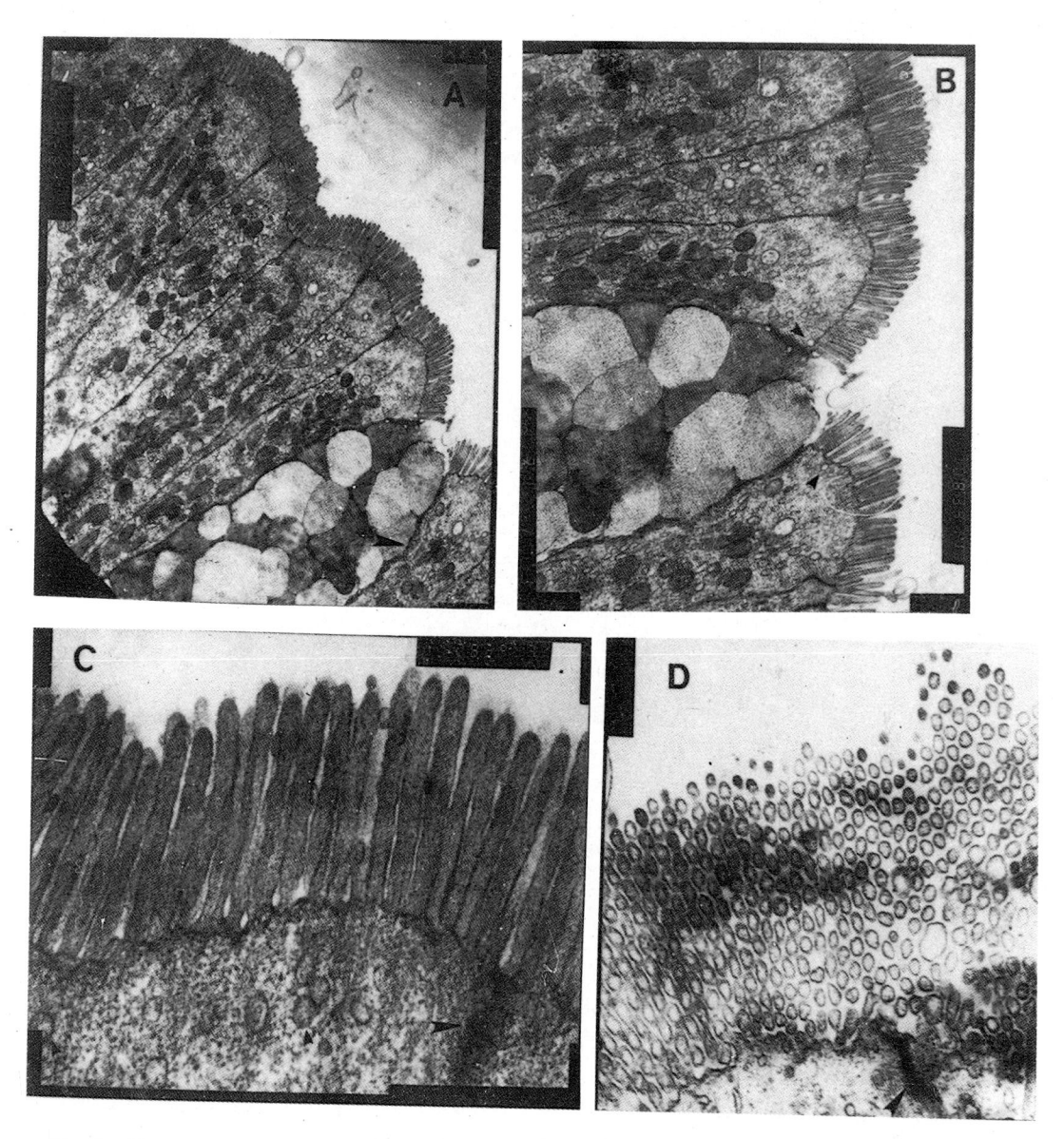

Fig. 8. Electron micrographs of absorbent cells of caècal mucosa: A—overview of several absorbent cells ($\times$ 6,000). B—apical part of some absorbent cells ($\times$ 10,000). C—longitudinal sectional view of microvilli at higher magnification ($\times$ 40,000). D—cross-sectional view of microvilli at higher magnification ($\times$ 30,000).

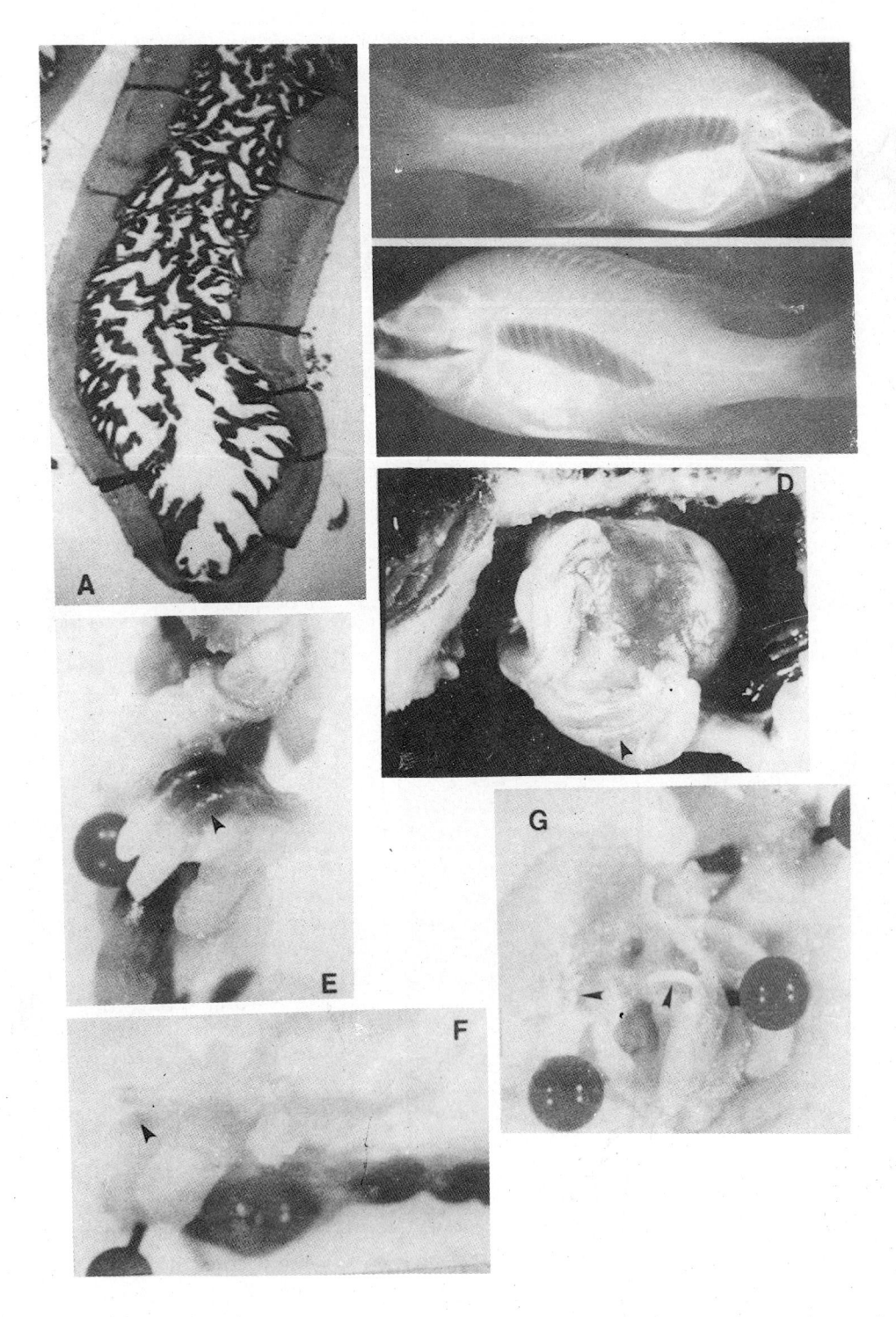

Fig. 9. Longitudinal section of caecum (A); x-radiograph of G.I. tract (B), one hour and fifteen minutes after feeding; radiograph, three hours after feeding. (C); caeca containing dyed food (D); plant materials in some caeca (E-F); and caeca harboring parasites (G).

possess more surface area, microvilli, and subcellular organelles than those of the caeca. The secretory and endocrine cells of the caeca also differ quantitatively from those of the intestine. These quantitative ultrastructural differences between the caecal and intestinal absorbent and secretory cells indicate that though absorption and secretion take place in both cases, the rate of absorption and secretion in unit area may not be the same in these two compartments.

We conducted both laboratory and field experiments to elucidate the process of movement/migration of semidigested food from stomach into caeca into intestine. Movement of food in the digestive tract was monitored by food mixed with a special dye or by radiography in the laboratory. Both methods confirmed that stomach contents enter the caeca in the same manner that food enters the intestine (Fig. 9).

Attempts were also made to verify whether fish use their caeca for digestive purposes while feeding under natural conditions. It was assumed that if the stomach contents of wild fish move to the caeca, a change in caecal contents will occur with variations in stomachal contents. Table 6 indicates that variation in caecal contents coincides with variation of stomachal contents. Variation in caecal contents can only be possible if there is a flow of food from the stomach into the caeca. In separate laboratory experiments, we detected plant materials in the caeca of plant-fed fish and animal materials in the caeca of animal-fed fish, which indicated that both plant and animal types of food enter the caeca. It would be valuable to determine the environmental impact on caecal contents.

PHYSIOLOGY: STUDY OF CAECAL ENZYMES AND MICROORGANISMS

Considerable attention has been given to detection of enzymatic activities in the fish gut and its diverticulae (Chesley, 1934; Ishida, 1936; George and Desai, 1947; Barrington, 1957; McBee and West, 1969; Kapoor et al., 1975; Travison, 1979; Kamoi et al., 1980; Hossain and Dutta, 1986), as well as caecal and intestinal enzymes (Chesley, 1934; Ishida, 1936). Lindsay and Harris (1980) determined some characteristics of the enzymes from the caeca of cod and haddock. These authors showed that lipase is much more common than any other enzyme in the caeca. It appears that acid phosphatase activity is highest in the stomach, and alkaline phosphatase activity highest in the caeca (Martin and

Table 6. Means and standard deviations of caeca contents of wild bluegills in two stomach groups and t-values testing significance of mean differences

Stomach groups	Mean of the % of poststomach content possessed by caeca	SD	Calculated t-value	Tabulated t-value
Stomach Group 1 (Stomach content <31.62% of G.I. tract contents)	7.78	5.20	3.41	2.052
Stomach Group 2 (Stomach content <31.62% of G.I. tract contents)	15.29	10.95		

Source: Hossain, 1988

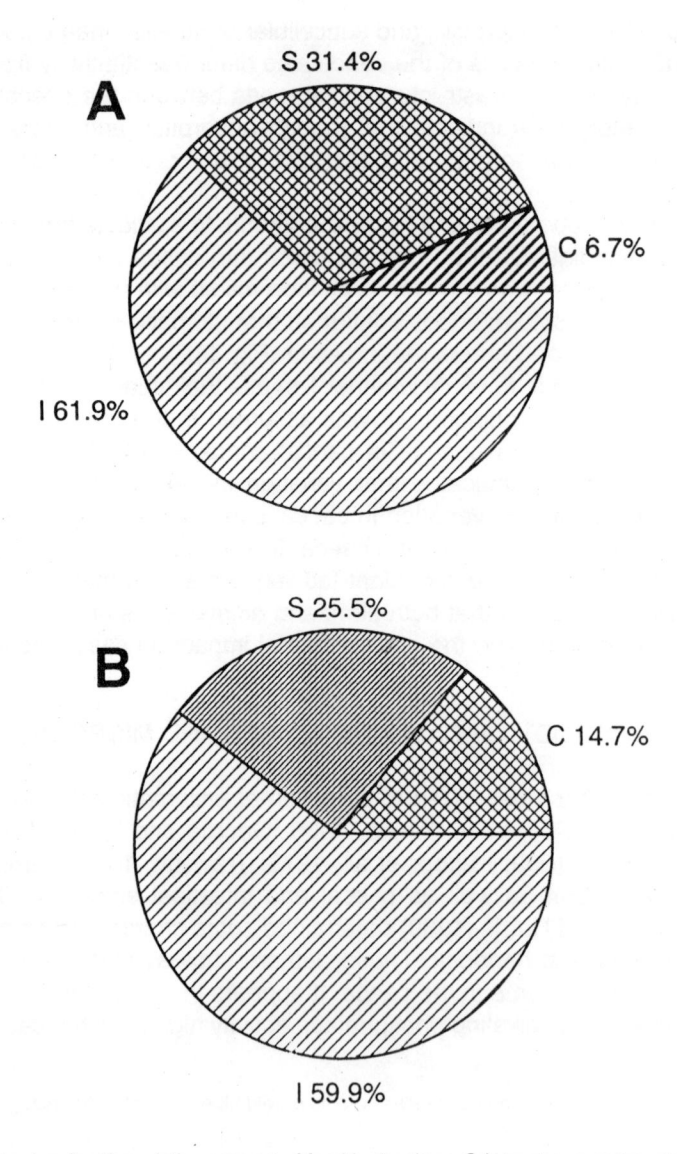

Fig. 10. Pie charts showing the relative amount of food in the three G.I. tract compartments (stomach S, caeca C, intestine I) in wild (A) and fasted (B) bluegills.

Sandercock, 1967). Effective peptic digestion on the striated border of the columnar cells of the caeca was reported by Janson and Olsson (1960). Protein-, fat-, and carbohydrate-digestive enzymes which are secreted into the lumen of the alimentary canal have also been traced to the caeca (Ishida, 1936; Idler, 1973; Sastry, 1974). Trypsin and endopeptidases were also detected in extracts of the caeca. In general, the pattern of various enzyme concentrations in the caeca is similar to that of the intestine (Idler, 1973; Kapoor et al., 1975; Watts and Lawrence, 1986).

The potential ability of caeca for synthesis of enzymes has also been investigated (Ishida, 1936; Jansson and Olsson, 1960). Significant quantities of DNA and RNA were found in the epithelial cells of the caeca, suggesting that synthesis of digestive enzymes occurs there. However, this view was not confirmed by further experimental work. As all living cells contain DNA and RNA, we do not consider the mere existence of these nucleic acids in cells of the caeca indicative of their ability to synthesize enzymes. Furthermore, the presence of enzymes or the ability of caeca to synthesize them does not necessarily mean that any digestive activity occurs there. For example, a large number of digestive enzymes are produced in the pancreas; but there is no evidence for digestive activities in this organ. More biochemical and histochemical investigations are needed for assessing whether enzyme biosynthesis occurs in the caeca and whether these enzymes are utilized by the caeca.

ENVIRONMENTAL EFFECT

The importance of caeca can also be evaluated by analysis of the interaction of caeca with the outside environment. The fish digestive tract has more direct contact with the external environment than any other internal organ system with the exception of the gills, mainly because food from outside comes directly into the gut. Hence, it may be expected that any significant alteration in the environment will induce some changes in the structure and function of the gut and its diverticulae (Hossain, 1988). Some studies have correlated environmental changes with structural and functional variability of the fish gut (Tyler, 1973; Singh and Bahuguna, 1983). A response of gut diverticulae to the external environment has yet to be intensively examined. Any difference in level of response of the caeca and the intestine to an environmental hazard would be an indirect estimate of the extent of their functional similarities and differences. In one of our studies we found that the relative amount of gut contents held by the caeca of starved fish was more than double that of wild ones (Fig. 10) (Hossain and Dutta, 1991a). This finding formed the basis for our assessment that the functional importance of caeca probably varies with the environmental conditions in which fish live.

References

Al-Hussaini, A.H. 1946. The anatomy and histology of the alimentary tract of the plankton feeder, *Atherinas forskali*. *J. Morph.* 80:251-286.

Al-Hussaini, A.H. 1949. On the functional morphology of the alimentary tract of some fishes in relation to differences in their feeding habits: Anatomy and Histology. *Quart. J. Micros. Sci.* 90:109-139.

Al-Hussaini, A.H. and A.A. Kholy. 1954. On the functional morphology of the alimentary tract of some omnivorous teleost fish. *Proc. Egypt. Acad. Sci.* 9:9-17.

Andrew, W. 1959. *Textbook of Comparative Histology*. Oxford University Press, New York.

Barnes, R.D. 1980. *Invertebrate Zoology*. Saunders Co. Publ., Philadelphia, 4th ed.

Barrington, E.J.W. 1957. The alimentary canal and digestion. *In* M.E. Brown (ed.), The *Physiology of Fishes*, pp. 109-161. Academic Press, New York.

Beklemishev, V.N. 1969. *Principles of Comparative Anatomy of Invertebrates*. University of Chicago Press, Chicago.

Berry, P.Y. and M.P. Low. 1970. Comparative studies on some aspects of the morphology and histology of *Ctenopharyngodon idellus, Aristichthys nobilis* and their hybrid (Cyprinidae). *Copeia* 4:708-726.

Blake, I.H. 1930. Studies on comparative histology of the digestive tube of certain teleost fishes. *J. Morph. and Physiol.* 50:39-70.

Blaxter, J.H.S. 1981. The rearing of larval fish. In A.D. Hawkins (ed.), The Aquarium Systems, pp. 301–323. Academic Press, New York.

Bond, C.E. 1979. Biology of Fishes. Saunders Co. Publ., Philadelphia, 1st ed.

Bucke, D. 1971. The anatomy and histology of the alimentary tract of the carnivorous fish, Esox lucius. J. Fish Biol. 3:421–431.

Buddington, R.K. and J.M. Diamond. 1987. Pyloric caeca of fish, a "new absorptive organ." Amer. J. Physio. 252:G66–G76.

Campbell, G. and G. Burnstock. 1968. Comparative physiology of gastrointestinal motility. In Handbook of Physiology, sec. 6, vol. IV:2213–2266. Amer. Physiol. Soc.

Chao, L.N. 1973. Digestive system and feeding habits of a marine stomachless fish. U.S. National Marine Fisheries Service Bull. 71:565–586.

Chesley, L.C. 1934. The concentrations of protease, amylase and lipase in certain marine fishes. Biol. Bull. 66:133–135.

Clarke, A.J. and D.M. Witcomb. 1980. A study of the histology and morphology of the digestive tract of the common eel. J. Fish Biol. 16:159–170.

Coetzee, D.J. 1981. Analysis of the gut contents of needlefish from Southern Cape, SA. J. Zool. 6:14–20.

Dawes, B. 1929. The histology of the alimentary tract of the plaice, Pleuronectes platessa. Quart. J. Micros. Sci. 73:243–273.

Dutta, H.M. and A. Hossain. 1985. Effects of fasting on the fish intestine and its diverticulae. Amer. Zool. 25:(4)130A.

Frange, R. and D. Grove. 1979. Digestion. In W.S. Hoar, D.J. Randall, and J.R. Brett (eds.), Fish Physiology, pp. 162–241. Academic Press, New York.

Geevarghese, C. 1983. Morphology of the alimentary tract in relation to diet among gobioid fishes. J. Nat. Hist. 17:731–741.

George, G.J. and N.S. Desai. 1947. Enzymes in the pyloric caeca of Scatophagus argus. J. Univ. Bombay 15:16–21.

Greene, C.W. 1912. The absorption of fats by the alimentary tract, with special reference to the function of the pyloric caeca in the king salmon. Trans. Amer. Fish. Soc. 32:93–100.

Groman, D.B. 1982. Histology of the Striped Bass. Amer. Fish. Soc. Monograph no. 3 (ISSN 0362-1715).

Hale, P.A. 1965. The morphology and histology of the digestive system of two freshwater teleosts, Poecilia reticulata and Gasterosteus aculeatus. J. Zool. 146:132–149.

Heinrichs, S.M. 1982. Ontogenetic changes in the digestive tract of the larval gizzard shad, Dorosoma cepedianus. Trans. Amer. Micros. Soc. 101:262–275.

Hossain, A. and H.M. Dutta. 1986a. Function and origin of fish caeca. The Ohio J. Sci. 87(2):1.

Hossain, A. 1988. Ontogenetic and morphological factors determining the origin and function of fish caeca. Ph.D. thesis, 172 pp. Kent State University, USA.

Hossain, M.A. and H.M. Dutta. 1968a. Methyl mercury induced alterations in the acid phosphatase activity in the intestine and caeca of bluegill fish. Bull. Environ. Contam. Toxicol. 36:460–467.

Hossain, M.A. and H.M. Dutta. 1983. Methyl mercury alterations in acid phosphatase activities in the intestine of the bluegill fish, Lepomis macrochirus (Teleostei). Ohio J. Science 83:71.

Hossain, M.A. and H.M. Dutta. 1986b. The intestinal loops of bluegill fish, Lepomis macrochirus. Acta Morphol. Neerl. Scand. 24:19–23.

Hossain, M.A. and H.M. Dutta. 1988. The embryology of the caeca in the digestive tract of bluegills, Lepomis macrochirus. Can. J. Zool. (4):988–1003.

Hossain, A. and H.M. Dutta. 1989. Assessment of functional discrepancy between fish intestine and intestinal caeca. Amer. Zool. 29(4):133A.

Hossain, A. and H.M. Dutta. 1991a. Food deprivation induces differential changes in contents and microstructures of digestive tract and appendages in bluegill fish, Lepomis macrochirus. Comp. Biochem. Physiol. 100A(3):769–772.

Hossain, A.M. and H.M. Dutta. 1991b. Assessment of structural and functional similarities and differences among the caeca of bluegill fish. Amer. Zool. 31(5):97A.

Hossain, A.M. and H.M. Dutta. 1991c. Fish caeca: A review of the history, evolution and functional morphology. Bull. Pol. Acad. Sci. Biol. Sci. 39(4):417–426.

Hossain, A. and H.M. Dutta. 1992. Role of caeca as food reservoirs in the digestive tract of bluegill sunfish, Lepomis macrochirus. Copeia 2:544–547.

Idler, D.R. 1973. Hormones in the life of the Atlantic Salmon. *In* The International Atlantic Salmon Foundation Publication, pp. 43-53.

Ishida, J. 1936. Distribution of the enzymes in the digestive system of stomachless fish. *Annot. Zool. Jap.* 15:263-284.

Jansson, B.O. and R. Olsson. 1960. The cytology of the caecal epithelial cells of *Perca perca*. *Acta Zool.* 41:267-276.

Jirge, S.K. 1971. Mucopolysaccharide histochemistry of the stomach of fishes with different food habits. *Folia Histochem et Cytochem.* 8:275-280.

Jobling, M. 1983. A short review and critique of methodologies used in fish growth and nutrition studies. *J. Fish Biol.* 23:685-703.

Kafuku, T.T. 1977. An ontogenetical study of the intestine coiling pattern of Indian major carps. *Bull. Freshw. Fish. Res. Lab.* (Tokyo) 27:1-20.

Kamoi, I., T. Suzuki, and T. Obara. 1980. Enzymatic properties of pyloric caeca in young yellowtail. *Bull. Jap. Soc. Sci. Fish.* 46:69-74.

Kapoor, B.G., H. Smith, and I.A. Verighina. 1975. The alimentary canal and digestion in teleosts. *Adv. Mar. Biol.* 13:109-239.

Kent, G.C. 1983. *The Comparative Anatomy of the Vertebrates*. Mosby C. Pub., St. Louis, MO, 5th ed.

Khanna, S.S. 1961. Alimentary canal of some teleostean fishes. *J. Zool. Soc. India* 13:206-219.

Khanna, S.S. and B.K. Mehrota. 1971. Morphology and histology of the teleostean intestine. *Anat. Anz.* 129:1-18.

Kimball, D.C. and W.T. Helm. 1971. A method of estimating fish stomach capacity. *Trans. Amer. Fish. Soc.* 3:572-575.

Kjorsvik, E., T. Van der Meeren, H. Kryvi, J. Arnfinnson, and P.G. Kvenseth. 1991. Early development of the digestive tract of cod larvae, *Gadus morhua* L., during start-feeding and starvation. *J. Fish. Biol.* 38(1):1-16.

Labhart, P. and P. Ziswiler. 1979. Comparative morphology of the alimentary tract of several cyprinodont fish. *Rev. Suisse Zool.* 4:843-854.

Lagler, K.F., J.E. Bardach, J.E. Miller, and D.R.M. Passino. 1977. *Ichthyology*. John Wiley and Sons, New York, 4th ed.

Lau, S.R. and P.L. Shafland. 1982. Larval development of the snook fish, *Centropomus undecimalis* (Pisces: Centropomidae). *Copeia* 3:618-627.

Lindsay, G.H. and J.E. Harris. 1980. Carboxymethylcellulase activity in the digestive tracts of fish. *J. Fish Biol.* 3:219-234.

Marias, J.F.K. 1980. Aspects of food intake, food selection and alimentary canal morphology in three mugil species. *J. Exp. Mar. Biol. Ecol.* 44:193-209.

Martin, N.V. and F.K. Sandercock. 1967. Pyloric caeca and gill raker development in lake trout. *J. Fish Res. Bd. Canada* 24:965-974.

McBee, R.H. and G.C. West. 1969. Cecal fermentation in the willow ptarmigan. *The Condor* 71:54-58.

Mohsin, S.M. 1962. Comparative morphology and histology of the alimentary canals in certain groups of Indian teleosts. *Acta Zool.* (Stockholm) 43:79-133.

Morton, J. 1967. Guts: The form and function of the digestive system. *In* The Institute of Biology Studies, no. 7. pp. 1-65. Arnold Publ., New York.

Nikolsky, G.V. 1963. *The Ecology of Fishes*. Academic press, London, 2nd ed.

Northcote, T.G. and R.J. Paterson. 1960. Relationship between number of pyloric caeca and length of juvenile rainbow trout. *Copeia* 3:248-250.

Pasha, K. 1964. The anatomy and histology of the alimentary canal of the omnivorous fish, *Mystus gulio*. *Proc. Indian Acad. Sci.* 59:211-223.

Rahimullah. M. 1945. A comparative study of the morphology, histology and probable functions of the pyloric caeca in Indian fishes together with discussion on their homology. *Proc. Ind. Acad. Sci.* 21:1-37.

Rahimullah, M. 1947. Disposition of the pyloric caeca in some freshwater and marine fishes of India. *J. Osmania Univ.* 13:21-52.

Reifel, C.W. and A.A Travill. 1978a. Structure and carbohydrate histochemistry of the stomach in eight species of teleost. *J. Morph.* 158:155-168.

Reifel, C.W. and A.A. Travill. 1978b. Gross morphology of the alimentary canal in ten teleostean species. *Anat. Anz.* 144:441-449.

Reifel, C.W. and A.A. Travill. 1979. Structure and carbohydrate histochemistry of the intestine in ten teleostean species. *J. Morph.* 162:343-360.

Ribble, D.O. and M.H. Smith. 1983. Relative intestine length and feeding ecology of freshwater fishes. *Growth* 47:292-300.

Richard, C.W. and L. Applegate. 1982. Alimentary canal development of muskellunge, *Esox masquinongy*. *Copeia* 3:717-719.

Romer, A.S. 1970. *The Vertebrate Body*. Saunders Publ., Philadelphia, 2nd ed.

Sastry, V.K. 1974. Histochemical localization of esterase and lipase in the digestive system of two teleost fishes. *Acta Histochem*. 51:18-23.

Scott, W.B. and E.J. Crossman. 1973. Freshwater Fishes of Canada. *Bull. Fish. Res. Bd. Canada*, no. 184.

Singh, H.R. and S.N. Bahuguna. 1983. Gross morphology of the alimentary canal and seasonal variation in feeding of *Noemacheilus montanus*. *Anat. Anz*. 154:119-124.

Sis, R.F., P.J. Ives, D.H. Lewis, and W.E. Haensly. 1979. The microscopic anatomy of the oesophagus, stomach and intestine of the channel catfish, *Ictalurus punctatus*. *J. Fish Biol*. 14:179-186.

Stroband, H.W.J. 1980. *Structure and function of digestive tract of the grass carp*. Ph.D. thesis, Van de Lanel bovwhogeschool te wageningen, The Netherlands.

Suyehiro, Y. 1942. A study on the digestive system and feeding habits of fish. *Jap. J. Zool*. 10:1-303.

Thorarensen, H., E. Mclean, E.M. Donaldson and A.P. Farrel. 1991. The blood vasculature of the gastrointestinal tract in chinook, *Oncorhynchus tshawytsche* (Walbaum) and coho, *O. kisutch* (Walbaum), salmon. *J. Fish. Biol*. 38(4):525-532.

Thurmond, T. 1979. *Histology and pathology of the alimentary canal of the American eel*. M.A. thesis. Univ. of Connecticut, Connecticut.

Travison, P. 1979. Histomorphological and histochemical researches on the digestive tract of the freshwater grass carp. *Anat. Anz*. 145:237-248.

Tyler, A.V. 1973. Alimentary Tract Morphology of Selected North Atlantic Fishes in Relation to Food Habits. Tech. Report Fish. Res. Bd. Canada, no. 361.

Vantue, V. 1980. Etude histologique de l'epithelium du tube digestif du bar, *Dicentrarchun labrax au cours du developpement postembryonnaire*. *Arch. Zool. Exp. Gen*. 121:191-206.

Wassersug, R.L. and R.K. Johnson. 1976. A remarkable pyloric caecum in the genus *Coccorella* with notes on gut structure and function in alepisauroid fishes. *J. Zool*. (London). 179:273-289.

Watts, S.A. and J.M. Lawrence. 1986. Seasonal changes in the activities of metabolic enzymes in the pyloric caeca of *Luidia clathrata*. (Echinodermata). *Amer. Soc. Zool*. 26:4-5.

Zihler, F. 1983. Gross morphology and configuration of digestive tracts of Cichlidae (Teleostei, Perciformes): Phylogenetic and functional significance. *The Nether. J. Zool*. 1:544-571.

6

The Structure and Function of Fish Liver

Jacques Bruslé and Gemma Gonzàlez i Anadon

Hepatology of vertebrates is largely based on knowledge gained from studying livers of mammals, especially rodents and humans. Although less known, because less studied, fish liver is of great interest. Indeed, it can be considered a starting point for comparative and phylogenetic studies among vertebrates. However, with a diversity of about 20,000 species, the description of any specific liver can hardly be used as a standard model for Teleostei, although common interorder morphologic features have been determined. In addition to this specific variability, some physiological characters of fish contribute to amplify their hepatic polymorphism. The fish liver appears, as does the liver of other vertebrates, as a key organ which controls many life functions and plays a prominent role in fish physiology (Plate 1), both in anabolism (proteins, lipids and carbohydrates) and catabolism (nitrogen, glycogenolysis, detoxication...). The fish liver plays an important role in vitellogenesis and, when compared with mammals, only a minor role in carbohydrate metabolism. On the other hand, the fish liver must be considered a target organ for many biological and environmental parameters that can alter liver structure and metabolism (Plate 2): food, pollutants, toxins, parasites, and microorganisms.

Two major features of fish physiology must be taken into account when studying the liver:
1. Fish are poikilothermal vertebrates, with substantial changes in metabolism related to temperature variations throughout the year. These changes will, of course, be reflected in the liver.
2. Fish spawn telolecithic eggs, rich in vitellus synthesized from precursor products (vitellogenin) through liver activity. There are, consequently, differences in the liver structures between males and females, immature and mature fish.

Fish are especially susceptible to environmental variations and respond more sensitively to pollutants than numerous mammals. Their liver is then a very interesting model for the study of interactions between environmental factors and hepatic structures and functions. Thus research on fish liver is expanding, especially in the field of troubles induced by aquaculture conditions or waterborne pollutants.

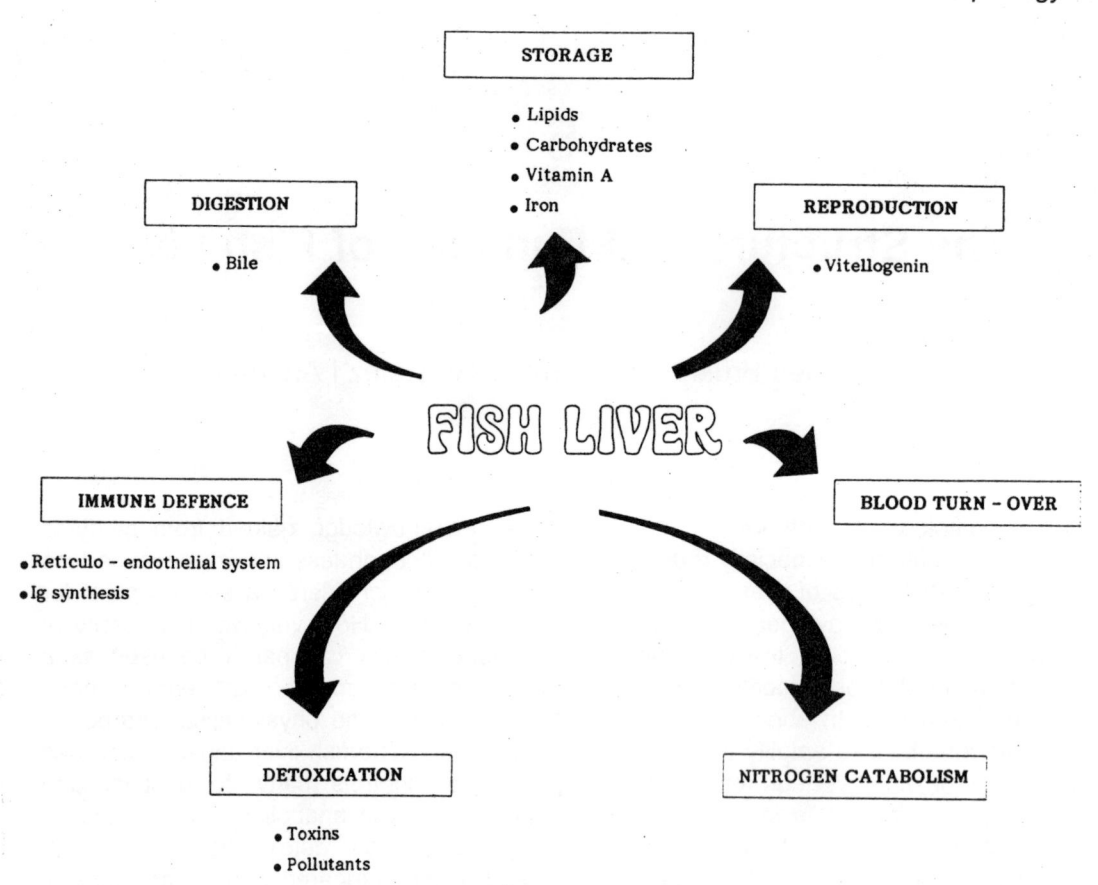

Fig. 1. Prominent functions of the fish liver.

GROSS ANATOMY

Shape

The liver in fish is a dense organ ventrally located in the cranial region of the general cavity. Its size, shape, and volume are adapted to the space available between other visceral organs (esophagus, stomach, spleen and intestine). It is divided into three lobes in many Teleostei species (two in Chondrichthyes and Dipnoi). However, no lobulation was recognized in some Teleostei, viz., *Oncorhynchus mykiss* (Robertson and Wexler, 1960), *Liza* spp. (Biagianti-Risbourg, 1991), *Lutjanus bohar* (Gonzalez, 1992) and *Serranus cabrilla* (Gonzalez et al., 1993), nor in the Agnath (Cyclostomata) *Lampetra* spp. (Shin,1977).

Color

Fish liver is generally reddish-brown because of its rich vascularization, tending towards yellow when fat storage is high. Liver of a yellowish color has been seen in *Anguilla anguilla, Dicentrarchus labrax,* and *Sparus aurata* (Bac et al., 1983) fed on artificial food responsible for lipid accumulation. A special case of green coloration was observed in the

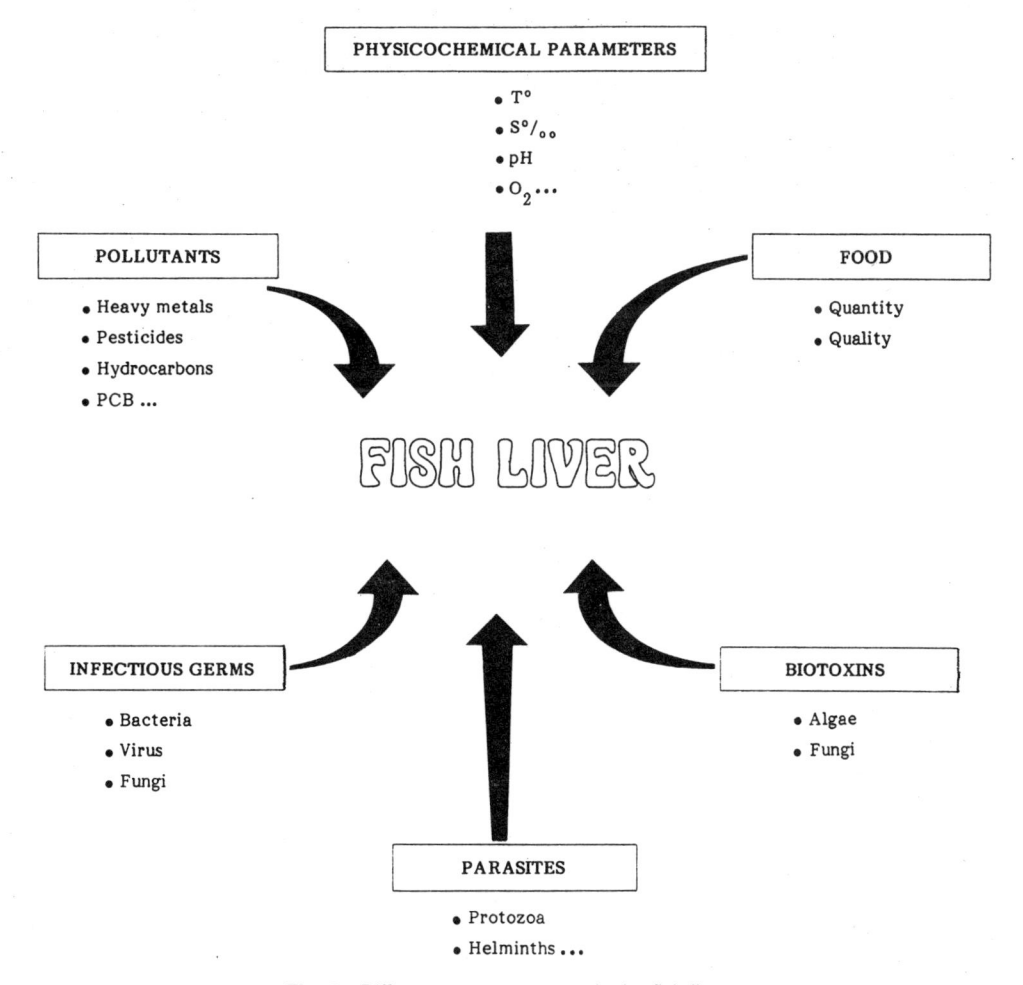

Fig. 2. Different stressors towards the fish liver.

sea lamprey *Petromyzon marinus* during metamorphosis, due to a biliary stasis following involution of the biliary tree (Youson et al., 1986).

Vascularization

As in other vertebrates, the vascular organization of this organ consists of two afferent blood vessels (hepatic artery and portal vein) and a single efferent vessel (hepatic vein) located at the hilum. Multiple small hepatic portal veins are seen in some flatfish. Detailed anatomical descriptions of the gross vascularization are few in number (Biagianti-Risbourg, 1990: review on *Liza* spp.; Mosconi-Bac, 1991: on *Dicentrarchus labrax*).

Hepatosomatic index (HSI = liver weight/body weight × 100)

The mean HSI value is species specific and correlates with the amount of fat deposition (higher in species which store high amounts of lipids) (Chiba and Honma, 1981; Oguri, 1985; Ando et al., 1993). In Osteichthyes the HSI is about 1–2% versus values as high as 10–20% in Elasmobranchs and Holocephalian fish (Oguri, 1978a, b). However, a high intraspecific variability occurs depending on differences in sex, season, age, and physiological condition regarding feeding, reproduction, or stress.

Sex. HSI is generally higher in females than in males, as shown in *Diodon holacanthus* (13.1 and 9.5 respectively; Chiba and Honma, 1981) and *Oncorhynchus mykiss* (2.6 and 1.5; Lincoln and Scott, 1984).

Sexual maturation. HSI decreases with maturation of gonads in many Teleostei such as ayu *Plecoglossus altivelis* (Aida et al., 1973) and Atlantic halibut *Hippoglossus hippoglossus* (Haug and Gulliksen, 1988). On the other hand, an experimental hormonal stimulation of estradiol induces an increase in this value (Aida et al., 1973; Korsgaard and Mommsen, 1993). Male steroids such as methyltestosterone (MT) induce a similar effect in goldfish *Carassius auratus* (Hori et al., 1979).

Feeding. HSI is highly sensitive to the nutritional status of the fish. It correlates with the quantity and quality of food (Hung et al., 1990), being higher in cultured than in wild fish, as observed in yellowtail *Seriola quinqueradiata* (1.7 and 1.4 respectively) (Shimeno et al., 1985). HSI can be also modified by the ingestion of anabolic steroids (MT: Lone, 1989).

Season and photoperiod. Fluctuations in HSI with the seasons have been shown in *Chaenogobius* spp. (Takahashi, 1974) and *Hippoglossus hippoglossus* (Haug and Gulliksen, 1988) depending on such factors as ambient temperature, feeding activity, and maturation period. Circadian variations of HSI (lower during the day) have also been found in *Oncorhynchus mykiss* held under different photoperiod regimes (Boujard and Leatherland, 1992).

Stress. Pollutants modify the HSI, as shown in bream *Abramis brama* (Sloof et al., 1983). Thus HSI may be a useful indicator of chemical water pollution.

LIGHT MICROSCOPY

Organization of hepatic parenchyma

The liver of vertebrates is a digestive gland of endodermic origin. It is made up of cellular plates, each of which separates several lacunae: the vascular (sinusoids) and biliary (canaliculi) network. Fish liver belongs to the "lower vertebrate category" in the classification of Ellias and Bengelsdorf (1952). It is constituted by highly anastomosed tubules that originate from the blind tubules organization of the Cyclostomata, the "myxine category". The hepatic parenchyma in fish is made of two cellular plates surrounded by sinusoids. Each plate shows polarized hepatocytes with a sinusoidal face for absorption and a biliary face for excretion. This "muralium duplex" is different from the "muralium simplex" in mammals (Mugnaini and Harboe, 1967). Indeed, the "mammal category" of liver is characterized by only one cellular plate which separates two sinusoids. However, Eurell and Haensly (1982) described a muralium restricted to only one cellular plate in the fish *Micropogon undulatus*. Such differences in fish-liver descriptions are due to the complexity of the hepatic organization which needs a tridimensional approach. Taking into account the planes of the histological sections, a monocellular cord located around the blood vessel will be seen when the sections are perpendicular to the sinusoids, in contrast to the bicellular cords found when the sections are parallel to the sinusoids (Plate 3).

In the lower Teleostei (Salmonidae: *Oncorhynchus mykiss*; Hampton et al., 1988; *Salmo salar*, Robertson and Bradley, 1991) the liver looks like an anastomosed tubular gland, in which four to nine hepatocytes surround a bile duct. In this pattern the sinusoidal face of the hepatocyte can be considered the basal region of the cell, and the opposite biliary face as apical. However, in higher Teleostei (Serranidae: *Serranus cabrilla*; Gonzalez et al., 1993), a similar number of hepatocytes surrounds each sinusoid. In this latter pattern the apical pole would be the sinusoidal face.

Hepatic architecture

The hepatic parenchyma of fish is very homogeneous and the hepatocytes are polygonal-shaped cells, often weakly basophilic (poor in organelles), compared to those of mammals (Plate 4, Fig. 1). The nucleus is spherical, with a single, central nucleolus. Regional or zonal enzymatic activity is not a classical feature of Teleostei liver (Hampton et al., 1985; Schär et al., 1985; Robertson and Bradley, 1991). Between two neighboring sinusoids, the hepatocytes are arranged as cords, usually two cells thick, but branching and anastomosing of cords can result in four or more cell layers.

Light-microscope observations show that it is not possible to distinguish hexagonal subdivisions of hepatic parenchyma (hepatic lobules), making it difficult to identify the acini of Rappaport as observed in mammals (Rappaport et al., 1954). The triads, constituted by a ramification of the portal vein, the hepatic artery, and a biliary duct, are indistinct, if not absent, in almost all Teleostei (Plate 4, Fig. 2). However, some triads are found in *Caranx* spp. and *Lutjanus bohar* (Gonzalez, 1992). Thus the term "portal region" is more correct than the "portal triads" of mammals when referring to fish liver. The hepatic

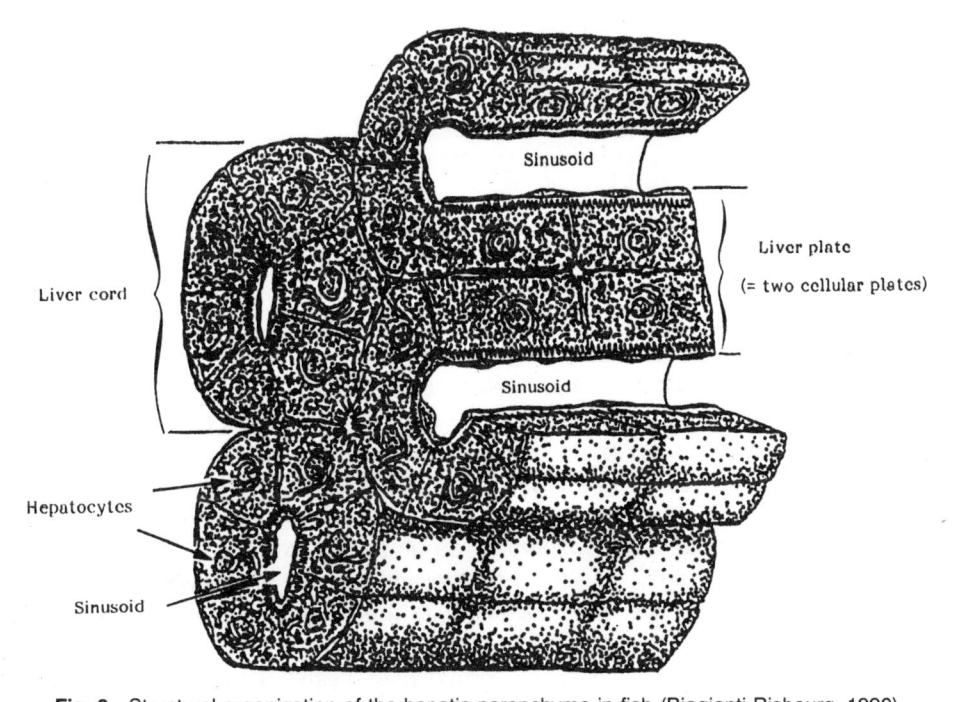

Fig. 3. Structural organization of the hepatic parenchyma in fish (Biagianti-Risbourg, 1990).

veins, classically located at the center of the hepatic lobules (also called center-lobular veins) are found randomly throughout the hepatic parenchyma of fish. Hence the terminal branches of the portal vein and the initial tributaries of the hepatic vein are difficult to identify. The portal venulae are differentiated from the hepatic venulae by a higher amount of periadventitial connective tissue in the former and because the lumen of the portal venula is devoid of blood cells. The hepatic artery differs from the veins by its narrow lumen, its thick wall with more elastic fibers, and its endothelial cells that are generally more voluminous than in veins. Communication between hepatic arteries and veins and also between portal and hepatic veins is assured by the capillary network of sinusoids (7–15 μm diameter). These capillaries are of the fenestrated type in fish but nonetheless are still named sinusoids, as in mammals. These sinusoids are radially disposed around the hepatic veins and constituted by a simple squamous endothelium without basal lamina.

The bile ducts consist of a simple cubic epithelium (Plate 4, Fig. 2), larger than the vascular one, with a PAS-positive brush border. In some cases "rodlet cells" are found among the epithelial cells. Under the epithelium a basal lamina and a wall consisting of both collagen and muscular fibers are observable. Some fibroblasts are present in the wall and some hematopoietic tissue is sometimes found at this location depending on the species (Ferguson, 1989).

The pancreatic exocrine tissue develops around the portal vein during ontogenesis. It remains extrahepatic or penetrates more or less deeply into the liver parenchyma depending on the species (*Ictalurus punctatus*, Hinton and Pool, 1976; *Micropogon undulatus*, Eurell and Haensly, 1982; *Diodon holacanthus*, Oguri, 1985; *Dicentrarchus labrax*, Mosconi-Bac, 1991; *Acanthurus blochii*, *Lutjanus bohar*, *Scarus* spp., Gonzalez, 1992; *Serranus cabrilla*, Gonzalez et al., 1993). Thus the existence of a hepatopancreas makes identification of the portal venula in these species relatively easy (Plate 4, Fig. 3). Pancreatic tissue can be differentiated from hepatic tissue by its acinar arrangement and its characteristic stain with hematoxylin-osin (basophilic basal pole and cytoplasm rich in eosinophilic zymogen granules). A thin septa of connective tissue separates the hepatocytes from the exocrine pancreatic cells.

Melano-macrophage centers (MMC) occur in the hepatic parenchyma of fish (Plate 4, Figs. 3 and 4). Their size, number, and content are highly variable, depending on the species, age, and health status (Agius, 1980). They are usually located in the vicinity of the hepatic arteries, portal veins, or bile ducts (portal regions). They concentrate heterogeneous materials such as lipofuscin (natural yellowish color), melanin (natural brown or black color), ceroid (PAS-positive) or hemosiderin (Perls-positive). Such products may play a role in neutralizing potentially toxic free radicals and cations produced during peroxydation of unsaturated lipids (Agius, 1985).

Differences in the parenchyma, related to hepatic energy stores (glycogen and lipids), are readily observed at the histological level. Glycogen is recognized by its PAS-positive affinity whereas lipid droplets are observable as empty vacuoles after histological treatment. The main factors responsible for the type and amount of storage are the quality of food and the feeding activity, both varying with the seasons (Quaglia, 1976). Glycogenic stores are abundant in some species of Cyprinidae, e.g. *Leuciscus idus* (Braunbeck et al., 1987), Salmonidae, e.g. *Oncorhynchus mykiss* (Vernier and Sire, 1976; Hacking et al., 1978) and in *Anguilla anguilla* (Barni et al., 1985). The glycogen reserves of fish hepatocytes are considered by some authors to be the result of metabolic deviation, induced by an unbalanced diet, related to metamorphosis or after long-term maintenance in an aquarium (Oguri, 1976; Ferguson, 1989; Segner and Witt, 1990). However, some authors have suggested that some fish may have the ability to synthesize or break down glycogen depending on their metabolic needs (Schär et al., 1985). In contrast, some other species, such as *Diodon* spp. (Oguri, 1985; Langdon, 1986), *Gadus morhua* (Fujita et al., 1986), *Lota lota* (Byczkowska-Smyk, 1968), or *Scarus* spp. (Gonzalez, 1992) accumulate large amounts of lipids (Plate 4, Figs. 5 and 6). They are mainly used as an energy resource and appear to store essential polyunsaturated fatty acids (PUFA) and vitamin A. The large amount of lipids accumulated in shark's liver is also a modulator of body density for maintenance of the fish's hydrostatic balance, with buoyancy control taking priority over lipid reserve energy (Baldridge, 1972).

ELECTRON MICROSCOPY

Hepatocytes

A high degree of interspecific variation occurs among Teleostei. This variation underlines the importance of determining hepatic ultrastructures for individual fish species. Liver diversity is a basic concept that must be highlighted whenever a "standard" hepatocyte description is presented.

The mean size of hepatocytes is about 20 μm in diameter. There are species-specific differences, with mean sizes ranging from 12 μm in *Micropogon undulatus* (Eurell and Haensly, 1982) to 30 μm in *Cyprinus carpio* (Kramar et al., 1974) and even 50 μm in *Gadus morhua* (Fujita et al., 1986). Such differences correlate with the size of energy stores (glycogen and/or lipids) depending on the specific metabolic activities related to seasonal changes, as observed in wintering of golden ide *Leuciscus idus* by Segner and Braunbeck (1990) and during gestation of gravid mosquito fish *Gambusia affinis* by Weis (1972).

The nucleus of fish hepatocytes is generally round to ovoid (Plate 5, Fig. 1) and centrally located except when cytoplasmic energy stores shift it to the periphery of the cell. The nucleolus, of considerable electron density, is single, central and fairly homogeneous. The chromatin is granular, with more condensed heterochromatin located at the periphery of the nucleus and also associated with the nucleolus. The classical nuclear envelope shows nuclear pores.

Fish hepatocytes are characterized by their prominent structural (sinusoidal-canalicular) and functional (vascular-biliary) polarity (Plate 5, Fig. 1), which is still retained in isolated cells (Bouche et al., 1979). Fish hepatocytes are poorer in organelles than those of mammals, suggesting low synthetic activity and hence low protein secretion. The rough endoplasmic reticulum (rER) is often arranged in an array parallel to the cellular membrane and nuclear envelope. The smooth endoplasmic reticulum (sER) is almost absent. The Golgi apparatus is poorly developed and located in the vicinity of the nucleus and the canalicular area where a few lysosomes and peroxisomes are present. Mitochondria display a relatively dense matrix, some cristae and electron-dense granules. They are distributed throughout the cytoplasm, often in close association with the rER. A single centriolar body is occasionally found. In some cases the organelles are concentrated around the nucleus and at the peripheral area of the cell (Plate 5, Fig. 2). This rather special arrangement, called "cytoplasmic segregation", has been described in some freshwater species, such as *Oncorhynchus mykiss* (Chapman, 1981), *Cyprinus carpio* (Saèz et al., 1984; Storch et al., 1984), and *Leuciscus idus* (Braunbeck et al., 1987) as well as in some marine species as *Dicentrarchus labrax* (Mosconi-Bac, 1990).

The hyaloplasm of fish hepatocytes includes a variable amount of store products, such as glycogen deposits in a typical rosette pattern (Plate 5, Fig. 2) and/or lipid globules of different electron densities (Plate 5, Fig. 3). Stored product distribution can be roughly divided into two patterns. In some cases they are widely distributed and scattered throughout the cytoplasm but they can also be concentrated at the center of the cell, in which case the nucleus and organelles are shifted to the periphery. Waste products, such as myelinic figures (Plate 5, Fig. 1), lipofuscin granules (Plate 5, Fig. 2), and residual bodies, which may have resulted from disturbances in the normal oxidative pathways of lipids, are quite common inside fish hepatocytes.

The plasmic membrane is differentiated into microvilli, on both the vascular and the biliary faces (Plate 5, Figs. 1 and 3). Cohesion between adjacent hepatocytes is assured by desmosomes, especially in the canalicular area. This physical barrier ensures the separation between blood and bile compartments. Except for these desmosomes, there are few connections between adjacent hepatocytes. Thus these junctional complexes around the biliary canaliculi may be considered important structures for supporting the hepatic tissue (Tanuma et al., 1982).

Sinusoids and their associated cells

In contrast with mammals, the sinusoidal lumen is limited by a fenestrated and continuous endothelium where no basal lamina occurs (Plate 6, Fig. 1). It consists of flattened cells with an ovoid, flat nucleus and a cytoplasm very poor in organelles but rich in microfilaments (a characteristic feature of flatfish according to Tanuma et al., 1982). In addition, endothelial cells display fenestrations (pores ranging from 0.05 to 0.2 μm in diameter in *Carassius auratus*; Nopanitaya et al., 1979). The presence of fenestrae in the endothelium and the lack of basal lamina allow plasmatic metabolites an easy crossing through the blood barrier to the hepatocytes, as revealed by extensive micropinocytic activity (Ferri and Sesso, 1981). Desmosomal junctions between endothelial cells and hepatocytes seem to make up a strong framework, supporting the architecture of the hepatic tissue (Plate 6, Fig. 1). The microfilaments of the cytoplasm allow the endothelial cells to play a contractile role in regulating the blood flow (Tanuma et al., 1982).

The perisinusoidal space, or space of Disse, lies between the sinusoidal endothelium and the hepatocyte. It is large and well developed in fish, in contrast to higher vertebrates. The hepatocyte microvilli project into this space and thus greatly amplify the blood-hepatocyte exchange surface area. The electron density of the space of Disse is low, although a loose network of microfilaments and some collagen fibrils may occur depending on the species. The amount of collagen in the space of Disse seems to be higher in species where no desmosomes occur between the endothelial cells (Kendall and Hawkins, 1975; Sakano and Fujita, 1982). Several categories of perisinusoidal cells, as in other vertebrates, have been recognized in the space of Disse in fish.

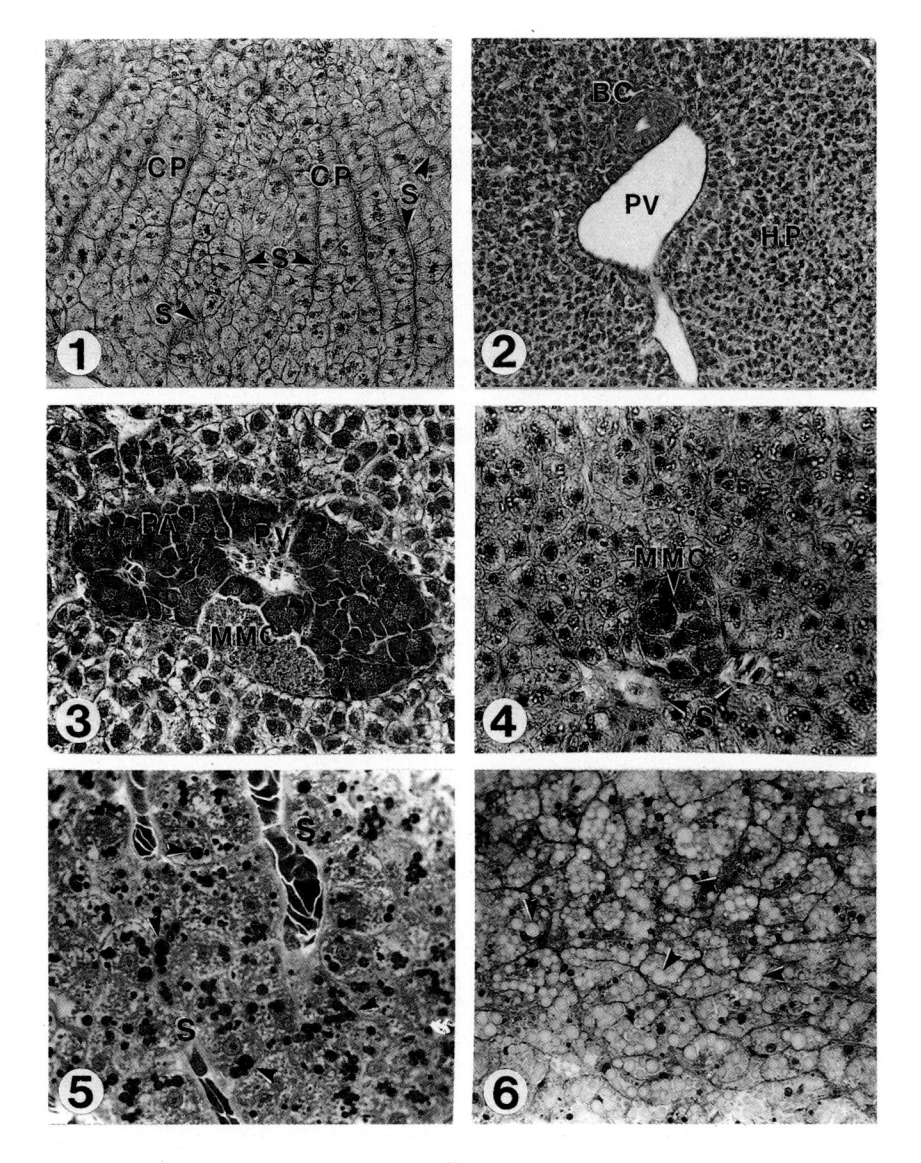

Plate 4. Histology of the fish liver.

Fig. 1. *Cephalopholis argus* (Serranidae; French Polynesia). RNS-PIC × 960. Hepatic parenchyma consisting of two cellular plates surrounded by sinusoids.

Fig. 2. *Serranus cabrilla* (Serranidae; Mediterranean Sea). H-E × 750. Association of a biliary canal and a portal vein at the portal region.

Fig. 3. *Lutjanus bohar* (Lutjanidae; Indian Ocean). RNS-PIC × 960. Pancreatic acini surrounding a portal vein. Lipofuscin granules are concentrated inside a melano-macrophage center.

Fig. 4. *Plectropomus leopardus* (Serranidae; French Polynesia). RNS-PIC × 1,200. An intrahepatic melano-macrophage center adjacent to a sinusoid.

Fig. 5. *Ctenochaetus striatus* (Acanthuridae; French Polynesia). Semithin section × 1,600. Details of hepatic parenchyma showing two cellar plates between two sinusoids. Hepatocytes contain some dense lipid droplets (arrowheads).

Fig. 6. *Cheilinus fasciatus* (Labridae; Indian Ocean). Semithin section × 1,200. Hepatic parenchyma characterized by a high amount of light lipid droplets (arrowheads).

BC—biliary canal; CP—cellular plate; HP—hepatic parenchyma; MMC—melano-macrophage center; PA—pancreatic acini; PV—portal vein; S—sinusoid.

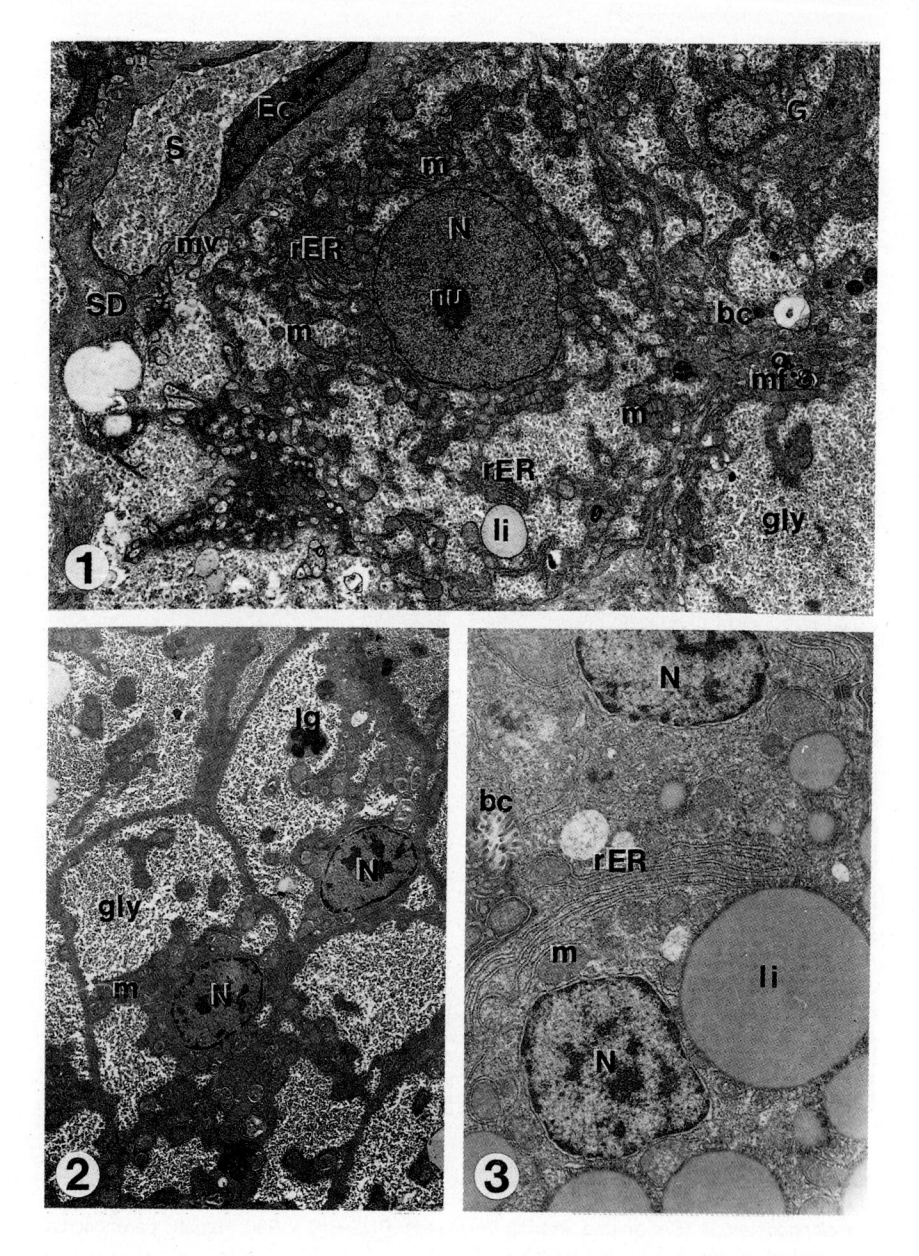

Plate 5. Cytology of the fish liver.

Fig. 1. *Anguilla anguilla* (Anguillidae; French Mediterranean Lagoon). TEM × 6,250. Hepatocyte showing a characteristic bipolarity: vascular pole (sinusoid) at the left and biliary pole (canaliculus) at the right of the micrograph (courtesy of Dr. Biagianti-Risbourg).

Fig. 2. *Lutjanus bohar* (Lutjanidae; New Caledonia). TEM × 4,500. Segregated hepatocytes characterized by a high amount of glycogen granules shifting the nucleus and the cytoplasmic organelles to one of the sides of the cell.

Fig. 3. *Serranus cabrilla* (Serranidae; Mediterranean sea). TEM × 9,000. Hepatocyte containing voluminous lipid droplets of low electron density.

bc—bile canaliculus; Ec—endothelial cell; G—Golgi apparatus; gly—glycogen; lg—lipid granule; li—lipid droplet; m—mitochondria; mf—myeline figure; mv—microvilli; N—nucleus; nu—nucleolus; rER—rough endoplasmic reticulum; S—sinusoid; SD—space of Disse.

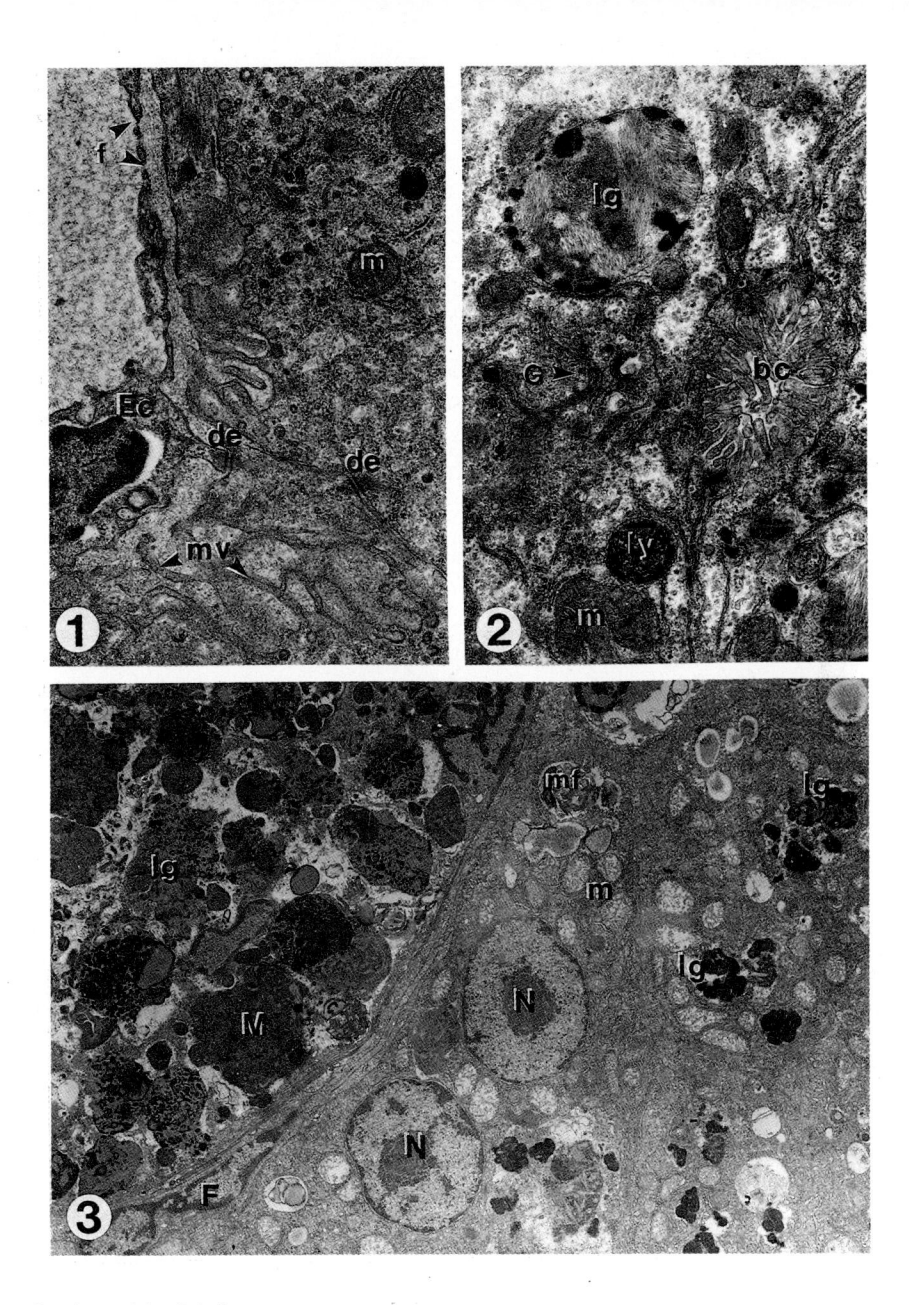

Plate 6. Cytology of the fish liver.

Fig. 1. *Liza aurata* (Mugilidae; French Mediterranean lagoon). TEM ×20,000. Vascular pole of the hepatocyte. Numerous microvilli project into the space of Disse. The endothelial cell display some fenestrae. Two desmosomes are observed between the endothelial cell and the hepatocyte, and between two adjacent hepatocytes (courtesy of Dr. Biagianti-Risbourg).

Fig. 2. *Liza ramada* (Mugilidae; French Mediterranean lagoon). TEM ×10,000. Biliary pole of the hepatocyte. Numerous microvilli project into the biliary lumen. At this location, several organelles such as lysosomes, Golgi apparatus and lipofuscin granules are commonly found (courtesy of Dr. Biagianti-Risbourg).

Fig. 3. *Cephalopholis argus* (Serranidae; New Caledonia). TEM ×6,000. A melano-macrophage center (left side), surrounded by a fibroblastic envelope, is adjacent to hepatocytes (right side) containing numerous dense lipofuscin granules.

bc—bile canaliculus; de—desmosome; Ec—endothelial cell; F—fibroblast; f—fenestrae; G—Golgi apparatus; lg—lipofuscin granule; ly—lysosome; M—macrophage; m—mitochondria; mf—myeline figure; mv—microvilli; N—nucleus.

Ito cells or fat-storing cells (FSC)

These cells of fibroblastic origin are variable in shape, mostly elongated, and rich in free ribosomes. They usually contain a large amount of lipid droplets that are the storage sites of vitamin A (Sakano and Fujita, 1982; Fujita et al., 1986; Wake et al., 1987; Robertson and Bradley, 1992). In some species, Ito cells are not free but connected by desmosomal complexes between them and with the endothelial cells and the hepatocytes. Such a tridimensional network might be the support of the liver tissue (hepatoskeletal system according to Fujita et al., 1986). The involvement of Ito cells in the fibrinogenesis in the space of Disse under pathological conditions has been speculated by Takahashi et al. (1978).

Macrophages and Kupffer cells

The hepatic perisinusoidal cells of the reticulo-endothelial system (Kupffer cells), which are typical of the mammal liver, have not been described in many Teleostei. These stellate and polymorphous cells are scarce and hardly ever observed in fish liver. When present, they are characterized by phagocytic activity, as revealed by injection of foreign molecules (Ferri and Sesso, 1981) and observed in wild Atlantic salmon (Robertson and Bradley, 1992). On the other hand, a special category of macrophage, the melano-macrophage (MM), is very common in fish. MM are located around the portal regions and are readily recognized by their cytoplasmic granules (melanin, hemosiderin, lipofuscin and/or ceroid). They are often concentrated into MMC (Plate 6, Figure 3) similar to those found in the spleen and kidney of fish (Agius, 1980).

Bile canaliculi

The organization of the biliary tree in fish is no different from that in higher vertebrates. It originates in many Teleostei as an intercellular canaliculus, formed by the close apposition of two hepatocytes (Plate 6, Fig. 2). The cell membrane projects numerous microvilli into the canalicular lumen and desmosomes and tight junctions assure the cohesion. However, the presence of intracellular bile canaliculi is a cytological feature of some freshwater fish. These intracellular canaliculi spring from the hepatocyte hyaloplasm in the vicinity of the nucleus (Tanuma, 1980). Golgi vesicles and lysosomes are concentrated in the cytoplasmic area near the canaliculi, suggesting bile secretory activity (Plate 6, Fig. 2). Abundant contractile microfilaments, concentrically located around the canaliculi, control bile flow through the bile ducts.

PHYSIOLOGICAL HEPATOCYTE POLYMORPHISM

Fish liver shows a high variability, both intra- and interspecific. Such differences from one species to another and from one individual to another were shown in the previous paragraphs concerning gross anatomy (size, color, HSI) and histology (pancreatic tissue, MMC, energy stores ...). However, a "standard" hepatocyte is described at the ultrastructural level, although there is high cellular diversity (Plate 7).

Several types of hepatocytes have been distinguished inside the same hepatic parenchyma. The main difference is a variation in the electron density and many

Fig 7. Diagram of fish liver tissue at the ultrastructural level, showing two cellular plates located between two sinusoids and a biliary canal. A melano-macrophage center is adjacent to the vascular and biliary structures.

BC—biliary canal; co—collagen fibers; d—desmosome; E—erythrocyte; EC—endothelial cell; F—fibroblast; f—fenestrae; g—Golgi; gl—glycogen; H—hepatocyte; Hl—hepatocyte rich in lipids; IC—Ito cell; L—lymphocyte; l—lipid droplet; li—lipofuscin granule; lu—lumen; Ma—macrophage; mv—microvilli; MMC—melano-macrophage center; S—sinusoids; SD—space of Disse; sH—segregated hepatocyte.

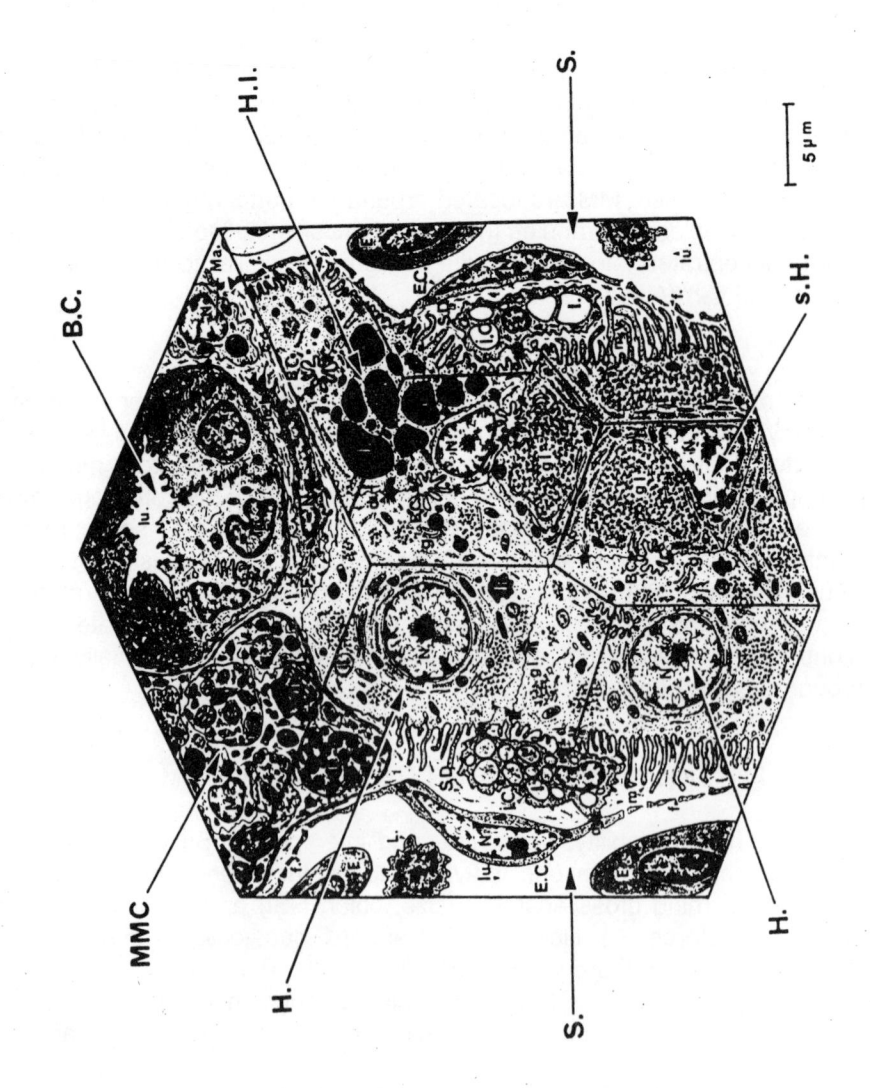

authors describe "dark" and "light" hepatocytes. The physiological turnover of hepatocytes probably explains the existence of dark cells that may be the result of the onset of a degenerative process. According to some authors (Saito and Tanaka, 1980), the various cytological treatments (fixatives, buffers . . .) may increase the electron density, which then must be considered an artifact. However, according to Leatherland and Sonstegard (1983; 1987), dark and light cells are not artifacts but a consistent feature of coho salmon liver that may result from different levels of metabolic activity. This theory could also explain the dark and light cells observed during ovogenetic maturation in Pacific herring (Gillis and McKeown, 1990).

On the other hand, the most conspicuous differences are observed between hepatocytes while comparing one liver with another. Many factors are involved in such hepatocyte variability, both endogenous (species, age and sex) and exogenous (temperature and feeding).

Species variability

This is seen in the size and the organization of both the nucleus and the cytoplasm. Such differences correlate with the amount of nucleic acids and energy stores respectively (Byczkowska-Smyk, 1973; 1981). Antarctic fish (such as Nototheniidae) exhibit a protein secretory system (Eastman and DeVries, 1981), which indicates a high degree of adaptation to low temperatures.

Age changes

These take place in the hepatocytes during liver ontogenesis and concern the number of mitochondria and the amount of glycogen, as observed in *Salmo trutta* by Byczkowska-Smyk (1967a and b) and in *Tilapia mossambica* by Jirge (1970). In *Anguilla anguilla* (Barni et al., 1985), the two distinct physiological situations of the trophic stage (yellow eel) and the reproductive stage (silver eel) result in two distinctive livers, the first storing glycogen and the second with a high number of lipid droplets. *Petromyzon marinus* (Sidon and Youson, 1983) also shows a special case of hepatocyte polymorphism during metamorphosis, where degenerative processes take place in correlation with the change of habitat. In a similar way, significant ultrastructural changes in hepatocytes are observed in the parr-smolt transformation of Atlantic salmon. One of the major functions of parr hepatocytes involves the storage of large amounts of glycogen. However, smolt hepatocytes exhibit an apparent shift away from such significant glycogen deposits whereas an increase in mitochondrial size occurs (Robertson and Bradley, 1991). Other examples of physiological changes in the hepatocytes are the accumulation of lipofuscin-like granules in old *Oryzias latipes* (Yamamoto and Egami, 1974) and the degenerative processes in sexually mature, spawning Pacific salmon (*Oncorhynchus* spp.) prior to death (Robertson and Wexler, 1960).

Sex

Sexual hepatocyte dimorphism is not apparent in juvenile and immature fish. Sexual differences in the liver occur during sexual maturation. They mainly pertain to the content of lipid and glycogen stores but are also reflected in the amount of rER and Golgi vacuoles, generally higher in females than in males, and correlate with exogenous vitellogenesis

(Ishii and Yamamoto, 1970; Van Bohemen et al., 1981). Vitellogenin, the egg-yolk protein precursor, is synthesized in the liver in response to estrogen. Ovarian hormones are then responsible for hepatocyte changes, as shown by experimental treatment with 17-β-estradiol (Aida et al., 1973; Ng et al., 1984), which are indicative of enhanced metabolic activity. Such activity has been induced in the liver of male zebra fish (Peute et al., 1985).

Temperature

Morphological changes in hepatocyte organization have been shown in the cyprinid *Leuciscus idus* (Braunbeck et al., 1987) to be an adaptation to temperature. Fish under cold acclimatization exhibited a proliferation of rER, Golgi and mitochondria, and glycogen accumulation; warm adaptation induced a decrease in glycogen and an increase in lipid stores and myelinic bodies. Such differences indicate thermal modifications in the lipid and carbohydrate metabolism, first observed in *Cyprinus carpio* (Storch et al., 1984). A segregation of fibrillar and granular nucleolar components can also be observed in carp (Saez et al., 1984) during thermal acclimatization.

Feeding

Lipid accumulation has been observed in the cytoplasm of hepatocytes of *Dicentrarchus labrax* fed on unbalanced diets (excess in oleic acid: Mosconi-Bac, 1987). Intranuclear lipid droplets have been observed in *Chanos chanos* (Storch and Juario, 1983) and in *Dicentrarchus labrax* fed on artificial lipid-rich feed (Mosconi-Bac, 1990). In addition, the size of the nuclei of hepatocytes correlates with the nutritional status, especially in larvae (*Odontesthes bonariensis*) which, under the stress of starvation, showed marked regressive changes (Strüssmann and Takashima, 1990).

Complex relationship

The complex relationship between all these factors makes it difficult to establish a simple classification. Indeed, the adaptative changes during seasonal acclimatization can be considered an adjustment of cellular and subcellular morphology, as revealed in *Leuciscus idus* (Segner and Braunbeck, 1990). Nutrition/starvation and temperature may be regarded as the most relevant factors for wintering in golden ide. Some other environmental factors are also able to modify the hepatocyte organization. For example, pollutants act as stressors that jeopardize the normal liver function and induce changes that must be considered pathological events. Interactions between factors are complex and difficult to define, leading to a concept of multifactorial etiology.

HIGHLIGHTS ON FISH LIVER MORPHOLOGY AND SUGGESTIONS FOR FURTHER INVESTIGATIONS

A full understanding of the fish liver might well constitute a good basis from which phylogenetic research should be approached. It could be considered the starting point for comparative studies among vertebrates. The high DNA content resulting from the manifold increase of the same genes, as observed in hepatocytes of Dipnoi (241 and 284 pg in *Lepidosiren* spp. and *Protopterus* spp. respectively), is considered a conservative character compared with more advanced Teleostei (3.4 pg in *Carassius auratus*; Byczkowska-Smyk, 1973). The structural organization of the liver could be used for valuable comparisons between primitive and more advanced vertebrates. The liver of Cyclostomata, with its simple tubular structure, is the most primitive liver in vertebrates. The physiological involution of the biliary tree in lampreys could be used as a good model for studying mammal pathologies, such as cholestasis and hemosiderosis, resulting from degeneration of the biliary ducts (Youson et al., 1983). The liver of Teleostei shows a more advanced organization than these primary vertebrates, since it comprises anastomosed tubules and thereby comes closer to the structure of mammal liver. However, some original features of fish liver are worth highlighting (muralium duplex, sinusoidal wall, intrahepatocyte bile canaliculi, and MMC). Despite these points of interest, studies on fish liver have up to now focused on only a small number of species. Indeed, among the 18,000 Teleostei, only 100 species have been investigated thus far. Freshwater fish, especially those of interest for culture (Salmonidae and Cyprinidae), are better known than brackish water and certainly marine fish. This shortage in fish liver research is especially evident with tropical species (Gonzalez, 1992), a large number of which have never been studied. For those investigated, studies were restricted for the most part to histological analyses. Cytological descriptions must be encouraged; findings with the light microscope are of limited value. It has been proved that the requisite histological fixatives definitely differ in ability to maintain liver morphology and glycogen deposits (Speilberg et al., 1993).

The morphological evolution of organs during ontogenesis is one of the main topics of embryology. Yet studies on the genesis of the liver and on the origin, differentiation, and onset of metabolic activity of hepatocytes in fish are few (Diaz et al., 1989; Diaz and Connes, 1990). The small size of fish larvae is a great handicap in this kind of study.

Interestingly, studies of fish liver have shown that it appears well suited to serve as a model for analysis of the interactions between natural environmental changes and hepatic morphology. Environmentally induced plasticity of hepatocyte structures was demonstrated by Braunbeck et al. (1987) and then by Segner and Braunbeck (1990) with reference to temperature acclimatization. In addition, many environmental stressors (natural biotoxins and xenobiotics of anthropogenic origin) act on the liver inducing metabolic perturbations and structural lesions. These hepatic alterations may reflect a variety of reactions of the whole organism to intoxication, leading to disease or kill. Such responses allow the fish liver to be considered a good indicator of fish health status (Bowser et al., 1990; Braunbeck et al., 1990; Vethaak and Rheinallt, 1990; Biagianti-Risbourg, 1992; Bruslé, 1993). On the same topic, cultured fish hepatocytes are also a good model for *in vitro* toxicology research. Dutta et al. (1993) observed changes in the hepatocyte diameter and other hepatic structures in fish upon exposure to malathion pesticide. The universality of many basic biological processes means that lower vertebrates such as fish can serve as useful research models for toxicity, xenobiotic metabolism, and genotoxicity studies on human and high vertebrate toxicity and also cancer research (Hightower and Renfro, 1988; Morimoto et al., 1988; Baksi and Frazier, 1990).

All these studies rely on a full understanding of the normal liver as revealed by fish presumed free of any pathology induced by various stressors. A problem arises when the liver of "control fish" from the field must be described, taking into account all the modifications resulting from their own history, regarding feeding, reproduction, and water quality. The difficulties are becoming all the greater as degradation of the aquatic environment and ecosystems progresses, making the liver of "true control fish" more and more scarce. Thus there is an urgent need to proceed in drawing up an extensive hepatic inventory of all the many species of fish that have not been investigated to date.

References

Agius, C. 1980. Phylogenetic development of melano-macrophage centres in fish. *J. Zool., London* 191:11–31.
Agius, C. 1985. The melano-macrophage centres of fish: A review. In M.J. Manning and M.F. Tatner (eds.), *Fish Immunology*, pp. 85–104. Academic Press, London.
Aida, K., K. Hirose, M. Yokote, and T. Hibiya. 1973. Physiological studies on gonadal maturation of fishes-II. Histological changes in the liver cells of ayu following gonadal maturation and oestrogen administration. *Bull. Jap. Soc. Sci. Fish*. 39:1107–1115.
Ando, S., Y. Mori, K. Nakamura, and A. Sugawara. 1993. Characteristics of lipid accumulation types in five species of fish. *Nippon Suisan Gakkaishi* 59:1559–1564.
Bac, N., S. Biagianti, and J. Bruslé. 1983. Etude cytologique ultrastructurale des anomalies hépatiques du loup, de la daurade et de l'anguille, induites par une alimentation artificielle. *Actes de Colloques* IFREMER 1:473–484.
Baksi, S.M. and G.M. Frazier. 1990. Review: Isolated fish hepatocytes—model systems for toxicology research. *Aquat. Toxicol.* 16:229–256.
Baldridge, H.D. 1972. Accumulation and function of liver oil in Florida sharks. *Copeia* 2:306–325.
Barni, S., G. Bernocchi, and G. Gerzeli. 1985. Morphohistochemical changes in hepatocytes during the life cycle of the European eel. *Tissue and Cell* 17:97–109.
Biagianti-Risbourg, S. 1990. *Contribution à l'étude du foie des muges (Téléostéens, Mugilidés) contaminés expérimentalement par l'atrazine (s-triazine herbicide)*. Thèse de Doctorat (spécialité Sciences). 451 pp. Université de Perpignan.
Biagianti-Risbourg, S. 1991. Fine structure of hepatocytes in juvenile grey mullets: *Liza saliens* Risso, *L. ramada* Risso and *L. aurata* Risso (Teleostei, Mugilidae). *J. Fish Biol.* 39:687–703.
Biagianti-Risbourg, S. 1992. Intérêt (éco)toxicologique d'une approche histo-cytologique pour la compréhension des réponses hépatiques de Mugilidés (Téléostéens) à des contaminations sub-aigues par l'atrazine. *Ichthyophysiol. Acta* 15:205–223.
Bouche, G., N. Gas, and H. Paris. 1979. Isolation of carp hepatocytes by centrifugation on a discontinuous ficoll gradient. A biochemical and ultrastructural study. *Biol. Cell.* 36:17–24.
Boujard, T. and J.F. Leatherland. 1992. Circadian pattern of hepatosomatic index, liver glycogen and lipid content, plasma non-esterified fatty acid glucose, T_3, T_4, growth hormone and cortisol concentrations in *Oncorhynchus mykiss* held under different photoperiod regimes and fed using demand-feeders. *Fish Physiol. Biochem.* 10:111–122.
Bowser, P.R., D. Martineau, R. Sloan, M. Brown, and C. Carusone. 1990. Prevalence of liver lesions in brown bullheads from a polluted site and a nonpolluted reference site on the Hudson River, New York. *J. Aquat. An. Health* 2:177–181.
Braunbeck, T., P. Burkhardt-Holm, and V. Storch. 1990. Liver pathology in eels (*Anguilla anguilla*, L.) from the Rhine River exposed to the chemical spill at Basle in November 1986. In G. Fisher (ed.), *Limnol. Aktuell,Biologie des Rheins*. Verlag, Stuttgart, 1:371–392.
Braunbeck, T., K. Gorgas, V. Storch, and A. Völkl. 1987. Ultrastructure of hepatocytes in golden ide (*Leuciscus idus melanotus*, L.; Cyprinidae: Teleostei) during thermal adaptation. *Anat. Embryol.* 175:303–313.
Bruslé, J. 1993. Etablissement du bilan sanitaire des poissons: un bon indicateur biologique de la qualité des eaux. In C.F. Boudouresque, M. Avon, and C. Pergent-Martini (eds.), *Qualité du Milieu Marin-Indicateurs Biologiques et Physico-chimiques*, pp. 215–235. GIS Posidonie Publ.
Byczkowska-Smyk, W. 1967a. The ultrastructure of the hepatic cells in the sea trout (*Salmo trutta*, L.) during ontogenesis. *Zool. Polon.* 17:105–119.
Byczkowska-Smyk, W. 1967b. The ultrastructure of the hepatic cells in the sea trout (*Salmo trutta*, L.) during ontogenesis. Part II. The nucleolus, the Golgi apparatus, the stored substances. *Zool. Polon.* 17:155–170.
Byczkowska-Smyk, W. 1968. Observations of the ultrastructure of the hepatic cells of the burbot (*Lota lota*, L.). *Zool. Polon.* 18:287–295.
Byczkowska-Smyk, W. 1973. Observations on the ultrastructure and size of the hepatocytes of the lung fish *Lepidosiren paradoxa* (Dipnoi, Pisces). *Acta Biol. Cracov.* 16:247–255.

Byczkowska-Smyk, W. 1981. The ultrastructure and size of hepatocytes in two species of the genus *Tilapia* (Perciformes, Cichlidae). *Acta Biol. Cracov.* 23:69-76.

Chapman, G.B. 1981. Ultrastructure of the liver of the fingerling rainbow trout *Salmo gairdneri* Richardson. *J. Fish Biol.* 18:553-567.

Chiba, A. and Y. Honma. 1981. Histological observations of some organs in the porcupine fish, *Diodon holacanthus*, stranded in Niigata on the coast of Japan Sea. *Jap. J. Ichthyol.* 28:287-294.

Diaz, J.P. and R. Connes. 1991. Development of the liver of the sea bass, *Dicentrarchus labrax* L. (Teleost, Fish): II. Hepatocyte differentiation. *Biol. Struct. Morphogen.* 3:57-65.

Diaz, J.P., R. Connes, P. Divanach, and G. Barnabe. 1989. Développement du foie et du pancréas du Loup, *Dicentrarchus labrax*: I. Etude de la mise en place des organes au microscope électronique à balayage. *Ann. Sc. Nat., Zoologie* 13:87-98.

Dutta, H.M., S. Adhikari, N.K. Singh, P.K. Roy, and J.S.D. Munshi. 1993. Histopathological changes induced by malathion in liver of a freshwater catfish *Heteropneustes fossilis* (Bloch). *Bull. Environ. Contam. Toxicol.* 51:895-900.

Eastman, J.T. and A.L. DeVries. 1981. Hepatic ultrastructural specialization in Antarctic fishes. *Cell Tissue Res.* 219:489-496.

Ellias, H. and H. Bengelsdorf. 1952. The structure of the liver of vertebrates. *Acta Anat.* 14:297-337.

Eurell, J.A. and W.E. Haensly. 1982. The histology and ultrastructure of the liver of Atlantic croaker *Micropogon undulatus* L. *J. Fish Biol.* 21:113-125.

Ferguson, H.W. 1989. *Systemic Pathology of Fish.* Iowa State University Press, Ames, Iowa, 263 pp.

Ferri, S. and A. Sesso. 1981. Ultrastructural study of the endothelial cells in teleost liver sinusoids under normal and experimental conditions. *Cell Tissue Res.* 219:649-657.

Fujita, H., H. Tatsumi, T. Ban, and S. Tamura. 1986. Fine ultrastructural characteristics of the liver of the cod (*Gadus morhua macrocephalus*), with special regard to the concept of a hepatoskeletal system formed by Ito cells. *Cell Tissue Res.* 244:63-67.

Gillis, D.J. and B.A. McKeown. 1990. Physiological and histological aspects of late oocyte provisioning, ovulation, and fertilization in Pacific herring (*Clupea harengus pallasi*). *Can. J. Fish Aquat. Sci.* 47:1505-1512.

Gonzalez, G. 1992. *Contribution à la connaissance des processus ciguatérigènes*. Thèse de Doctorat (spécialité Oceanologie), 335 pp. Université de Perpignan.

Gonzalez, G., S. Crespo, and J. Bruslé. 1993. Histo-cytological study of the liver of the cabrilla sea bass, *Serranus cabrilla* (Teleostei, Serranidae), an available model for marine fish experimental studies. *J. Fish Biol.* 43:363-373.

Hacking, M.A., J. Budd, and K. Hodson. 1978. The ultrastructure of the liver of the rainbow trout: normal structure and modifications after chronic administration of a polychlorinated biphenyl Aroclor 1254. *Canad. J. Zool.* 56:477-491.

Hampton, J.A., P.A. McCuskey, R.S. McCuskey, and D.E. Hinton. 1985. Functional units in rainbow trout (*Salmo gairdneri*) liver: I. Arrangement and histochemical properties of hepatocytes. *Anat. Record* 213:166-175.

Hampton, J.A., R.C. Lantz, P.J. Goldblatt, D.J. Lauren, and D.E. Hinton. 1988. Functional units in rainbow trout (*Salmo gairdneri*, Richardson) liver: II. The biliary system. *Anat. Record* 221:619-634.

Haug, T. and B. Gulliksen. 1988. Variations in liver and body condition during gonad development of Atlantic halibut, *Hippoglossus hippoglossus* (L.). *Fisk Dir. Skr. Ser. Hav Unders.* 18:351-363.

Hightower, L.E. and J.L. Renfro. 1988. Recent applications of fish cell culture to biomedical research. *J. Exper. Zool.* 248:290-302.

Hinton, D.E. and C.R. Pool. 1976. Ultrastructure of the liver in channel catfish *Ictalurus punctatus* (Rafinesque). *J. Fish Biol.* 8:209-219.

Hori, S.H., T. Kodama, and K. Tanahashi. 1979. Induction of vitellogenin synthesis in goldfish by massive doses of androgens. *Gen. Comp. Endocrinol.* 37:306-320.

Hung, S.S.O., J.M. Groff, P.B. Lutes, and F.K. Fynn-Aikins. 1990. Hepatic and intestinal histology of juvenile white sturgeon fed different carbohydrates. *Aquaculture* 87:349-360.

Ishii, K. and K. Yamamoto. 1970. Sexual differences of the liver cells in the goldfish, *Carassius auratus*, L. *Bull. Fac. Fish., Hokkaido Univ.* 21:161-168.

Jirge, S.K. 1970. Changes in the distribution of glycogen in the liver at various stages of development of *Tilapian* larvae. *Ann. Histochim.* 15:283-287.

Kendall, M.W. and W.E. Hawkins. 1975. Hepatic morphology and acid phosphatase localization in the channel catfish (*Ictalurus punctatus*). *J. Fish. Res. Bd. Canada* 32:1459-1464.

Korsgaard, B. and T.P. Mommsen. 1993. Glyconogenesis in hepatocytes of immature rainbow trout (*Oncorhynchus mykiss*): Control by estradiol. *Gen.Comp. Endocrinol.* 89:17-27.

Kramar, R., H. Goldenberg, P. Bock, and N. Klobukar. 1974. Peroxisomes in the liver of the carp (*Cyprinus carpio* L.). Electron microscopic, cytochemical and biochemical studies. *Histochemistry* 40:137-154.

Langdon, J.S. 1986. Haemosiderosis in *Platycephalus bassensis* and *Diodon nicthemerus* in South-east Australian coastal waters. *Aust. J. Mar. Freshw. Res.* 37:587-593.

Leatherland, J.F. and R.A. Sonstegard. 1983. Interlake comparison of liver morphology and *in vitro* hepatic monodeiodination of L-thyroxine in sexually mature coho salmon, *Oncorhynchus kisutch* Walbaum, from Lakes Erie, Ontario, Michigan and Superior. *J. Fish Biol.* 22:519-536.

Leatherland, J.F. and R.A. Sonstegard. 1987. Ultrastructure of the liver of lake Erie coho salmon from posthatching until spawning. *Cytobios* 54:195-208.

Lincoln, R.F. and A.P. Scott. 1984. Sexual maturation in triploid rainbow trout, *Salmo gairdneri* Richardson. *J. Fish Biol.* 25:385-392.

Lone, K.P. 1989. The effect of feeding three anabolic steroids in different combinations on the growth, food conversion efficiency and protein and nucleic acid levels of liver, kidney, brain and muscle of mirror carp (*Cyprinus carpio*). *Fish Physiol. Biochem.* 6:149-156.

Morimoto, T., H. Asakawa, M. Nakano, and T. Watanabe. 1988. Establishment and some biological characteristics of a cell line derived from yamame embryonal liver. *Nippon Suisan Gakkaishi* 54:1881-1887.

Mosconi-Bac, N. 1987. Hepatic disturbances induced by an artificial feed in the sea bass (*Dicentrarchus labrax*) during the first year of life. *Aquaculture* 67:93-99.

Mosconi-Bac, N. 1990. Reversibility of artificial feed-induced hepatocyte disturbances in cultured juvenile sea bass (*Dicentrarchus labrax*): an ultrastructural study. *Aquaculture* 88:363-370.

Mosconi-Bac, N. 1991. *Effets de l'alimentation sur les caractères structuraux et métaboliques du foie de loup, Dicentrarchus labrax, en aquaculture*. Thèse de Doctorat (spécialité Sciences), 193 pp. Université de Perpignan.

Mugnaini, E. and S.B. Harboe. 1967. The liver of *Myxine glutinosa*; a true tubular gland. *Z. Zellforsch. Mikrosk. Anat.* 78:314-369.

Ng, T.B., N.Y.S. Woo, P.P.L. Tam, and C.Y.W. Au 1984. Changes in metabolism and hepatic ultrastructure induced by estradiol and testosterone in immature female *Epinephelus akaara* (Teleostei, Serranidae). *Cell Tissue Res.* 236:651-659.

Nopanitaya, W., J. Aghajanian, J.W. Grisham, and J.L. Carson. 1979. An ultrastructural study on a new type of hepatic perisinusoidal cell in fish. *Cell Tissue Res.* 198:35-42.

Oguri, M. 1976. On the enlarged liver in 'cobalt' variant of rainbow trout. *Bull. Jap. Soc. Sci. Fish.* 42:823-830.

Oguri, M. 1978a. Histochemical observations on the interrenal gland and liver of European spotted dogfish. *Bull. Jap. Soc. Sci. Fish.* 47:703-707.

Oguri, M. 1978b. On the hepatosomatic index of Holocephalian fish. *Bull. Jap. Soc. Sci. Fish.* 44:131-134.

Oguri, M. 1985. On the liver tissue of freshwater stingrays and balloon fish. *Bull. Jap. Soc. Sci. Fish.* 51:717-720.

Peute, J., R. Huiskamp, and P.G.W.J. van Oordt. 1985. Quantitative analysis of estradiol-17-β-induced changes in the ultrastructure of the liver of the male zebrafish, *Brachydanio rerio*. *Cell Tissue Res.* 242:377-382.

Quaglia, A. 1976. Seasonal variations in the ultrastructure of the liver of the pilchard *Sardina pilchardus* Walb. *Archo Oceanogr. Limnol.* 18:525-530.

Rappaport, A.M., Z.J. Borowy, W.M. Lougheed, and W.N. Lotto. 1954. Subdivision of hexagonal liver lobules into a structural and functional unit. *Anat. Rec.* 119:11-33.

Robertson, J.C. and T.M. Bradley. 1991. Hepatic ultrastructure changes associated with the parr-smolt transformation of Atlantic salmon (*Salmo salar*). *J. Exper. Zool.* 260:135-148.

Robertson, J.C. and T.M. Bradley. 1992. Liver ultrastructure of juvenile Atlantic salmon (*Salmo salar*). *J. Morphol.* 211:41-54.

Robertson, O.H. and B.C. Wexler. 1960. Histological changes in the organs and tissues of migrating and spawning Pacific salmon (genus *Oncorhynchus*).*Endocrinology* 66:222-239.

Saez, L., T. Zuvic, R. Amthauer, E. Rodriguez, and M. Krauskopf. 1984. Fish liver protein synthesis during cold acclimatization: seasonal changes of the ultrastructure of the carp hepatocyte. *J. Exper. Zool.* 230:175-186.

Saito, Y. and Y. Tanaka. 1980. Glutaraldehyde fixation of fish tissues for electron microscopy. *J. Electron Microsc.* 29:1-7.

Sakano, E. and H. Fujita. 1982. Comparative aspects on fine structure of the Teleost liver. *Okajima Fol. Anat. Jap.* 58:501-520.

Schär, M., I.P. Maly, and D. Sasse. 1985. Histochemical studies on metabolic zonation of the liver in the trout (*Salmo gairdneri*). *Histochemistry* 83:147–151.

Segner, H. and T. Braunbeck. 1990. Adaptative changes of liver composition and structure in golden ide during winter acclimatization. *J. Exp. Zool.* 255:171–185.

Segner, H. and U. Witt. 1990. Weaning experiments with turbot (*Scophtalmus maximus*): electron microscopic study of liver. *Marine Biology* 105:353–361.

Shimeno, S., H. Hosokawa, M. Takeda, H. Kajiyama, and T. Kaisho. 1985. Effect of dietary lipid and carbohydrate on growth, feed conversion and body composition in young yellowtail. *Bull. Jap. Soc. Sci. Fish.* 51:1893–1898.

Shin, Y.C. 1977. Some observations on the fine structure of lamprey liver as revealed by electron microscopy. *Okajima Fol. Anat. Jap.* 54:25–60.

Sidon, E.W. and J.H. Youson. 1983. Morphological changes in the liver of the sea lamprey, *Petromyzon marinus* L., during metamorphosis. II. Canalicular degeneration and transformation of the hepatocytes. *J. Morphol.* 178:225–246.

Sloof, W., C. Van Kreijl, and A. Baars. 1983. Relative liver weights and xenobiotic-metabolizing enzymes of fish from polluted surface waters in the Netherlands. *Aquatic Toxicology* 4:1–14.

Speilberg, L., O. Evensen, B. Bratberg, and E. Skjerve. 1993. Evaluation of five different immersion fixatives for light microscopic studies of liver tissue in Atlantic salmon *Salmo salar*. *Dis. Aquat. Org.* 17:47–55.

Storch, V. and J.V. Juario. 1983. The effect of starvation and subsequent feeding on the hepatocytes of *Chanos chanos* (Forsskal) fingerlings and fry. *J. Fish Biol.* 23:95–103.

Storch, V., U. Welsch, M. Schünke, and E. Wodtke. 1984. Einfluss von Temperatur und Nahrungsentzug auf die Hepatocyten von *Cyprinus carpio* (Cyprinidae, Teleostei). *Zool. Beitr. N.F.* 28:253–269.

Strüssmann, C.A. and F. Takashima. 1990. Hepatocyte nuclear size and nutritional condition of larval pejerrey, *Odontesthes bonariensis* (Cuvier et Valenciennes). *J. Fish Biol.* 36:59–65.

Takahashi, S. 1974. Sexual maturity of isaza (*Chaenogobius isaza*)-I. The seasonal changes of growth and sexual maturation. *Bull. Jap. Soc. Sci. Fish.* 40:847–857.

Takahashi, Y., H. Tsubouchi, and K. Kobayashi. 1978. Effects of vitamin A administration upon Ito's Fat-Storing Cells of the liver in the carp. *Arch. Histol. Jap.* 41:339–349.

Tanuma, Y. 1980. Electron microscope observations on the intra-hepatocytic bile canaliculus and sequent bile ductules in the crucian, *Carassius carassius*. *Arch. Histol. Jap.* 43:1–21.

Tanuma, Y., M. Ohata, and T. Ito. 1982. Electron microscopic study on the sinusoidal wall of the liver in the flatfish, *Kareis bicoloratus*: demonstration of numerous desmosomes along the sinusoidal wall. *Arch. Histol. Jap.* 45:453–471.

Van Bohemen, C.G., J.G.D. Lambert, and J. Peute. 1981. Annual changes in plasma and liver in relation to vitellogenesis in the female rainbow trout, *Salmo gairdneri*. *Gen. Comp. Endocrinol.* 44:94–107.

Vernier, J.M. and M.F. Sire. 1976. Evolution of the glycogen content and of glucose-6-phosphate activity in the liver of *Salmo gairdneri* during development. *Tissue and Cell* 8:531–546.

Vethaak, D. and T. Rheinallt. 1990. A review and evaluation of the use of fish diseases in the monitoring of marine pollution in the North Sea. *Int. Council Expl. Sea*, CM/E11: 62 pp.

Wake, K., K. Motomatsu, and H. Senoo. 1987. Stellate cell storing retinol in the liver of adult lamprey, *Lampetra japonica*. *Cell Tissue Res.* 249:289–299.

Weis, P. 1972. Hepatic ultrastructure in two species of normal, fasted and gravid Teleost fishes. *Am. J. Anat.* 133:317–332.

Yamamoto, M. and N. Egami. 1974. Sexual differences and age changes in the fine structure of hepatocytes in the medaka, *Oryzias latipes*. *J. Fac. Sci. Univ. Tokyo* 13:199–209.

Youson, J.H., P.A. Sargent, and E.W. Sidon. 1986. Iron loading in the liver of parasitic adult lampreys, *Petromyzon marinus* L. *Amer. J. Anat.* 168:37–49.

Youson, J.H., P.A. Sargent, D. Ogilvie, and R.R. Shivers. 1983. Morphology of the green livers in upstream migrants of *Petromyzon marinus* L. *J. Morphol.* 188:347–361.

7

Ultrastructural Diversity of the Biliary Tract and the Gallbladder in Fish

J. Gilloteaux, C.K. Oldham and S. Biagianti-Risbourg

The fish gallbladder, like the mammalian one, is an accessory organ of the digestive system that stores and secretes concentrated bile. This bile has several functions, such as facilitating several digestive functions, eliminating conjugated metabolites in the liver (including xenobiotics), and participating in the enterohepatic bile circulation. The quality of the bile itself and its precursors in fish have been surveyed and studied by Haslewood (1978). It was observed that diet plays a major role in determining the type of bile salts an animal has. Bile salts can have a deleterious action on the gallbladder wall, but this potential threat is balanced by the beneficial presence of bound cholesterol (Jacyna et al., 1986) as well as mucus. The main bile salts have been characterized in fishes according to systematic and dietary habits but only in a limited number of species (Haslewood, 1978; Cornelius, 1986). In consequence, it is difficult to discuss whether the fine structures of gallbladder and bile duct epithelium reflect the make-up or alteration of the bile composition due to changes in diet. One can just assume that, as in the mammalian gallbladder, the epithelium of the cystic duct and of the gallbladder in fish is able to concentrate the bile by altering its ionic and water permeability. Little is known about the regulation of bile acids and their impact on the liver and biliary tract ultrastructure and metabolism resulting from xenobiotics. The same is true of the gallbladder. The general microscopic anatomy and the production of mucinous compounds by the surface epithelium are some features common to the fish and the mammalian gallbladder (Gilloteaux et al., 1992, 1993a–c; Karkare and Gilloteaux, 1995; Karkare et al., 1995).

The first description of fish gallbladder was given by Aristotle (Suyehiro, 1942). Since then only a handful of publications have focused on this organ. Most of the contributions concern the feeding habits of fish and it is quite difficult to glean descriptions of the fine structures of this organ in fish from the literature. Most of the textbooks treating the topic of the fish digestive system lack the elementary histology and ultrastructure of the gallbladder and the associated biliary tract.

The dietary demands and adaptive strategy of quality of bile production are probably reflected in the ultrastructural diversity of these accessory organs of the digestive system. A few examples are presented here which demonstrate that if the gallbladder shows an apparent simple histological structure, ultrastructural observations (transmission and scanning electron microscopy) reveal in better detail the differences existing between species. Comparative ultrastructural aspects could also provide some insight into the mechanism of gallstones or histopathologic changes of the same organs of higher vertebrates, including man, since pathologic aspects often offer ontogenetic homologies.

Certain types of algae and protozoan (*Myxosporidae*) found as exo- and endoparasites release natural toxicants, which can threaten many species of fish. From our survey of the literature it was concluded that more studies on the aforementioned topics would be most welcome, considering their economical impact.

MORPHOLOGICAL INTERRELATIONSHIP BETWEEN LIVER, THE BILIARY SYSTEM AND THE GALLBLADDER: DEVELOPMENT AND HISTOLOGY

The ontogeny of the gallbladder, biliary system, and liver in fish differs somewhat in various taxa (Barrington, 1957; Bertin, 1958).

Agnatha: class Cyclostoma, Myxinidae (Hagfishes) and Petromyzontidae (lampreys)

As reported by Haslewood (1978), "hagfishes are known to have a very large liver, containing cholesterol in free and esterified form, and a relatively enormous gallbladder." The liver of the larval sea lamprey is composed of hepatocytes arranged in a tubular pattern around bile canaliculi (Peek et al., 1979) and a gallbladder which may best be described as intrahepatic, situated within the anterior portion of the liver, separated from the liver cells by connective tissue, blood vessels, and muscle fibers (Youson and Sidon, 1978).

Gnatha: class Chondrichtyes, subclass Elasmobranchii

In sharks (order Selachii) development shows that two embryonic evaginations arise from the hepatic placode of the endoderm and fuse on the medial line to form the gallbladder and the cystic duct. The lateral portions constitute the biliary canals and the hepatic tissue (Scammon, 1913). The development of one may precede the other in any order of fish. For example, in the genus *Squalus* the gallbladder develops independent of the liver and connects the liver only secondarily. In Chondrichthyes the liver is horseshoe shaped, with equal right and left lobes linked by an isthmus. The gallbladder is elongated and more or less embedded in the right lobe, and sometimes can be completely surrounded by hepatic tissues.

Gnatha: Osteichthyes (teleosts or bony fishes)

In subclass Sarcopterygii, order Dipnoi (lungfishes), some species, such as *Neoceratodus*, demonstrate only one evagination arising from the primitive endoderm, although the same medial portion becomes the cystic duct and gallbladder. Again, the lateral portion provides the hepatic tissue (Neumayer, 1904). In *Neoceratodus* and other Dipnoi the gallbladder is associated with the anterior lobe of the liver (Bluntschli, 1904).

In the subclass Actinopterygii, order Acipenseriformes (sturgeon), the hepatic placode evaginates into several epithelial buds which anastomose by engulfing connective tissue and blood spaces. These branching tissues ultimately form a lumen which becomes the biliary channels and hepatic tissue, as is found in *Acipenser*.

The subclass Teleostei generally shows an extreme variability in liver morphology. The liver may have one, two, three, or even more lobes. The gallbladder may be spherical, ovoid, elongated, or differently shaped. Histological studies of the gallbladders of eight freshwater teleosts showed considerable morphological variability (Chakrabarte et al., 1973). It is interesting to speculate whether feeding habits play any part in this variability. Bottom-feeders, plankton-eaters, omnivores, and others show differences in their digestive systems which may have developed according to their feeding habits. It may be expected that the gallbladder, as an accessory organ of the digestive system, may also differ in morphology. Histologically, the features described by light microscopy of the gallbladder differ mainly in the mucosal layer and the distribution of pancreatic tissue surrounding the gallbladder. In some species pancreatic tissue is noted within the gallbladder wall tissue while it may not be present in other species (Chakrabarte et al., 1973). Occasionally, the gallbladder may be missing, as in *Lota* (Hyrtl, 1868). The number of cystic ducts present may vary (Chakrabarte et al., 1973). The cystic duct of teleosts is often sinuous and may contain swellings along its length where bile ducts abut. Sometimes a certain number of these ducts may connect directly into the intestine.

BILIARY SYSTEM

Descriptions in literature of the biliary tracts of fish reveal some confusion regarding the nomenclature to be used in describing the morphology of the elements of the biliary system from its proximal region (canaliculi) to its distal region (ducts). Indeed, various authors have used different terminology to describe identical regions of this system. We have chosen to use the terminology of Hampton et al. (1985), which is similar to that used in a recent work by Biagianti-Risbourg (1990), as it appears to be the most complete and precise.

ULTRASTRUCTURAL CHARACTERISTICS AND NOMENCLATURE

It can be seen at first glance that the hepatic stroma is not as clearly arranged into cords as in the classic mammalian system (Fig. 1). The bile canaliculi are intercellular canals communicating with preductules ("terminal intraparenchymental bile ductules" (Fig. 2) according to Tanuma, 1980). The lumen of these canaliculi is delineated by the plasma membrane of the hepatocyte (canalicular pole) and by preductal biliary cells. These preductal cells can be linked to the hepatocytes by tight junctions and desmosomes in the vicinity of the biliary lumen. Even though several papers have described them (Hampton et al., 1985), the existence of preductal cells has been refuted by Tanuma (1980), who claimed they were simply fibroblasts. However, according to Biagianti-Risbourg (1990), preductal cells exist because: (1) they are connected to hepatocytes by tight junctions and desmosomes, as mentioned previously; (2) no collagen fibers are associated with these cells nor are there any collagen fibers in the immediate vicinity; and (3) they contain numerous microtubules in the area of the biliary lumen (Hinton and Pool, 1976) and therefore demonstrate a sort of functional polarity.

According to an efferent gradient, the number of biliary cells increases and the biliary lumen is then entirely circumscribed by cells which belong to the biliary system. The ductules (or cholangioles) possess their own epithelium, composed of ductal cells with, apparently, no basal lamina and showing no clear polarity (Fig. 3). The ductal cells have a high nucleocytoplasmic ratio. They vary in morphology, from spherical to ovoid, and present no apparent polarity. However, they contain numerous microfilaments (actin) surrounding the ductal lumen. The ductules are also distinguished from preductules by the presence of microvilli. The diameter of the ductules increases correspondingly with the number of biliary cells constituting the ductule walls. In general, the ductules are canals with a diameter between 5.5 and 10 μm. Preductule fibroblasts also increase concurrently along the efferent gradient.

Coalescence of these efferent ductules concludes in the formation of bile ducts or biliary ducts with a diameter between 10–40 μm. These ducts are lined by simple stratified cuboidal to cylindrical epithelial cells (Fig. 4). These cells, possessing few microvilli, have a nucleocytoplasmic ratio smaller than that of the ductal cells and present a clear polarity, due to the fact that they are attached to a basement membrane. These cells are also rich in organelles and display basolateral interdigitations. The number of cells increases progressively with the diameter of the duct and the epithelial lining is eventually surrounded by fibroblasts and smooth muscle cells.

The common bile ducts are of an even larger diameter (more than 40 μm in diameter) and their walls composed of very tall, simple columnar epithelial cells (e.g. 20.2 ± 1.7 μm × 6.3 ± 1.8 μm for a canal of 46.7 ± 3.3 μm in diameter; Biagianti-Risbourg, 1990). This epithelium shows characteristics of a transport epithelium, possessing microvilli, vesicles and vacuoles, lysosomal bodies, and basolateral interdigitations between adjacent cells, named "lamellar structures" by Sire et al. (1981).

Similar studies of the biliary system were conducted by Hampton et al. (1988) in the rainbow trout *Salmo gairdneri*. The proximal region of the intrahepatic biliary tract (preductal and ductal) has only been described in a small number of teleosts belonging to Cyprinidae (*Carassius auratus*) by David (1961), Yamamoto (1962, 1965), Nopanitaya et al. (1979) and *Carassius carassius* by Tanuma (1980); Ictaluridae (*Ictalurus punctatus*) by Hinton and Pool (1976); Salmonidae (*Salmo gairdneri*) by Hacking et al. (1978); Pimelomidae (*Pimelodus maculatus*) by Ferri (1982); and Morinidae (*Dicentrarchus labrax*) by Diaz and Connes (1988).

The liver in telostean fish is similar to that of higher vertebrates in that it contains typical interhepatocytic biliary canaliculi formed by junctional complexes of the plasma membranes of the hepatocytes (Hampton et al., 1988). These canaliculi are associated with preductules; the canaliculi are joined directly to the preductules and are very short, not lengthy, as in higher vertebrates. Indeed, they are so short that the name "biliary pockets" has been proposed by Diaz and Connes (1988) to name these canaliculi in *Dicentrarchus labrax*. Biliary pockets such as these have also been observed during ontogenesis of mammals (Ellias, 1955). They would thus seem to represent a characteristic stage in many vertebrates, including ontogenetic development in the human fetus (Koga, 1971).

In young Mugilidae (Biagianti-Risbourg, 1990) preductules are observed with difficulty; ductules or cholangioles, however, are very long and sinuous and readily observable in longitudinal and transverse sections. These long ductules may be a result of the general tubular architecture of a fish liver. On the other hand, in the mammalian liver the hepatocytes are organized in plates or cords, and as a result of this organization, one finds unusually short ductules. In some cases, preductules may be absent in fish. For example, in a few Cyprinidae, the bile canaliculi directly abut the ductules (Yamamoto, 1965; Nopanitaya et al., 1979; Tanuma, 1980).

The bile ducts of teleosts have been described by several authors, especially in Salmonidae (Hampton et al., 1988) and Mugilidae (Biagianti-Risbourg, 1990); in the gray mullet *Liza* the bile duct cells present numerous interdigitations similar to those described in Agnatha. Between the duct cells, basolateral spaces may be found. They vary in size, according to their functional activity, associated as in higher vertebrates with bile concentration by removal of water and regulation of some electrolyte flows. These spaces are substantially reduced in juveniles, as shown in freshwater teleosts (*Salmo gairdneri* by Hampton et al., 1988) and marine teleosts (*Dicentrarchus labrax*, by Diaz and Connes, 1988).

The common bile duct is circumscribed by a wall formed of a layer of cylindrical epithelial cells; the height of the cells increases with the diameter of the duct. In both Salmonidae and Mugilidae these cells show properties of a transport epithelium. In Salmonidae the cells produce a mucus readily observable in the apical cytoplasm.

The branches of the hepatic artery run parallel to the biliary ducts; this association of arteriole and bile ducts is consistently observed in Mugilidae and Salmonidae (Hampton et al., 1988). The portal spaces or triads, so characteristic of mammals, are seldom found in fish. The codistribution of the biliary ducts and hepatic arteries allows for electrolyte and metabolic exchanges to occur and constitutes a passageway for transport to the bile (Sternlieb and Quintana, 1985).

GALLBLADDER ULTRASTRUCTURE

Ultrastructural investigations of the gallbladder and biliary system in fish have only been undertaken recently. Two large groups of fish have been investigated so far and the research on these groups, Agnatha and Osteichthyes, is presented below. In addition, some of the research done by our laboratories on an Elasmobranch and on the ultrastructure of marine teleosts gallbladders (unpublished) is also reported.

The successive layers found in mammalian gallbladders can be differentiated in fish using light (LM) and electron microscopy (TEM). They are the surface epithelium, the lamina propria and subjacent connective tissue which constitutes the submucosal layer, the fibromuscular layer, the subserosal layer and the serosal layer, usually covered by a mesothelium, unless in areas adjacent to other organs (pancreas, liver) where adventitial connective tissue will join both organs.

Agnatha: class Cyclostoma

A comparison of the biliary system of fish should properly begin with a most primitive class of fish, the jawless fish. Very detailed studies of the biliary system, including the biliary ducts, have been done in Agnatha: in *Myxine glutinosa* (Mugnaini and Harboe, 1967) and in *Petromyzon marinus* (Sidon et al., 1980; Sidon and Youson, 1983 a, b).

The bile ducts of the larval lamprey are composed of cells bearing lateral folds and large numbers of microvilli on the apical surface, forming a brush border nearly filling the lumen. Intracytoplasmic cisternae are located at the periphery; the cisternae connect with the intercellular spaces via pores (Sidon et al., 1980). Zonulae occludentes (or tight junctions) at the bile canaliculi, although "leaky", form a bile-blood barrier (Youson et al., 1987). The bile ducts have a large amount of surface area exposed to the blood vessels due to their convoluted nature; this arrangement of bile ducts and blood vessels has been postulated to function in bile modification (Yamamoto et al., 1986). The bile duct epithelium functions as a transport epithelium in *Myxine glutinosa* (Mugnaini and Harboe, 1967) and in *Petromyzon marinus* (Sidon et al., 1980).

The larval gallbladder epithelial cells are similar to those of the bile ducts, although lacking intracytoplasmic cisternae and pores and containing a less developed brush border. Gallbladder cells contain large amounts of glycogen, as may be found in some mammals. The cells contain smaller numbers of mitochondria and have narrower intercellular spaces than are found in mammals, possibly signifying a reduced transport function (Sidon et al., 1980).

The metamorphosis of a larval sea lamprey into an adult includes the loss of the intrahepatic gallbladder and bile ducts. The cells of the extrahepatic common bile duct dedifferentiate and eventually give rise to the adult pancreas (Youson et al., 1987; Elliott and Youson, 1993). The nonparenchymal cells of the young adult liver are similar to those of other vertebrates but, unlike those of mammals and teleosts, include fenestrae which are not grouped into sieve plates; this does not seem to be involved in bile stasis (Youson et al., 1985). The parenchymal cells of the adult liver show a loss of the zonulae occludentes after biliary atresia; gap junctions increase however, allowing for greater cell contact during bile stasis (Youson et al., 1987).

Elasmobranch gallbladders

The gallbladders of two elasmobranchs were studied in our laboratories but only one is illustrated here since the two did not appear to differ morphologically. In both the dogfish (*Scyliorhynus canicula*) and the electric ray (*Torpedo marmorata* Risso), the gallbladder is composed of the five characteristic histological layers mentioned above.

In *Torpedo* both LM (Fig. 5) and TEM (Figs. 6, 9–10) showed the wall of the gallbladder to be generally 300–500 μm thick. The surface epithelial cells are tall and cylindrical, approximately 20–50 μm in height and 5 μm or less in width. This is a typical epithelium, showing folds and with cells linked to each other by junctional complexes, as seen in Fig. 6. Bulging apices are a common observation at the epithelium luminal surface (Figs. 5–6, 9–10). The cytoplasm of the cells contains numerous mitochondria, although the apical portion of the cells does not; instead fine cytoplasmic granulation is present in the apical portion showing no peculiar organelle. Closer examination of the epithelium by TEM showed a diversification of pale and dark cells, not necessarily apparent at the light microscopic level. The pale cells do not appear to have a bulging apex and are scattered in small numbers throughout the most abundant dark cells. They also contain smoother ER than the mitochondria, but further investigations are planned to characterize and differentiate between these cell types. Nuclei occupy the lowest third volume of the cells (Fig. 6). The basolateral spaces of this epithelium can also be invaded by cells resembling lymphocytes and/or macrophages. The submucosal layer is about 100–150 μm in thickness and contains loose connective tissue. Many blood vessels can be detected in the subepithelial zone in the 1-μm-thick sections because they demonstrate wavy empty spaces lined by endothelium, in which sometimes entrapped nucleated hemocytes can be distinguished (Fig. 5). These small blood vessels usually have a fenestrated endothelium on the side facing the epithelium. The fibromuscular layer (± 100 μm thick) is formed by loosely arranged smooth muscle fibers which appear scalloped or sinuous due to fixation procedures. No muscle bundles or layer appear to be organized as in all other vertebrates. A narrow subserosal layer (± 50–70 μm thick) is limited by a loose connective layer of about 50–80 μm covered by a serosal mesothelium.

Scanning electron microscopy or SEM (Fig. 9) showed and confirmed the existence of bulging apices, as seen in Fig. 6. They are covered by short, fine microvilli. Among them one can detect minute bulging excrescences. However, among the depressions or crypts of folds of the epithelium, smooth-surface bulging apices can be seen. TEM observations (summarized in Fig. 10) show the beginning and final stage of extrusion of these apical excrescences. Some contain an apical electron density that appears as a fine granular cytoplasm (Fig. 6). Following apical decapitation or extrusion the cytoplasmic content appears more heterogeneous because it shows as electron-dense and lucent, pale mucus-containing vesicles (Fig. 10). Although no histochemical or cytochemical tests have yet been conducted to confirm the assumption of mucus production, these features resemble those described in other teleosts (Viehberger, 1982, 1983; also Figs. 11, 12, 16) and in mammals (Gilloteaux et al., 1992, 1993a–c). Other epithelial cells show a somewhat paler cytoplasm because of abundant smooth endoplasmic reticulum (Fig. 6) and no decapitation.

Osteichthyes gallbladders, class Teleost

The gallbladder of teleosts is composed of the five typical layers described in mammals. In the rainbow trout (*Salmo gairdneri*) the mucosal surface epithelium is composed of cells covered with numerous microvilli; these form a true brush border in rainbow trout but are less regularly arranged in trench. Glycogen has been observed in the gallbladder surface epithelial cells of some teleosts, such as the tench (Viehberger, 1982). Cytoplasmic protrusions or excrescences are also common; these have not been observed in Agnatha but have been described in mammals during fetal development (Laitio and Nevalainen, 1972) and under the effects of sex steroids (Gilloteaux et al., 1993 a–b; Karkare and Gilloteaux, 1995; Karkare et al., 1995; and others) and detected in Elasmobranchs, such as *Torpedo marmorata*, as well as several teleosts (this review).

Among the twelve teleost species studied in our laboratories, three species—a selected group of three—are used as examples in this report, all of them marine, bottom-feeders, predatory and/or omnivorous.

EXAMPLES OF TELEOST GALLBLADDER ULTRASTRUCTURE

In the Triglydae family the species *Triglosporum lastoviza* Brünnich has a small gallbladder and portions of it are covered by pancreatic lobes. The epithelial cells are very tall (50–85 μm) and clavate, being up to 5 μm in diameter at the apex and much narrower at the basal membrane (Figs. 7–8). The nuclei usually appear in the lower third of the cell and may be found situated at various levels, giving the appearance of a pseudostratified epithelium (Fig. 7). The epithelial cells have bulging apices, almost as if secretory, demonstrating features similar to those detected in the *Torpedo* anteriorly. LM of 1-μm-thick sections stained with toluidine blue viewed at higher magnification (Fig. 8) showed that the apical region stained least. The perinuclear and basal cytoplasm contains many stained granules, which correspond to lysosomal and/or complex mucopolysaccharidic content. TEM observations demonstrate that the apical region is less electron dense than the rest of the cell. Apices may also appear extensively elongate, which burst to release their content in the gallbladder lumen (Fig. 11). However, with careful scrutiny, one can detect mucus-containing vesicles released from some decapitated apices (Fig. 12). The morphology of these vesicles (0.6–1.5 μm in diameter) suggests a content of mucoproteinaceous material which takes on a fine fibrillar or striated appearance. Depending on the plane of sectioning, some vesicles contain a densely contrasted core, indicating the presence of an anionic proteinaceous material, similar to that found in other mucous glands. This teleost and elasmobranch gallbladder are illustrated together as they appear to have similar morphology and the demonstrated differences, detected only by TEM, concern cell apices, mucus production and release.

In the Acanthopterygian *Uranoscopus*, LM light microscopy shows a thick gallbladder wall ranging from 400 μm to 1 mm in thickness. The surface epithelium is very darkly stained with toluidine and thin, from 25–45 μm in height (Fig. 13). The cells are 4–5 μm in width. The submucosal layer is very thick and can occupy more than half of the wall width of the organ. It is constituted by a loose connective tissue represented by a narrow network of subepithelial blood vessels and in the remaining portion a loose connective tissue matrix is invaded by elongated fibroblasts containing occasionally large macrophagic cells filled by cholesterol inclusions (diffraction crosses, not shown here). A 200–300-μm-thick muscular layer is organized in obliquely oriented bundles invaded by blood vessels. Finally, the serosal mesothelium surrounds the major organ outer surface. In addition, some lobes of the pancreas can be attached to the gallbladder. Higher magnification of the epithelium (Fig. 14) shows that the cells are filled by numerous stained granules and pale nuclei.

The apical surface is peculiar, showing an irregularly spaced, jagged appearance, as though cut by pinking shears. Fine extensions of the apical cells also contain toluidine blue-stained granules (Fig. 14). The cell surface is usually covered by long, extended microvilli, which can attain more than 10 μm in length; some even appear extremely long and branching within the gallbladder lumen and its slightly more electron dense content. Numerous vacuoles as well as other granulations are also present throughout the cells and many appear perinuclear. An SEM view of the epithelial surface (Fig. 15) revealed irregular, cone-shaped excrescences resembling those seen in Fig. 14 (histology) among widely distributed microvilli. The microvilli were probably blunted at some stage of the SEM preparative procedure, especially that associated with the critical-point drying step. Similar structures are confirmed when viewed by TEM (Fig. 16), including the tall microvilli and irregular pyramid-like to conical extrusions showing heterogeneous vesicles, suggesting the presence of a mucoid material. The presence of such structures confirms the positive toluidine blue staining seen by light microscopy in the same cell apical extensions as viewed in Fig. 14. Notice also the views from Figures 15 and 16 (arrowed) of a dilated apex. The cytoplasm of these cells also shows minute electron-dense elongated exocytotic vesicles (100–200 nm length diameter, 50 to 75 nm diameter) of an unknown content (probably anionic mucus product). Large lipidic inclusions, 0.2–1.5 μm in diameter, can also be found throughout the cytoplasm among angular, electron-dense bodies (0.3 to 1 μm in diameter) containing fine electron-lucent curved and straight crystals resembling

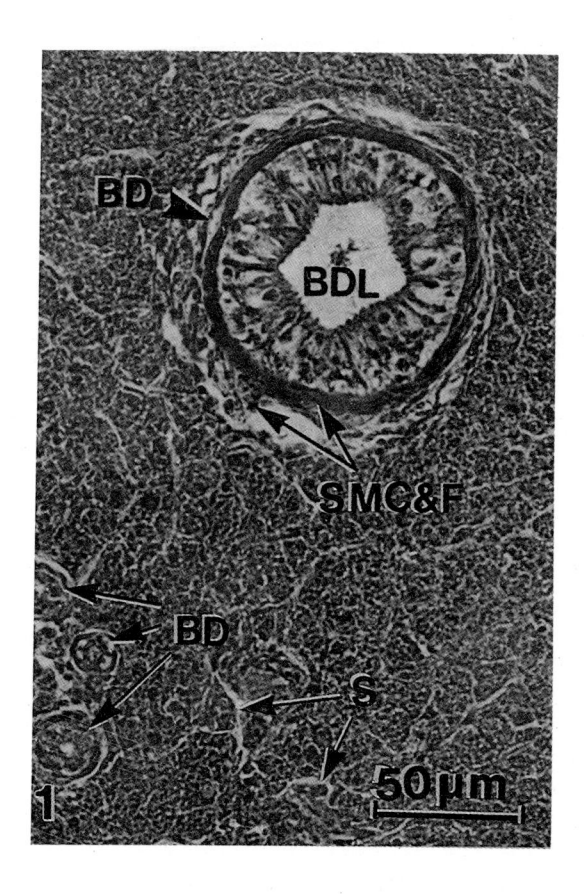

Fig. 1. Illustrations of *Liza* sp. (Teleostei, Mugilidae):.
LM micrograph of a large bile duct lumen (BDL) depicted among the liver parenchyma (Fig. 1) and small diameter bile ducts (BD). S—sinusoid; SMC & F—smooth muscle cells and fibrocytes

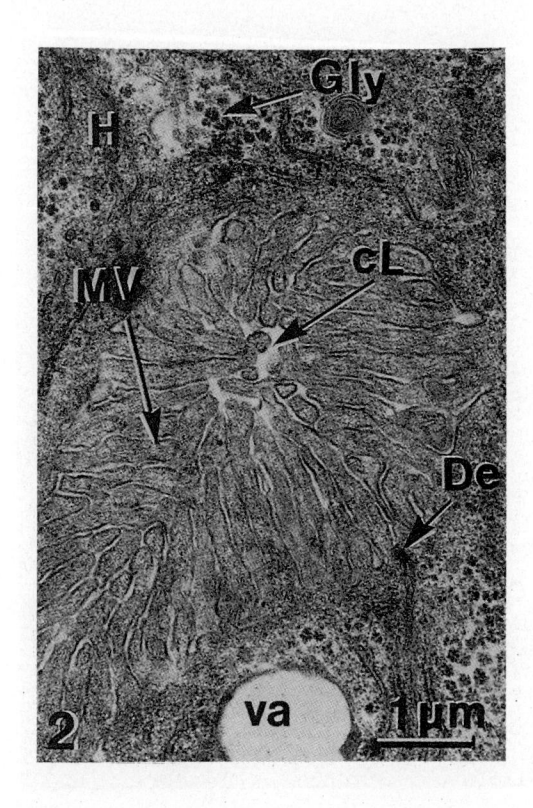

Fig. 2. Illustrations of *Liza* sp. (Teleostei, Mugilidae):.
TEM of bile canaliculus. cL—canalicular lumen; De—desmosome; Gly—glycogen; H—hepatocyte; Va—vacuole

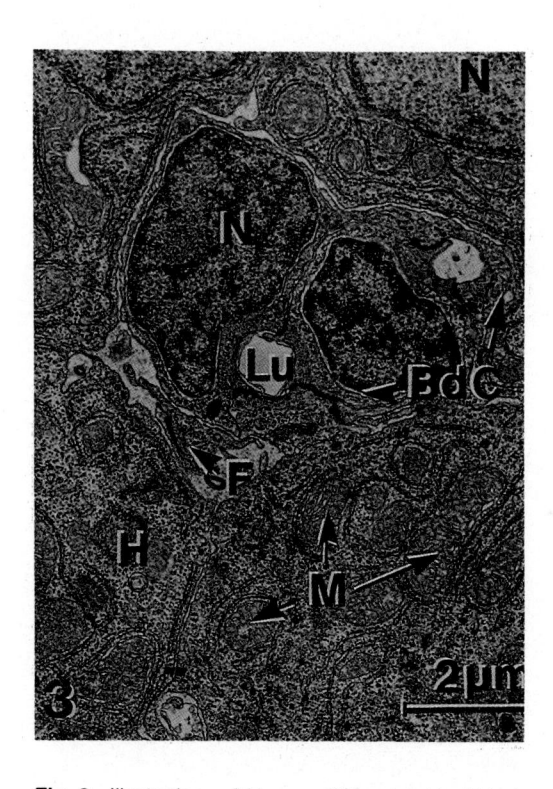

Fig. 3. Illustrations of *Liza* sp. (Teleostei, Mugilidae):.
TEM view of a bile ductule. BdC—bile ductule cell; F—fibrocyte; H—hepatocyte; Lu—ductule lumen; M—mitochondria; N—nucleus

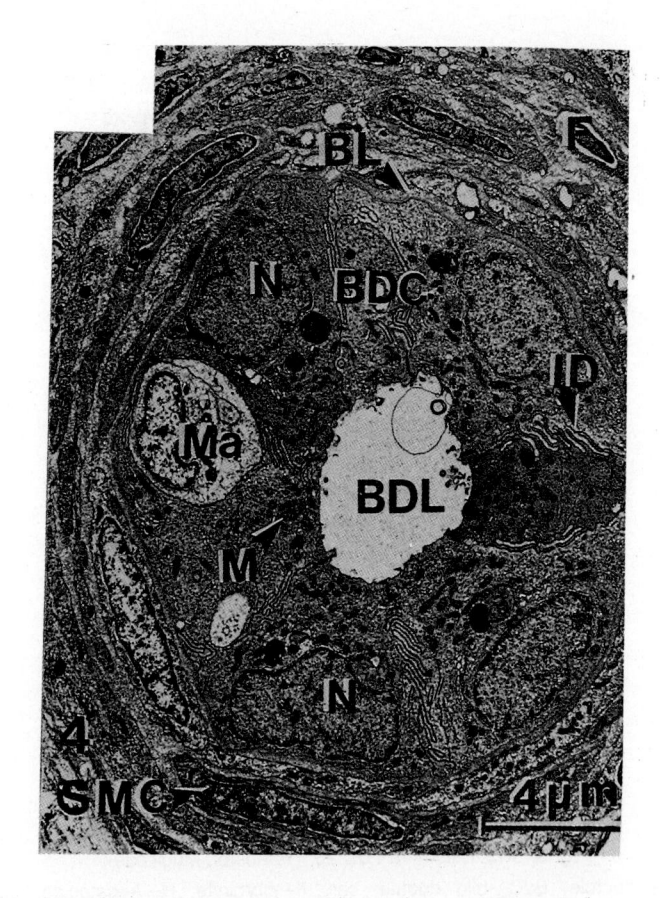

Fig. 4. Illustrations of *Liza* sp. (Teleostei, Mugilidae):.
TEM of bile duct. BDC—bile duct cell; BDL—bile duct lumen; F—fibrocyte; ID—interdigitations or basolateral digitations; M—mitochondria; Ma—macrophage; N—nucleus; SMC—smooth muscle cell

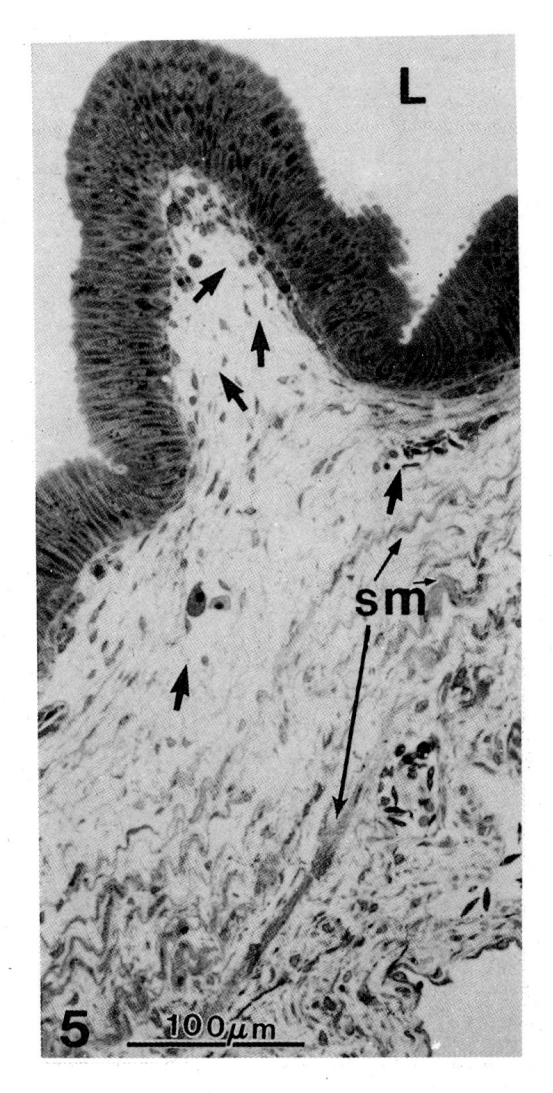

Fig. 5. LM micrographs of *Torpedo marmorata* Risso gallbladder.
Overall LM view of the gallbladder wall from 1-μm-thick section, toluidine blue stained. Folding depicted. L—lumen; sm—smooth muscle cells; small thick arrows outline subepithelial blood vessels

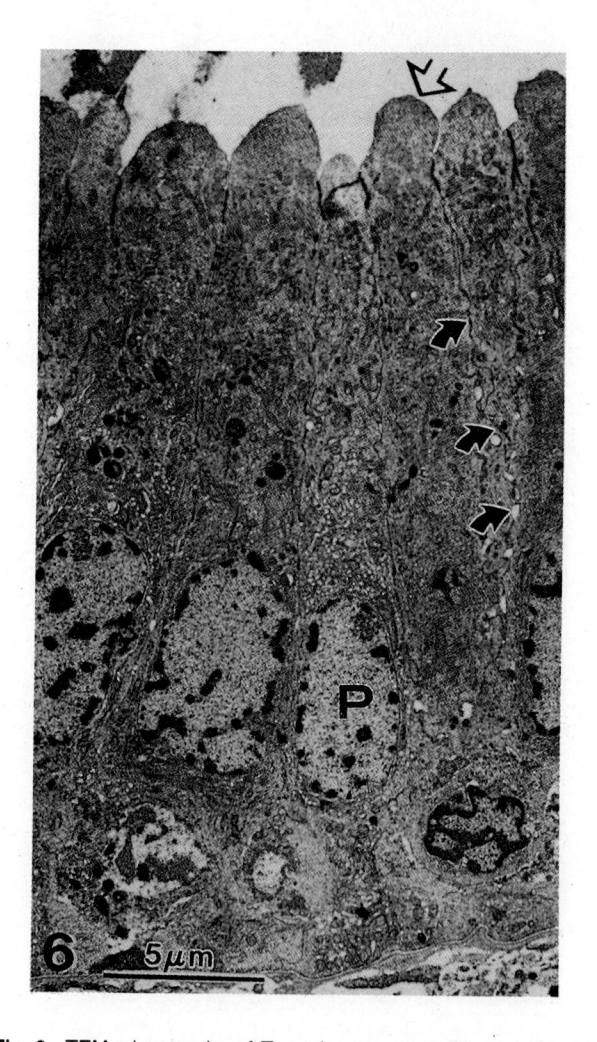

Fig. 6. TEM micrographs of *Torpedo marmorata* Risso gallbladder. Detailed TEM view of surface epithelial cells where bulging apices are shown (open arrow) and basolateral digitations are minimal by showing intercellular cell junctions (curved arrows). P—pale cell

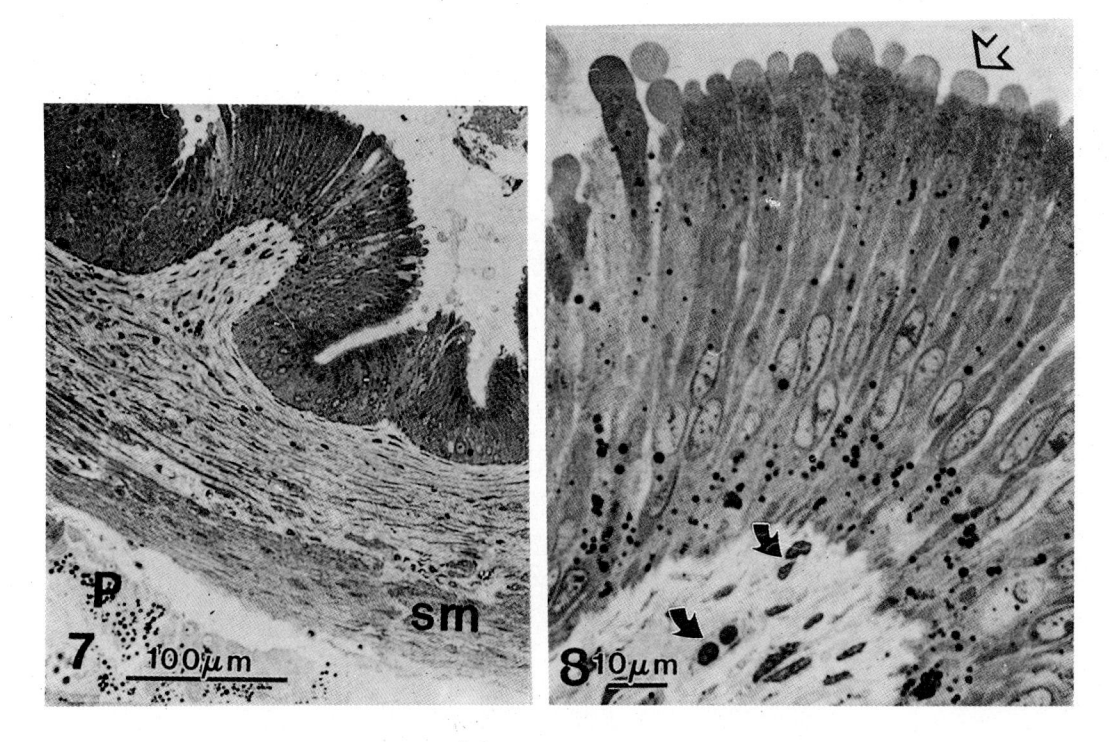

Figs. 7 and 8. LM (7) and (8) views of *Triglosporum lastoviza* gallbladder surface epithelial apices. Small (curved arrows) and large (stars) apical excrescences depicted in Fig. 7. Notice the microvilli when no apical bulging occurs. Enlarged view of apical excrescences (*) and decapitations indicating their heterogeneity in (Fig. 8). Among the excrescences, pale regions correspond to mucin-like secretory product. L—lumen.

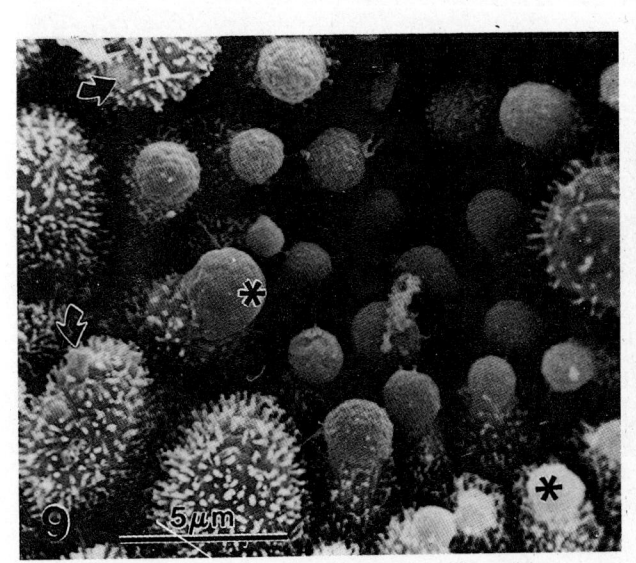

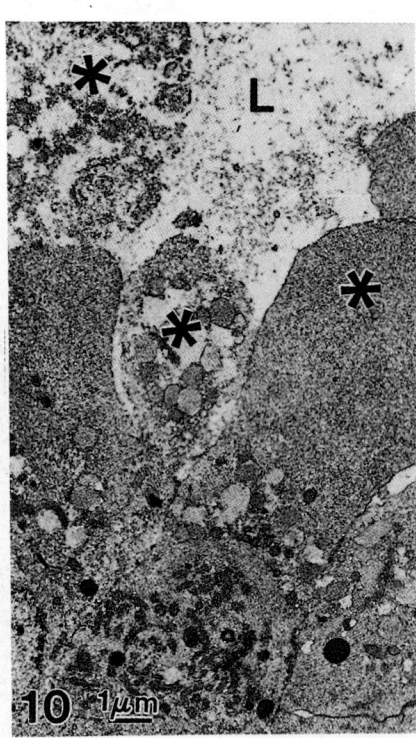

Figs. 9 and 10. LM and TEM views of *Torpedo marmorata* Brünnich gallbladder wall (9) and surface epithelium (10). Notice foldings of the mucosal layer, the well-defined fibromuscular layer with dense bundling of the smooth muscles fibers (sm); P—pancreas. Notice the clavate tall cells marking the surface epithelium with apical excrescences (open arrow). Subepithelial blood vessels also depicted (curved arrows).

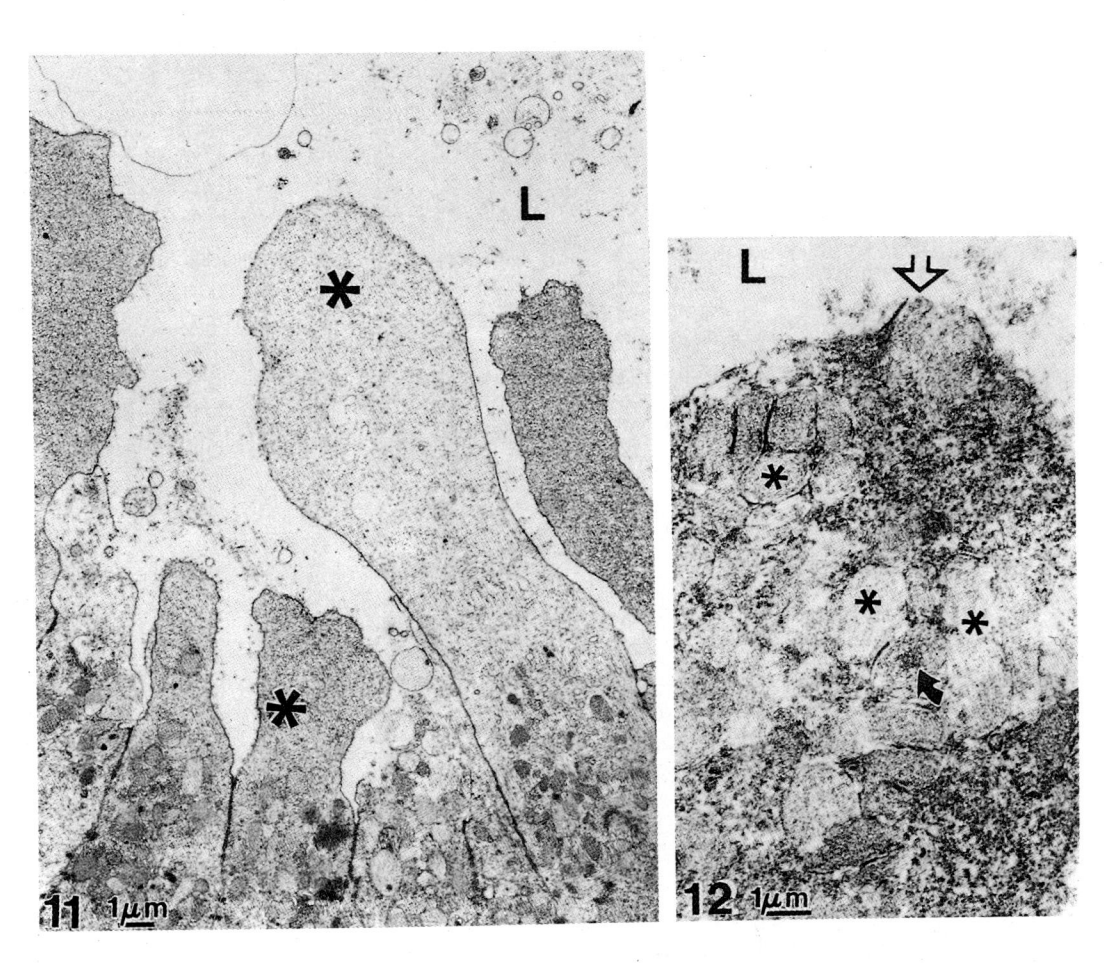

Figs. 11 and 12. TEM of *Triglosporum lastoviza* gallbladder surface epithelium. Small to large and extended excrescences (*) of the apical regions. L—gallbladder lumen in Fig. 8. Detailed view of mucoid discharges in the lumen (L) (Fig. 12). Stars indicate some mucus-containing vesicles. Curved arrow indicates intravesicular, condensed anionic material.

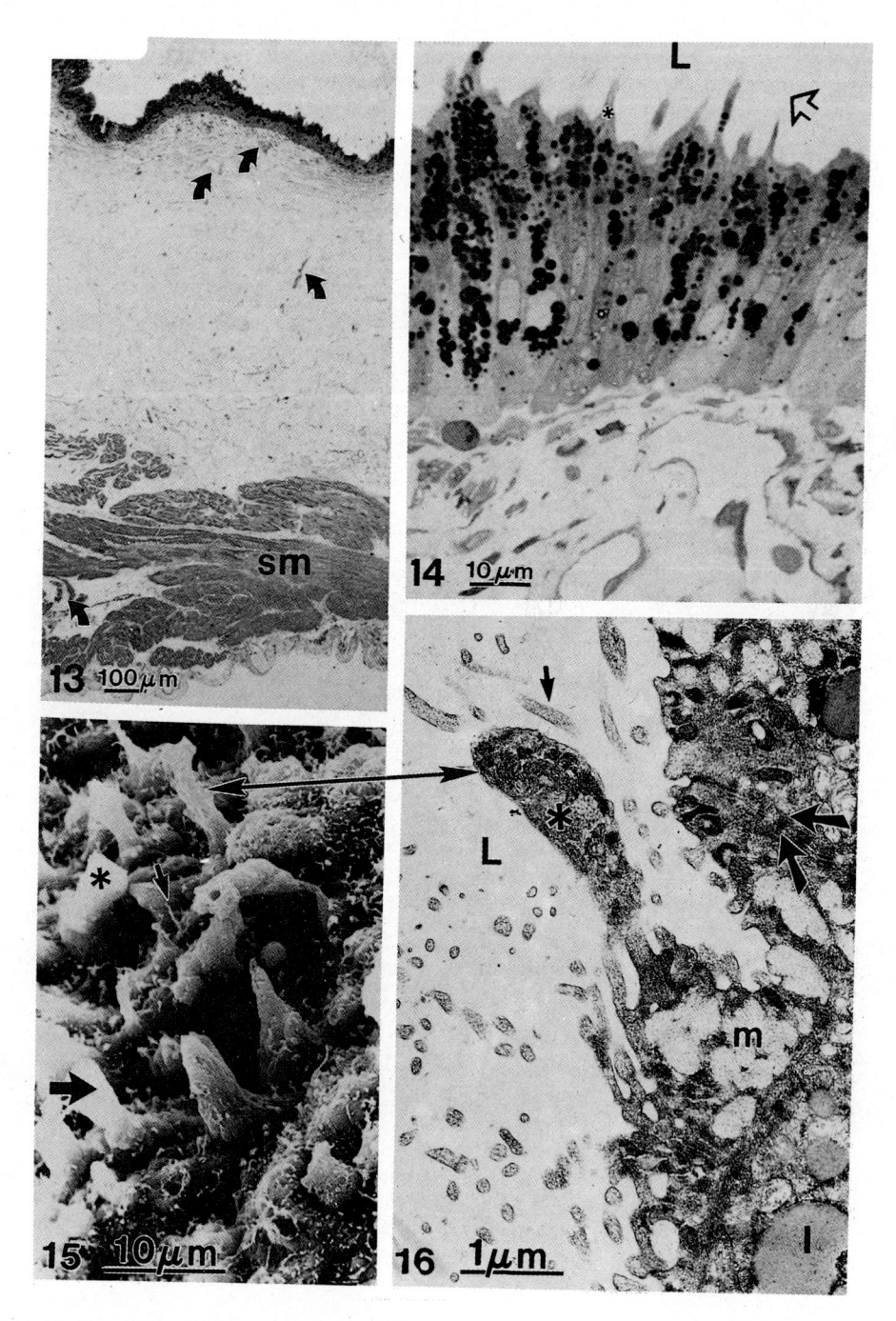

Figs. 13–16. LM (13 and 14), SEM (15) and TEM (16) views of *Uranoscopus* sp. gallbladder epithelium. Overall wall shown with densely stained (toluidine) surface epithelium in 13. Subepithelial and other blood vessels indicated by curved arrows. Sm—smooth muscle bundles of the fibromuscular layer. Detailed LM view of the epithelial cells demonstrating their abundant content of heavily stained inclusions and some subjacent blood vessel lumena in 14. SEM view of the luminal surfaces showing the irregular profiles of cell apices (* indicates a small cholesterol-like calculus) in 15. The double-arrowed line across this Figure and Fig. 16 suggests the similar structure being viewed by TEM in the latter (*) where apices contain heterogeneous filling by organelles, but mainly mucus-containing vesicles (m). Tall microvilli (arrow) discernible among these excrescences. L—lumen; l—lipidic inclusion. Curved arrows indicate junctional complex.

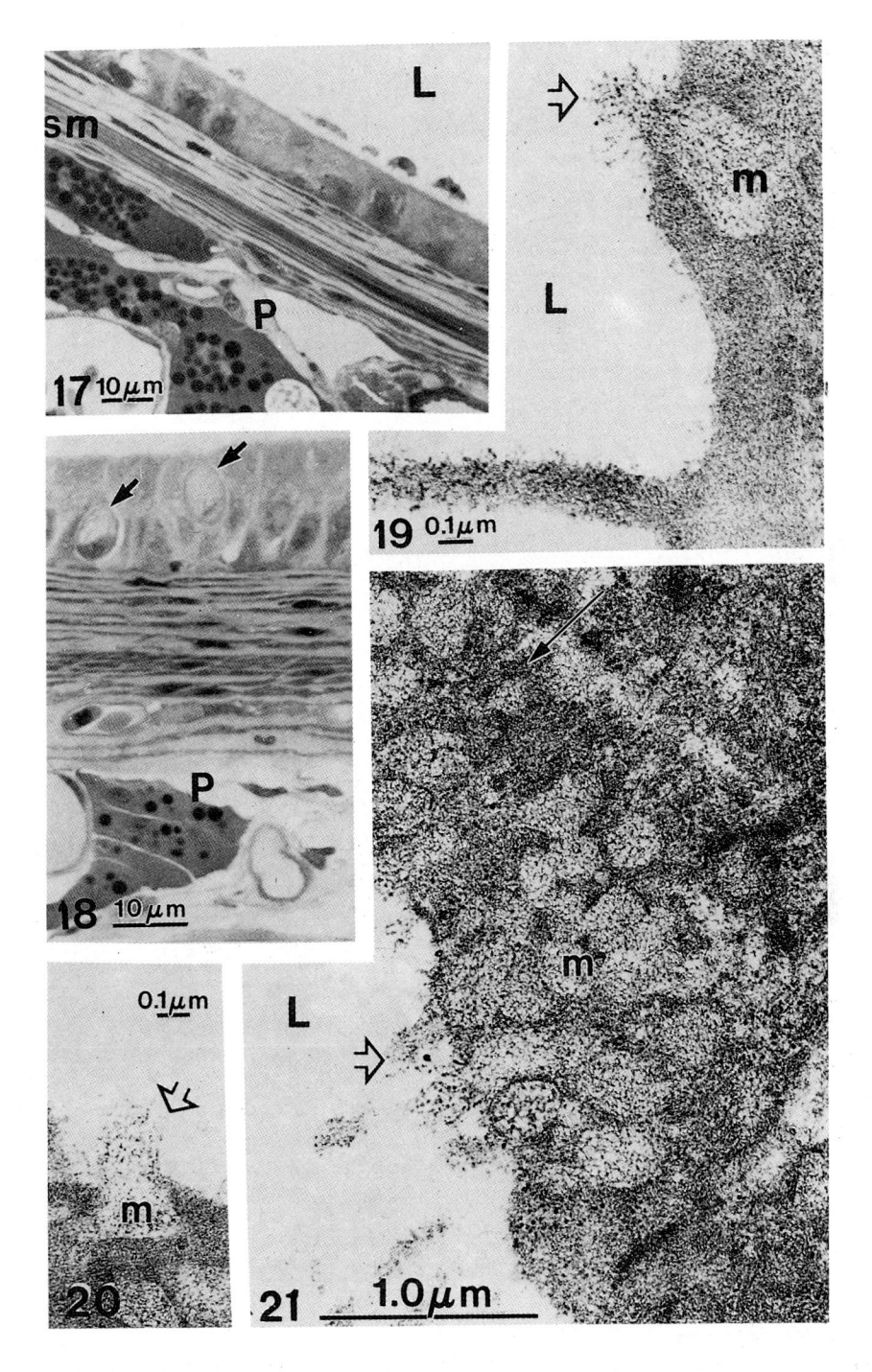

Figs. 17–21. LM (17–18) and TEM (19–21) views of *Scorpena scrofa* gallbladders. Overall view of the gallbladder wall showing some pancreatic tissues (P) surrounded by dilated blood vessels. In the lumen (L) a trail of peculiar wandering cells discernible (Myxosporidia ?) in 17 and 18. Sm—smooth muscles of the fibromuscular layer. In 18, intraepithelial spores are arrowed. In 19–21, TEM views of the apical aspects of the surface epithelial cells showing mucous exocytotic events (open arrows) from isolated apical vesicles (in 19 and 20) or aggregates of vesicles (in 21). Note the long slender microvillus in 19, in the lumen (L).

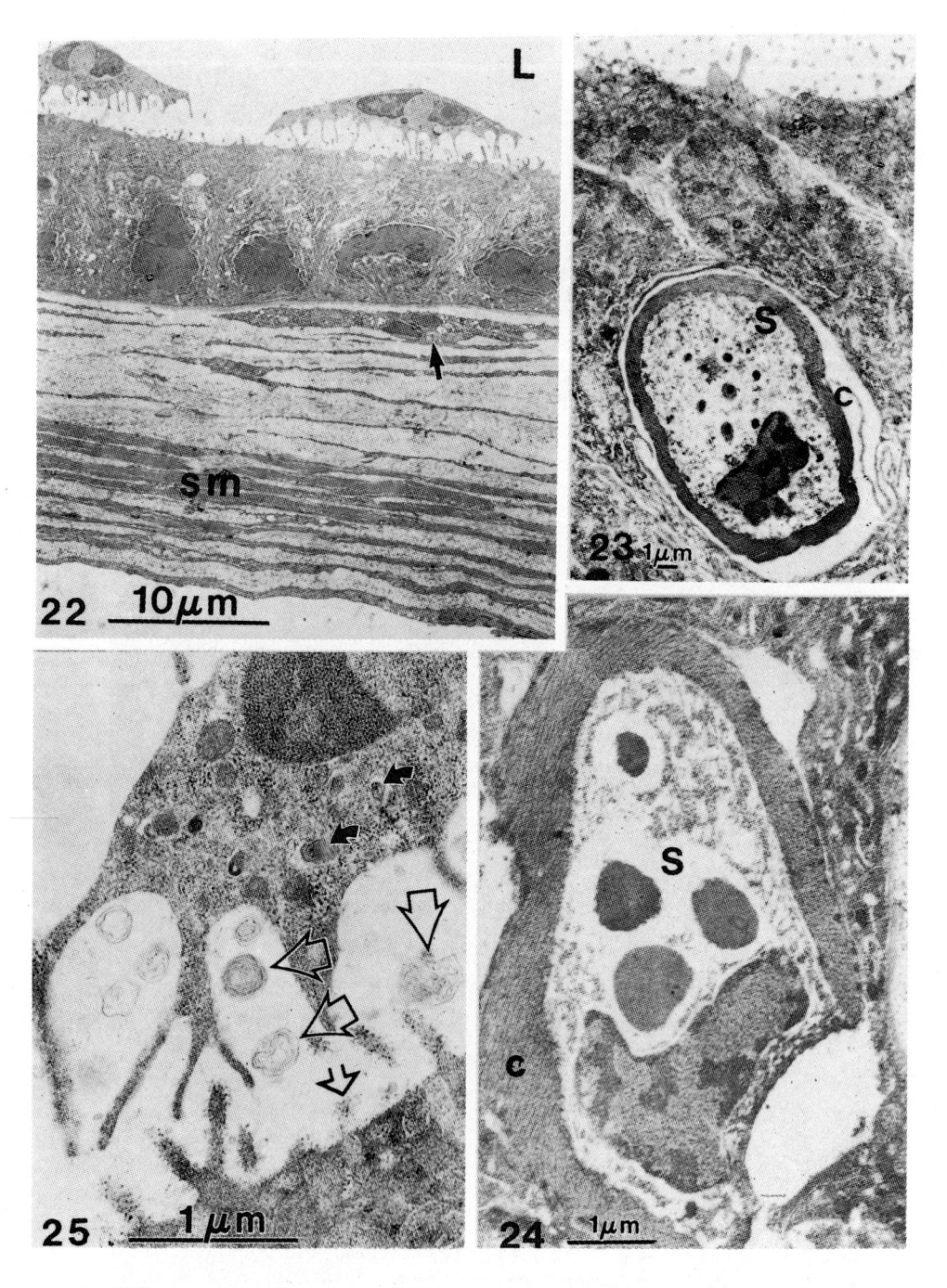

Figs. 22–25. TEM micrographs of *S. scrofa* gallbladder (22) and associated parasites (23–25). In 22, an overall view of the gallbladder wall, including some cells of the trail shown in Fig. 17. Arrow depicts a subepithelial blood vessel; sm—smooth muscle of fibromuscular layer. In 23 and 24, spores (S) and fibrous, cyst-like, finely fibrous wall (c). Note the intercellular space created by the densely contrasted inclusions, often containing a crystalloid density. In 25, a parasitic cell is shown with its elongated cell extensions above cell surfaces. Note the phospholipidic whorls (open arrows) also present in the cytoplasm of these bizarre cells (curved arrows) and among its mitochondria.

cholesterol. The remaining cytoplasmic spaces between organelles are extremely reduced because filled with numerous matrix-pale mitochondria. Often the outlines of the epithelial cells are difficult to delineate but a careful observation reveals well-defined junctional complexes and cell junctions extending between cells (Fig. 16).

The submucosal layer is fairly thick and can occupy over half the thickness of the wall. This layer is invaded by a subepithelial network of blood vessels and the remainder is made of slender fibroblasts and a loose matrix of connective tissue. The fibromuscular layer (200–300 µm thick) shows longitudinal, oblique, and cross-sectional bundles of smooth muscles. The outer subserosa is less than 15 µm thick and is covered by a mesothelium. Some lobes of the pancreas are attached in some regions of the gallbladder, (Gilloteaux and Gilloteaux, 1996).

In a Scorpenidae (*Scorpena scrofa* L.), one can see at first glance that the gallbladder wall is fairly thin and only attains between 30 to 50 µm in thickness. It is covered in many places by lobes of the pancreatic tissue, recognizable by their chromaticity and pattern of staining of secretory granules contained in the glandular cells (Figs. 17–18 and 22). The organ presents the usual histological layering. The surface epithelium is a simple but low columnar to cuboidal epithelium (± 15 µm), which appears fairly smooth surfaced, with no folding. The apical region is usually darkly stained by toluidine blue in the 1-µm thick sections (Figs. 17–18), indicating the presence and production of a mucoid secretory product, as determined by TEM (Figs. 19–20 and 25). A narrow, 8–10-µm-thick submucosal layer made of dense irregular connective tissue containing scattered elongated fibrocytes and a loose network of blood vessels is sandwiched between the epithelium and a 5 or 7-µm-thick fibromuscular layer containing 6 or 7 interconnected smooth muscle fibers (Figs. 17–18 and 22). A narrow or barely distinguishable subserosal layer is covered by a serosal mesothelium. The mesothelial covering coats the overall organ with or without pancreatic tissues. When a lobe of the pancreas is adjacent to the fibromuscular layer, a thicker stratum of loose connective tissue is shared by both organs (Figs. 17–18).

With the TEM, the apical surfaces of the epithelial cells can be detailed as containing single large (± 2.5–3 µm length diameter by ± 1–1.5 µm diameter) or rows and/or aggregates of small (± 0.2–0.5 µm) mucus-containing apical vesicles or undergoing diverse stages of exocytosis (Figs. 19–21, 25). Within the mucoid secretory product, a more electron-dense material is also present (Fig. 21). The content appears finely fuzzy and is shown to empty at the epithelial surface of the cells, covered by only a few slender and fine microvilli (base ± 0.2 µm and length 1–1.2 µm, Figs. 19, 22, 23 and 25). These microvilli cannot be detected by LM because they appear at the limit of its resolution power; they are only detectable with TEM. They are coated by an abundant glycocalyx (Figs. 19, 23, and 25). The nuclei are extremely chromatic and cells display abundant RER and SER; the mitochondria are detected with difficulty but are usually small and located at the basal side of the cells.

GALLBLADDER PARASITES

Among protozoans, the subphylum of Cnidospora includes the order Myxosporidia, common parasites of fishes, amphibians, and some reptiles but not known for birds and mammals. The same species can be observed in freshwater or marine fishes, even in widely separated regions. This cosmopolitan dispersion could be the result of migrations made by some species and to the trashing at sea or along coasts of contaminated freshwater viscera altered or improper for human consumption (Poisson, 1953). Some species can be endemic. In the most recent edition of the famous *Traité de Zoologie*, under the editorship of the late Professor P.P. Grassé, Poisson (1953) fully describes and discusses many species of these protozoans. They are placed in a well-named section of the text, "Parasites of Uncertain Affinities", since these organisms are usually classified with difficulty even by specialists in the field and hence data describing the complete life cycle is lacking for numerous species. Most of the descriptions, if not all, rely on interpretation of old LM observations (Kudo, 1966; Manwell, 1968). Only recently has some ultrastructural data been published on membrane specializations of the spores (Desportes-Livage and Nicolas, 1990) and their cytology (Uspenskaya, 1987, 1988). For the relevance of our survey, it is interesting to note that Myxosporidia can be detected as diffuse infiltrates in tissues and can form voluminous tumors. Some species live free in the lumen of the gallbladder and other viscera, viz. the urinary bladder, testes, and ovaries, but for many species the gallbladder is the most frequently infected organ or, even, the specific organ infected. The origin and site of penetration of these infectious organisms are not yet known.

The interesting survey of the literature reported by Poisson (1953) reveals how widely spread across species these parasites can be. Even so, it does not appear to be complete. Among marine fishes, the gallbladder of *Lophius*, *Pleuronectes*, *Drepanopsetta*, *Hippoglossus*, and *Maena vulgaris* can be infected by a specific species of the family Ceratomixidae. The *Leptotheca agilis* Thélohan parasitizes *Trygon pastinaca* L. and *Scorpena* sp. gallbladders. Another species, *L. vikrami* Tripathi, lives in the gallbladder of *Zeus faber*. In the suborder Sphaerosporea members of the family Chloromyxidae are parasites of the gallbladder of the following species: *Torpedo*, *Raia*, and *Cestracion* (*Chloromyxum leydigi* Mingazzini); *Merluchius gayi* (*C. rosenbushi* Gelormini); *C. truttae* Léger infects the gallbladder of *Trutta fario* and can kill the trout by diarrhea, liver infection, and progressive muscular wasting. *C. histolyticum* Pérard affects *Scomber scomber* L. and *C. trijugum* Kudo has been located in *Xenotis megalotis* and *Pomoxis sparoides* gallbladders. Recently, several new species were discovered along the Atlantic coast of Africa (Fomena et al., 1990) and the collective character of *C. leydigi* has been confirmed (Kovaleva, 1988) as well as in Russian seas (Aseeva, 1992; Shul'Man, 1989). In the family Myxididae, *Mixidium gasterostei* Noble parasitizes *Gasterosteus aculeatus* gallbladder and *M. kudoi* Meglitsch *Ictarulus furcatus* gallbladder. From the same family, *Sphaeromyxa balbianii* Trélohan has been reported in the gallbladders of *Motella*, *Clupea*, *Cepola*, etc.; *S. sabrazesi* Laveran and Mesnil in *Hippocampus brevirostris* Cuv. and *H. guttulatus* Cuv. develops and evolves in the biliary tract. *S. incurvata* Doflein in *Blennius ocellatus*; *S. hollandi* in *Molva vulgaris*, *Centronotus gunellus*, *Brosmius brosme*; *S. exneri* Awerinzew in *Thysanophris japonicus*, and *S. gasterostei* Georgévitch in *Gastrosteus spinachia* parasitize the gallbladder. The genus *Zschokella* parasitizes the gallbladders of *Gaidropsarus cirratus* (*Z. russelli* Tripathi) and *Acipenser sturio* (*Z. sturionis* Tripathi). Davies and Stenkowski (1988) showed *Z. russelli* Tripathi in the hepatic ducts of *Ciliata mustela* L. (*Gadidae*). The family Trilosporidae prevails in the gallbladder of several marine fishes while Myxobolidae are mainly found in those of freshwater fish. Interesting recent studies by Kazubski and El-Tantawy (1989) and El-Tantawy (1989a) conducted in a lake of Poland confirms this freshwater distribution and the potential infection by another parasite as a result of the introduction of a new fish in this lake (El-Tantawy,1989b). These studies and Shul'Man's (1989) show that these parasites have a yearly cycle of population growth which fluctuates with seasonal changes. This was also confirmed in American myxobolid parasites (*Unicauda* sp.) infesting Montana Columbia River fishes (Mitchell, 1989). Finally, among the Coccomyxidae, *C. morovi* Léger and Hesse is a parasite of the gallbladder of species of sardines.

Other large parasites can be found in the liver bile ducts and the gallbladder. Again, without claiming this review to be complete, we would like to mention the potential finding of Nemathelminth worms (Dorier, 1953) i.e., Gordiacea V. Siebold Nematomorpha (Vejdovsky, 1886) since one of us (J.G.) found several of them filling the gallbladder cavity of a *Uranoscopus* caught in the Banyuls area of France. Parasitic trematodes can also affect the liver and indirectly the biliary tracts (Biagianti, 1984).

Example of gallbladder parasite ultrastructure

From our experience, one Myxosporidia parasite is associated with the gallbladder surface epithelium of *S. scrofa*, identified as a possible stage of the plasmode of *Leptotheca agilis*. In fact, several individuals of this species (3 of 5 collected along the Banyuls Mediterranean coast mid-July) contained these gallbladder parasites. The location of ovoid spore-like structures among the epithelial cells is illustrated in Figure 18 (LM view from random sections across the gallbladder) while Figures 23 and 24 depict them from observations done by TEM. They range from 10–15 μm in length to 6 to 8 μm in diameter, having a fibrillar wall which can attain up to 1 μm in thickness. These spores are extracellular because following TEM examination a rim of cytoplasm always encircled the epithelial cell nuclei from which slender basolateral processes free from contact between adjacent cells constituted a "space" for this spore. These spores contain a cytoplasmic structure with a lobed, chromatic nucleus, fine granular ribonucleoprotein-like particles, and other electron-dense saccules containing intensely contrasted crystalloid densities or inclusions which could be sporonts (Fig. 24), as described by Kudo (1966). Other illustrations show a bizarre trail (Fig. 17, LM); Figure 22 gives a TEM view of these cells with an enlarged view in Figure 25. These cells are binucleated and show elongated, slender processes which can attain up to 2 μm in length and 80–125 nm in width and appear as if they were tiptoeing filopodia around and above the epithelial microvilli engulfing whorls of phospholipids (Fig. 25). One can detect small vacuoles where the membranous whorls appear endocytosed and with many small mitochondria. From these observations, one can ask whether the existence of densely stained epithelial nuclei and the abundant mucus could not be the result of this parasitic "infection". However, other intact gallbladders show similar preservation for samples fixed in the best conditions soon after capture.

OTHER PECULIAR STRUCTURES: THE "RODLET CELLS"

In order to complete our survey, the presence of "rodlet cells," described in cyclostomes, elasmobranchs, and teleosts, must also be mentioned. These cells can attain a diameter ranging up to 7 μm and can be detected ultrastructurally among epithelial cells, including the liver, the biliary tract, and the gallbladder epithelium and many other tissues of many teleost species (Bielek and Viehberger, 1983; Dawe et al., 1964; Leino, 1974; Mattey et al., 1979; Mayberry et al., 1979, Viehberger and Bielek, 1982; and others). Their nature is still controversial as they have been considered parasitic by some, with this interpretation rejected by others. Several ultrastructural studies on these "cells" have been conducted (Leino, 1974; Mattey et al., 1979 and 1980; Morrison and Odense, 1978), including one by one of us (Biagianti-Risbourg, 1990). Using histochemistry (Bielek and Viehberger, 1983; Inagawa et al., 1990) and combined molecular biology techniques, including *in situ* hybridization and ultrastructure (Barber and Westermann, 1975, 1983 and 1986a, b; Barber et al., 1979; Leino, 1974), it has been demonstrated that these "rodlet cells" are foreign to whatever fish host tissues, and hence are probably also a kind of parasitic host, as claimed by old zoologists, viz. Laguesse (1897) and Labbé (1897) cited in textbooks.

EFFECT OF NATURAL AND MAN-MADE TOXINS ON THE LIVER, BILIARY TRACT, AND GALLBLADDER OF FISHES

Environmental contaminants are of increasing concern. Many fish live in water in which a number of toxins may have been introduced naturally by man. Much of this water may be destined for human consumption. In addition, human consumption of fish subjected to toxic substances is proving a hazard as biological magnification of these toxins occurs. Toxic effects on fish have so far been studied primarily in terms of mortality figures, but some studies have focused on the pathological effects of toxins. *Microcystis aeroginosa* (a Cyanobacteria or bluegreen alga) has recently been shown to produce a toxic hepatotoxin which affects the liver and biliary structures (Rabergh et al., 1992). Cockell and Bettger (1993) studied the effect on the gallbladder of disodium arsenate heptahydrate (DSA) fed to juvenile rainbow trout (*Oncorhynchus mykiss*). The fish fed DSA showed inflammation and swelling of submucosal tissues. Sloughing of the epithelium occurred during the first day of exposure. Long-term exposure resulted in chronic inflammation and depression of growth.

CONCLUSION

To conclude this survey of the gallbladder ultrastructure and the many aforementioned parasites specific for the gallbladder (and the biliary tract, i.e., the liver also), let us note that they remain poorly studied ultrastructurally. Our survey should stimulate more investigations using the combined imaging facilities of LM, TEM and SEM to enable a better understanding of these parasites, especially in terms of the damage inflicted on the target organ and their respective life cycle, often damaging or lethal for fish. The fishing industry is currently limited in many regions of the world as a result of either a temporary exhaustion of natural stocks or regional pollution. As aquaculture progressively replaces these traditional modes of capture, the threat of these pollutants will be more harmful for mankind. Hence it is important to study the impact of such pollutants and parasites to prevent their adverse economic influence in zones where fish constitute the major or only source of protein intake for human beings.

Acknowledgments

This report has been supported in part by the Summa Health Foundation, Akron OH. The capture and collection of fish gallbladders reported in this survey was done during a short field study by J.G. at the Laboratoire Océanologique Arago of Banyuls, Station of Research of the Université de Paris, France during the summer of 1993. The authors thank Ms. M. Vijayalaksmi for reviewing the manuscript.

References

Aseeva, N.L. 1992. Myxosporidia from *Lepidosetta bilineata* of the Avacha Bay. *Parazitologiya* (St Petersburg) 26:161-165.

Barber, D.L. and J.E. Westermann. 1975. Rodlet cells in *Castostomus commersoni* (Teleostei, Pisces) secretory cell or parasite. *Experientia* (Basel) 31:924-925.

Barber, D.L. and J.E. Westermann. 1985. Reappraisal of nuclear DNA content of rodlet cells compared to several cell types from some freshwater teleosts using two methods of microdensitometry. *J. Fish Biol.* 27:817-826.

Barber, D.L. and J.E. Westermann. 1986a. Comparison of the DNA of nuclei of rodlet cells and other cells in the chub *Semotilus atromaculatus*. Hybridation *in situ*. *Can. J. Zool.* 64:801-804.

Barber, D.L. and J.E. Westermann. 1986b. The rodlet cells of *Semotilus atromaculatus* and *Catostomus commersoni*, Teleostei. Studies on its identity using histochemistry and DNase I-gold, RNase A-gold, and S-1 nuclease-gold labeling techniques. *Can. J. Zool.* 64:805-813.

Barber, D.L., J.E. Westermann, and D.N. Jensen. 1979. New observations on the rodlet cell *Rhabdospora thelohani* in the white sucker *Catostomus commersoni*. Light microscopic and electron microscopic studies. *J. Fish Biol.* 14:277-284.

Barrington, E.J.W. 1957. The alimentary canal and digestion. *In* M.E. Brown (ed.), *Physiology of Fishes*. Academic Press, London.

Bertin, L. 1958. Agnathes et Poissons. *In* P.-P. Grassé (ed.), *Traité de Zoologie*, 13, fasc. 2:1248-1302. Masson & Cie, Paris.

Biagianti-Risbourg, S. 1990. *Contribution à l'étude du foie de juvéniles de muges (Téléostéens, Mugilidés) contaminés expérimentalement par l'atrazine (S-triazine herbicide): approche ultrastructurale et métabolique: intérêt en écotoxicologie*. Ph.D. thesis, Université de Perpignan, France.

Biagianti, S. 1984. Etude de l'action pathogène de parasites infestant le foie des poissons d'intérêt aquacole. I. Etude des altérations induites par un Trématode: Labratrema minimus dans le foie de deux espèces de Muges. Rap. CNEXO nb. 82/2719, 1-15.

Bielek, E. and G. Viehberger. 1983. New aspects of the rodlet cell in Teleosts. *J. Submicr. Cytol.* 15:681-694.

Bluntschli, H. 1904. Der feinere Bau der Leber von *Ceratodus*, züglich ein Beitrag zur vergleichenden Histologie der Fishleber. *Jena Denkschr.* 4:335-375.

Chakrabarte, J., R. Saharya, and D.K. Belsare. 1973. Structure of the gallbladder in some freshwater teleosts. *Z. Mikrosk.-anat. Forsch.* 87:23-32.

Cockell, K.A. and W.J. Bettger. 1993. Investigations of the gallbladder pathology associated with dietary exposure to disodium arsenate heptahydrate in juvenile rainbow trout (*Oncorhynchus mykiss*). *Toxicology* 77:233-248.

Cornelius, C.E. 1986. Comparative bile pigment metabolism in Vertebrates. *In* J.D. Ostrow (ed.), *Bile Pigments and Jaundice*, pp. 601-647. Marcel Dekker, Inc. New York.

David, H. 1961. Zur submikroskopischen Morphologie intrazellularer Gallenkapillaren. *Acta Anat.* 47:216-224.

Davies, A.J. and I.K. Sienkowski. 1988. Further studies on *Zschokkella russelli* Tripathi, *Myxozoa myxosporea* from *Ciliata mustela* L. (Teleostei, Gadidae) with emphasis on ultrastructural pathology and sprorogenesis. *J. Fish Dis.* 11:325-336.

Dawe, C.J., M.F. Stanton, and F.J. Schwartz. 1964. Hepatic neoplasms in native bottom-feeding fish of Deep Creek Lake, Maryland. *Cancer Res.* 24:1194-1201.

Desportes-Livage, I. and G. Nicolas. 1990. The plasma membrane of myxosporidian valve cells: freeze-fracture data. *Protozool.* 37:243-249.

Diaz, J.P. and Connes, R. 1988. Particularités de l'organisation ultrastructurale du foie du loup *Dicentrarchus labrax* L. (Poisson Téléostéen). *Ann. Sci. Nat. Zool.* 9:123-141.

Dorier, A. 1953. Classe des *Gordiacea* V. Siebold 1843 (*Nematomorpha* Vejdovsky 1886). *In* P.P. Grassé (ed.), *Traité de Zoologie*, 4:1201-1222. Masson & Cie, Paris.

Ellias, H. 1955. Origin and early development of the liver in various vertebrates. *Acta Hepatol*. 3:1-56.

Elliott, W.M. and J.H. Youson. 1993. Development of the adult endocrine pancreas during metamorphosis in the sea lamprey, *Petromyzon marinus* L. II. Electron microscopy and immunocytochemistry. *Anat. Rec.* 237:271-290.

El-Tantawy, S.A.M. 1989a. Myxosporidian parasites in fishes in lakes Dgal Wielki and Warniak (Mazurian Lakeland, Poland) I. Survey of parasites. *Acta Parasitol. Pol.* 34:203-220.

El-Tantawy, S.A.M. 1989b. Myxosporidian parasites in fishes in lakes Dgal Wielki and Warniak (Mazurian Lakeland, Poland). II. Infection of fishes. *Acta Parasitol. Pol.* 34:221-234.

Ferri, S. 1982. Fine structure of a freshwater teleost (*Pimelodus maculatus*). Intrahepatic biliary pathways. *Anat. Anz.* 151:187-196.

Fomena, A., Bouix, G., and E. Birgi. 1990. Contribution to the study of myxosporidia of freshwater fishes in Cameroon. II. New species of *Myxobolus* Butschli 1882. *Bull. Inst. Fondam. Afr. Noire*. Ser. A. *Sci. Nat.* 40:167-192, 1984/1985.

Gilloteaux, J. and L.C. Gilloteaux. 1996. Morphological aspects of a Perciform Teleost gallbladder: LM, TEM and SEM of *Uranoscopus*. *Belg. J. Zool.* (submitted).

Gilloteaux, J., S. Karkare, and T.R. Kelly. 1993a. Apical excrescences in the gallbladder epithelium of the female Syrian hamster in response to medroxyprogesterone. *Anat. Rec.* 236:479-485.

Gilloteaux, J., S. Karkare, W. Ko, and T.R. Kelly. 1992. Female sex steroid induced epithelial changes in the gallbladder of the ovariectomized Syrian hamster. *Tissue and Cell* 25:527-536.

Gilloteaux, J., E. Kosek, and T.R. Kelly. 1993b. Epithelial surface changes and gallstone formation in the Syrian hamster gallbladder as a result of sex steroid treatment. *J. Submicr. Cytol. Pathol.* 25:157-172.

Gilloteaux, J., E. Kosek, and T.R. Kelly. 1993c. Epithelial surface changes and induction of gallstones in the male Syrian hamster gallbladder as a result of a two-month sex steroid treatment. *J. Submicr. Cytol. Pathol.* 25:519-533.

Hacking, M.A., J. Budd, and Hodson, K. 1978. The ultrastructure of the liver of the rainbow trout: normal structure and modification after chronic administration of a polychlorinated biphenyl aroclor 1254. *Can. J. Zool.* 56:477-491.

Hampton, J.A., P.A. McCuskey, R.S. McCuskey, and D.E. Hinton. 1985. Functional units in rainbow trout (*Salmo gairdneri*, Richardson) liver. I. Histochemical properties and arrangement of hepatocytes. *Anat. Rec.* 213:166-175.

Hampton, J.A., R.C. Lantz, P.J. Goldblatt, D.J. Lauren, and D.E. Hinton. 1988. Functional units in rainbow trout (*Salmo gairdneri*, Richardson) liver. II. The biliary system. *Anat. Rec.* 221:619-634.

Haslewood, G.A.D. 1978. The biological importance of bile salts. *In* A. Neuberger and E.L. Tatum (eds.), *Frontiers in Biology*, pp. 79-182. North-Holland Publ. Co., Amsterdam.

Hinton, D.E. and C.R. Pool. 1976. Ultrastructure of the liver in channel catfish *Ictalurus punctatus* (Rafinesque). *J. Fish Biol.* 8:209-219.

Hyrtl, C.J. 1868. Uber Ampulla am *Ductus cysticus* der Fisches. *Denkschr. K. Akad. Wiss. Wien* (*Math. Naturw.*) 28:185-190.

Inagawa, T., Y. Hashimoto, Y. Kon, and M. Sugimura. 1990. Lectin histochemistry as special markers for rodlet cells in carp *Cyprinus carpio* L. *J. Fish Dis.* 13:537-540.

Jacyna, R., P. Ross, and A.D. Bouchier. 1986. Biliary cholesterol, friend or foe? *Quart. J. Med. New Ser.* 65:991-996.

Karkare, S. and J. Gilloteaux (1995). Gallstone induced by sex steroids in the female Syrian hamster: duration effects. *J. Submicr. Cytol. Pathol.* 27:53-74.

Karkare, S., T.R. Kelly, and J. Gilloteaux (1995). Morphological aspects of female Syrian hamster gallbladder induced by one-month steroid treatment. *J. Submicr. Cytol. Pathol.* 27:35-52.

Kazubski, S.L. and S.A.M. El-Tantawy. 1989. *Myxobolus waniakiensis*, new species (Myxosporidia, Bivalvulea, Myxobolidae), new parasite of *Lota lota* from Lake Waniak (Mazurian Lakeland, Poland). *Acta Parasitol. Pol.* 34:199-202.

Koga, A. 1971. Morphogenesis of intrahepatic bile ducts of the human fetus. Light and electron microscopic study. *Z. Anat. Entwicklgesch.* 135:156-184.

Kovaleva, A.A. 1988. Myxosporidia of the genus *Chloromyxum* (Cnidospora, Myxospora) of cartilaginous fishes from the Atlantic coast of Africa. *Parazitologiya* (*Leningr.*) 22:384-388.

Kudo, R.R. 1966. *Protozoology* Ch.C. Thomas Publ. Springfield IL., (5th ed.), pp. 32-33, 774-806.

Laitio, M. and T. Nevalainen. 1972. Scanning and transmission electron microscope observations on human gallbladder epithelium. *Z. Anat. Entwickgesch.* 136:319-325.

Leino, R.L. 1974. Ultrastructure of immature developing and secretory rodlet cells in fish. *Cell Tissue Res.* 155:367-381.

Manwell, R.D. 1968. *Introduction to Protozoology.* Dover Publ. Inc. New York, (2nd ed.), pp. 469-476.

Mattey, D.L., M. Morgan, and D.E. Wright. 1979. Distribution and development of rodlet cells in the gills and pseudobranch of the bass *Dicentrarchus labrax. J. Fish Biol.* 15: 363-370.

Mattey, D.L., M. Morgan, and D.E. Wright. 1980. A scanning electron microscope study of the pseudobranchs of 2 marine Teleosts. *J. Fish Biol.* 16:331-343.

Mayberry, L.F., A.A. Marchiondo, J.E. Ubelaker, and D. Kazic. 1979. *Rhadospora thelohani* new record, Apicomplexa new host and geographic records with taxonomic considerations. *J. Protozool.* 26:168-178.

Mitchell, L.G. 1989. Mixobolid parasites *Myxozoa myxobolidae* infecting fishes of Western Montana USA with notes on histopathology, seasonability, and intraspecific variation. *Can. J. Zool.* 67:1915-1922.

Morrison, C.M. and P.H. Odense. 1978. Distribution and morphology of the rodlet cell in fish. *J. Fish Biol.* 35:101-116.

Mugnaini, E. and S.B. Harboe. 1967. The liver of *Myxine glutinosa*: a true tubular gland. *Z. Zellforsch.* 78:341-369.

Neumayer, L. 1904. Die Entwicklung des Darmkanales, von Lunge, Leber, Milz und Pankreas bei Ceratodus. *Jena Denkschr.* 4:379-422.

Nopanitaya, W., J. Aghajanian, J.W. Grisham, and J.L. Carson. 1979. An ultrastructural study on a new type of hepatic perisinusoidal cell in fish. *Cell Tissue Res.* 198:35-42.

Peek, W.D., E.W. Sidon, J.H. Youson, and M.M. Fisher. 1979. Fine structure of the liver in the larval lamprey, *Petromyzon marinus* L.: hepatocytes and sinusoids. *Amer. J. Anat.* 156:231-250.

Poisson, R. 1953. Sous-Embranchement des Cnidosporidies. In P.P. Grassé (ed.), *Traité de Zoologie. Protozoaires: Rhizopodes, Actinopodes, Sporozoaires, Cnidosporides,* pp. 1006-1041. Masson & Cie, Paris.

Rabergh, C.M.I., G. Byland, and J.E. Eriksson. 1992. Histopathological effects of microcystin-LR, a cyclic peptide from the cyanobacterium (blue-green alga) *Microcystis aeruginosa*, on common carp (*Cyprinus carpio* L.). *Aquatic Toxicology* 20:131-146.

Scammon, R.E. 1913. The development of the elasmobranch liver. *Amer. J. Anat.* 14:33-390.

Shul'Man, B.S. 1989. Life cycle of some Myxosporidia from fishes of the Kola Peninsula. *Parazitologiya (Leningr.)* 23:216-221.

Sidon, E.W. and J.H. Youson. 1983a. Morphological changes in the liver of the sea lamprey, *Petromyzon marinus* L., during metamorphosis. I. Atresia of the bile ducts. *J. Morphol.* 177:109-124.

Sidon, E.W. and J.H. Youson. 1983b. Morphological changes in the liver of the sea lamprey, *Petromyzon marinus* L., during metamorphosis. II. Canalicular degeneration and transformation of the hepatocytes. *J. Morphol.* 178:225-246.

Sidon, E.W., W.D. Peek, J.H. Youson, and M.M. Fisher. 1980. Fine structure of the liver in the larval lamprey, *Petromyzon marinus* L.: bile ducts and gall bladder. *J. Anat.* 131(3):501-519.

Sire, M.F., C. Lutton, and J.M. Vernier 1981. New views on intestinal absorption of lipids in teleostean fishes: an ultrastructural and biochemical study in the rainbow trout. *J. Lipid Res.* 22:81-94.

Sternlieb, I. and N. Quintana. 1985. Biliary proteins and ductular ultrastructure. *Hepatology* 5:139-143.

Suyehiro, Y. 1942. A study of the digestive system and feeding habits of fish. *Jap. J. Zool.* 10:1-303.

Tanuma, Y. 1980. Electron microscope observations on the intrahepatocytic bile canalicules and subsequent bile ductules in the crucian, *Carassius carassius. Arch. Histol. Jap.* 43:1-21.

Uspenskaya, A.V. 1987. Evolutionary aspects of Myxosporidia, cytological investigation. *Tsitologiya* 29:867-873.

Uspenskaya, A.V. 1988. Peculiarities of host-parasite relationship of some intracellular Myxosporidia. *Parasitologiya (Leningr.)* 22:196-200.

Viehberger, G. 1982. Apical surface of the epithelial cells in the gallbladder of the rainbow trout and the tench. *Cell Tissue Res.* 224:449-454.

Viehberger, G. 1983. Ultrastructural and histochemical study of the gallbladder epithelia of rainbow trout and tench. *Tissue and Cell* 15:121-135.

Viehberger, G. and E. Bielek. 1982. Rodlet cells, gland cell or protozoan? *Experientia (Basel)* 38:1216-1218.

Yamamoto, T. 1962. Some observations on the fine structure of terminal biliary passages in the goldfish liver. *Anat. Rec.* 142:293-303.

Yamamoto, T. 1965. Some observations of the fine structure of the intrahepatic biliary passages in goldfish (*Carassius auratus*). *Z. Zellforsch.* 65:319-330.

Yamamoto, K., P.A. Sargent, M.M. Fisher, and J.H. Youson. 1986. Convoluted bile ducts in the liver of the larval lamprey, *Petromyzon marinus* L. *Anat. Embryol.* 173:355–359.

Youson, J.H. and E.W. Sidon. 1978. Lamprey biliary atresia: First model system for the human condition? *Experientia (Basel)* 34:1084–1086.

Youson, J.H., K. Yamamoto, and R.R. Shivers. 1985. Nonparenchymal liver cells in a vertebrate without bile ducts. *Anat. Embryol.* 172:89–96.

Youson, J.H., L.C. Ellis, D. Ogilvie, and R.R. Shivers. 1987. Gap junctions and zonulae occludentes of hepatocytes during biliary atresia in the lamprey. *Tissue and Cell* 19(4):531–548.

8

Recent Advances in the Functional Morphology of Follicular Wall, Egg-Surface Components, and Micropyle in the Fish Ovary

Sardul S. Guraya

A follicular wall envelops the growing oocytes in the fish ovary. It consists of a zona pellucida (or chorion), follicular epithelium, basement membrane or basal lamina, and theca—all of which have been extensively subjected to techniques of electron microscopy, histochemistry, and biochemistry in elasmobranchs and teleosts as these fish are of great economic and evolutionary interest. The results obtained with such techniques have demonstrated a great diversity in the development, structure and function of various components of the follicular wall, egg surface, and micropyle(s). These components are discussed here from a comparative point of view for a better understanding of their functional morphology in relation to reproductive processes of fish. Our knowledge of the functional morphology of egg envelopes of fish which are oviparous, ovoviviparous, and viviparous and live in diverse aquatic conditions has greatly increased in the twentieth century. The cellular processes of oocyte growth, maturation, and ovulation are not discussed here as they have been reviewed by Guraya (1979, 1982, 1986, 1994a, b) and Riehl (1991). For endocrinological aspects, reference may be made to Idler et al. (1987) and to Jalabert et al. (1991).

FOLLICULAR EPITHELIUM

Morphology and histochemistry

Either a few follicle cells associate with very young oocytes of fish ovaries (Jollie and Jollie, 1964; Chaudhry, 1956; Anderson, 1967; Flügel, 1967a; Rastogi, 1970; Guraya et al., 1975; Shackley and King, 1977; Guraya, 1978, 1994b; Brusle, 1980a), or a single layer of squamous follicle cells with the young oocytes as in *Cynolabias melanotaenia* (Wourms, 1976). With oocyte growth, the follicle cells multiply in number, apparently by mitosis, to form a continuous follicular epithelium which constitutes a single-layered structure throughout oocyte growth in teleosts (Kraft and Peters, 1963; Guraya, 1965, 1976, 1978, 1979; Hurley and Fisher, 1966; Flügel, 1967a, b; Anderson, 1967; Götting, 1967, 1970, 1974,

1976; Erhardt and Götting, 1970; Nicholls and Maple, 1972; Gabaeva and Ermolina, 1972; Chinareva, 1973; Busson-Mabillot, 1973; Azevedo, 1974; Wourms, 1976; Kapoor, 1976; Shackley and King, 1977; Tesoriero, 1977a; Riehl, 1977, 1978a–c; Riehl and Schulte, 1977a; Gabaeva, 1977; Emel'Yanova, 1979; Toshimori and Yasuzumi, 1979a, b; Selman and Wallace, 1982; Guraya and Kaur, 1979, 1982; Guraya, 1986, 1994b; Selman et al., 1991; Riehl, 1991). But a follicular epithelium consisting of two cell layers is observed in *Fundulus heteroclitus* (Anderson, 1966). The oocytes of some fish species (e.g. *Tilapia thollori, Arius thalassinus,* dogfish *Scoliodon sarrakowah*) may develop a pseudostratified follicular epithelium for some stages of oocyte growth due to the placement of nuclei in three to four layers (Kraft and Peters, 1963; Gabaeva and Ermolina, 1972; Guraya, 1978, 1986). This pseudostratified morphology is of transitory nature as the follicular epithelium again develops a monolayer structure in the advanced stages of folliculogenesis, indicating that the pseudostratified structure may be indicative of cellular reserve to be utilized during the rapid growth of the oocyte.

The follicle cells undergo morphological changes during oocyte growth. To start with, they are generally spindle shaped or flattened parallel to the oocyte surface or squamous and show little cytoplasmic differentiation. But just prior to or at the initiation of yolk formation, they undergo gradual changes in morphology from squamous to cuboidal or columnar. With stretching of the follicular epithelium, the amount of cytoplasm in the follicle cells is reduced. During follicle growth the morphology of the follicle cell nucleus simultaneously changes from oblong and ellipsoid to round. Nuclei with nucleoli lie toward the basal region. The follicle cells in some elasmobranchs are uniform in size but in others two types of cell develop—a large cell with a reticular nucleus and abundant cytoplasm, and a small columnar cell with a densely staining nucleus (Chieffi, 1961; Guraya, 1978).

During the later stages of oocyte growth the follicular epithelium generally stretches, developing some wide intercellular spaces except at zones where the cells are connected by attachments (Flügel, 1967a; Azevedo, 1974; Nagahama et al., 1978; Kagawa et al., 1981). An amorphous material often forming a network-like structure usually fills the spaces (Nagahama et al., 1978; Kagawa et al., 1981). The chemical composition of this intercellular matrix during follicle growth has yet to be detailed. Wide intercellular spaces between follicle cells have been observed in the teleost *Plecoglossus altivelis* except for two focal membrane appositions near the basal lamina (Toshimori and Yasuzumi, 1979a, b). The space lying between the follicle cells shows a great expansion near the free surface of the epithelium, where it apparently opens directly toward the chorion or zona pellucida. Kobayashi (1985) observed communication of oocyte-follicle cells in the chum salmon ovary with electron microscopy. Intercellular junctions in the form of gap junctions and desmosomes have been reported in the follicular epithelium of different teleost species (Yamamoto, 1963; Flügel, 1964; Anderson, 1967; Kessell et al., 1985; Riehl, 1991). Riehl (1977) reported that the follicle cells in *Noemacheilus barbatulus* are toothed and develop few desmosomes. The boundaries of the follicular epithelium in *Gobio gobio* are straight and more desmosomes develop in this fish than in *N. barbatulus*. Toshimori and Yasuzumi (1979a, b) reported the presence of tight junctions in the follicular epithelium of *Plecoglossus*, which are believed to be "leaky" or "intermediate" regarding the number of strands. Intercellular junctions may perform some important functions in the transport of substances through the follicle cells but the mechanisms of transport have yet to be detailed at the molecular level. However, fundamental functions of cell junctions have been related to the follicle cell-to-follicle cell as well as to the follicle cell-to-oocyte transport of small molecules (either ionic, nutritional, metabolic, or regulatory in nature) in the developing ovarian follicles of mammals and in other tissues (Larsen and Wert, 1988). Such functions may also be performed by cell junctions of the follicular wall in the fish ovary. Further investigations are required to determine the synthetic events and regulatory mechanisms involved in the cell junctions in the follicular wall of the fish ovary.

Gabaeva and Boglazova (1977) investigated the morphodynamics and proliferative activity of the follicular epithelium in the oogenesis of *Xiphophorus helleri*. Its maximum proliferative activity is related to the period of its morphological transformation. During oocyte growth, the follicular epithelium of *Hemichromis* alters from a flattened prismatic into a high prismatic structure (Gabaeva, 1977). Therefore, it can be called highly prismatic with secondary specialization. A cross section of the spindle-shaped follicle cells appears quadratic in stages I and II and quadrilateral in stage III of oocyte growth in the goby *Pomatoschistus minutus* (Riehl, 1978b). Very little cytoplasmic differentiation is seen in the flattened or squamous follicle cells enveloping primary oocytes but it increases as the morphology of the follicle cells changes from squamous to cuboidal or columnar.

Histochemical investigations have revealed lipid droplets consisting mainly of phospholipids and mitochondria, a Golgi complex, and an abundant RNA-containing basophilic substance (or ergastoplasm) in the follicle cells (Guraya, 1965, 1976, 1978, 1979; Hurley and Fisher, 1966; Ramadan et al., 1979a, b; Guraya and Kaur, 1982; Sun and Xizai, 1983; see reviews by Guraya, 1986, 1994b; Riehl, 1991). The presence of cholesterol in the follicle cells has been demonstrated using polarizing microscopy (Sun and Xizai, 1983). Carbohydrates and proteins were also revealed besides the various enzyme activities such as LDH, MDH, and nonspecific esterase in the follicle cells (Sun and Xizai, 1983).

Various electron microscopic investigations have demonstrated the presence of various cell organelles, such as elements of granular endoplasmic reticulum, several free ribosomes, a well-developed Golgi complex, smooth coated vesicles, mitochondria with lamellar cristae, centrioles, dense bodies (secretory or lysosomal), lipid droplets, etc. (Jollie and Jollie, 1964; Hurley and Fisher, 1966; Anderson, 1966, 1967; Flügel, 1967a; Götting, 1967; Nicholls and Maple, 1972; Hirose, 1972; Busson-Mabillot, 1973, 1977; Azevedo, 1974; Chinareva and Kirchinskaja, 1975; Wourms, 1976; Wourms and Sheldon, 1976; Shackley and King, 1977; Nagahama et al., 1976, 1978; Riehl, 1977, 1978a–c; Riehl and Schulte, 1977a; Hoar and Nagahama, 1978; Emel'Yanova, 1979; Kagawa and Takano, 1979; Kagawa et al., 1981; Stehr, 1982; Ohta and Teranishi, 1982; Stehr and Hawkes, 1983; Hart et al., 1984; Thiaw and Mattei, 1991, 1992). Fine filaments and microtubules have also been reported in the follicle cells (Anderson, 1967). With previtellogenic oocyte growth, ribosomes, elements of endoplasmic reticulum, mitochondria, and the Golgi complex progressively increase in amount and concomitantly undergo changes in morphology. The most important demonstrated alterations are found in the endoplasmic reticulum, mitochondria, and Golgi apparatus. The endoplasmic reticulum may proliferate and change into secretory cavities (Wourms and Sheldon, 1976; Busson-Mabillot, 1977; Stehr and Hawkes, 1983; Hart et al., 1984). Mitochondrial changes were mostly related to number, size, electron density of the matrix, and aspect of the matrix (Yamamoto, 1964; Hirose, 1972; Busson-Mabillot, 1977; Iwamatsu et al., 1988; Nakashima and Iwamatsu, 1989; Cruz-Landin and Cruz-Höfling, 1989). However, Thiaw and Mattei (1992) have demonstrated degenerating ultrastructural evolution of the mitochondria in the follicle cells of a species of Cyprinodontidae, *Epiplatys spilargyreus*, during development of its ovarial follicles. Mitochondria with dense matrix and well-developed cristae are few in number until the end of previtellogenesis. They proliferate during vitellogenesis and then are modified by deterioration of their matrix, leading to the formation of multilamellar structures in the vacuolized mitochondria. During postvitellogenesis these modifications are advanced and the mitochondria degenerate, leaving vacuoles that contain heterogeneous structures. At the end of these mitochondrial changes, the follicle cells degenerate and release the heterogeneous structures which will participate in forming the secondary envelope of the oocyte. The saccules of the Golgi complex develop a rather dense matrix. Secretion material produced by the proliferating Golgi apparatus of the follicle cells during previtellogenesis goes on to make up the secondary envelope of the eggs in *Aphyosemion splendopleur* (Thiaw and Mattei, 1991). Various ultrastructural alterations in follicle cells are also indicative of their maturation during follicle growth under the influence of hormones (gonadotrophin and possibly steroids), which are known to influence the proliferation and metabolism of follicle (granulosa) cells of developing follicles in mammals as they develop hormone receptors (Guraya, 1985; Greenwald and Roy, 1994). Similar investigations on the follicle cells of developing and maturing follicles of fish ovary would prove interesting as very little work has been carried out on this aspect to date (Van der Kraak et al., 1983; Bhattacharya, 1992; Bhattacharya et al., 1994).

Functions

Protein and lipid synthesis and their transport into the oocyte. Electron microscope and histochemical investigations have revealed that the follicular epithelium synthesizes proteins and lipids during oocyte growth. The presence of various organelles, such as many ribosomes, elements of granular endoplasmic reticulum, a well-developed Golgi complex, coated vesicles, and dense bodies in the follicle cells is indicative of protein synthesis. These proteins may be required partly for growth and maturation processes of the follicular epithelium itself, and partly for their transport into the oocyte as well as for formation of the zona pellucida (Guraya, 1986). Chinareva and Kirchinskaja (1975) reported the synthesis and accumulation of substances in the follicular epithelium of *Coregonus peled*, which are released to constitute the two outer layers of the zona pellucida (see also Chinareva, 1975; Stehr, 1982). The material of the outer layer (or secondary envelope) of the zona pellucida in the growing follicles of the fishes *Cynolebias melanotaenia* and *C. ladigesi* also accumulates in the granular endoplasmic reticulum of the follicle cells during stage V (Wourms, 1976; Wourms and Sheldon, 1976). It is deposited in stage VI. Donato et al. (1980) demonstrated protein inclusions in the follicle cells of *Chromis chromis*, and their transportation into the oocyte by means of microvilli.

Some phospholipid bodies developed in the follicle cells appear to be transported into the ooplasm of growing oocytes (Guraya, 1965, 1976, 1978, 1986). Corresponding to the deposition of the yolk, lipid droplets form large accumulations in follicle cells of the ovary of the dogfish, *Scoliodon*. Simultaneously the follicle cells develop triglycerides besides the phospholipids (Guraya, 1978). Lance and Callard (1969) also observed accumulation of lipid droplets during maturation of the follicle walls in the ovary of *Squalus acanthias*. It remains to be investigated whether this increase in lipid droplets of the follicular epithelium in elasmobranchs is related to steroidogenesis for providing steroid hormone precursors or simply to yolk deposition through providing lipid yolk precursors. The exact mechanisms of synthesis and release of various substances from the follicular epithelium remain to be revealed more precisely by correlative electron microscope, autoradiographic, and molecular probes to ascertain their precise functions.

Corresponding to oocyte growth, the follicle cells form numerous cytoplasmic processes which traverse the zona pellucida to a variable distance, depending on the stage of oocyte growth as well as on the fish species (Guraya, 1986; Riehl, 1991). They interdigitate with the oocyte microvilli, which maintain close contact with the follicle cells either by lying next to short microvillar extensions on the surface of the follicle cells within the subfollicular space, or one microvillus from a follicle cell will develop contact with one microvillus of the oocyte within the pore canal, as will be described later. By employing freeze fracture replicas, many small accumulations of intramembranous particles have been reported on the cleavage faces of cytoplasmic membranes of follicle cells. Sometimes the follicle cell processes lie close to the oolemma forming the macula adherens between the follicle cells and oocytes of *Lebistes reticulatus*. Toshimori and Yasuzumi (1979a, b) reported the presence of gap junctions between the oocyte and follicle cells, especially on the surface of follicle cells in *Plecoglossus altivelis*. Shorter and longer processes from the follicular epithelium project into the radial canals of the cortex radiatus (zona pellucida) in the follicles of freshwater teleosts (*Noemacheilus barbatulus* and *Gobio gobio*) (Riehl, 1977, 1978b, 1991). The plugs of the cortex radiatus externus (outer layer of the zona pellucida) attach to the follicular epithelium of stage III oocytes which are anchored to the follicle by the follicle cell processes (Riehl, 1976, 1991).

The follicle cell processes and microvilli of the oocyte carry out specific functions during oocyte growth especially in relation to transport of substances, cell-to-cell communication, provision of mechanical support etc. Hurley and Fisher (1966) indicated the possibility of protoplasmic continuity between follicle cell and oocyte via some microvilli, as judged from the absence of intervening membranes between them. This observation still remains to be extended and confirmed as no cytoplasmic continuity at the points of contact between the microvilli and follicle cell processes has been demonstrated in other investigations (Azevedo, 1974; Erhardt, 1976; Riehl, 1977, 1978a–c; Stehr, 1982). Thus the transport of substances across their membranes appears to be facilitated by diffusion, active transport, and pinocytosis (Hurley and Fisher, 1966; Flügel, 1967a; Nicholls and Maple, 1972; Erhardt, 1976). Azevedo (1974) reported numerous pinocytotic vesicles in the periphery of oocytes in a viviparous teleost (*X. helleri*) suggesting the transfer of substances from the follicle cells into the oocyte. An intense formation of vesicles below the primary oocyte membrane is observed in early stage I oocytes of *N. barbatulus* and *G. gobio* (Riehl, 1977a). Simultaneously the primary oocyte membrane partially dissolves and establishes an intimate contact between oocyte cytoplasm and follicular epithelium in *N. barbatulus*. This observation remains to be examined and confirmed in future investigations. The peripheral ooplasm in *Pomatoschistus minutus* also shows many polysomes in addition to intense pinocytotic activity in stage II (Riehl, 1978a). The possible roles of the pinocytotic vesicles and the polysomes are described. Na^+, K^+ activated ATPase localized on the oocyte and follicular microvilli in *Heterandria formosa* (Riehl, 1980a), possibly play an important role in the exchange of substances between the follicle and oocyte. Hurley and Fisher (1966) correlated the diameter of the microvilli and the surface area of microvilli with the changing diameter of growing oocytes in the trout *Salvelinus fontinalis*, suggesting a close correlation between morphological changes of the membranes of the follicular wall and transport of substances to the growing oocyte.

The interfollicular cell spaces containing amorphous material and developing some junctional complexes during follicle growth may also be involved in the selective transport of substances from outside the follicular epithelium (Yamamoto, 1963; Flügel, 1964; Riehl, 1977, 1991; Anderson, 1967; Azevedo, 1974; Nagahama et al., 1978; Toshimori and Yasuzumi, 1979a, b; Kagawa et al., 1981). The regulation and mechanisms of transport of substances through the follicular epithelium and oocyte surface remain to be determined at the molecular level.

Steroid hormone synthesis. In vitro biochemical studies using various hormone precursors have revealed that ovaries of various fish species can synthesize a variety of steroids including progestins, corticosteroids, androgens, and estrogens by the Δ^5 (17α-hydroxypregnenolone and dehydroepiandosterone) or the Δ^4 (progesterone and 17α-hydroxyprogesterone) pathways (Guraya, 1976; Colombo et al., 1982; Fostier et al., 1983; Young et al., 1983a, b; Bhattacharya, 1992; Bhattacharya et al., 1994). Piscine and mammalian gonadotropins stimulate the synthesis of ovarian steroids to a variable degree. Regulation of steroidogenesis and steroid activity by various hormonal and external environmental factors has already been discussed in earlier reviews (Fostier et al., 1983; Bhattacharya, 1992; Bhattacharya et al., 1994). Species variation in this regard are also reported and the molecular mechanisms involved in the action of gonadotropin(s) and other hormones for regulating steroid biosynthesis in various ovarian cell types are currently of great interest. Salmon et al. (1984) investigated the binding of gonadotropin to thecal and follicular cells from amago salmon ovary in radioactive iodine labeled salmon. Gonadotropin receptors are reported for the fish ovary (Van der Kraak et al., 1983; Bhattacharya, 1992). As in mammals (Guraya, 1985; Greenwald and Roy, 1994), cAMP also acts as the intracellular mediator of gonadotropin because it relates to investigations demonstrating increased cAMP levels or adenyl cyclase activity in ovarian homogenates from several teleosts after incubation with gonadotropin (Young et al., 1983b; Bhattacharya, 1992; Bhattacharya et al., 1994). The absence of adenylate cyclase in the fish ovary has been shown cytochemically (Mester et al., 1980).

Ovarian steroid hormones play an important role in the regulation of various reproductive processes of fish (Fostier et al., 1983; Idler et al., 1987). Although the cellular sites of the synthesis of ovarian steroid hormones in fish are not fully resolved, ovarian follicular layers (follicular epithelium and theca) enveloping oocytes form the major cellular sites of ovarian steroidogenesis (see reviews by Guraya, 1976, 1978, 1986; Hoar and Nagahama, 1978; Lambert, 1978; Nagahama et al., 1982; Nagahama, 1983; Fostier et al., 1983; Bhattacharya, 1992; Bhattacharya et al., 1994). Most of the earlier *in vitro* biochemical studies of steroid biosynthesis pertained to whole ovaries (Fostier et al., 1983); attempts to isolate cell types of follicle wall were made later (Nagahama, 1983, 1984). By employing electron microscopy and histochemical techniques for the localization of various hydroxysteroid dehydrogenases involved in the biosynthesis of steroid hormones, very divergent views were expressed about the roles of follicular epithelium in the synthesis of 17β-estradiol, now well known to be the secretory product of the follicle in the fish ovary (Nagahama et al., 1982; Nagahama, 1983, 1984; Fostier et al., 1983; Bhattacharya, 1992; Bhattacharya et al., 1994).

The most conspicuous changes which occur in the follicle cells of the teleost during the maturation stage just before ovulation, include dilation of vesicular cisternae of rough ER, the presence of abundant round, oval or rod-shaped mitochondria with lamellar/tubular cristae, and formation of lysosome-like bodies of variable inner structure with lipid droplets (Anderson, 1967; Yamamoto and Onozato, 1968; Nicholls and Maple, 1972; Nagahama et al., 1976, 1978, 1982; Hoar and Nagahama, 1978; Kagawa and Takano, 1979; Iwamatsu and Ohta, 1981; Kagawa et al., 1981; Ohta and Teranishi, 1982; Nagahama, 1983). An amorphous material fills the dilated granular endoplasmic reticulum. The Golgi complex forms a prominent structure and generally lies in the apical cytoplasm towards the oocyte surface; it consists of stacks of several flattened cisternae associated with numerous small vesicles. Electron-dense membrane-bound secretory granules, approximately 100–200 nm in diameter, frequently lie in association with the Golgi field; some of these granules appear to form a contact with the plasma membrane or lie in the intercellular space. Various electron microscopic investigations have indicated that the follicle cells of preovulatory follicles do not develop ultrastructural features related to steroidogenesis, as also suggested by Ohta and Teranishi (1982). Rather, their features are indicative of protein synthesis as already discussed. But some researchers suggest that development of elements of agranular endoplasmic reticulum that are actually intermediate between the smooth and granular forms and of mitochondria with tubular or vesicular cristae in the follicle cells of preovulatory and postovulatory follicles in teleosts, are indicative of their possible involvement in steroid production. Wallace and Selman (1980) also reported specific ultrastructural changes in follicle cells during final oocyte maturation of *Fundulus heteroclitus*. These changes consist of (1) the development of enormous Golgi complexes with accumulated secretory material and (2) an increase in the number of cisternae of granular ER and free ribosomes. These ultrastructural alterations are believed to be related to the production of a maturation-inducing steroid. This suggestion remains to be confirmed by further investigations.

Δ^5-3β-hydroxysteroid dehydrogenase (3β-HSDH) activity is present in the follicle cells of *Squalus acanthias* (Lance and Callard, 1969) which also react positively for glucose-6-phosphate dehydrogenase (G-6-PDH) (Lance, 1968). Both enzyme systems also show an increase in their activity in the follicular epithelium of growing oocytes which is believed to form the possible site for synthesis or metabolism of steroid hormones. But very contradictory observations exist about the enzyme systems related to steroid hormone synthesis in the follicle cells of teleost ovaries (Guraya, 1976, 1978, 1979, 1986; Nagahama, 1983; Fostier et al., 1983). With some reservations, the species differences must be taken into consideration as the various investigations may not have been made at every stage of ovarian development and maturation. Strong 3β-HSDH and G-6-PDH activities are reported in the follicle cells of the developing follicles in the guppy *Poecilia reticulata* ovary (Lambert, 1966) as also observed for *Tilapia nilotica* (Sun and Xizai, 1983) G-6-PDH participating in the hexose monophosphate shunt forms a necessary process in the synthesis of steroid hormones and its presence in the follicle cells indicates their steroid synthesizing activity. 17β-HSDH activity can also be readily demonstrated in the follicle cells of the guppy ovary, showing that these most likely form the site of estrogen synthesis in the fish (Lambert, 1970; Lambert and van Oordt,

1974). Strong 3β-HSDH activity is also seen in the follicular epithelium during vitellogenesis, especially at the exogenous phase in the swordtail *Xiphophorus helleri* (Lambert and van Oordt, 1974), *Brachydanio rerio* (Lambert et al., 1973; Lambert and van Oordt, 1974), *Sarotherodon aureus, Tilapia aurea* and the mullet *Mugil capito* (Livni et al., 1969; Livni, 1971), *Trachurus mediterraneus* (Bara, 1974), *Acanthobrama terrae-sanctae* (Yaron, 1971), medaka (Iwasaki, 1973; Kagawa and Takano, 1979), *Monopterus albus* (Tang et al., 1978), and loach (Ohta and Teranishi, 1982). In addition, 17β-HSDH activity is also present in the follicle cells of *M. capito* (Livni et al., 1969; Livni, 1971), guppy (Lambert, 1970), and goldfish (Khoo, 1975). The follicle cells of vitellogenic and ripe follicles in the ovary of *Channa gachua* show 3β-HSDH, 17β-HSDH, 11β-HSDH, and G-6-PDH activities, revealing their steroidogenic potential (Shanbag and Nadkarni, 1981). Van den Hurk and Peute (1979) observed a weak but definite 3β-HSDH activity in the follicle cells of *Salmo gairdneri* during the phase of exogenous vitellogenesis. In the ovary of the loach, *Misgurnus anguillicaudatus*, the follicle cells show a weak but distinct positive response for 3β-HSDH only during the periods from prematuration to spawning (Ohta and Teranishi, 1982). These histochemical investigations as well as others (Iwasaki, 1973; Tang et al., 1978; Khoo, 1975) indicate that the follicle cells form the possible source of estrogens in some teleosts (Nagahama, 1983; Fostier et al., 1983; Bhattacharya, 1992; Bhattacharya et al., 1994). In *Brachydanio rerio*, Lambert (1978) confirmed the correlation between development of 3β-HSDH activity in the follicular epithelium and capacity of the ovary to produce 17β-estradiol and estrone *in vitro*. The cells become active only during the spawning period in *Oryzias latipes* (Iwasaki, 1973) or after ovulation induced with hCG in *Clarias lazera* (Van den Hurk and Richter, 1980). But 3β-HSDH activity is observed in the follicle cells of the preovulatory follicles in *Brachydanio rerio* (Yamamoto and Onozato, 1968), *Carassius auratus* (Nagahama et al., 1976) and white-spotted char *Salvelinus leucomaenis* (Kagawa et al., 1981). Kagawa et al. (1981) suggested that these cells form the principal source of estradiol in the *S. leucomaenis* ovary. These workers could not relate any cell type of the follicle to a preovulatory peak of 17β-estradiol secretion. Further correlative *in vitro* biochemical and physiological experiments are recommended to reveal the exact steroidogenic function of the follicle cells from the ovaries of different species. But recent *in vitro* biochemical studies have indicated that both thecal and follicular epithelium are required for production of 17β-estradiol. However, Onitake and Iwamatsu (1986) established steroid hormones in the follicle (granulosa cells) of the medaka, *Oryzias latipes* through immunocytochemical studies.

Lambert (1970) reported the presence of 17β- and 3α-HSDH activities in the peripheral ooplasm of older, yolk-laden oocytes, suggesting the possibility that steroids produced by the follicular epithelium may also be transported to the oocyte. In this regard the absence of 3β-HSDH in the oocytes indicates that the latter do not synthesize steroid hormones and that the enzymes 3α- and 17β-HSDH found at the periphery of yolk-laden oocytes are possibly involved in the intermediate metabolism of steroids produced by the follicle cells. Lambert (1970) stated that the physiological meaning of steroid metabolism in oocytes is difficult to understand. Steroids are possibly needed as building material for the oocyte but they may also play a role in the disappearance of the germinal vesicle as well as during embryogenesis (Jalabet et al., 1991; Selcer and Leavitt, 1991).

THECA AND SURFACE EPITHELIUM

Morphology and histochemistry

Surface epithelium (or serosa) lying outer to the thecal layer constitutes the outermost layer of the follicles in the teleost ovary. The thin and flattened serosal cells are connected with one another by tight junctions and desmosomes (Nagahama et al., 1978). Their lateral plasma membranes are highly convoluted and cytoplasm shows both agranular and granular ER; the elements of granular ER are relatively more developed. The serosal cells contain small mitochondria, microfilaments, and free ribosomes. Micropinocytotic vesicles are often seen at both apical and basal surfaces. Toshimori and Oura (1979) observed the usual architecture of occluding junctions between surface epithelial cells of ovarian follicles in teleost (*Plecoglossus altivelis*). Double strands of intramembranous particles are present on the face. A narrow furrow-like gap is seen between two rows of particles in these strands. Two groove types are present on the E-face. On the P-face the double strands are evidently registered with the grooves (type I or type II) on the complementary E-face and a row of particles on the E-face is registered with a furrow-like region between two rows in the double strands on the P-face.

Thecal layer. A thick layer of theca envelops the follicular epithelium outside the basal lamina which separates the thecal layer from the follicular epithelium; the basal lamina stains for carbohydrates and proteins (Guraya, 1978, 1986) and shows collagenous fibers (Anderson, 1967). Electron microscope investigations revealed that it comprises a series of membranes and electron-dense material (Hurley and Fisher, 1966; Anderson, 1967; Azevedo, 1974; Götting, 1967, 1970, 1974, 1976; Guraya, 1986, 1994b; Flügell, 1967a; Erhardt and Götting, 1970).

The thecal cell layer consists of collagenous fibers, capillary loops and fibroblast-like (or stromal) cells (Hurley and Fisher, 1966). Electron microscope investigations have revealed the presence of some enlarged or hypertrophied cells designated as the thecal gland cells or special thecal cells (Yamamoto and Onozato, 1968; Nicholls and Maple, 1972; Nagahama et al., 1976, 1978, 1982; Kagawa et al., 1981; Guraya and Kaur, 1982; Nagahama, 1983; Guraya, 1986, 1994b; Fostier et al., 1983; Nakamura et al., 1993).

The vascularized theca interna of follicles in the dogfish *Scoliodon* is composed of large spherical or polygonal cells with abundant cytoplasm (containing RNA, some diffuse lipoproteins, and granular mitochondria), and a vesicular nucleus with a large nucleolus rich in RNA (Guraya, 1978). The thecal tissue also shows lipid droplets consisting of phospholipids which show sparse distribution. The theca interna is surrounded by the broad theca externa which consists of fibroblast-like cells with very little cytoplasmic differentiation. The fibroblast-like cells in the thecal layer of fish follicles show slight development of cytoplasm with some elements of granular endoplasmic reticulum, free ribosomes, a few mitochondria of small size, and a very small Golgi zone (Hurley and Fisher, 1966; Yamamoto and Onozato, 1968; Azevedo, 1974). The stromal cells of the thecal layer form an association with each other by maculae adherens (Anderson, 1976; Azevedo, 1974).

Various enzymes demonstrated histochemically in the thecal layer relate either to permeability or to steroid biosynthesis (Guraya, 1976, 1978, 1979, 1986; Nagahama et al., 1982; Colombo et al., 1982; Nagahama, 1983; Fostier et al., 1983). The strong alkaline phosphatase activity observed in the thecal layer of the follicle appears to be related to the transport of substances across its various membranes (Varma and Guraya, 1968). In contrast to the studies of Lambert (1966, 1970) on *Poecilia reticulata* ovaries, demonstrating enzyme activity related to steroidogenesis in the follicle cells, Bara (1965) observed that 3β-HSDH activity is greatest in some of the thecal cells and not in the follicular epithelium in the ovary of the mackerel *Scomber scomber* at different stages of its reproductive cycle. Activity is most intense at the start of vitellogenesis and decreases as the follicle matures. Localization of glucose-6-phosphate dehydrogenase (G-6-PDH) corresponds to the distribution of 3β-HSDH except that it is also seen at a low level in the follicle cells (Bara, 1965). 3β-HSDH activity in the zebra fish (*Brachydanio rerio*) also occurs in the thecal layer, where it is seen in enlarged cells (Yamamoto and Onozato, 1968; Van Ree et al., 1977) and has also been demonstrated for the goldfish (Nagahama et al., 1976) and carp, *Cyprinus carpio* (Colombo et al., 1982). Yaron (1971) demonstrated 3β-HSDH activity in both follicular and thecal cells of *Acanthobrama terrae-sanctae* and *Tilapia nilotica*. Saidapur and Nadkarni (1976) also observed 13β-17β-, and 11β-HSDH, and G-6-PDH activities in the thecal cells of the ovary in *Mystus cavasius*, as also reported in stromal cells of *Channa gachua* (Shanbag and Nadkarni, 1981). 17β-HSDH activity was also reported in the thecal cells of the ovaries of *Trachurus mediterraneus* (Bara, 1974). The thecal and interstitial cells of the adult trout ovary showed maximum 3β-HSDH activity at the time of meiotic maturation and ovulation (Van den Hurk and Peute, 1979) suggesting that these cells may produce progestins and/or corticosteroids, which are now known to be involved in the regulation of these processes (Jalabert et al., 1991). Kagawa et al. (1981) did not find 3β-HSDH activity in the special thecal cells at any stage in the preovulatory follicles of *Salvelinus leucomaenis*, suggesting that the special thecal cells at this stage of preovulatory growth of the follicle show low steroidogenic activity.

Various electron microscope investigations produced strong evidence for the presence of isolated special thecal cells lying close to capillaries in the thecal layer of ovarian follicles in various teleosts, such as *Brachydanio rerio* (Yamamoto and Onozato, 1968), *Cichlasoma nigrofasciata* and *Haplochromis multicolor* (Nicholls and Maple, 1972), *Carassius auratus* (Nagahama et al., 1976), *Oncorhynchus kisutch* (Nagahama et al., 1978), *S. gairdneri* (Van den Hurk and Peute, 1979), and *Salvelinus leucomaenis* (Kagawa et al., 1981); these cells may also form clusters in some teleosts (Bara, 1965, 1974; Livni, 1971; Nagahama et al., 1978; Nakamura et al., 1993) and could derive from the same stromal cells, the interstitial cells (Yamamoto and Onozato, 1968; Lambert and van Oordt, 1974; Saidapur and Nadkarni, 1976; Guraya, 1976, 1979; Van den Hurk and Peute, 1979). Yamamoto and Onozato (1968) observed that special thecal cells in the zebra fish *Brachydanio rerio* develop from the ovarian interstitial cells at an early stage and later localize in the thecal layer (Guraya and Kaur, 1982). The hypertrophied cells develop the ultrastructural characteristics of well-established steroid gland cells (Christensen and Gillim, 1969; Guraya, 1971, 1974; Christensen, 1975; Neaves, 1975) as reported for various teleosts (Yamamoto and Onozato, 1968; Nicholls and Maple, 1972; Nagahama et al., 1976, 1978; Van den Hurk and Peute, 1979; Iwamatsu and Ohta, 1981). Special thecal cells contain a central nucleus with a large nucleolus. Agranular endoplasmic reticulum forms the conspicuous feature throughout the cytoplasm. Granular endoplasmic reticulum, relatively less developed, often forms several stacked cisternae. The moderately developed Golgi complex often lies adjacent to the nucleus. Round, oval, or rod-shaped mitochondria with tubular or vesicular cristae are seen. Membrane-bound dense bodies of variable size occur frequently near the Golgi apparatus.

Steroid hormone synthesis

Various cytological and histochemical features of the special thecal cells are certainly indicative of steroidogenesis, as discussed in detail in previous reviews (Guraya, 1976, 1979, 1986; Nagahama et al., 1982; Nagahama, 1983; Fostier et al., 1983; Nakamura et al., 1993). From a review of both histochemical and ultrastructural data it can be concluded that the special thecal cells of the follicular wall constitute the major cellular site for steroidogenesis in the teleost ovary (see also Guraya, 1976, 1978, 1979, 1986; Hoar and Nagahama, 1978; Nagahama et al., 1982; Nagahama, 1983, 1984; Fostier et al., 1983; Nakamura et al., 1993). These cells are believed to secrete either 17β-estradiol or progesterone (Nagahama et al., 1978; Kagawa et al., 1981; Nagahama, 1983; see reviews by Guraya, 1986, 1994b). But the nature of steroid hormones synthesized by the special thecal cells *in vivo* remains to be determined more precisely. Kagawa et al. (1981) discussed the functional significance of preovulatory rise in plasma progesterone levels in relation to the induction of oocyte maturation (Jalabert et al., 1991; Selcer and Leavitt, 1991). But more precisely timed sampling before or during final maturation would be helpful in locating the cellular sites of progesterone synthesis in the follicle wall. However, 17α, 20β-dihydroxy-4-pregnen-3-one is secreted *in vitro* by ovarian tissue of amago salmon (*Oncorhynchus rhodurus*), and constitutes one of the most potent inducers of oocyte maturation (Suzuki et al., 1981). It is known to be synthesized by the follicle in response to gonadotropin (Fostier et al., 1983; Nagahama et al., 1983b; Young et al., 1983b; Nagahama 1983, 1984; Ueda et al., 1984; Yamauchi et al., 1984; Hirose et al., 1985; Guraya, 1986) and its increased concentrations are observed in the plasma of females undergoing final oocyte maturation (Nagahama, 1983; Fostier et al., 1983; Young et al., 1983b; Ueda et al., 1984; Yamauchi et al., 1984; Hirose et al., 1985; Jalabert et al., 1991). *In vitro* biochemical experiments have demonstrated that under the influence of gonadotropin the thecal layer secretes 17β-hydroxyprogesterone, which is transported to the follicular epithelium and converted to 17α, 20β-dihydroxy-4-pregnen-3-one (the maturation inducing steroid) (Nagahama, 1983, 1984; Bhattacharya, 1992; Bhattacharya et al., 1994); 20β-hydroxysteroid dehydrogenase, the key enzyme involved in this conversion occurs in the follicular epithelium.

Further correlative histochemical, electron microscopic, autoradiographic and biochemical methods including *in vitro* experiments are required to more precisely ascertain the cellular sites of steroidogenesis during maturation of the follicular wall in the fish ovary, which may vary with the species and its mode of reproduction. However, *in vitro* experiments have produced direct evidence for the presence of 3β-HSDH in amago salmon (*Oncorhynchus rhodurus*) follicles and demonstrated that this enzyme is required for stimulation of 17β-estradiol by mammalian and fish gonadotropins (Young et al., 1982; Kagawa et al., 1982a, b; Fostier et al., 1983; Bhattacharya, 1992; Bhattacharya et al., 1994). These data indicate that fish gonadotropin (SG-G 100) directly regulates 17β-estradiol production in amago salmon follicles, although the cellular site of its production and the mechanism of gonadotropin stimulation of steroidogenesis remains to be defined (see also Fostier et al., 1983; Bhattacharya, 1992; Bhattacharya et al., 1994). But investigations on the amago salmon using isolated thecal and follicle cells have revealed that both cell types are essential for 17β-estradiol production *in vitro* in response to salmon gonadotropin (Nagahama et al., 1982; Kagawa et al., 1982b, 1983; Nagahama, 1983, 1984; Bhattacharya, 1992; Bhattacharya et al., 1994). In response to hCG, Kagawa et al. (1984) obtained *in vitro* 17β-estradiol and testosterone production by ovarian follicles at different stages of their development and maturation in the goldfish, *Carassius auratus*. The decrease in 17β-estradiol production in tertiary yolk stage follicles is related in part to a decrease in aromatase activity at this stage.

It can be concluded from the correlation of various data that the function of the vascularized thecal layer during the major part of vitellogenesis appears to be the production of estrogen precursors (androgens, especially testosterone), which are converted to 17β-estradiol in the follicular epithelium (follicle cells) by its aromatase system (Young et al., 1982; see reviews by Guraya, 1986; Bhattacharya, 1992; Bhattacharya et al., 1994). Thecal layers having no aromatizing enzyme system cannot produce 17β-estradiol. High concentrations of plasma testosterone in vitellogenic females function as a precursor for estrogen production (Kagawa et al., 1982b, 1983, 1984; Bhattacharya, 1992; Bhattacharya et al., 1994), suggesting that both layers of the follicle are required for gonadotropin stimulated 17β-estradiol production. Ultrastructural, histochemical, and *in vitro* biochemical experiments have shown that the thecal layer produces androgen, involving numerous biosynthesis steps, and the follicular epithelium forms the site of aromatization of androgens, involving only a few biosynthesis steps. The thecal cells are also ultrastructurally more steroidogenic than the follicle cells, which show fewer features of steroidogenesis, as already described in Section I. A similar two-cell type model for the production of follicular estrogens *in vitro* is also well established for the rat and some other mammals in which granulosa (or follicle) cells also show fewer signs of steroidogenesis (Guraya, 1985). In mammals, FSH plays an important role in the aromatization of androgens to form estrogen (or stimulates granulosa cell aromatase activity) (Greenwald and Roy, 1994). But detailed studies are required to determine the involvement of gonadotropin(s) in the conversion of androgens to 17β-estradiol by the follicle (granulosa) cells of various teleost species. In other words, the mechanism of induction or activation of the granulosa cell aromatase system remains to be defined at the molecular level. Nevertheless, it is known that aromatase activity increases during vitellogenesis and thereafter rapidly decreases in the postvitellogenic period (Young et al., 1983a). The mechanism by which fish gonadotropin stimulates testosterone production by the thecal layer also remains to be defined more precisely (see reviews by Bhattacharya, 1992; Bhattacharya et al., 1994). Further investigations are needed to establish whether this pattern also occurs *in vivo*. When the vitellogenesis phase ends and the oocytes have fully grown, the estradiol 17-β level suddenly drops and concomitantly 17α.20β-di OH prog production by the follicular wall increases in response to GtH (see Guraya, 1986; Jalabert et al., 1991; Bhattacharya, 1992; Bhattacharya et al., 1994). A two-cell type model is also proposed for this shift in steroidogenesis. The thecal layer produces large quantities of 17α-hydroxyproesterone in response to GtH, which then enters follicle cells where GtH activates 20β-hydroxy steroid dehydrogenase, the key enzyme involved in the conversion of 17α-hydroxyprogesterone to 17α.20β-di OH prog (see Bhattacharya, 1992; Bhattacharya et al., 1994). However, precise mechanism of the regulation of this shift in steroidogenic pattern as well as the receptor quality and quantity in relation to specific time of follicle wall maturation, remains to be determined at the molecular level (Ishii, 1991; Jalabert et al., 1991; Selcer and Leavitt, 1991; Bhattacharya, 1992; Bhattacharya et al., 1994; Manna and Bhattacharya, 1993).

ZONA PELLUCIDA (OR CHORION)

Between the surface of the growing oocyte and follicular epithelium an acellular layer is formed which is called the zona pellucida. It shows great diversity in structure and chemistry in various groups of fish (Guraya, 1986; Johnson and Werner, 1986; Cotelli et al., 1988; Groat and Alderdica, 1986; Begovac and Wallace, 1989; Hamazaki et al., 1989; Oppen-Bernstein, 1990; Riehl, 1991). This structure has also been designated the chorion, vitelline membrane, zona radiata, cortex radiatus, etc. (Guraya, 1978; Laale, 1980; Renard et al., 1987; Riehl, 1991) and common synonyms include coat, covering, pellicle, sac, and envelope—none of which give much information about its origin and structural organization (reviewed by Laale, 1980). It is now well established that the substances of the zona pellucida are deposited between the oocyte surface and follicular epithelium during follicle growth, which develop microvilli and follicle cell processes respectively. The development, structure, and function of the follicle processes have already been discussed. Here developmental, structural, chemical, and functional aspects of the microvilli and zona material are presented for various fishes.

Structure

Microvilli. The plasma membrane of young oocytes forms a close association with that of follicle cells, being separated from it by a narrow space (Jollie and Jollie, 1984; Flügel, 1967a; Anderson, 1967; Erhardt and Götting, 1970; Tsukahara, 1971; Hirose, 1972; Erhardt, 1976, 1978; Busson-Mabillot, 1973; Azevedo, 1974; Riehl and Schulte, 1977a; Flegler, 1977; Riehl, 1978a–c; Upadhyay et al., 1978; Emel'Yanova, 1979; Stehr, 1982; Selman and Wallace, 1982; Guraya, 1986, 1994b; Riehl, 1991). With the growth of oocyte the oolemma forms microvilli covering the entire oocyte surface. The microvilli elongate and lie in the space between the oocyte and follicle cells as the latter become displaced from the former (see Hurley and Fisher, 1966; Flegler, 1977; Riehl and Schulte, 1977a; Erhardt, 1978; Upadhyay et al., 1978; Riehl, 1978a–c, 1991; Stehr, 1982; Bruslé, 1980a, b; Riehl and Greven, 1990; Abraham et al., 1991; see review by Riehl, 1991). Some of the microvilli become so long that they extend into the intercellular space of the follicle cells where they constitute conspicuous groups but do not branch (Azevedo, 1974). Some microvilli also indent the faces of follicle cells lying adjacent to the oocyte surface. The gap junctions enable the microvilli to establish contact with the follicle cells in the teleost (*Plecoglossus altivelis*) (Toshimori and Yasuzumi, 1979a, b). These contact zones appear to be a seven-layered membrane with an overall thickness of 18 µm in standard fixation.

Microvilli of the large or mature oocytes form thin slender structures and show a denser interior than those of growing oocytes (Jollie and Jollie, 1964; Hurley and Fisher, 1966; Flügel, 1967a; Anderson, 1967; Götting, 1967, 1974, 1976; Hirose, 1972; Busson-Mabillot, 1973; Azevedo, 1974; Wourms and Sheldon, 1976; Wourms, 1976; Flegler, 1977; Erhardt, 1976, 1978; Riehl and Schulte, 1977a; Selman and Wallace, 1982; Stehr, 1982; Hart and Donovan, 1983; Guraya, 1986; Riehl, 1991). These lie in the pore canals of the zona pellucida as a space always develops between its components and the microvilli. The material enveloping the pore canals forms the continuous phase of the zona pellucida, whereas the pore canals themselves constitute the discontinuous phase. But very definite striations develop in the walls of the pore canals (Hurley and Fisher, 1966). These show a "ribbing" arranged spirally around the pore canals.

The microvilli correspond to the striations of light microscopic studies (see references in Kraft and Peters, 1963; Guraya, 1965, 1986; Guraya et al., 1975, 1977). Due to the presence of these numerous striations or microvilli the layer of zona pellucida which lies adjacent to the oocyte, was generally termed the zona radiata (or cortex radiatus internus) and the outer, relatively homogeneous layer as the zona pellucida proper (or cortex radiatus externus or zona radiata externa) in some earlier studies (Hurley and Fisher, 1966; Flügel, 1967a; Riehl, 1978a-c; Flegler, 1977; Riehl and Schulte, 1977a, 1978; Erhardt, 1976). Microvilli during all stages of oocyte growth have a core filament extending into the microvilli from the ooplasm.

Zona material. The microvilli during the early stages of oocyte growth do not show an intermicrovillous substance which, however, begins to deposit on the oocyte surface during its further growth to form the zona pellucida (Jollie and Jollie, 1964; Hurley and Fisher, 1966; Droller and Roth, 1966; Anderson 1966, 1967; Flügel, 1967a; Götting, 1967; Erhardt and Götting, 1970; Caloianu-Iordachel, 1971a, b; Busson-Mabillot, 1973; Azevedo, 1974; Chinareva and Krichinskaja, 1975; Wourms, 1976; Erhardt, 1976, 1978; Flegler, 1977; Shackley and King, 1977; Riehl and Schulte, 1977a; Guraya, 1978, 1986, 1994b; Stehr, 1982; Selman and Wallace, 1982; Stehr and Hawkes, 1983; Schmehl and Graham, 1987; Riehl and Greven, 1990; Riehl, 1991). Depending on the species as well as on the stage of oocyte growth in the same species, the zona pellucida develops a variable number of layers (or zones), thus giving rise to either a monopartite, bipartite, or tripartite acellular envelope between the oocyte surface and follicular epithelium (see Guraya, 1986, 1994b; Riehl and Greven, 1990; Riehl, 1991). The major zones can also be distinguished in light microscopic studies (Chaudhry, 1956; Kraft and Peters, 1963; Guraya, 1965, 1978, 1986; Anderson, 1967; Riehl and Schulte, 1977a) but details of their development and structure cannot. Under an electron microscope, however, the width, structure, and texture of the different zones show considerable diversity among teleosts (Guraya, 1986; Kobayashi and Yamamoto, 1987; Riehl, 1991). Götting (1967, 1974) reported that formation of the zona pellucida differs in ovoviviparous and viviparous fishes. This difference in oviparous species lies in a decrease in zona layers for establishing the close relationship between maternal and fetal blood (see also Erhardt and Götting, 1970; Azevedo, 1974; Götting, 1976; Riehl and Greven, 1990; Riehl, 1991).

The tripartite structural organization of the zona pellucida is generally found in oviparous fishes (Guraya, 1965; Hurley and Fisher, 1966; Anderson, 1967; Götting, 1967, 1970, 1974; Busson-Mabillot, 1973; Chinareva and Kirchinskaja, 1975; Wourms, 1976; Erhardt, 1976; Treasurer and Holiday, 1981; Stehr, 1982; Hart and Donovan, 1983; see reviews by Guraya, 1986, 1994b; Riehl, 1991). Erhardt (1978) also found that the cortex radiatus externus (zona pellucida proper) of the oviparous fish *Lutianus analis* develops a triple-layered structure while the cortex radiatus internus (zona radiata proper) gets encircled by bundle structures. Three zones, labeled Z-1, Z-2, and Z-3, were demonstrated in the zona pellucida enveloping the ovarian oocytes of *Hippocampus erectus* and *Syngnathus fuscus* (Anderson, 1967) and also for *Brachydanio rerio* (Hart and Donovan, 1983). Of these, zone 1 disappears upon maturation of the egg cell. Zone 1 of the zona pellucida, present in the oocytes of seahorse (*Hippocampus erectus*), was not found in the sunfish (*Lepomis microchirus*), brown trout (*Salmo trutta*), killfish (*Fundulus heteroclitus*) and other kinds of fish investigated by Fisher (1963) and Flügel (1967b). Nor does it develop in the C-O sole oocyte (Stehr, 1982). The morphological organization of zones 2 and 3 in the zona pellucida of the seahorse and pipefish is the same as that reported for the zona pellucida of the teleost *Cynolebias belotti* (Müller and Sterba, 1963) and *Agonus cataphractus* (Götting, 1965). The zones of zona pellucida in the latter two teleost species are also altered in a manner similar to that reported for the highly ordered zone 3 in the seahorse and pipefish (Anderson, 1967). In the white sturgeon *Acipenser transmontanus*, the thick zona pellucida of its mature oocyte consists of four distinct layers (Cherr and Clark, 1982).

Electron microscope studies have also revealed that each layer of zona pellucida may also consist of as many as three zones (Anderson, 1967; Wourms, 1976; Erhardt, 1976). It has also been shown that the inner layer of zona pellucida in *Crenilabrus melops, Crenilabrus cinercus, Crenilabrus mediterraneus, Ctenolabrus exoletus, Ctenolabrus rupestris, Limanda limanda* (= *Pleuronectes limanda*), *Hypoglossoides platessoides, Platichthys flesus, Pleuronectes flesus*, and *Gadus marrhua* is composed of 6, 7, 9, 17–18, 5–6, 6, 6, and 5 membranous lamellae respectively (Lönning and Solemdal, 1972). An interesting observation by Lönning and Solemdal (1972) is that the species *Pleuronectes platessa* from Bergen and Tromoso showed variation in ultrastructure, suggesting either a lack of species specificity in regard to morphological organization of the zona pellucida or a possible effect of the environment in the modification of zona pellucida morphology. Manner et al. (1977) reported the presence of a middle multilayered structure composed of 19 lamellae in the zona pellucida of *Pimephales promelas*.

The layer next to the oocyte forms a reticular network of lamellae in *Hippocampus erectus* and *Syngnathus fuscus* (Anderson, 1967). *Oryzias latipes* (Hirose, 1972; Tesoriero, 1977a, 1978), *Blennius pholis* (Shackley and King, 1977), *Fundulus heteroclitus* (Kemp and Allen, 1956; Flügel, 1967b), *Cynolebias melanoteania* (Wourms, 1976), *Gobio gobio* and *Noemacheilus barbatulus* (Riehl, 1977), *Pleuronichthys coenosus* (Stehr, 1982; Stehr and Hawkes, 1983); complex tubular systems or crossbanded fibrils in *Dermogenys pusillus* (Sterba and Müller, 1962; Müller and Sterba, 1963; Flegler, 1977) and *Cynolebias belotti* (Götting, 1965); lamellae in *Cichlasoma nigrofasciatus* (Busson-Mabillot, 1977); 16 horizontal electron-dense lamellae alternating with 15 interlamellae of lower electron density in *Brachydanio rerio* (Hart and Donovan, 1983), and densely packed in *Lebistes reticulatus* (Jollie and Jollie, 1964). The alternating arrangement of obliquely placed fibrillar lamellae constituting ribs is seen in the C-O sole (Stehr, 1982; Stehr and Hawkes, 1983) and *Salvelinus fontinalis* (Hurley and Fisher, 1966). Hurley and Fisher (1966) reported that the ribs are placed spirally around the pore canals. In those teleost eggs having more than one layer, the outer layer shows a different ultrastructure from the inner layer, varying from stratified and fibrous in *Lebistes reticulatus* (Jollie and Jollie, 1964), homogeneous in *Blennius pholis* (Shackley and King, 1977), and densely packed in *Oryzias latipes* (Hirose, 1972).

In the starry flounder (*Platichthys stellatus*), Stehr and Hawkes (1979) reported that the zona pellucida forms 0.22–0.50% of the egg diameter, consisting of six continuous horizontal lamellae, covered by a thin triple-layered border and pierced by numerous regularly spaced pore canals. The thicker membrane of the pink salmon (*Oncorhynchus gorbuscha*) egg constitues 0.80–1.0% of the egg's diameter and consists of numerous short discontinuous lamellae traversed by pore canals and enveloped by a coating of irregular thickness.

During embryogenesis of these two species, contrasting environmental conditions may be reflected by the thin membrane and simple lamellar structure in the pelagic egg of the starry flounder and the thick membrane complex lamellar structure in the demersal egg of the pink salmon.

During ovum maturation and ovulation the zona pellucida undergoes changes. A rapid increase in size occurs during maturation and ovulation of teleost oocytes (Kuo et al., 1974; Hirose et al., 1976; Hirose, 1972; Wallace and Selman, 1981; Stehr, 1982; Guraya, 1986). Major changes in the zona pellucida of *Dermogenys pusillus* have been detailed by Flegler (1977). In the zona pellucida of seahorse and pipefish, zone 1 develops first and disorganizes during later stages of oocyte maturation; the significance of its disappearance remains to be determined in future investigations (Anderson, 1967). Zones 2 and 3 are greatly altered. Similarly, the different zones of the zona pellucida in the teleost *Cichlasoma nigrofasciata* are greatly altered structurally and chemically during vitellogenesis, preovulation and ovulation (Busson-Mabillot, 1973), as also observed for the medaka *O. latipes* (Hirose, 1972). When yolk vesicle fusion is completed and the nucleus moves from the center of the oocyte, the microvilli withdraw from the pore canals. The zona pellucida decreases in thickness and simultaneously loses its striated appearance. This type of change is also observed in *Blennius pholis* (Shackley and King, 1977), *Fundulus heteroclitus, Pleuronectes platessa* (Flügel, 1967a), *Cynolebias melanotaenia* (Wourms, 1976) and C-O sole (Stehr, 1982). Thinning of the zona pellucida after ovulation appears to be partially a result of the envelope stretching when the egg increases in diameter. However, the pore canals in salmonids are plugged with an electron-dense material at the outer surface of the envelope and the striated appearance of the zona pellucida persists (Flügel, 1964, 1967a, 1970; Hirose, 1972).

Origin

Divergent views still continue to be expressed about the origin of zona material. Its synthesis continues to be attributed either to the oocyte itself or to the follicular epithelium or to both (Chaudhry, 1956; Kemp and Allen, 1956; Hurley and Fisher, 1966; Anderson, 1967; Rastogi, 1970; Chinareva and Kirchinskaja, 1975; Wourms, 1976; Wourms and Seldon, 1976; Shackley and King, 1977; Tesoriero, 1977a, b, 1978; Sobhana and Nair, 1977; Stehr, 1982; Lopes et al., 1982). The extent to which such a role might differ among species as well as between follicles at different stages of growth still remains to be determined in detail. Our knowledge about the origin of the fish zona pellucida is mostly the result of various studies carried out with ultrastructural and histochemical methods; very little attempt has been made to apply immunocytochemical methods and autoradiography for this purpose. Although each of these techniques is limited in scope by itself, their correlative application can be of great help in obtaining a better insight into the cellular and developmental regulation of zona pellucida formation during follicle growth.

Anderson (1967) observed ultrastructural evidence for involvement of organelles of the oocyte in the synthesis and secretion of zona material. Vesicular and elongated structures originating from the endoplasmic reticulum and Golgi complex were observed in the peripheral ooplasm during the formation of each zone of the zona pellucida. The lumina of these components were confluent with the material forming the egg coat during its development. Tesoriero (1977a, b, 1978) also indicated that the mechanism of development of zona pellucida in *Oryzias latipes* appears to involve the transfer of precursor glycoprotein substance from the Golgi bodies of the oocyte to the zona pellucida by means of populous dense-cored vesicles. In the early stages of zona pellucida formation in *Cynolebias melanotaenia* oocytes, smooth-surfaced, dense-cored vesicles are also observed to fuse with the oolemma and deposit the zona material on the exterior of the oocyte (Wourms, 1976). Tesoriero (1977a) made a distinction between pinocytotic-coated vesicles and exocytotic smooth-surfaced vesicles, yolk is transported into the oocyte by coated vesicles. C-O sole oocytes at stages 5-7 show only a few Golgi complexes but contain abundant rough endoplasmic reticulum and smooth dense-cored vesicles lying next to the developing striated zona pellucida (Stehr, 1982), suggesting that the oocyte itself may synthesize the zona material of the striated layer and transport it in dense-cored vesicles which fuse with the oolemma and release their contents into the surface of the oocyte.

Chinareva and Kirchinskaja (1975) reported that the materials of the outer two layers of the zona pellucida, consisting of three layers, are formed by the follicular epithelium in the fish *Coregonus peled* (see also Chinereva, 1975). Wourms (1976) and Wourms and Sheldon (1976) also found that the substance of the outer layer (the secondary envelope) of the zona pellucida in the maturing follicles of annual fishes *Cynolebias melanotaenia* and *C. ladigesi* is produced by the follicle cells at stage 5, when they also show the ultrastructural characteristics of protein-synthesizing cells. The formation of granular endoplasmic reticulum correlates with the initial formation of tubular material. Shackley and King (1977) described alterations in the follicular epithelium of marine teleost (*Blennius pholis*), together with development of the zona, believed to be follicular in origin. They report two types of follicle cells, which appear to play different roles in the process of zona formation. The nature of the forces that regulate all variation in the developmental processes of zona layers of teleosts remains to be determined more precisely (Laale, 1980; Riehl, 1991). The layered morphology of the zona pellucida indicates that substances of variable physicochemical properties are possibly formed at different stages of oocyte growth by the follicle cells and oocyte (Anderson, 1967; Chinareva and Kirchinskaja, 1975; Erhardt, 1976; Wourms, 1976; Wourms and Sheldon, 1976; Riehl, 1978b; Stehr, 1982). Immunocytochemical investigations may reveal biological variations in the cellular origin of zona components among different fish species. The application of specific techniques for zona components at the mRNA level would be helpful in solving this question. Actually, studies on the synthesis of zona proteins at the transcriptional and translational level are required for demonstration of how the macromolecular composition of its unique glycoprotein has diverged, whereas the basic physiological functions are highly conserved.

From the recent data discussed here, it appears that little progress has been made in regard to problems of origin and naming of egg envelopes. Therefore, further correlative electron microscopic, autoradiographic, and biochemical investigations related to synthesis and secretion, and morphological information on the subcellular organization of both egg and follicle cells are required to resolve the extant difficulties. Monoclonal antibody methods would also be very useful in determining the role of different cell types in the formation of zona proteins. However, the morphological and chemical differences of zona pellucida, as discussed in different species, may be indicative of adaptations of the fish to diverse ecological conditions. Morphological investigations have shown that formation of the zona pellucida is developmentally regulated, as evidenced from its development during specific stages of oogenesis. But the nature of the signals that initiate the zona pellucida formation remains to be determined. These may be hormonal, genetic, or the result of cell contact and establishment of intercellular communication between the oocyte and follicle cells (Guraya, 1986). Whatever the signals may be, regulation of zona pellucida development could occur at the level of postranslational modification, translation, transcription and/or genome organization. The molecular mechanism of regulation remain to be determined in future investigations using recombinant DNA and monoclonal antibody techniques. Since intraspecies, biochemical, histochemical, and morphological differences are reported in the fish zonae (Guraya, 1986; Riehl, 1991), the site of synthesis and the postranslational modification possibly also vary among fish species. The cellular origin of zona pellucida materials also needs to be investigated, using *in vitro* biosynthesis of proteins and carbohydrates.

To summarize, it may be said that the zona pellucida (chorion) is the primary envelope as it mainly originates from the oocyte. However, we cannot exclude a prior contribution of the follicle cells (Begovac and Wallace, 1989) or of the liver (Hamazaki et al., 1989), which, however, could be interpreted as a complement in the acquisition of the supramolecular architecture. In teleostean fishes the follicle cells participate in producing the secondary envelope of the egg, as judged from the formation of its pressure in the endoplasmic reticulum (Wourms and Sheldon, 1976; Busson-Mabillot, 1977; Stehr and Hawkes. 1983), or in the Golgi apparatus (Thiaw and Mattei, 1991). In the case of *Epiplatys spilargyreus* zona pellucida is formed by the destruction of the follicle cells that release the elements that participate in forming its outer layer secondary envelope (Thiaw and Mattei, 1992). Further investigations are required to define more precisely the complementary roles of follicle cells in various fish species inhabiting variable water bodies.

Chemistry

Most previous studies have focused primarily on the morphological architecture of the structural components of zona pellucida (or chorion) as already discussed (see also reviews by Guraya, 1986; Riehl and Greven, 1990; Riehl, 1991). Zona material in elasmobranchs and teleosts is composed of mucopolysaccharides, glycoproteins, carbohydrate-protein matrix or protein and polysaccharide combinations, as revealed through histochemical techniques (Arndt, 1960a, b; Stahl and Leray, 1961; Guraya, 1965, 1978; Anderson, 1967; Nakano, 1969; Tesoriero, 1977b; Pelizaro et al., 1981; Lopes et al., 1982; Stehr, 1982; Guraya, 1986, 1994b). The chemical characteristics of fish egg zona pellucida differ between its individual layers or zones. In the flounder (*Liopsetta*) the zona pellucida consists mainly of polysaccharides in the early phases of its oogenesis but later the polysaccharide component decreases and the protein component predominates (Nakano, 1969). However, the zona pellucida in cyprinoids shows only protein throughout oogenesis, and the thin hyaline layer outside the zona pellucida consists of polysaccharides (Nakano, 1969). The inner layer of zona pellucida in teleosts, such as *Cichlosoma nigrofasciata* (Busson-Mabillot, 1977) and *Noemacheilus barbatulus* (Riehl, 1978b), is composed only of protein, whereas the outer layer (which is jelly-like in *Cichlosoma nigrofasciata*) shows a combination of proteins and polysaccharides. The entire zona pellucida in *Blennius pholis* is composed only of basic proteins and no polysaccharide material while the outer layer also reacts positively for S-H groups and the inner layer for S-S groups (Shackley and King, 1977). Both zones of zona pellucida in the C-O sole (Stehr, 1982) contain mucopolysaccharides. Relatively more mucopolysaccharide material is seen in the inner layers than in the hexagon walls. Riehl (1977) reported that the cortex radiatus internus (zona radiata) of *Noemacheilus barbatulus* and *Gobio gobio* is composed of proteins and neutral lipids; the cortex radiatus externus (zona pellucida proper) consists mainly of polysaccharides while acid mucopolysaccharides predominate in *G. gobio*. In both species, the cortex radiatus externus shows some protein. After coming into contact with water, the polysaccharides and acid mucopolysaccharides of the cortex radiatus externus become adhesive, thus facilitating attachment of the eggs to the substrate. In lambari *Astyanax bimaculatus*, the zona radiata (zona pellucida) consists of neutral polysaccharides and protein radical-a amino groups, cystine, arginine, cystein, tyrosine and tryptophan (Lopes et al., 1982).

Makeeva and Mikodina (1977) found that the zona pellucida (chorion) in different species of Cyprinidae showing differences in reproductive characteristics always reveals protein and mucopolysaccharides (both acid and neutral). As a rule, the acidic mucopolysaccharides are present along the outer edge of the zona pellucida or in the ends of villi. A thin membrane and a hyaline layer are always present with a villous zona pellucida. Tesoriero (1977b) used silver methenamine to demonstrate ultrastructural localization of polysaccharides in *Oryzias latipes* oocytes. He observed nonspecific deposition of silver grains and that the microvilli, attaching filaments, zona pellucida, and dense-cored vesicles were the most heavily labeled. He also used other more specific staining techniques which revealed polysaccharides on these structures.

In oviparous fish species, the eggs are shed in very diverse aquatic habitats; their zona pellucida has, therefore, attained a great structural and chemical diversity during evolution in order to adapt to variable physicochemical conditions of water. The acidic mucopolysaccharides of the zona pellucida (chorion) appear to help the eggs stick to the substratum in Cyprinidae (Makeeva and Mikodina, 1977). Further investigations are required to determine the comparative solubility of proportion, molecular organization, and composition of zona pellucida in various fish species from different water bodies to obtain an insight into the functional roles of zona molecules in the biology of the fish egg. Recent studies have increased our knowledge about the macromolecular composition of its protein and carbohydrate components by revealing the presence of glycosidically bound sugars among the polypeptide components of the egg zona pellucida (chorion) of some fish species, such as *Carassius auratus, Salmo gairdneri* (Cotelli et al., 1986, 1988), *Oncorhynchus mykiss* (Brivio et al., 1991), and *Chum salmon* (Kobayashi, 1982) as well as major structural proteins in cod (*Gadus morhua*) eggshells (Oppen-Berntsein et al., 1990). In all cases chorion components showed considerable heterogeneity on two-dimensional electrophoresis, which was attributed to the heterogeneous glycosylation level and possibly to some other postranslational modifications of the polypeptides (Brivio et al., 1991). These authors have identified and characterized trout chorion glycoproteins by analyzing mature forms of the molecules isolated from ovulated eggs. Their data support the evidence that the two mature forms of glycoconjugates showing an rm of 129 and 47 kD contain asparagine-linked (N-linked) oligosaccharides of a complex (or hybrid) type. The four major components (129, 62, 54 and 47 kD), representing about 80% of the total chorion, are devoid of O-linked oligosaccharides but 54 and 62 kD components appear to be insensitive to any enzymatic deglycosylation. The solubility properties of fish zona pellucida can be investigated from its sensitivity to both chemical and enzymatic dissolution as well as to pH (Kügel et al., 1990). Antigenic correlation has also been observed among the polypeptides of the chorion (Cotelli and Brivio, 1989). Further analysis of the immunological behavior as well as the amino acid sequencing in various fish species would provide more precise answers with respect to the suggested presence of blocks of homologies common to the main chorion components.

SURFACE STRUCTURES OF EGGS

The morphological characteristics of fish egg envelopes are highly adapted to the environmental conditions in which the embryo develops (Ivankov and Kurdyayeva, 1973; Guraya, 1978, Stehr and Hawkes, 1979; Laale, 1980; Brummett and Dumont, 1981). Laale (1980) and Riehl (1991) have provided critical reviews of earlier results about the great diversity of membrane adhesiveness and modifications in the eggs of various teleost species, which have also been described in other recent studies on the surface structure of pelagic and demersal eggs in various teleost species (Hart et al., 1984; Kobayakawa, 1985; Markle and Frost, 1985; Howe, 1987; Mikodina, 1987; Lönning et al., 1988; Howe et al., 1988; Mooi, 1990; Mooi et al., 1990). Demersal eggs, commonly subjected to abrasive forces, generally form thick envelopes with complex lamellae (Hurley and Fisher, 1966; Flügel, 1967a; Ivankov and Kurdyayeva, 1973; Osanai, 1977; Manner et al., 1977; Stehr and Hawkes, 1979, 1983). Such eggs may also develop a complex surface structure that carries out specialized functions. For example, the eggs of various teleost species develop some type of adhesive device for attachment to the substratum. Fibrils on the surface of *Fundulus heteroclitus* eggs (Tsukahara, 1971; Dumont and Brummett, 1980; Brummett and Dumont, 1981) help them adhere to each other and to the substrate (Anderson, 1974) as well as to retain moisture when they become exposed at low tide (Kuchnow and Scott, 1977). Hollow spines develop on the surface of *Cynolebias melanotaenia* eggs (Wourms and Sheldon, 1976; Wourms, 1976; Sponaugle, 1980), which may also provide them partial protection upon exposure to air during the dry season (Wourms, 1976). In contrast to demersal eggs, pelagic eggs with a thin zona pellucida generally have a smooth surface. Their thin zona pellucida is made up of horizontal lamellae (Ginsburg, 1968; Hagstrom and Lönning, 1968; Lönning and Solemdal, 1972; Ivankov and Kurdyayeva, 1973; Lönning and Hagstrom, 1975; Stehr and Hawkes, 1979, 1983; Guraya, 1986; Riehl, 1991). The pelagic eggs of a few teleost species develop complex structures, such as regularly spaced bristles, protuberances, or raised hexagonal patterns (Ahlstrom and Moser, 1980), but we know little about their origin during oogenesis and function in the mature egg (Robertson, 1981; Stehr, 1982). As already mentioned, the formation of teleost egg zona pellucida has been extensively investigated for fish with smooth egg surfaces. But the formation of complex surface structures of teleost egg zona pellucida during oogenesis has been investigated in only a few teleost species with demersal eggs, which include *Cynolebias melanotaenia* (Wourms, 1976; Wourms and Sheldon, 1976; Sponaugle, 1980), *Cichlasoma nigrofasciata* (Busson-Mabillot, 1977) and *Fundulus heteroclitus* (Kemp and Allen, 1956; Anderson, 1966). Information about the fine morphology of the zona pellucida of fish eggs during and after fertilization is meager (Hart, 1990; Riehl, 1991). However, Perry

(1984), using scanning electron microscopy did demonstrate that the unfertilized eggs of the winter flounder *Pseudopleuronectes americanus* has a cross pattern of depressions which radiate in all directions across the zone surface giving them a wrinkled look. After fertilization, the zona pellucida surface attains a regular smoother appearance. The pores of the unfertilized eggs first become flush with the zona surface, then thicken and rise above it after fertilization. The zona pellucida of the fertilized eggs takes on a granular appearance in contrast to the smooth and uniformly textured zona pellucida of unfertilized eggs.

The development, structure, chemical nature, and function of adhesive devices during fish oogenesis were studied in the last decade (Patzner, 1984; Groat and Alderdice, 1985; Hart et al., 1984; Mikodina, 1987). These authors as well as others (Markle and Frost, 1985; Howe, 1987; Lönning et al., 1988; Howe et al., 1988; Riehl and Kock, 1989; Riehl and Ekau, 1990; Mooi, 1990; Mooi et al., 1990; Riehl, 1991) have discussed the problems of egg surface morphology, development, and evolution in various teleost fish species inhabiting different niches of fresh and marine water bodies. The egg surface structures show great adaptations to the physicochemical characteristics of the water and substratum as supported by comparative observations on pelagic and demersal eggs (review by Riehl, 1991). Rubstov (1978) investigated the structural features of the egg envelope in *Cyprinus carpio* and *Clupea harengus* in the zone of mucilage attachment of substrate (vegetation). The envelopes of the two fish show differences according to individual structures in spite of the fact that the features of spawning grounds, substrates for egg laying, and adaptive characteristics of their embryogenesis are very similar. Busson-Mabillot (1977) reported the development, structure, chemistry, and function of the adhesive apparatus of the teleost *Cichlasoa nigrofasciata* egg, whereby they are attached to the substratum after laying. The adhesive apparatus is composed of two distinct elements; one, filaments connected to the zona pellucida surface and the other, a mucous jelly coat which forms the outer envelope of the zona pellucida. Follicle cells successively produce these two elements during the course of vitellogenesis. The filaments constitute bundles of protein tubules about 20 mm in diameter. At the beginning of vitellogenesis, the tubular proteins produced by the rough endoplasmic reticulum are released directly into the extracellular space where they polymerize to form continuous tubules. Later, the glycoproteins of the jelly coat are formed and accumulate as voluminous granules in the rough endoplasmic reticulum of the follicle cells. Release of these glycoproteins occurs during ovulation and proceeds as a type of apocrine secretion. Abraham et al. (1991) have demonstrated the presence of microfollicle cells of the jelly coat in the oocyte envelope of the sheatfish (*Silurus glanis* L.). Anderson (1966) also correlated development of fibrilla on *Fundulus* eggs to an increase in microtubules, mitochondria, rough endoplasmic reticulum, ribosomes, and Golgi complexes in the follicle cells. Kemp and Allen (1956) found extracellular strands of filaments among *Fundulus* follicle cells and concluded that the fibrils developed from the filaments and intercellular matrix of the follicle cell processes. The attaching filaments in the oocytes of *Cyprinodon variegatus* and *Oryzias latipes* are of two kinds and of different thickness. In cross section they seem to consist of tubular units.

Up to 220 attaching filaments develop in stage II of oocyte growth in the goby *Pomatoschistus minutus* and are connected with the egg envelope at the animal pole

(Riehl, 1978a). These filaments constitute a modified part of the cortex radiatus externus (zona pellucida proper) and are composed of two substances that differ in electron density. The attaching filaments are situated between the follicle cells and, possibly, are produced by these cells. They help in attaching eggs to the substrate. Scanning electron microscope investigation of eggs in Acipenseridae has demonstrated a sticky surface on their gelatinous layer which can perform special movements enabling attachment to a substrate (Markov, 1978). In the eggs of isaza (*Chaenogobius isaza*) the adherent filaments are formed during the tertiary yolk globule stage between the zona pellucida and follicular epithelium (Takahashi, 1978). The filaments have a reticular structure around the micropyle, running almost parallel to the meridional lines on the wall of the egg surface. After the egg was released in water, it stuck to the wall of the vessel; meanwhile the adherent filaments detached from the egg envelope and turned inside out except around the micropyle. The stalk of the egg was thus formed. The structure and origin of adherent filaments are essentially the same in other species of gobiid fish. The eggs of osmerids and salangids, though only distantly related to gobiids, also develop adherent envelopes (filaments) of similar structure. The complex zona pellucida of *Cynolebias melanotaenia* is covered with uniformly spaced, hollow, conical projections which terminate as a crown of recurved spikes (Wourms, 1976). These constitute equilateral triangles originating from a pentagonal base, which is part of a surface raised in a pattern of pentagon and hexagon interconnected ribs. Small vesicles containing 25-nm tubules appear within the rough endoplasmic reticulum and are then transported to the exterior of the cell and assembled into the ornamentation of the zona pellucida (Wourms and Sheldon, 1976; Sponaugle, 1980). The C-O sole (*Pleuronichthys coenosus*) is one of five species within the genus *Pleuronichthys* that produce eggs with raised hexagonal surface structures (Sumida et al., 1979). The species in this genus are unusual in that they are among the very few teleosts in Pacific Northwest waters that produce pelagic eggs with ornate surface structures. Ultrastructural investigations of C-O sole eggs have shown that ornate hexagonal surface structures are produced from glycoproteins secreted by the follicle cells (Stehr, 1982; Stehr and Hawkes, 1983). They first develop as a thin osmiophilic layer evenly distributed on the surface of the young oocyte, averaging 0.12 nm in diameter. The vertical walls or struts of the hexagons begin as a discrete thickening in the osmiophilic layer that forms stout extensions which lengthen and thin to a final height averaging 0.09 nm. The oocyte surface is enveloped by closely placed follicle cells and the walls of the hexagons develop in the space between these cells. The hexagonally patterned zona pellucida (chorion) detailed by Stehr and Hawkes (1983) possibly provides the egg with additional protection, resiliency, and buoyancy during embryogenesis. The functional meaning of pelagic egg surface extensions including their possible role in gas exchange remains to be determined more precisely in future investigations.

Ovulation in most fishes breaks the relationship of the developing ovum and its follicular attachment (Guraya, 1986). In various fish species the contribution of the follicular epithelium to the external egg layer is a secretion from the follicle cells as already described. However, Shelton (1978) reported the development and transformation of the follicular epithelium into a functional egg envelope has been reported in *Dorosoma petenense*. The highly adhesive egg capsule of *D. petense* develops from the transformed ovarian follicular epithelium, which in growing oocytes develops from a loosely organized layer of cuboidal cells into a simple columnar cell layer, then into a granular layer, and finally into a layer with a radially striated structure. This changed membrane is not ruptured at ovulation but persists throughout incubation and is the source of tenacious egg adhesiveness.

MICROPYLE(S)

The notable developmental, morphological, and chemical complexity of the zona pellucida or chorion seem to be responsible for the formation of micropyles during oogenesis in the eggs of various teleost species (Ginsburg, 1968; Rastogi, 1970; Riehl and Götting, 1974, 1975; Szöllösi and Billard, 1974; Kuchnow and Scott, 1977; Riehl, 1978c, 1979, 1980b; Riehl and Schulte, 1977b, 1978; Stehr and Hawkes, 1979; Brummett and Dumont, 1979; Hosokawa, 1979; Dumont and Brummett, 1980; Kudo, 1980; Laale, 1980; Hosokawa et al., 1981; Iwamatsu and Ohta, 1981; Hart and Donovan, 1983; Guraya, 1986, 1994b; Kobayashi and Yamamoto, 1981, 1987; Takano and Ohta, 1982; Hirai, 1988; Nakashima and Iwamatsu, 1989; Pithawalla et al., 1993). Various workers have also reviewed the results of light and electron microscopic studies on the development, structure and function of micropyles in various fish species (Laale, 1980; Hosaja and Luczynski, 1984; Guraya, 1986; Riehl, 1991). The structure and development of micropyles, which characterize the animal pole of the oocyte, vary greatly among teleosts. The ultrastructure of zona pellucida surrounding the distal and proximal micropyle openings, the size of the openings, and the shape of the canal differ greatly between species. Micropyle development is related to the formation of the surrounding layers of the follicular wall. Riehl (1977) reported that the micropyle of *Noemacheilus barbatulus* appears in late stage I and its further development is completed in the late stage II of oocyte growth. The micropyles of *N. barbatulus* and *Gobio gobio* are developed by a modified follicle cell which appears at the late perinucleolus stage in isaza, *Chaenogobius isaza* (Takahashi, 1981). This modified follicle cell forms a thin apophysis, which leaves open the micropyle canal. The micropyle of *N. barbatulus* is developed by the cortex radiatus externus (zona radiata) while the micropyle of *G. gobio* is produced only by the cortex radiatus internus. Osanai (1977) investigated the envelopes surrounding the chum salmon (*Oncorhynchus keto*) oocytes in the preovulatory and postovulatory stages. The follicle cells enclosing the zona pellucida of ovarian oocytes get dispersed at ovulation. Then micropyles can be seen in the ovulated ripe eggs. Sobhana and Nair (1977) reported some observations on the origin and fate of the micropyle during oogenesis of the olive carp (*Puntius sarana*). In the oocytes of *N. triangularis*, the micropyle is developed by the coalescence of three enlarged follicle cells (Kumari and Nair, 1979). The micropyle canal in the eggs of starry flounder (*Psettodes stellatus*) measures 8 μm at the opening and tapers to 3.6 μm as it traverses the membrane (Stehr and Hawkes, 1979). The 15–16-μm micropyle opening in the pink salmon (*Oncorhynchus gorbuscha*) egg is developed by an area of protrusion and the funnel-shaped canal tapers to 3 μm at its terminal structure. Numerous micropyles (average 7) are confined to a 100–200-μm zone at the animal pole and traverse the zona pellucida of eggs in the white sturgeon *Acipenser transmontanus* (Cherr and Clark, 1982). Their outer opening measures 15 μm in diameter. The micropylar canal tapers twice, ultimately ending at the oolemma with an inner opening diameter of 1.2 μm. The micropyles of the white sturgeon egg show more morphological complexity than the micropyles in other fish eggs. The micropylar apparatus in the egg of *Brachydanio rerio* is composed of a cone-shaped vestibule and a tapered canal penetrating the zona pellucida (Hart and Donovan, 1983). The outer diameter of the canal measures 7.5–8.5 μm and the inner diameter 2.3 μm. The inner micropylar aperture forms a circular cluster of 10–20 microvilli-like projections. No cortical granules are seen in the cytoplasm directly below the sperm entry site.

Riehl (1978b) investigated the problems of development, structure, and function of the micropyle in the oocytes of the goby *Pomatoschistus minutus*. In its egg the micropyle is placed at the animal pole between the points of departure of attaching filaments and composed of a channel perforating the cortex radiatus (zona pellucida). The micropyle pit is absent and thus the micropyle of *P. minutus* belongs to type 3 of micropyles reported by Riehl and Götting (1974, 1975) and Riehl (1980b). This micropyle, the smallest among the teleost eggs investigated to date, is produced by a modified cell of the follicular epithelium, the so-called plug cell, which differs from other follicle cells by its shape and size. The eggs of *P. minutus* adhere to the substrate by the micropyle hole. Problems of insemination due to this fact are discussed. The development and structure of micropyles have also been reported in pelagic eggs of some marine fishes (Hirai, 1988), chum salmon *Oncorhynchus keto* (Kobayashi and Yamamoto, 1981, 1987) and medaka *Oryzias latipes* (Nakashima and Iwamatsu, 1989).

The micropyle is occupied by the highly specialized micropylar cell until matured oocytes are released from their surrounding egg envelopes at the time of ovulation. The large triangular micropylar cells contain granular ER, mitochondria, small vesicles, Golgi complex, filaments, and microtubules, which have a specific localization within the cytoplasm (Ohta and Teranishi, 1982; Nagahama, 1983). The apical cytoplasm shows numerous microtubules placed parallel to the long axis of the micropylar cell. The cisternae of granular ER with amorphous substance are visible in the basal cytoplasm as is their secretory function. The micropylar cells reveal no ultrastructural features indicative of steroidogenesis in the ovary of loach *Misgurnus anguillicaudatus* but these cells react intensely to 3β-HSDH at the yolk globule stage (Ohata and Teranishi, 1982). This intense reaction of the micropylar cells persists throughout the year, in sharp contrast to the follicle cells in which enzyme activity disappears during the postspawning and sexually quiescent period. Pithawalla et al. (1993) investigated the structural and metabolic events at the micropylar region during the late stages of oogenesis in the hagfish.

Riehl (1980b) used SEM to investigate the micropyle of salmonids and four coregonids and suggested that the micropyles and the surface pattern of fish eggs form important specific criteria for identification of various teleosts (Riehl, 1979, 1991; Riehl and Schulte, 1978). The micropyles reported belong to two types: (1) micropyle with flat pit and long canal (type II) and (2) micropyle with canal but without pit (type III). Riehl and Götting (1974) described the third type with a hollow pit leading into a short canal. The sides of the micropyle canal are generally reinforced by annular and helicoid thickenings (Szöllösi and Billard, 1974; Riehl, 1979, 1991; Stehr and Hawkes, 1979). These concentric ridges within the micropyle canal appear to correspond to edges of the horizontal lamellae of the zona pellucida. SEM investigations of the funnel-shaped micropyle in mature eggs have demonstrated that the wider distal opening enters a canal which tapers to an aperture just large enough to allow entry of a single sperm (Kuchnow and Scott, 1977; Stehr and Hawkes, 1979; Ohta and Teranishi, 1982). Since the diameter of the inner micropylar aperture in the egg of *Brachydanio rerio* is slightly larger than the size of its sperm head, the block to polyspermy is believed to be mechanical and guaranteed by the morphological design of the micropyle (Hart and Donovan, 1983). Thus, micropyles of teleost eggs facilitate fertilization (Ginsburg, 1968, 1987; Jaffe and Gould, 1985; Hart and Donovan, 1983; Kobayashi and Yamamoto, 1987; Hart, 1990; Amanze and Iyengar, 1990; Riehl, 1991). The micropyle is believed to form a sperm guidance system in teleost fertilization (Amanze and Iyengar, 1990). But in *Esox lucius* the eggs are fertilized even when the micropyle is blocked (Riehl and Götting, 1975). Riehl and Schulte (1977b) reported that ripe eggs can be fertilized after their micropyles are clogged, indicating that the egg envelope can be traversed by sperm at other sites. Riehl (1991) has detailed the role of the teleost micropyle.

SUMMARY

The functional morphology of the follicular wall (composed of follicular epithelium, basal lamina, theca and zona pellucida or chorion), egg surface components, and micropyle in the fish ovary has been reviewed and discussed in detail by correlating the results of electron microscopic, histochemical, and biochemical studies carried out during the twentieth century. The great diversity in development, structure, and function in various fish species is revealed. The follicular epithelium synthesizes mainly proteins, as evidenced by the presence of numerous ribosomes, elements of granular endoplasmic reticulum, well-developed Golgi complex, coated vesicles, and dense bodies. The possible synthesis of lipids and steroid hormone (estrogen) also occurs. The isolated special thecal cells lying close to capillaries in the thecal layer synthesize estrogen precursors which are converted to 17β-estradiol in the follicular epithelium, as shown by cytological and histochemical features of well-established steroid gland cells. Changes in morphological interrelationships between the follicle cell processes and oocyte microvilli during oocyte growth reveal great diversity among fish species. The possible mechanisms of transport of substances across their cellular membranes have been described. The zona pellucida or chorion, consisting mainly of carbohydrates and proteins, lies between the oocyte surface and follicular epithelium and shows a great diversity in structure and chemistry in various groups of fishes. Its material is produced mainly by the oocyte itself but the contribution of some materials by the follicle cells is also indicated as the detailed ultrastructure and chemistry of the different zones of the zona pellucida vary greatly among fish species. The surface structures of eggs comprise adhesive filaments, hollow spines or regularly spaced bristles and protuberances, or raised hexagonal patterns which show many variations in their development, structure, and function in pelagic and demersal eggs of various fish species. The shape, structure, and origin of micropyles, which play an important role in fertilization of teleosts, show species variation. The surface pattern of eggs and the micropyles constitute important characteristic criteria for the identification of teleosts.

Acknowledgment

Prepared and supported under the program of CSIR-Emeritus Scientist Scheme.

References

Abraham, M., V. Hilge, S. Lison, and Tibika. 1984. The cellular envelope of oocytes in teleosts. *Cell Tissue Res.* 235:403–410.

Abraham, M., V. Hilge, R. Riehl, and V. Iyer. 1991. The muco-follicle cells of the jelly coat in the oocyte envelope of the sheatfish (*Silurus glanis* L.). *Cell Tissue Res.* 266:231–236.

Ahlstrom, E.H. and H.B. Moser. 1980. Characters useful in identification of pelagic marine fish eggs. *Calcofi Rep.* 21:121–131.

Amanze, D. and A. Iyengar. 1990. The micropyle: a sperm guidance system in teleost fertilization. *Development* 109:495–500.

Anderson, E. 1966. A study of the fibrillar appendages associated with the surface of eggs of the killifish, *Fundulus heteroclitus*. *Anat. Rec.* 154:308–309.

Anderson, E. 1967. The formation of the primary envelope oocyte differentiation in teleosts. *J. Cell Biol.* 35:193–212.

Anderson, E. 1974. Comparative aspects of the ultrastructure of the female gamete. *Int. Rev. Cytol.* 4: suppl., pp. 1–70.

Arndt, E.A. 1960a. Die Aufgaben des Kerns während der Oogenese der Teleosteer. *Z. Zellforsch.* 51:356–378.

Arndt, E.A. 1960b. Untersuchungen über die Eihüllen von Cypriniden. *Z. Zellforsch.* 52:315–327.

Azevedo, C. 1974. Evolution des envelopes ovocytaires, au cours de l'ovogenese, chez un teleosteen vivipare, *Xiphophorus helleri*. *J. Microscopie*. 21:43–54.

Bara, G. 1965. Histochemical localization of 3β-hydroxysteroid dehydrogenase in the ovaries of a teleost fish, *Scomber scomber* L. *Gen. Compar. Endocr.* 5:284–296.

Bara, G. 1974. Location of steroid hormone production in the ovary of *Trachurus mediterraneus*. *Acta Histochem.* 51:90–101.

Begovac, P.C. and R.A. Wallace. 1988. Stages of oocyte development in the pipe fish, *Syngnathus scovelli*. *J. Morphol.* 197:353–369.

Begovac, P.C. and R.A. Wallace. 1989. Major vitelline envelope proteins in pipe fish oocytes originate within the follicle and are associated with Z_3 layer. *J. Exp. Zool.* 251:56–73.

Bhattacharya, S. 1992. Endocrine control of fish reproduction. *Current Sci.* 63:135–139.

Bhattacharya, S., S. Halder, and P.R. Manna. 1994. Current status of endocrine aspects of fish reproduction—an overview. *Proc. Indian Nat'l. Sci. Acad.* B60:33–44.

Brivio, M.F., R. Bassi, and F. Cotelli. 1991. Identification and characterization of the major components of the *Oncorhynchus mykiss* egg chorion. *Mol. Reprod. Develop.* 23:85–93.

Brummett, A.R. and J.N. Dumont. 1979. Initial stages of sperm penetration into the egg of *Fundulus heteroclitus*. *J. Exp. Zool.* 210:417–434.

Brummett, A.R. and J.N. Dumont. 1981. A comparison of chorions from eggs of northern and southern populations of *Fundulus heteroclitus*. *Copeia* 1981:602–614.

Bruslé, S. 1980a. Fine structure of early previtellogenic oocytes in *Mugil (Liza) auratus*, Risso. 1810 (Teleostei: Mugilidae). *Cell Tiss. Res.* 207:123–134.

Bruslé, S. 1980b. Fine structure of the oocytes and their envelopes in *Chelon labrosus* and *Liza aurata* (Teleostei, Mugilidae). *Zool. Sci.* 2:681–693.

Busson-Mabillot, S. 1973. Evolution des enveloppes de l'ovocyte et de l'oeuf chez un poisson teleosteen. *J. Microscopie* 18:23–44.

Busson-Mabillot, S. 1977. Un type particulier de sécretion-étocribe: celui de l'appareil adhésif de l'oeuf d'un poisson téléostéen. *Biol. Cell.* 30:233–244.

Caloianu-Iordachel, M. 1971a. Cytochemical and ultrastructural data on the cytoplasm of the young oocytes in the sturgeon *Huso huso* L. *Rev. Roum. Biol. Ser. Zool.* 16:165–169.

Caloianu-Iordachel, M. 1971b. Ovogenesis in acipenserid fish. The morphogenesis and histochemical composition of the external membranes. *Rev. Roum. Biol. Ser. Zool.* 16:113–120.

Chaudhry, H.S. 1956. The origin and structure of the zona pellucida in the ovarian eggs of teleosts. *Z. Zellforsch.* 43:478–485.

Cherr, G.N. and W.H. Clark, Jr. 1982. Fine structure of the envelope and micropyles in the eggs of the white sturgeon, *Acipenser transmontanus*. *Develop. Growth Different.* 24:341–352.

Chieffi, G. 1961. La luteogenesi nei salacei ovovivipari: Ricerche istologiche e istochimiche in *Torpedo marmorata* e *Torpedo ocellata*. *Publ. Staz. Zool., Napoli* 32:145.

Chinareva, I.D. 1973. Changes in the follicular epithelium and formation of egg membranes in oocytes of the last year of development in *Coregonus peled* Arkh. Anat. Gistol. Embriol. 64:79-87 (in Russian).

Chinareva, I.D. 1975. Formation of a pectinate sheet of the *Coregonus peled* oocyte egg membrane (based on electron microscopic data). *Vest Leningr. Univ. Biol.* 1:28-34 (in Russian).

Chinareva, I.D. and E.B. Kirchinskaja. 1975. Electron microscopic study of formation of egg membranes of oocytes in the last year of development in peleds. *Arkh. Anat. Gistol. Embriol.* 68:94-99 (in Russian).

Christensen, A.K. 1975. Leydig cells. In R.D. Greep and E.B. Astwood (eds.), *Handbook of Physiology*, vol. 5, Endocrinology, pp. 57-94. Amer. Physiolo. Soc., Washington.

Christensen, A.K. and S.W. Gillim. 1969. The correlation of fine structure and function in steroid secreting cells with emphasis on those of the gonads. In K.W. McKerns (ed.), *The Gonads*, pp. 415-488. Amsterdam, North Holland.

Colombo, L., P.C. Belvedere, K. Bieniarz, and P. Epter. 1982. Steroid hormone biosynthesis by ovary of the carp *Cyprinus carpio* during wintering. *Compar. Biochem. Physiol.* 72B:367-376.

Cotelli, F. and M. Brivio. 1989. Biochemical analysis of fish egg chorion components. Paper presented in 19th FEBS Meeting.

Cotelli, F., F. Androntico, M. Brivio, and C.L. Lamia. 1988. Structure and competition of fish egg chorion (*Carassius auratus*). *J. Ultrastructure. Mol. Struct. Res.* 99:70-78.

Cotelli, F., F. Androntico, R. Bassi, M. Brivio, C. Cecagno, S. Denis Donini, M.L. La Rosa, and C. Lara Lamina. 1986. Studies on the composition, structure and differentiation of fish egg chorion. *Cell Biol. Int. Rep.* 10:471.

Cruz-Landin, C. and M.A. Cruz-Höfling. 1989. Electron microscopic studies on the development of the chorion of *Astyanax bimaculatus* (Teleost. Characidae). *Zool. Jb. Anat.* 119:241-249.

Donato, A., A. Contini, A. Maugeri, and S. Fasulo. 1980. Structural and ultrastructural aspects of the growing oocytes of *Chromis chromis* (Teleostei, Labridae). *Riv. Biol. Norm. Patol.* 6:31-66.

Droller, M.J. and T.F. Roth. 1966. An electron microscopic study of yolk formation during oogenesis in *Lebistes reticulatus* Guppy L. *J. Cell Biol.* 28:209-232.

Dumont, J.N. and A.R. Brummett. 1980. The vitelline envelope, chorion and micropyle of *Fundulus heteroclitus* eggs. *Gamete Res.* 3:24-44.

Emel'Yanova, N.G. 1979. Electron microscopic study of the development of egg membranes and follicle of the silver carp *Hypophthalmichthys molitrix*. *Vopr. Ikthiol.* 19:302-312.

Erhardt, H. 1976. Licht- und elektronenmikroskopische Untersuchungen an den Eihüllen des marinen Teleosteers *Lutjanus sgnagris*. *Helgolander Wiss. Meeresunters* 28:90-105.

Erhardt, H. 1978. Elektronenmikroskopische Untersuchungen an den Eihüllen von *Lutjanus analis* (Cuvier & Valenciennes, 1828) (Lutjanidae: Perciformis, Pisces). *Biol. Zbl.* 97:181-187.

Erhardt, H. and M.J. Götting. 1970. Light and electron microscopic studies in egg cells and envelopes of *Platypoecilus maculatus*. *Cytobiologie* 2:249-440 (in German).

Fisher, K.C. 1963. The formation and properties of the external membrane of the trout egg. *Trans. R. Soc. Can. J.* 1:323-332.

Flegler, C. 1977. Electron microscopic studies on the development of the chorion of the viviparous teleost *Dermogenys pusillus* (Hemirhamphidae). *Cell Tiss Res.* 179:155-279.

Flügel, H. 1964. Electron microscopic investigations on the fine structure of the follicular cells and the zona radiata of trout oocytes during and after ovulation. *Naturwissenschaften* 51:564-565.

Flügel, H. 1967a. Licht- und elektronenmikroskopische Untersuchungen an Oozyten und Eiern einiger Knochenfische. *Z. Zellforsch.* 83:82-116.

Flügel, H. 1967b. Elektronenmikroskopische Untersuchungen an den Hüllen der Oozyten und Eiern des Flussbarsches *Perca fluviatilis*. *Z. Zellforsch.* 77:245-256.

Flügel, H. 1970. Zur funktion der Golgi-apparatas in den Follikelzellen von *Gorgonus ulbala* L. *Cytobiologie* 4:450-459.

Fostier, A., B. Jalabert, R. Bilard, B. Breton, and Y. Zona. 1983. The gonadal steroids. In W.S. Hoar, D.J. Randall, and E.M. Donaldson (eds.), *Fish Physiology*, vol. IX. Reproduction, pt. A, pp. 227-372. Academic Press, New York.

Gabaeva, N.S. 1977. Characteristics of the follicular epithelium in the fishes with demersal roe. Classification of unilaminate monomorphous follicular epithelium of vertebrates. *Arkh. Anat. Gistol. Embriol.* 73:48-55 (in Russian).

Gabaeva, N.S. and N.O. Ermolina. 1972. On changes in the follicular epithelium of ovum membranes during oogenesis of the sheatfish *Arius thalassinus*. *Arkh. Anat. Gistol. Embriol.* 63:97-106 (in Russian).

Gabaeva, N.S. and E.K. Boglazova. 1977. The morphodynamics and proliferative activity of the follicular epithelium in the oogenesis of the sword-tailed minnows *Xiphophorus helleri. Biol. Nauk. Moscow* 20:59–65.
Ginsburg, A.S. 1968. *Fertilization in Fishes and the Problem of Polyspermy*. Akad. Nauk. SSSR (in Russian).
Ginsburg, B.S. 1987. Egg cortical reaction during fertilization and its role in block to polyspermy. *Sov. Sci. Rev. F. Physiol. Gen. Biol.* 1:307–375.
Götting, K.J. 1965. Die Feinstruktur der Hullschichten reifender Oocyten von. *Agonus cataphractus* L. (Teleostei: Asonidae). *Z. Zellforsch.* 66:405–414.
Götting, K.J. 1967. Der Follikel und die peripheren Strukturen der Oocyten der Telecosteer and Amphibien. *Z. Zellforsch.* 79:481–491.
Götting, K.J. 1970. Zur Darstellung der Ultrastruktur des Teleosteer-Follikels mittels der Gefrierätztechnik. *Micron* 1:356–372.
Götting, K.J. 1974. Oocyte ultrastructure of oviparous and ovoviviparous teleosts as revealed by freeze etching. *Int. Congr. Electron. Microsc. Canberra* 8:668–669.
Götting, K.J. 1976. Fortpflanzung und Oocyten-Entwicklung bei der Aalmuter (*Zoarces viviparus*) (Pisces, Osteichthyes). *Heigoländer Wiss Meeresuniers* 28:71–89.
Greenwald, G.S. and S.K. Roy. 1994. Follicular development and its control. In E. Knobil and J.D. Neil (eds.), *The Physiology of Reproduction*, 2nd ed., pp. 629–724. Raven Press, New York.
Groat, E.P. and D.F. Alderdica. 1985. Fine structure of the external egg membrane of five species of pacific salmon and steelhead trout. *Can. J. Zool.* 63:352–366.
Guraya, S.S. 1965. A comparative histochemical study of fish (*Channa marulius*) and amphibian (*Bufo stomaticus*) oogenesis. *Z. Zellforsch.* 65:662–700.
Guraya, S.S. 1971. Morphology, histochemistry and biochemistry of human ovarian compartments and steroid hormone synthesis. *Physiol. Rev.* 5:785–807.
Guraya, S.S. 1974. Gonadotropins and functions of granulosal and thecal cells *in vivo* and *in vitro*. In N.R. Moudgal (ed.), *Gonadotropins and Gonadal Function*, pp. 220–236. Academic Press, New York.
Guraya, S.S. 1976. Recent advances in the morphology, histochemistry and biochemistry of steroid-synthesizing cellular sites in the nonmammalian vertebrate ovary. *Int. Rev. Cytol.* 44:365–409.
Guraya, S.S. 1978. Maturation of the follicular wall of nonmammalian vertebrates. In R. Jones (ed.), *The Vertebrate Ovary*, pp. 261–329. Plenum Press, New York.
Guraya, S.S. 1979. Recent advances in the morphology and histochemistry of steroid synthesizing cellular sites in the gonads of fish. *Proc. Indian Nat'l. Sci. Acad.* 45B:452–461.
Guraya, S.S. 1982. Recent progress in the structure, origin, composition and function of cortical granules in animal egg. *Int. Rev. Cytol.* 78:257–360.
Guraya, S.S. 1985. *Biology of Ovarian Follicles in Mammals*. Springer, Heidelberg.
Guraya, S.S. 1986. The cell and molecular biology of fish oogenesis. *Mongr. Devel. Biol.* 18:1–229. Karger, Basel.
Guraya, S.S. 1994a. Nucleolar extrusions and their functional significance in fish oogenesis. In H.R. Singh (ed.), *Advances in Fish Biology*, pp. 155–166. Hindustan Pub. Corp., Delhi.
Guraya, S.S. 1994b. Gonadal development and production of gametes in fish. *Prod. Indian Nat'l. Sci. Acad.* B. 60:15–32.
Guraya, S.S. and S. Kaur. 1979. Morphology of the postovulatory follicle (or corpus luteum) of teleost (*Cyprinus carpio* L.) ovary. *Zool. Beitr.* 25:381–390.
Guraya, S.S. and S. Kaur. 1982. Cellular sites of steroid synthesis in the oviparous teleost fish (*Cyprinus carpio* L.): a histochemical study. *Proc. Indian Acad. Sci.* (*Anim. Sci.*) 91:587–597.
Guraya, S.S., S. Kaur, and P.K. Saxena. 1975. Morphology of ovarian changes during reproductive cycle of fish, *Mystus tengara* (Ham). *Acta Anat.* 91:222–260.
Guraya, S.S., H.S. Toor, and S. Kumar. 1977. Morphology of ovarian changes during the reproductive cycle of the *Cyprinus carpio communis* (Linn.). *Zool. Beitr.* 23:405–437.
Hagstrom, B.A. and S. Lönning. 1968. Electron microscopic studies of unfertilized and fertilized eggs from marine teleosts. *Sarasia* 33:73–80.
Hamazaki, T.S., Y. Nagahama, I. Iuchi, and K. Yamagami. 1989. A glycoprotein from the liver constitutes the inner layer of the egg envelope (zona pellucida interna) of the fish, *Oryzias latipes*. *Dev. Biol.* 133:101–110.
Hart, N.H. 1990. Fertilization in teleost fishes: mechanisms of sperm-egg interactions. *Int. Rev. Cytol.* 181:1–66.
Hart, N.H. and M. Donovan. 1983. Fine structure of the chorion and site of sperm entry in the egg of *Brachydanio rerio*. *J. Exp. Zool.* 227:277–296.

Hart, N.H., R. Pietri, and M. Donovan. 1984. The structure of the chorion and associated surface filaments in *Oryzias*—evidence for the presence of extracellular tubules. *J. Exp. Zool.* 230:273–296.

Hirai, A. 1988. Fine structure of the micropyles of pelagic eggs of some marine fishes. *Jap. J. Ichthyol.* 35:351–357.

Hirose, K. 1972. The ultrastructure of the ovarian follicle of medaka, *Oryzias latipes*. *Z. Zellforsch.* 123:316–329.

Hirose, K., Y. Machida, and E.M. Donaldson. 1976. Induction of ovulation in the Japanese flounder (*Limanda yokohamae*) with human chorionic gonadotropin and salmon gonadotropin. *Bull. Jap. Soc. Sci. Fish.* 42:13–20.

Hirose, K., S. Adachi, and Y. Nagahama. 1985. Changes in plasma steroid hormone levels during sexual maturation in the ayu *Plecoglossus altivelis*. *Bull. Jap. Soc. Sci. Fish.* 51:399–403.

Hoar, W.S. and Y. Nagahama. 1978. The cellular sources of sex steroids in teleost gonads. *Ann. Biol. Anim. Biochim. Biophys.* 18:893–898.

Hosaja, M. and M. Luczynski. 1984. Micropyle in three Coregoniae species (Teleostei). *Z. Angew. Zool.* 71:21–27.

Hosokawa, K. 1979. Scanning electron microscopic observations of the micropyle in *Oryzias latipes*. *Jap. J. Ichthyol.* 26:94–99.

Hosokawa, K., T. Fusimi, and T. Matsusato. 1981. Electron microscopic observation of the chorion and micropyle apparatus of the porgy, *Fagrus major*. *Jap. J. Ichthyol.* 27:339–343.

Howe, E. 1987. Breeding behaviour, egg surface morphology and embryonic development in four Australian species of the genus *Pseudomugil* (Pisces, Melanotaenidae). *Austr. J. Mar. Freshwat. Res.* 38:385–395.

Howe, E., E. Howe, and S. Doyle. 1988. The structure of the egg in blueyes pseudomugils sp. *J. Austr. New Guinea Fish. Assn.* 5:205–211.

Hurley, D.A. and K.C. Fisher. 1966. The structure and development of the external membrane in young eggs of the brook trout, *Salvelinus fontinalis* (Mitcheill). *Can. J. Zool.* 44:173–189.

Idler, D.R., L.W. Crim, and J.M. Walsh (eds.). 1987. *Reproductive Physiology of Fish*. Marine Sciences Research Laboratory, St. Johns, Newfoundland.

Ishii, S. 1991. Gonadotropins. In P.K.T. Pang and M.P. Schreibman (eds.), *Vertebrate Endocrinology: Fundamentals and Biomedical Implications*, vol. 4, pt. B, *Reproduction*, pp. 33–66. Academic Press, San Diego.

Ivankov, V.N. and V. Kurdyayeva. 1973. Systematic differences and the ecological importance of the membranes in the fish eggs. *J. Ichthyol.* 13:864–873.

Iwamatsu, T. and T. Ohta. 1981. On a relationship between oocyte and follicle cells around the time of ovulation in the medaka (*Oryzias latipes*). *Annot. Zool. Jap.* 54:17–29.

Iwamatsu, T., T. Ohta, E. Oshima, and N. Sakai. 1988. Oogenesis in the medaka, *Oryzius latipes*—stages of oocyte development. *Zool. Sci.* 5:353–373.

Iwasaki, V. 1973. Histochemical detection of $\Delta^5 - 3\beta$-hydroxysteroid dehydrogenase in the ovary of medaka, *Oryzias latipes*, during annual reproductive cycle. *Bull. Fac. Fish Hokkaido Univ.* 23:177–184.

Jaffe, L.A. and M. Gould. 1985. Polyspermy preventing mechanism. In C.B. Motz and A. Morrey (eds.), *Biology of Fertilization*, vol. 1, pp. 223–350. Academic Press, New York.

Jalabert, B., A. Fostier, B. Breton, and C. Weil. 1991. Oocyte maturation in vertebrates. In P.K.T. Pang and M.P. Schreibman (eds.), *Vertebrate Endocrinology. Fundamentals and Biomedical Implications*, vol. 4, pt. A, *Reproduction*, pp. 23–90. Academic Press, San Diego.

Johnson, E.Z. and R.G. Werner. 1986. Scanning electron microscopy of the chorion of selected fresh water fishes. *J. Fish Biol.* 29:257–265.

Jollie, W.P. and L.C. Jollie. 1964. The fine structure of the ovarian follicle of the ovoviviparous poeciliid fish, *Lebistes reticulatus* I. Maturation of follicular epithelium. *J. Morph.* 114:479–502.

Kagawa, H. and K. Takano. 1979. Ultrastructure and histochemistry of granulosa cells of pre- and post-ovulatory follicles in the ovary of the medaka *Oryzias latipes*. *Bull. Fac. Fish. Hokkaido Univ.* 30:191–204.

Kagawa, H., K. Takano, and Y. Nagahama. 1981. Correlation of plasma estradiol-17β and progesterone levels with ultrastructure and histochemistry of ovarian follicles in the white-spotted char, *Salvelinus leucomaenis*. *Cell Tiss. Res.* 218:315–329.

Kagawa, H., G. Young, S. Adachi, and Y. Nagahama. 1982a. Estradiol-17β production in amago salmon (*Oncorhynchus rhodurus*) ovarian follicles; role of the thecal and granulosa cells. *Gen. Compar. Endocrin.* 47:440–448.

Kagawa, H., G. Young, and Y. Nagahama. 1982b. Estradiol-17β production in isolated amago salmon (*Oncorhynchus rhodurus*) ovarian follicles and its stimulation by gonadotropins. *Gen. Compart. Endocr.* 47:361–365.

Kagawa, H., G. Young, and Y. Nagahama. 1983. Relationship between seasonal plasma estradiol-17β and testosterone levels and *in vitro* production by ovarian follicles of amago salmon (*Oncorhynchus rhodurus*). *Biol. Reprod.* 29:301-309.

Kagawa, H., G. Young, and Y. Nagahama. 1984. In vitro estradiol-17β and testosterone production by ovarian follicles of the goldfish, *Carassius auratus*. *Gen. Compar. Endocr.* 54:139-143.

Kapoor, C. 1976. Studies on the oogonial cycle in *Puntius ticto* (Ham.). II. Formation of mature follicles. *Z. Mikrosk. Anat. Forsch.* 9:1049-1055.

Kemp, N.E. and M.D. Allen. 1956. Electron microscopic observations on the development of the chorion of *Fundulus*. *Biol. Bull. Mar. Biol. Lab. Woods Hole* 11:293.

Kessel, R.G., H.N. Tung, R. Roberts, and H.W. Beams. 1985. The presence and distribution of gap junctions in the oocyte-follicle cell complex of the zebra fish, *Brachydanio rerio*. *J. Submicrosc. Cytol.* 17:239-253.

Khoo, K.H. 1975. The corpus luteum of goldfish (*Carassius auratus* L.) and its function. *Can. J. Zool.* 53:1306-1313.

Kobayakawa, M. 1985. External characteristics of the eggs of Japanese catfishes (Silurus). *Jap. J. Ichthyol.* 32:104-106.

Kobayashi, W. 1982. The structure and amino acid composition of the envelope of the chum salmon egg. *J. Fac. Sci. Hokkaido Univ.* Ser. VI. *Zool.* 23:1-12.

Kobayashi, W. 1985. Communication of oocyte-granulosa cells in the chum salmon ovary detected by transmission electron microscopy. *Develop. Growth Differ.* 27:553-556.

Kobayashi, W. and T.S. Yamamoto. 1981. Fine structure of the micropylar cell and its change during oocyte maturation in the chum salmon, *Onchorynchus kata*. *J. Morphol.* 184:263-276.

Kobayashi, W. and T.S. Yamamoto. 1987. Light and electron microscopic observations of sperm entry in the chum salmon egg. *J. Exp. Zool.* 243:311-322.

Kraft, A.V. and H.M. Peters. 1963. Vergleichende-Studien über die Oogenese in der *Gattung tilpala* (Cichlidae, Teleostei). *Z. Zellforsch.* 61:434-485.

Kuchnow, K.P. and J.R. Scott. 1977. Ultrastructure of the chorion and its micropyle apparatus on the mature *Fundulus heteroclitus* (Walbaum) ovum. *J. Fish. Biol.* 10:197-201.

Kudo, S. 1980. Sperm penetration and the formation of a fertilization cone in the common carp egg. *Develop. Growth Different.* 22:403-414.

Kugel, B., R.W. Hoffmann, and A. Friess. 1990. Effects of low pH on the chorion of rainbow trout, *Oncorhynchus mykiss*, and brown trout, *Salmo trutta* f. *fario*. *J. Fish Biol.* 37:301-310.

Kumari, S.D.R. and N.D. Nair. 1979. Oogenesis in the hill stream loach *Noemacheilus triangularis*. *Zool. Anz.* 203:259-271.

Kuo, C.M., C.E. Nash, and Z.H. Shedadeh. 1974. A procedural guide to induce spawning in grey mullet (*Mugil cephalus* L.). *Aquaculture* 2:1-14.

Laale, H.W. 1980. The perivitelline space and egg envelopes of bony fishes: a review. *Copeia* 1980:210-226.

Lambert, J.G.D. 1966. Location of hormone production in the ovary of the guppy, *Poecilia reticulata*. *Experientia* 22:476.

Lambert, J.G.D. 1970. The ovary of the guppy *Poecilia reticulata*. The granulosa cells as the sites of steroid biosynthesis. *Gen. Compar. Endocr.* 15:464-476.

Lambert, J.G.D. 1978. Steroidogenesis in the ovary of *Brachydanio rerio* (Teleostei). *In* P.G. Gaillard and H.H. Boer (eds.), *Comparative Endocrinology*, pp. 65-68. Elsevier/North Holland Biomedical Press, Amsterdam.

Lambert, J.G.D. and P.G.W.J. van Oordt. 1974. Ovarian hormones in teleosts. *Fortschr. Zool.* 22:340-349.

Lambert, J.G.D., J.A.M. Mattheij, and P.G.W.J. van Oordt. 1973. The ovary and hypophysis of the zebra fish *Brachydanio rerio* during the reproductive cycle (abstract). *Gen. Compar. Endocr.* 18:602.

Lance, V.A. and K.P. Callard. 1969. A histochemical study of ovarian function in the ovoviviparous elasmobranch, *Squalus acanthias*. *Gen. Compar. Endocr.* 13:255-267.

Larsen, W.J. and S.E. Wert. 1988. Roles of cell junctions in gametogenesis and in early embryonic development. *Tissue and Cell* 20:809-840.

Livni, B.N., M. Abraham, C. Leray, and A. Yashouv. 1969. Histochemical localization of hydroxysteroid dehydrogenases in the ovary of an euryhaline teleost fish, *Mugil capito*. *Gen. Compar. Endocr.* 13:493.

Livni, N. 1971. Ovarian histochemistry of the fishes, *Cyprinus carpio*, *Mugil capito* and *Tilapia aurea* (Teleostei). *Histochem. J.* 3:304-414.

Lönning, S. and P. Solemdal. 1972. The relation between thickness of chorion and specific gravity of eggs from Norwegian and Baltic flat fish populations. *Fisk Dir. Skr. Ser. Hayunders*, 16:77-87.

Lönning, S. and B.E. Hagstrom. 1975. Scanning electron microscope studies of the surface of the fish egg. *Asarte* 8:17-22.

Lönning, S., E. Kjorsvik, and J.B. Falk-Petersex. 1988. A comparative study of pelagic and demersal eggs from common marine fishes in northern Norway. *Sarsia* 73:49-60.

Lopes, R.A., H.S. Leme Dos Santos, J.R.V. Costa, M.G. Pelizaro, and N. Costagnoli. 1982. Histochemical study of oocyte zona radiata of the lambari *Astyanax bimaculatus lacustris* (Osteichthyes: Characidae). *Zool. Anz.* 208:265-268.

Makeeva, A.P. and E.V. Mikodina. 1977. Structure of egg membrane in Cyprinidae and some data on their chemical nature. *Biol. Nauk., Moscow* 20:60-64.

Manna, P.R. and S. Bhattacharya. 1993. [^{123}I] gonadotropin binding to the ovary of an Indian major carp, *Catla catla*, at different stages of reproduction cycle. *J. Bio. Sci.* 18:361-372.

Manner, H.W., M. Vancura, and C. Muehleman. 1977. The ultrastructure of the chorion of the fathead minnow *Pimephales promelas*. *Trans. Amer. Fish. Soc.* 106:110-114.

Markle, D.F. and L.A. Frost. 1985. Comparative morphology, seasonality and a key to planctonic fish eggs from Nova Scotian shelf. *Can. J. Zool.* 63:246-257.

Markov, K.P. 1978. Stickiness of egg membranes in Acipenseridae. *Vopr. Ikhtiol*. 18:483-493 (in Russian).

Mester, R., D. Scripcariu, and M.I. Varo. 1980. Cytochemical localization of adenylate cyclase in fish ovary. *Rev. Roum. Biol. Ser. Biol. Anim.* 25:117-120.

Mikodina, E.V. 1987. Surface structure of the egg membranes of teleostean fishes. *J. Ichthyol.* 27:106-113.

Mooi, R.D. 1990. Egg surface morphology of Pseudochromeids (Perciformes, Percoidei), with comments on its phylogenetic implications. *Copeia* 1990:455-475.

Mooi, R.D., R. Winterbottom, and M. Burridge. 1990. Egg surface morphology, development and induction in the Congrogadinae (Pisces, Perciformes,Pseudochromidae). *Can. J. Zool.* 68:923-934.

Müller, H. and G. Sterba. 1963. Electronenmikroskopische Untersuchungen über Bildung und Struktur der Eithüllen bei Knochnfischen. H. Die Eihüllen jüngerer und alterer Oozyten von *Cynolebias belotti* Steindachner (Cyprinodontidae). *Zool. Jb. Anat.* 80:469-488.

Nagahama, Y. 1983. The functional morphology of teleost gonads. In W.S. Hoar, D.J. Randall, and E.M. Donaldson (eds.), *Fish Physiology*, vol. IX: *Reproduction*, pt. A. pp. 223-276. Academic Press, New York.

Nagahama, Y. 1984. Mechanism of gonadotropic control of steroidogenesis in the teleost ovarian follicle. In *Gunma Symp. on Endocrinol*. vol. 21, pp. 167-182. Center for Academic Publications, Tokyo.

Nagahama, Y., K.C. Chan, and W.S. Hoar. 1976. Histochemistry and ultrastructure of pre- and post-ovulatory follicles in the ovary of the goldfish, *Carassius auratus*. *Can. J. Zool.* 54:1128-1139.

Nagahama, Y., W.C. Clarke, and W.S. Hoar. 1978. Ultrastructure of putative steroid-producing cells in the gonads of coho (*Oncorhynchus kisutch*) and pink salmon (*Oncorhynchus gorbuscha*). *Can. J. Zool.* 56:2508-2519.

Nagahama, Y., H. Kagawa, and G. Young. 1982. Cellular sources of sex steroids in teleost gonads. *Can. J. Fish. Aquat. Sci.* 39:56-64.

Nagahama, Y., G. Young, and H. Kagawa. 1983a. Steroidogenesis in the amago salmon (*Oncorhynchus rhodurus*) ovarian follicle: a two cell-type model. *Proc. 9th Int. Symp. Comp. Endocrinol.*, Hong Kong.

Nagahama, Y., K. Hirose, G. Young, S. Adachi, K. Suzuki, and B.I. Tamaoki. 1983b. Relative *in vitro* effectiveness of 17α-20β-dihydroxy-4-pregnen-3-one and other pregnene derivative on germinal vesicle breakdown in oocytes of ayu (*Plecoglossus altivelis*), amago salmon (*Oncorhynchus rhodurus*), rainbow trout (*Salmo gairdneri*) and goldfish (*Carassius auratus*). *Gen. Compar. Endocr*. 51:15-23.

Nakamura, J., J.L. Specker, and V. Nagahama. 1993. Ultrastructural analysis of developing follicle during early vitellogenesis in tilapia, *Oreochromes niloticus*, with special reference to the steroid producing cells. *Cell Tissue Res*. 272:33-39.

Nakano, E. 1969. Fishes. In C.B. Metz and A. Monroy (eds.), *Fertilization*, vol. II, pp. 295-324. Academic Press, New York.

Nakashima, S. and T. Iwamatsu. 1989. Ultrastructural changes in micropylar cells and formation of the micropyle during oogenesis in the medaka *Oryzias latipes*. *J. Morphol*. 202:339-349.

Neaves, W.B. 1975. Leydig cells. *Contraception* 2:571-504.

Nicholls, T.O. and G. Maple. 1972. Ultrastructural observations on possible sites of steroid biosynthesis in the ovarian follicular epithelium of two species of cichlid fish, *Cichlasoma nigrofasciata* and *Haplochromis multicolor*. *Z. Zellforsch*. 128:317-335.

Ohta, H. and T. Teranishi. 1982. Ultrastructure and histochemistry of granulosa and micropylar cells in the ovary of the loach, *Misgurnus anguillicaudatus* (Cantor). *Bull. Fac. Fish Hokkaido Univ*. 33:1-8.

Onitake, K. and T. Iwamatsu. 1986. Immunocytochemical demonstration of steroid hormones in the granulosa cells of the medaka, *Oryzias latipes. J. Exp. Zool.* 239:97-103.

Oppen-Bernstein, D.O., J.V. Helvik, and B.T. Waltker. 1990. The major structural proteins of cod (*Gadus morhua*) eggshells and protein cross binding during teleost egg hardening. *Develop. Biol.* 137:258-265.

Osanai, K. 1977. Scanning electron microscopy of the envelopes surrounding the chum-salmon oocytes. *Bull. Mar. Biol. St. Asamushi* 16:21-25(1973).

Patzner, R.A. 1984. The reproduction of *Blennius pavo* (Teleostei, Blenniidae). II. Surface structures of the ripe egg. *Zool. Anz.* 213:44-50.

Pelizaro, M.G., H.S. Leme Dos Santos, R.A. Lopes, and N. Castagnoli. 1981. Rhythm of development in the oocyte of the tambiu *Astyana bimaculatus* (Pisces: Characidae). A morphometric and histochemical study. *Arch. Biol.* 92:415-432.

Perry, J.M. 1984. Post-fertilization changes in the chorion of winter flounder, *Pseudopleuronectes americanus* Walbaum; eggs observed with scanning electron microscope. *J. Fish Biol.* 25:83-94.

Pithawalla, R.B., F.A. Castillos, and L.J. Wilson. 1993. The hagfish oocyte at late stages of oogenesis. Structural and metabolic events at the micropylar region. *Tissue and Cell* 25:259-274.

Ramadan, A., A. Ezza, and S. Hafey. 1979a. Studies on fish oogenesis. 2. Cytological studies on developing oocytes of *Merluccius merluccius. Folia Morph.* 27:18-24.

Ramadan, A., A. Azzat, and S. Hafez. 1979b. Cytochemical studies on developing oocytes of *Merluccius merluccius. Folia Morph.* 27:182-186.

Rastogi, R.K. 1970. Studies on the fish oogenesis. IV. Origin, structure and fate of the egg membranes in some freshwater teleosts. *Acta Biol. Hung.* 21:35-42.

Renard, P., R. Billard, and R. Christen. 1987. Formation of the chorion in carp oocyte. An analysis of the kinetics of its elevation as a function of oocyte ageing, fertilization and the composition of the dilution medium. *Aquaculture* 62:153-168.

Riehl, R. 1976. Eine besondere Haftvorrichtung zwischen Cortex radiatus externus und Follikel-epithel der Oocyten von *Gobio gobio* (L.) (Teleostei: Cyprinidae). *Z. Naturforsch.* 31C:62B.

Riehl, R. 1977. Konzentrische "Lamellen" in jungen Oocyten von *Noemacheilus barbatulus* (L.) (Teleostei, Cobitidae). *Biol. Zool.* 86:523-528.

Riehl, R. 1978a. Die Oocyten der Grundel *Pomatoschistus minutus* I. Licht- und elektronenmikroskopische Untersuchungen an Eihülle and Follikel. *Helgoländer Wiss. Meeresunters* 31:314-332.

Riehl, R. 1978b. Licht- und elektronenmikroskopische Untersuchungen an den Oocyten der Süsswasser-Teleosteer *Noemacheilus barbatulus* (L.) und *Gobio gobio* (L.) (Pisces. Teleostei). *Zool. Anz. Jena* 201:199-219.

Riehl, R. 1978c. Feinbau, Entwicklung und Bedeutung der Eihüllen bei Knochenfischen. *Riv. Ital. Piscic. Ittiop.* 13A: 113-121.

Riehl, R. 1979. Ein erweiterter und verbesserter Bestimmungschlüssel für die Eier deutscher Süsswasser-Teleosteer. *Z. Angew Zool.* 66:199-216.

Riehl, R. 1980a. Ultracytochemical localization of Na^+, K^+-activated ATPase in the oocytes of *Heterandria formosa* Agassiz, 1853 (Pisces, Poecillidae). *Reprod. Nutr. Dev.* 20:191-196.

Riehl, R. 1980b. Micropyle of some salmonids and coregonids. *Env. Biol. Fishes*, 5:59-66.

Riehl, R. 1988d. Die Oocyten der Grundel *Pomatoschistus minutus*. II. Licht mikroskopische Untersuchungen zur Kenntnis der Mikropyle. *Microscopia Acta* 80:287-291.

Riehl, R. 1991. Structure of oocytes and egg envelopes in oviparous teleosts. An overview. *Acta Biol. Benrodis* 3:27-65.

Riehl, R. and K.J. Götting. 1974. Zu Struktur und Vorkommen der Mikropyle an Eizellen und Eiern von Knochenfischen (Teleostei). *Arch. Hydrobiol.* 74:393-402.

Riehl, R. and K.J. Götting. 1975. Bau ünd Entwicklung der Mikropylen in den Oocyten einiger Süsswasser-Teleosteer. *Zool. Anz. Jena* 195:363-373.

Riehl, R. and E. Schulte. 1977a. Licht- und elektronenmikroskopische Untersuchungen an den Eihüllen der Elritze (*Phoxinus phoxinus* (L.), Teleostei, Cyprinidae). *Protoplasma* 92:147-162.

Riehl, R. and E. Schulte. 1977b. Vergleichende raster elektronenmikroskopische Untersuchungen an den Mikropylen ausgewählter Süsswasser-Teleostei). *Arch. Fisch. Wiss.* 28:95-107.

Riehl, R. and E. Schulte. 1978. Bestimmungsschlüsel der wichtigsten deutschen Süsswasser-Teleosteer anhand ihrer Eier. *Arch. Hydrobiol.* 83:200-212.

Riehl, R. and K.H. Kock. 1989. The surface structure of Antarctic fish eggs and its use in identifying fish eggs from the southern ocean. *Polar Biol.* 9:197-203.

Riehl, R. and W. Ekau. 1990. Identification of Antarctric fish eggs by surface structures as shown by eggs of *Trematomus eulepidotus* (Teleostei, Nototheniidae). Validation of the method. *Polar Biol.* 11:27–31.

Riehl, R. and H. Greven. 1990. Electron microscopical studies on oogenesis and development of egg envelope in the viviparous teleosts, *Heterandria formosa* (Poeciliidae) and *Ameca splendens* (Goodeidae). *Zool. Beitrage N.F.* 3:247–252.

Robertson, D.A. 1981. Possible functions of surface structure and size in some planktonic eggs of marine fishes. *N.Z.J. Mar. Freshw. Res.* 15:147–153.

Rubtsov, V.F. 1978. Membrane mucilage of carp (*Cyprinus carpio* L.) and herring (*Clupea harengus marisalter* Bert.) roe. *Vopr. Khitio* 18:305–313 (in Russian).

Saidapur, S.K. and V.B. Nadkarni. 1976. Steroid synthesizing cellular sites in the ovary of catfish, *Mystus cavasius*: a histochemical study. *Gen. Compar. Endocr.* 30:457–461.

Salmon, C., H. Kagawa, S. Adachi, Y. Nagahama, and Y.A. Fontaine. 1984. Mise en évidence de liaison spécifique de la gonadotropine de Salmon chum (*Oncorhynchus keta*) dans des preparations membranaires de granulosa d'ovaire due Salmon amago (*Oncorhynchus thodurus*). *C.R. Hedb. Séanc. Acad. Sci. Paris* 298:337–340.

Schmehl, M.K. and E.F. Graham. 1987. Comparative ultrastructure of the zona radiata from eggs of six species of salmonids. *Cell Tissue Res.* 250:513–519.

Selcer, K.W. and W.W. Leavitt. 1991. Estrogens and progestins. In P.K.T. Pang and M.P. Schreibman (eds.), *Vertebrate Endocrinology: Fundamentals and Biomedical Implications*, vol. 4, pt. B. Reproduction, pp. 67–114. Academic Press, San Diego.

Selman, K. and R.A. Wallace. 1982. The inter- and intracellular passage of proteins through the ovarian follicle in teleosts. In C.J.J. Richter and H.J.T. Goos (eds.), *Reproductive Physiology of Fish*, pp. 151–154. Proc. Int. Symp. on Reproductive Physiology of Fish, Wageningen.

Selman, K., R.A. Wallace, and D. Player. 1991. Ovary of the sea horse *Hippocampus erectus*. *J. Morphol.* 209:285–304.

Shackley, S.E. and P.E. King. 1977. Oogenesis in a marine teleost, *Blennius pholis* L. *Cell Tiss. Res.* 181:105–128.

Shackley, S.E. and P.E. King. 1978. Protein yolk synthesis in *Blennius pholis* L. *J. Fish Biol.* 13:179–193.

Shanbag, A.B. and V.B. Nadkarni. 1981. Identification of steroidogenic tissues in the ovary of a fish, *Channa gachua*, during different phases of reproductive cycle. *J. Anim. Morphol. Physiol.* 28:128–134.

Sobhana, B. and N.B. Nair. 1977. Oogenesis in the olive carp, *Puntius sarana submasutus*. *Zool. Anz.* 198:373–379.

Sponaugle, D.L. 1980. *Annual fish oogenesis: morphology and pattern formation of the secondary egg envelope of Cynolebias melanotaensis*. M.S. thesis, Clemson University SC.

Shelton, W.L. 1978. Fate of the follicular epithelium in *Dorósoma peteneuse* (pisces: Clupeidae). *Copeia* 1978:237–244.

Stahl, P.A. and C. Leray. 1961. L'ovogenese chez les poissons téléostéens. I. Origine et signification de la zona radiata et de ses annexes. *Arch. Anat. Microsc. Morph. Exp.* 30:215–261.

Stehr, C.M. 1982. *The development of the hexagonally structured egg envelope of the C-O sole (Pleuronichthys coenosus)*. M.S. thesis, University of Washington, Wash.

Stehr, C.M. and J.W. Hawkes. 1979. The comparative ultrastructure of the egg membrane and associated pore structures in the starry flounder, *Platichthys stellatus* (Pallas) and pink salmon, *Oncorhynchus gorbuscha* (Walbaum). *Cell Tiss. Res.* 202:347–356.

Stehr, C.M. and J.W. Hawkes. 1983. The development of the hexagonally structured egg envelope of the C-O sole (*Pleuronichthys coenosus*). *J. Morph.* 178:267–284.

Sterba, G. and H. Muller. 1962. Elektronenmikroskopische Untersuchungen über Bildung und Struktur der Ethüllen Bei Knochenfischen. I. Die Hüllen junger Oozyten von *Cynolebias belotti* staindachner (Cyprinodontidae). *Zool. J. Anat.* 80:65–80.

Sumida, B.Y., E.H. Ahlstrom, and H.G. Moser. 1979. Early development of seven flatfishes of the eastern north Pacific with heavily pigmented larvae (Pisces, Pleuronectiformes). *Fishery Bull.* 77:105–145.

Sun, J. and W. Xizai. 1983. A preliminary study on the histological chemistry of the ovary of *Tilapia nilotica* during its development. *Zool. Res.* 4:279–286.

Suzuki, K., B.-I. Tamaoki, and Y. Nagahama. 1981. In vitro synthesis of an inducer for germinal vesicle breakdown of fish oocytes, 17α-20β-dihydroxy-4-pregnen-3-one by ovarian tissue preparation of amago salmon (*Oncorhynchus rhodurus*). *Gen. Compar. Endocr.* 45:533–535.

Szöllösi, D. and R. Billard. 1974. The micropyle of trout eggs and its reaction to different incubation media. *J. Microscopie* 21:55-62.

Takahashi, S. 1978. The adherent filaments of eggs of isaza (*Chaenogobius isaza* Tanaka, Pisces). *Zool. Mag. Tokyo* 87:216-220.

Takahashi, S. 1981. Sexual maturity of the isaza, *Chaenogobius isaza*. 2. Gross morphology and histology of the ovary. *Zool. Mag. Tokyo* 90:54-61.

Takano, K. and H. Ohta. 1982. Ultrastructure of micropylar cells in the ovarian follicles of the pond smelt, *Hypomesus transpacificus nipponensis*. *Bull. Fac. Fish. Hokkaido Univ.* 33:65-78.

Tang, F., B. Lofts, and S.T.H. Chan. 1978. Δ^5-3β-hydroxysteroid dehydrogenase activities in the ovary of the rice-field eel, *Monopterus albus*. *Experientia* 30:316-317.

Tesoriero, J.V. 1977a. Formation of the chorion (zona pellucida) in the teleost, *Oryzias latipes* I. Morphology of early oogenesis. *J. Ultrastruct. Res.* 59:282-291.

Tesoriero, J.V. 1977b. Formation of the chorion (zona pellucida) in the teleost, *Oryzias latipes*. II. Polysaccharide-cytochemistry of early oogenesis. *J. Histochem. Cytochem.* 25:1376-1380.

Tesoriero, J.V. 1978. Formation of the chorion (zona pellucida) in the teleost, *Oryzias latipes*, III. Autoradiography of (^{3}H)-proline incorporation. *J. Ultrastruct. Res.* 64:315-326.

Thiaw, O.T. and X. Mattei. 1991. Morphogenesis of the secondary envelope of the oocyte in a teleostean fish of the family Cyprinodontidae, *Aphyosemion splendopleur*. *J. Submicrosc. Cytol. Pathol.* 23:419-426.

Thiaw, O.T. and X. Mattei. 1992. Natural degenerating mitochondria in ovarian follicles of a cyprinodontid fish, *Epiplatys spilargyreus* (Teleost). *Mol. Reprod. Develop.* 32:67-72.

Toshimori, K. and C. Oura. 1979. An unusual architecture of occluding junctions between surface cells in teleost ovarian follicles (*Plecoglossus altivelis*). *Archim. Histol. Jap.* 42:543-550.

Toshimori, K. and F. Yasuzumi. 1979a. Tight junctions between ovarian follicle cells in the teleost (*Plecoglossus altivelis*). *J. ultrastruct. Res.* 67:73-78.

Toshimori, K. and F. Yasuzumi. 1979b. Gap junctions between microvilli of an oocyte and follicle cells in the teleost (*Plecoglossus altivelis*). *Z. Mikrosk-anat. Forsch.* 93:459-464.

Treasurer, J.W. and F.G.T. Holiday. 1981. Some aspects of the reproductive biology of perch *Perca fluviatilis*: a histological description of the reproductive cycle. *J. Fish. Biol.* 18:359-376.

Tsukahara, J. 1971. Ultrastructural study on the attaching filaments and villi of the oocyte of *Oryzias latipes* during oogenesis. *Develop. Growth Differ.* 13:173-180.

Ueda, H., O. Hiroi, A. Hara, K. Yamauchi, and Y. Nagahama. 1984. Changes in serum concentrations of steroid hormones, thyroxine and vitellogenin during spawning migration of the chum salmon, *Oncorhynchus keta*. *Gen. Compar. Endocr.* 53:203-211.

Upadhyay, S.N., B. Breton, and R. Billard. 1978. Ultrastructural studies on experimentally induced vitellogenesis in juvenile rainbow trout (*Salmo gairdneri* R.). *Ann. Biol. Anim. Biochim. Biophys.* 18:1019-1025.

Van den Hurk, R. and J. Peute. 1979. Cyclic changes in the ovary of the rainbow trout, *Salmo gairdneri* with special reference to sites of steroidogenesis. *Cell Tiss. Res.* 199:289-306.

Van den Hurk, R. and C.J.J. Richter. 1980. Histochemical evidence for granulosa steroids in follicle maturation in the African catfish *Clarias lazera*. *Cell Tiss. Res.* 211:345-348.

Van den Kraak, G., L. Hoa-Ren, E.M. Donaldson, H.M. Dyte, and G.A. Hunter. 1983. Effects of LHRH and (des-10-glycine, 6-D-alanine) LHRH ethylamide on plasma gonadotropin levels and oocyte maturation in adult female coho salmon (*Oncorhynchus kisutch*). *Gen. Compar. Endocr.* 49:470-476.

Van Ree, G.E., D. Lok, and G. Bosman. 1977. *In vitro* induction of nuclear breakdown in oocytes of the zebrafish *Brachydanio rerio* (Ham. Buch). Effects of the composition of the medium and of protein and steroid hormones. *Proc. K. Ned. Akad. Wet. Ser. C.* 80:353-371.

Varma, S.K. and S.S. Guraya. 1968. The localization and functional significance of alkaline phosphatase in the vertebrate ovary. *Experientia* 24:398-399.

Wallace, R.A. and K. Selman. 1980. Oogenesis in *Fundulus heteroclitus*. II. The transition from vitellogenesis into maturation. *Gen. Compar. Endocr.* 42:345-354.

Wallace, R.A. and K. Selman. 1981. Cellular and dynamic aspects of oocyte growth in teleosts. *Amer. Zool.* 21:325-343.

Wourms, J.P. 1976. Annual fish oogenesis. I. Differentiation of the mature oocyte and formation of the primary envelope. *Devl. Biol.* 50:338-354.

Wourms, J.P. and H. Sheldon. 1976. Annual fish oogenesis. II. Formation of the secondary egg envelope. *Devl. Biol.* 50:355-366.

Yamamoto, K. and H. Onozato. 1968. Steroid-producing cells in the ovary of the zebrafish, *Brachydanio rerio*. *Annot. Zool. Jap.* 41:119–128.

Yamamoto, M. 1963. Electron microscopy of fish development. II. Oocyte-follicle cell relationship and formation of chorion in *Oryzias latipes*. *J. Fac. Sci. Tokyo Univ.*, sect. IV. *Zool.* 10:123–127.

Yamamoto, M. 1964. Electron microscopy of fish development. III. Changes in the ultrastructure of the nucleus and cytoplasm of the oocyte during its development in *Oryzias latipes*. *J. Fac. Sci. Tokyo Univ.*, sect. II, 10:335–346.

Yamauchi, K., H. Kagawa, M. Ban, N. Kasahara, and Y. Nagahama. 1984. Changes in plasma estradiol-17β and 17α,20β-dihydroxy-4-pregnen-3-one levels during final oocyte maturation of the masu salmon *Oncorhynchus masou*. *Bull. Jap. Soc. Sci. Fish.* 50:21–37.

Yaron, Z. 1971. Observations on the granulosa cells of *Acanthobrama terrae-sanctae* and *Tilapia nilotica* (Teleostei). *Gen. Compar. Endocr.* 17:247–252.

Young, G., H. Kagwa, and Y. Nagahama. 1982. Inhibitory effect of cyanoketone on salmon gonadotropin-induced estradiol-17β production by ovarian follicles of the amago salmon (*Oncorhynchus rhodurus*) *in vitro*. *Gen. Comp. Endocr.* 47:357–360.

Young, G., H. Kagwa, and Y. Nagahama. 1983a. Evidence for a decrease in aromatase activity in the ovarian granulosa cells of amago salmon (*Oncorhynchus rhodurus*) associated with final oocyte maturation. *Biol. Reprod.* 29:310–315.

Young, G., H. Oeda, and Y. Nagahama. 1983b. Estradiol-17β and 17α,20β-dihydroxy-4-pregnen-3-one production by isolated ovarian follicles of amago salmon (*Oncorhynchus rhodurus*) in response to mammalian pituitary and placental hormones and salmon gonadotropin. *Gen. Compar. Endocr.* 52:329–335.

9

Atretic Follicles and Corpora Lutea in the Ovaries of Fishes: Structure-Function Correlations and Significance

Bhagyashri A. Shanbhag and Srinivas K. Saidapur

It is a common feature of any vertebrate ovary that the oocyte that is born is destined either to grow, mature, and ovulate and leave behind a corpus luteum (postovulatory follicle) or to undergo atresia at some stage of its growth and development. While the corpus luteum is formed by the cells of the follicular envelope after expulsion of the mature ovum, the atretic follicles are formed by degeneration and resorption of oocyte and follicle cells *in situ*. There are excellent reviews on both the corpora lutea and atretic follicles, which deal with such various aspects as cytomorphology, histochemistry, biochemistry, and physiology in various vertebrate groups (Barr, 1968; Lofts and Bern, 1972; Browning, 1973; Guraya, 1973, 1976, 1979, 1985, 1986; Byskov, 1978; Saidapur, 1978, 1982; Xavier, 1987). It is apparent from these that atretic and luteal bodies are common features of the piscine ovary. Here, we review the various facets of follicular atresia, luteogenesis, and luteolysis in the fish ovary based on histological, histochemical, ultrastructural, and biochemical findings of recent years. The endocrine capacity of the corpora lutea, their role in gestation and other reproductive functions, and atretic follicles, their degeneration and the possible hormonal and/or environmental factors affecting/governing follicular atresia are discussed.

ATRETIC FOLLICLES

Follicular atresia has been studied in some species of elasmobranchs and several species of teleosts (Lofts and Bern, 1972; Byskov, 1978; Saidapur, 1978; Nagahama, 1983; Guraya, 1979, 1986). It is known to occur more commonly in fishes during prespawning, spawning, and postspawning phases (Saidapur, 1978). It is seen more frequently in the vitellogenic follicles. Atresia of small previtellogenic follicles is rare.

Process of Atresia: Elasmobranchs

Follicular atresia has been reported in the ovaries of oviparous, ovoviviparous, and viviparous elasmobranchs (Table 1). Though atresia has been observed in all the stages of follicular development (Hisaw and Hisaw, 1959), it is less frequent in previtellogenic follicles than in vitellogenic follicles. Further, the nature of the involution process of the degenerating follicles is influenced by the extent of vitellization of the ovum. Small previtellogenic follicles are seen to undergo simple resorption. In these follicles the clear fluid of the ovum decreases, the wall of the follicle collapses, and the granulosa atrophies and disintegrates (Hisaw and Hisaw, 1959). On the other hand, the degenerative changes of vitellogenic follicles are more complex. The various histological features of different stages of atresia of vitellogenic follicles have been described in some species of elasmobranchs (Chieffi, 1962; Lance and Callard, 1969; Te Winkel, 1972), the salient features of which are as follows:

Stage 1: Characterized by enlargement of small cells of the granulosa followed by invagination of these cells along with the thecal cells into the follicular lumen to give rise to well-vascularized villi-like structures. Phagocytosis of the yolk is underway.

Stage 2: Yolk phagocytosis is completed and anastomosis of hypertrophied follicular cells more evident.

Stage 3: Follicular cavity reduces in size due to anastomosis of follicular elements and more vascular elements appear between the follicular villi. Yolk granules are no longer present. According to Chieffi (1962) the atretic follicles at this stage exhibit morphological features of an epithelial gland. In *Squalus acanthias*, stage 3 atretic follicles may appear very similar to stage 3 of corpora lutea (Lance and Callard, 1969).

Stage 4: Marked by sclerosis of the vascular network of villi resulting in pigmentary degeneration of atretic follicles. This final stage of degeneration is not distinguishable from later stages of corpora lutea in *S. acanthias* (Lance and Callard, 1969).

There appear to be no studies on the ultrastructure of the follicles undergoing atresia in the elasmobranch ovary. A few histochemical studies carried out show considerable species variation with respect to various histochemical tests. The granulosa cells of stage 3 atretic follicles of *Torpedo marmorata* gave positive results in the Schultz test for cholesterol and/or its esters and in the Asbel-Seligman test for the carbonyl group (Chieffi, 1962). Also, atretic follicles increase in number during pregnancy in *Torpedo* and exhibit Δ^5-3β-hydroxysteroid dehydrogenase (Δ^5-3β-HSDH) activity. These observations led Chieffi (1962) to propose that true corpora lutea arise from preovulatory follicles undergoing atresia in the ovoviviparous species, *Torpedo marmorata* and *T. ocellata*. Subsequently, *in vitro* studies on the isolated atretic follicles of *Torpedo* species, they were shown to possess steroidogenic capacity (Lupo di Prisco, 1968). However, atretic follicles in *Scylliorhinus stellaris* and *Scylliorhinus canicula* did not exhibit a positive reaction to the Schultz and Asbel-Seligman tests (Chieffi, 1962). Later, Lance and Callard (1969) also reported negative tests for the Schultz reaction and Δ^5-3β-HSDH activity in all stages of atresia in an ovoviviparous species, *Squalus acanthias*. Thus there appear to be species variation in the histochemical features of atretic follicles in elasmobranchs. While Hisaw and Hisaw (1959) and Lance and Callard (1969) consider them to be simply abortive elements in the ovary of *Raja binoculata*, *R. erinacea*, *Squalus acanthias*, *S. suckleyi*, and *Mustelus canis*, at least in the ovary of *Torpedo* (Chieffi, 1962) there is evidence for their secretory activity during gestation.

Table 1. Fishes in which atretic follicles have been studied

Species (1)	Histochemical features (2)	Ultra-structure (3)	In vitro steroid synthesis (4)	Reference (5)
Elasmobranchs				
Cetorhinus maximus	–	–	–	Mathews, 1950
Raja binoculata	–	–	–	Hisaw and Hisaw, 1959
Raja erinacea	–	–	–	Hisaw and Hisaw, 1959
Squalus acanthias	–	–	–	Hisaw and Hisaw, 1959
	–ve Schultz –ve Δ^5-3β-HSDH	–	–	Lance and Callard, 1969
Mustelus canis	–	–	–	Hisaw and Hisaw, 1959 Te Winkel, 1972
Scylliorhinus canicula	–ve Schultz –ve Ashbel Seligman	–	–	Chieffi, 1962
Scylliorhinus stellaris	–ve Schultz –ve Ashbel Seligman	–	–	Chieffi, 1962
Torpedo marmorata	+ve Schultz	–	–	Chieffi, 1962
Torpedo ocellata	+ve Ashbel Seligman +ve Δ^5-3β-HSDH –ve Δ^5-3β-HSDH	–	–	Chieffi, 1962
			Progesterone	Lupo di Prisco, 1968
Scoliodon sorrokowah	–ve Δ^5-3β-HSDH	–	–	Guraya, 1972
Bony fishes				
Gobius paganellus	–	–	–	Vivien, 1939
Rhodeus amerus	–	–	–	Bretschneider and Duvyene de Wit, 1947; Polder, 1964
Lebistes reticulatus	–	–	–	Stolk, 1951
Gadus merlangus	–	–	–	Gokhale, 1957
Carassius auratus	–	–	–	Beach, 1959
	–ve Δ^5-3β-HSDH	–	–	Khoo, 1975
	–ve Δ^5-3β-HSDH	+	–	Yamamoto and Yamazaki, 1961
Mugil cephalus	–	–	–	Stenger, 1959
Gasterosteus aculeatus	–	–	–	Tromp-Blom, 1959
Scomber scomber	–ve Δ^5-3β-HSDH	–	–	Bara, 1960, 1965
Plecoglossus altivelis	–	–	–	Honma, 1961
Mystus seenghala	–	–	–	Sathyanesan, 1961
Glyptothorax nectinopterus	–	–	–	Pant, 1968
Ophiocephalus punctatus	–	–	–	Belsare, 1962
Pleuronectes platessa	–	–	–	Barr, 1963
Heteropneustes fossilis	–	–	–	Nair, 1963; Sunderaraj and Goswami, 1968
Tor tor	–	–	–	Rai, 1966
Gobius giuris	–	–	–	Rajalakshmi, 1966
Xenentodon cancila	–	–	–	Rastogi, 1966
Eucalia inconstans	–	–	–	Brakevelt and McMilan, 1967

(Contd.)

Table 1 Continued

Species (1)	Histochemical features (2)	Ultra-structure (3)	In vitro steroid synthesis (4)	Reference (5)
Monopterus albus	–	–	–	Chan et al., 1967
Clarias batrachus	–	–	–	Lehri, 1968
Mystus tengara	–	–	–	Rastogi, 1968
				Guraya et al. 1975
Amphipnous cuchia	–	–	–	Rastogi, 1969
Brachydanio rerio	–ve Δ^5-3β-HSDH	–	–	Van Ree, 1976
	–ve G-6-PDH			
	+ve SDH			
	+ve acid PT			
Notopterus notopterus	–	–	–	Shrivastava, 1969
Poecilia reticulata	–ve Δ^5-3β-HSDH	+	–	Lambert, 1970
	+ve acid PT			
	cholesterol			
Acanthobrama terrae-sanctae	–ve Δ^5-3β-HSDH	–	–	Yaron, 1971
Tilapia nilotica	+ve acid PT	–	–	Yaron, 1971
Channa gachua	–	–	–	Sanwal and Khanna, 1972
	+ve Δ^5-3β-HSDH			Shanbhag and Nadkarni, 1981
	–ve G-6-PDH			
Mystus cavasius	+ve Δ^5-3β-HSDH			Saidapur and Nadkarni, 1976
	+ve 17β-HSDH			
	+ve 11β-HSDH			
	+ve G-6-PDH	–	–	
Salmo gairdneri	–ve Δ^5-3β-HSDH	–	–	Van den Hurk and Peute, 1979
Perca fluviatilis	–ve Δ^5-3β-HSDH	+	–	Lang, 1981
Cyprinus carpio	–ve sudanophilia	–	–	Guraya and Kaur, 1982
Amblypharyngodon chakaiensis	–	–	–	Babu and Nair, 1983
Puntius sarana	–ve Δ^5-3β-HSDH	–	–	Bhat and Dutt, 1989a
	+ve G-6-PDH			
	+ve Acid PT			

Atresia of previtellogenic follicles: Bony fishes

Under normal circumstances, atresia of previtellogenic follicles is rarely encountered in the ovary of teleosts and has been described in only a few species (Rajalakshmi, 1966; Guraya et al., 1975, 1977; Mani and Saxena, 1985). The histogenesis of previtellogenic atretic follicles has been excellently reviewed by Guraya (1986). The salient features are described below:

Stage 1: Shrinkage of oocyte is accompanied by the appearance of clear spaces in the peripheral ooplasm and between the nucleus and ooplasm.

Stage 2: Disorganization of nuclear and cytoplasmic organelles takes place. The nucleoli are distributed irregularly.

Stage 3: The follicle cells hypertrophy, invade the disorganized nucleocytoplasmic mass, and phagocytize the latter. Invasion of oocytes by follicle cells is observed in a number of species [e.g., *Mystus tengara, Cyprinus carpio, Xenontodon cancila* (Guraya et al., 1975, 1977; Guraya, 1986); *Channa gachua* (Sanwal and Khanna, 1972; Shanbhag and

Nadkarni, 1981); *Amphipnous cuchia* (Rastogi, 1969)]. Rajalakshmi (1966), on the other hand, observed that atresia of previtellogenic follicles is accomplished without invasion of the oocyte by granulosa cells in *Gobius giuris*. In large previtellogenic atretic follicles the thecal cells also hypertrophy.

Stage 4: This stage is marked by complete degeneration of the oocyte which is occupied by phagocytic and blood cells. After complete resorption, the follicle remains as a nodule in the stroma. These nodules are highly sudanophilic in the case of large previtellogenic follicles.

Histochemical studies on the atretic previtellogenic follicles are limited to a few species. No Δ^5-3β-HSDH activity could be detected in the atretic follicles in *Scomber scomber* (Bara, 1965a), *Acanthobrama terrae-sanctae*, *Tilapia nilotica* (Yaron, 1971), and *Salmo gairdneri* (Van den Hurk and Peute, 1979). A slight enzymatic activity was observed in the early stages of atresia in *Poecelia reticulata* (Lambert, 1970), *Mystus cavasius* (Saidapur and Nadkarni, 1976), *Channa gachua* (Shanbhag and Nadkarni, 1981), and *Puntius sarana* (Bhat and Dutt, 1989a). In *M. cavasius* in early atretic follicles, slight 17β-hydroxysteroid dehydrogenase (17β-HSDH), and 11β-hydroxysteroid dehydroginase (11β-HSDH) and intense G-6-PDH activities were also reported (Saidapur and Nadkarni, 1976). Acid phosphatase was seen in all the stages of atretic follicles of *P. reticulata* (Lambert, 1970), *T. nilotica* (Yaron, 1971), *B. rerio* (Van Ree, 1976) and *P. sarana* (Bhat and Dutt, 1989b). Though cholesterol or its esters (steroid hormone precursor) could not be detected in stage 1 atretic follicles in *P. reticulata*, cholesterol could be detected in the later stages (stage 3) of atresia (Lambert, 1970).

Atresia of vitellogenic (yolky) follicles

Atresia of vitellogenic follicles has been reported in a number of species of teleosts (Saidapur, 1978; Guraya, 1986). It is more frequent in the ovaries of teleosts during the prespawning, spawning, and postspawning phases. The histological features of follicles undergoing atresia have been variously described by different researchers. The salient histological features are summarized below:

Stage 1: Initially, the oocyte shrinks due to loss of fluid and loses contact with the granulosa; folding and dissolution of the zona pellucida are also seen.

Stage 2: Disorganization and liquefaction of yolk and ooplasm.

Stage 3: Granulosa cells hypertrophy, invade the oocyte, and commence phagocytization. Ingrowth of thecal cells and blood vessels may also occur.

Stage 4: Intense yellowish orange pigment appears in the degenerating follicular cells. Histochemically, the yellow granules are due to aging lipids.

Stage 5: Marked by disintegration of these structures forming fibrous nodules with pyknotic nuclei.

Cystic/fibrous and bursting atresia have been reported in the ovaries of amphibians, reptiles, and birds (Bhujle et al., 1979; Gouder et al., 1979; Saidapur et al., 1982). Occasionally, bursting and cystic atresia occur in the ovary of teleosts. In bursting atretic follicles, the granulosa and yolk are thrown out through an opening in the follicular wall. The thecal cells then invade the follicular lumen and phagocytize the residual degenerating granulosa and yolk (Sathyanesan, 1961; Rastogi, 1968). In cystic atretic follicles, the theca hypertrophies and becomes fibrous. Degeneration of the granulosa and vitelline membrane is accompanied by liquefaction and resorption of yolk platelets. The degenerating follicle now resembles a cyst with the fibrous theca surrounding the empty follicular cavity. The cyst wall ultimately breaks and mingles with the stroma (Shanbhag and Nadkarni, 1982).

Histochemical studies on the atretic vitellogenic follicles in teleosts are scarce. Development of sudanophilic lipids in the region of cortical alveoli has been reported in the early stages of atretic follicles in some teleosts (Guraya and Kaur, 1982). With the advancement of atresia these lipid masses coalesce, filling the follicular cavity. The dense cholesterol positive lipids have been attributed to degenerative changes in the vitellogenic follicles (Guraya, 1986). No Δ^5-3β-HSDH activity has been reported in the vitellogenic atretic follicles of teleost species studied so far (Saidapur, 1978; Nagahama, 1983; Guraya, 1986). A strong alkaline and acid phosphatase activity in vitellogenic atretic follicles is reported in *Brachydanio rario* (Van Ree, 1976). Electron microscopic studies on the atretic follicles of *Perca fluviatilis* (Lang, 1981) revealed that the granulosa cells of atretic follicles show only features related to phagocytosis and lysosome formation. There is no smooth endoplasmic reticulum in these cells.

Causes of follicular atresia

The mechanisms which regulate ovarian follicular atresia in fishes are not yet properly understood. Various factors influence follicular atresia: reproductive state, inadequate amounts of hormones (extraovarian and intraovarian), environmental conditions, food availability, water pollution, captivity, etc. Indeed, a number of studies point out that poor supply of gonadotropins or imbalance between gonadotropins and other hormones, especially steroid hormones, may cause follicular atresia in fishes (Saidapur, 1978; Fostier et al., 1983; Goetz, 1983; Guraya, 1986). The atresia of yolky follicles during the spawning period has been related to inadequate levels of gonadotropins and subsequently to the low levels of gonadotropins during the postspawning period (Saidapur, 1978). Studies involving hypophysectomy or treatment with antigonadotropic agents have demonstrated an increased incidence of atresia, thereby suggesting that gonadotropins are needed for follicular growth (Hoar, 1965; Barr, 1968; Sunderaraj and Goswami, 1968; Donaldson, 1973; Mackay, 1973; Anand and Sunderaraj, 1974; Van Ree, 1976). Also, antisera to vitellogenic hormone (fish gonadotropin with low carbohydrate) is known to arrest vitellogenesis in flounder (Ng et al., 1980). The incidence of atresia is found to be greater when fishes are exposed to a short photoperiod (Vlaming, 1974; Saxena and Anand, 1977; Borg and Van Veen, 1982; Bhat and Dutt, 1989c) and presumed to be due to increased release of melatonin. Further, melatonin treatment of fish is reported to inhibit gonadal recrudescence and to increase follicular atresia (Vlaming, 1974; Sunderaraj and Keshavnath, 1976; Saxena and Anand, 1977). *In vitro* studies on oocyte maturation have pointed out that denuded oocytes (oocyte without follicular wall) do not respond to gonadotropin, suggesting that maturation of follicles by gonadotropin is mediated by a maturation-inducing substance, a steroid, produced by the follicular wall (Goetz, 1983). All these studies lend evidence to the view that improper gonadotropin delivery to the follicles or imperfect balance between gonadotropins and steroids may cause follicular atresia in fishes, as suggested for other vertebrates (Saidapur, 1978; Guraya, 1985).

The majority of fishes are seasonal breeders and timing of reproduction may be influenced by a number of factors, such as photoperiod, temperature, salinity of water, rainfall, abundance of food, nest availability (Wooten, 1982; Lam, 1983; Bye, 1984) and so on.

The nonbreeding season is generally marked by unfavorable conditions and small ovaries showing extensive atresia.

Studies carried out on captive fishes revealed that the ovarian follicles do not advance beyond the early yolk stage and need exogenous hormones for further growth. In gray mullet, *Mugil cephalus*, held in captivity, the ovarian follicles did not undergo final maturation and underwent atresia (Kuo, 1982). Apparently, stress is responsible for causing follicular atresia in captive fishes.

In recent years fish have been exposed to increased levels of pollutants in the water in which they live. Several studies have been carried out on the effect of biocides on fish under laboratory conditions (Singh and Singh, 1980a-c; Eller, 1981; Ram and Sathyanesan, 1983; Saxena and Bhatia, 1983; Shukla and Pandey, 1984a, b; Mani and Saxena, 1985), which were recently reviewed (Lam, 1983; Patil and Saidapur, 1989). Chronic exposure to endrin (chlorinated hydrocarbon) is known to result in the formation of atypical oocytes, increased follicular atresia, and inhibition of ovulation in *Salmo clarki* (Eller, 1981). Similarly, increased follicular atresia is observed in *Colisa fasciatus* treated with arsenic (Shukla and Pande, 1984a, b). The responses of the ovary to different biocides may also vary. In *Channa punctatus* treated with two different biocides, carbofuran and fenitrothion at toxicologically safe concentrations, relatively more oocytes underwent atresia following fenitrothion treatment. The ovary had only atretic follicles following 120 days exposure to fenitrothion. On the other hand, about 60% atretic follicles were found in the ovaries of the carbofuran-treated fish following the 120-day exposure as against 12% in controls (Mani and Saxena, 1985). The mechanism of action of biocides on the fish ovary is not yet properly understood. Guraya (1986) suggested that biocides might affect follicular growth in fish by (1) inhibiting secretion of gonadotropins, (2) affecting liver metabolism, thereby interfering with vitellogenin synthesis which is needed for the growth of vitellogenic follicles, and (3) acting directly on the developmental process of the follicles at subcellular and molecular levels.

It is apparent from the aforesaid discussion that hormones as well as environmental factors control the initiation and also the rate of follicular atresia. Adverse environmental stress may induce ovarian follicular atresia and reduce fecundity in fishes in adverse environmental conditions.

Origin of interstitial gland cells

In the mammalian ovary the interstitial gland cells are known to arise from ordinary connective tissue of the ovarian stroma and also from the theca interna of the atretic follicles (Guraya, 1985). However, opinions differ about their origin and distribution in the ovary of fish. In the past, based on histological studies, the presence of interstitial gland cells in the fish ovary was considered doubtful. But histochemical tests for localization of lipids and enzymes involved in steroidogenesis have not only shown their presence in the fish ovary, but also that they arise from the hypertrophied thecal cells of large previtellogenic atretic follicles (Saidapur and Nadkarni, 1976; Saidapur, 1978; Guraya, 1986). Guraya (1972), by localizing lipids in the frozen sections of the dogfish (*Scoliodon sorrokowah*) ovary, showed the presence of interstitial gland cells and suggested that thecal cells of large previtellogenic follicles contribute to these cells. Similarly, Saidapur and Nadkarni (1976) showed that interstitial gland cells arise from thecal cells of large previtellogenic atretic follicles in *Mystus cavasius* and showed their steroidogenic ability based on the localization of various steroid dehydrogenases (Δ^5-3β-HSDH, 17β-HSDH, and 11β-HSDH) in these cells. In the immature rainbow trout, *Salmo gairdneri*, the interstitial gland cells of the ovary show ultrastructural features such as smooth endoplasmic reticulum and mitochondria with tubular cristae as in other steroid secreting cells (Upadhaya, 1977). Also, the interstitial gland cells in the developing ovary of platyfish, *Xiphophorus maculatus* (Schreibman et al., 1982), exhibit Δ^5-3β-HSDH and G-6-PDH enzyme activity, suggesting their steroidogenic potential. Guraya (1986) suggested that these cells in the immature ovary could be designated as primary interstitial cells in contrast to the secondary interstitial cells arising from thecal cells of atretic follicles, comparable to those in mammals. Histological and histochemical studies on the ovary of *Cyprinus carpio* (Guraya and Kaur, 1982) have also shown that special thecal cells of the follicle may eventually contribute to the interstitial gland cells following atresia. The special thecal cells, larger than other thecal cells, show the ultrastructural features of steroid secreting cells in a number of species of teleosts (Yamamoto and Onozato, 1968; Nicholls and Maple, 1972; Nagahama et al., 1976, 1978; Van den Hurk and Peute, 1979; Kagawa et al., 1981). Such special thecal cells are often grouped in clusters close to blood vessels. In zebrafish the special thecal cells are differentiated from ovarian interstitial gland cells at an early stage and later become incorporated in the thecal layer (Yamamoto and Onozato, 1968). In the catfish *M. cavasius*, contrarily, during follicular atresia the special thecal cells remain in the stroma as interstitial cells (Saidapur and Nadkarni, 1976). Thus, it seems that in the fish ovary, as in mammals, during follicular development the interstitial gland cells become incorporated in the theca interna and if the follicles become atretic, these cells once again contribute to the interstitial gland cells of the ovarian stroma.

The presence of Δ^5-3β-HSDH, 17β-HSDH, 11β-HSDH activities and also the ultrastructural features of steroid-producing cells, indicate that the interstitial gland cells are involved in steroid production. However, the nature of steroids secreted by interstitial gland cells of the fish ovary remains to be determined.

Significance of atresia

Extensive follicular atresia is a general feature of the ovary in the nonbreeding season. This may be an adaptation which temporarily suspends breeding activity during the months that are unfavorable for breeding and raising the young. Therefore, it appears that extensive follicular atresia in some fishes during the nonbreeding season might be nature's rule for prevention of breeding during unfavorable conditions.

Whether or not atretic follicles have any endocrine role in fish is a matter of controversy. Conclusions are drawn from histochemical studies, mainly tests for cholesterol and localization of Δ^5-3β-HSDH activity in the atretic follicles, and *in vitro* incubation studies with steroids. Chieffi (1962) and Lupo di Prisco (1968) suggested that the atretic follicles in two species of elasmobranchs, *Torpedo marmorata* and *T. ocellata* possess steroidogenic potentiality. As a large number of preovulatory follicles were found to undergo atresia during early gestation in these species, Chieffi (1962) referred to them as "preovulatory corpora lutea." However, their role in gestation is not clear. On the contrary, the atretic follicles of other elasmobranch species studied, *Scylliorhinus canicula, S. stellaris* (Chieffi, 1962), *Squalus acanthias* (Lance and Callard, 1969) and *S. sorrokowah* (Guraya, 1972) showed no steroidogenic potentiality.

Among teleosts also there is a controversy regarding the endocrine capacity of the atretic follicles. Based on histochemical and ultrastructural studies, Bara (1965a), Yaron (1971), Van den Hurk and Peute (1979), and Lang (1981) suggested that it is unlikely that atretic follicles have any steroidogenic function. On the other hand, atretic follicles of the guppy, *P. reticulata* (Lambert, 1970), the catfish, *M. cavasius* (Saidapur and Nadkarni, 1976), *C. gachua* (Shanbhag and Nadkarni, 1981), and *P. sarana* (Bhat and Dutt, 1989a) show weak Δ^5-3β-HSDH activity in early stages. Lambert (1970) and Bhat and Dutt (1989a) attribute this weak enzyme activity in the granulosa of early atretic follicles as merely continuation of the preexisting activity of the granulosa cells of normal follicles and do not consider the atretic follicles to have any endocrine role. However, Saidapur and Nadkarni (1976) are of the opinion that the granulosa of early atretic follicles in *Mystus cavasius* attain steroidogenic potential as Δ^5-3β-HSDH activity is not present in the granulosa of normal follicles but appears during early stages of atresia. Saidapur (1978) also proposed that even though there may be only weak activities of enzymes involved in steroidogenesis at any given time, there may be a large number of atretic follicles so that their combined contribution to hormone production cannot be neglected. Interestingly, the vitellogenic atretic follicles do not give positive results to any of the above referred histochemical tests, suggesting that they are merely in the process of degeneration. Apparently atresia of vitellogenic follicles limits the number of eggs that could be supported for vitellogenesis, maturation, and ovulation under inadequate hormone availability and/or energy budget of the animals.

CORPORA LUTEA

Luteogenesis and luteolysis

Development of corpora lutea (postovulatory follicles) in the ovaries following ovulation from the follicular envelope is a ubiquitous feature of the ovaries in fish, with the exception of viviparous forms with follicular gestation. Fish exhibit diverse reproductive patterns such as oviparity, ovoviviparity, and viviparity. There is even diversity in the pattern of gestation in viviparous forms. Therefore, it is not surprising to find a wide variety of difference in the development of corpora lutea, their lifespan and functions among fish. Our current knowledge of morphology, luteogenesis, and luteolysis of the corpora lutea and their possible functional significance in fish reproduction is reviewed here.

Elasmobranchs

In most elasmobranchs studied so far, in oviparous, ovoviviparous, and viviparous species, the corpus luteum forms a glandular structure (Browning, 1973; Guraya, 1976; Saidapur, 1982; Xavier, 1987). The histogenesis of corpora lutea has been described in some species of elasmobranchs (Hisaw and Hisaw, 1959; Chieffi, 1962; Lance and Callard, 1969; Te Winkel, 1972) (Table 2). The salient features of histological changes during the development and regression of corpora lutea in elasmobranchs are as follows:

Stage 1: Immediately after ovulation, the granulosa cells become folded but do not fill the cavity of CL. The phagocytic "foam cells" ingest the yolk and tissue detritus remain in the follicular lumen. The theca is extremely thick.

Table 2. Fishes in which corpora lutea have been studied

Species (1)	Histochemical features (2)	Ultra-structure (3)	In vitro steroid synthesis (4)	Reference (5)
Elasmobranchs				
Rhinobatus granulatus	–	–		Samuel, 1943
Squalus acanthias*	–	–		Hisaw and Albert, 1947
				Hisaw and Hisaw, 1959
	+ve Δ^5-3β-HSDH	–	–	Lance and Callard, 1969
	+ve G-6-PDH			
	+ve cholesterol			
Cetorhinus maximus*	–	–	–	Mathews, 1950
Squalus suckley*	–	–	–	Hisaw and Hisaw, 1959
Raja binoculata	–	–	–	Hisaw and Hisaw, 1959
Hydrolagus colloeri	–	–	–	Hisaw and Hisaw, 1959
Mustelus canis*	–	–	–	Hisaw and Hisaw, 1959; Te Winkel, 1972
Torpedo marmorata*	+ve Δ^5-3β-HSDH	–	–	Chieffi, 1961, 1962
			17β-estradiol	Lupo di Prisco, 1968
Torpedo ocellata*	+ve Δ^5-3β-HSDH	–	–	Chieffi, 1961, 1962
	–ve Schultz			
Scylliorhinus stellaris	+ve Schultz	–	–	Chieffi and Botte, 1961;
	+ve Δ^5-3β-HSDH			Lupo et al., 1965
	+ve Ashbel Seligman		Progesterone	Lupo di Prisco, 1968
Scylliorhinus canicula	+ve Schultz	–	–	Chieffi and Botte, 1961;
	+ve Ashbel Seligman			Chieffi, 1962
	+ve Δ^5-3β-HSDH			
Several Raja spp.	+ve Schultz	–	–	Botte, 1963
Bony fishes				
Gasterosteus aculeatus	–	–	–	Craig-Bennet, 1931; Lam et al., 1978a
Fundulus heteroclitus	–	–	–	Mathews, 1938
Scomber scomber	+ve Δ^5-3β-HSDH	–	–	Bara, 1965a, b
Mystus seenghala	–	–	–	Sathyanesan, 1962
Pleuronectes platessa	–	–	–	Barr, 1963
Heteropneustes fossilis	–	–	–	Nair, 1963
Tor tor	–	–	–	Rai, 1966
Gobius giuris	–	–	–	Rajalakshmi, 1966
Xenentodon cancila	–	–	–	Rastogi, 1966
Eucalia inconstans	–	–	–	Braekevelt and McMilan, 1967
Sebastodes paucispinis*	–	–	–	Moser, 1967
Clarias batrachus	–	–	–	Lehri, 1968
Notopterus notopterus	–	–	–	Shrivastava, 1969
Cyprinus carpio	–	–	–	Blanc-Livni, 1971; Guraya and Kaur, 1979
Merluccius merluccius hubbsi	–	–	–	Christiansen, 1971
Acanthobrama terrae-sanctae	–	–	–	Yaron, 1971
Cichlasoma nigrofasciatum	–	–	–	Nicholls and Maple, 1972

Glossogobius giuris	−	−	−	Saksena and Bhargava, 1972
Channa gachua	−	−	−	Sanwal and Khanna, 1972
Gillichthys mirabilis	−	−	−	Vlaming, 1972
Brachydanio rerio	+ve Δ^5-3β-HSDH	−	−	Lambert et al., 1972
	+ve G-6-PDH			Lambert and Van Oordt, 1974
	+ve Alk.PT			Van Ree, 1976
	+ve Acid PT			Lambert, 1978
Oryzias latipes	+ve Δ^5-3β-HSDH	−	−	Iwasaki, 1973;
				Kagawa and Takano, 1979
Hypseleotris galii	−ve Schultz	−	−	Mackay, 1973
Trachurus mediterraneus	+ve Δ^5-3β-HSDH	−	−	Bara, 1974
Mystus tengara	−	−	−	Guraya et al., 1975
Carassius auratus	+ve Δ^5-3β-HSDH	−	−	Khoo, 1975
		+	·	Nagahama et al., 1976
	+ve Δ^5-3β-HSDH	−	−	Hoar and Nagahama, 1978
Puntius sophore	−	−	−	Agarwala and Dixit, 1977
Oncorhynchus kisutch	−	+	−	Nagahama et al., 1976
Oncorhynchus gorbuscha	−	+	−	Nagahama et al., 1978
Salmo gairdneri	+ve Δ^5-3β-HSDH	+ ·	−	Van den Hurk and Peute, 1979
	+ve G-6-PDH			
Oncorhynchus keta	+ve Δ^5-3β-HSDH	−	−	Sufi et al., 1980
	+ve 17β-HSDH			
Oncorhynchus masou	+ve Δ^5-3β-HSDH	−	−	Sufi et al., 1980
Clarias lazera	+ve Δ^5-3β-HSDH	−	−	Van den Hurk and Richter, 1980
Salvelinus leucomaenis	+ve Δ^5-3β-HSDH	−	−	Kagawa et al., 1981
Oncorhynchus rhodurus	+ve Δ^5-3β-HSDH	−	17α-20β-diOH-progesterone	Nagahama and Kagawa, 1982 Young et al., 1983

*Live bearing (ovoviviparous/viviparous).

Stage 2: Granulosa cells completely fill the lumen and this central mass of granulosa is surrounded by theca, forming a compact structure.

Stage 3: Characterized by vacuolization of the granulosa cells and infiltration of connective tissue, suggesting the onset of involution of the corpus luteum.

Stage 4: Advanced stage of involution of the corpus luteum, which is reduced in size. Degeneration of granulosa cells and more infiltration by connective tissue take place.

Hisaw and Hisaw (1959) studied the ovary in five species of elasmobranchs (oviparous, ovoviviparous and viviparous) and found little difference in the process between atresia of large preovulatory follicles and the changes that occur in corpora lutea following ovulation. In both cases extensive involution of the granulosa cells to form villi-like structures and phagocytosis of the yolk and cellular debris from the lumen of the follicle were observed. Ultimately, the corpus luteum becomes organized into a compact structure and is not distinguishable from atretic follicles derived from large preovulatory follicles (Hisaw and Hisaw, 1959; Lance and Callard, 1969). In the ovoviviparous *Torpedo marmorata*, the corpus luteum does not luteinize but rapidly undergoes sclerosis and degenerates (Chieffi, 1962). In this species the corpora lutea appear to be derived only from the granulosa cells and the theca interna contributes to the stroma. In *Scylliorhinus* (oviparous) the corpus luteum develops from the granulosa cells and the cuboidal cells of the theca interna though both remain separated by connective tissue (Chieffi and Botte, 1961; Chieffi, 1962; Dodd, 1983; Dodd and Sumpter, 1984). The theca cells also contribute to luteal cell mass in *Rhinobatus* (Samuel, 1943) and several species of the genus *Raja*

(Botte, 1963). The fully developed corpora lutea of *Cetorhinus maximus* (Mathews, 1950) are 4–5 mm in diameter and their central cavity contains lymphocytes and miscellaneous cellular debris.

Steroidogenic potential of corpus luteum

Histochemical studies on the corpora lutea in elasmobranchs are limited and there exists considerable species variation with regard to their response to different histochemical tests. The corpora lutea of *Torpedo marmorata* (Chieffi, 1962) exhibited for a weak sudanophilia whereas the granulosa lutein cells of *Scylliorhinus stellaris* and *S. canicula* gave positive results in the Schultz test for cholesterol and/or its esters and in the Aschbel-Seligman test for the carbonyl group. Similarly positive tests for cholesterol in the corpora lutea were seen in several oviparous species of the genus *Raja* (Botte, 1963). A positive Δ^5-3β-HSDH activity was seen in the corpora lutea of *S. stellaris* but the enzyme activity could not be localized in the case of *T. marmorata* (Lupo di Prisco et al., 1965). In *Squalus acanthias* the granulosa lutein cells of stage 1 corpus luteum exhibited an intense Δ^5-3β-HSDH activity. The enzyme activity decreased in stages 2 and 3 as gestation advanced. G-6-PDH enzyme activity also showed a similar pattern of distribution (Lance and Callard, 1969). Further, an intense staining with Fattrot B for lipids in the granulosa cells of corpora lutea in this species is reported. The granulosa cells contain discrete round lipid droplets. Also, these cells contained cholesterol in the corpora lutea of stages 1–4.

In vitro studies have shown that the isolated corpora lutea in *S. stellaris* yielded progesterone but no estrogen whereas those from *T. marmorata* did not produce progesterone but could synthesize estrogen (Lupo di Prisco, 1968).

Bony fishes. Corpora lutea have been observed in the ovary of both oviparous and viviparous fishes (Table 2). The corpora lutea degenerate more rapidly in most oviparous forms but in some they develop into glandular structures. In species such as *Pleuronectus platessa* (Barr, 1963), *Gobius giuris* (Rajalakshmi, 1966), *Mystus seenghala* (Sathyanesan, 1962), *Heteropneustes fossilis* (Nair, 1963), and *Hypseleotris galii* (Mackay, 1973) the ruptured follicle shrinks rapidly after ovulation and does not form a corpus luteum. Also there is no hypertrophy of follicular membranes after ovulation. In *Tor tor*, on the other hand, following ovulation, hypertrophy of the granulosa cells occurs and these along with theca cells fill the follicular lumen (Rai, 1966). In *Clarias batrachus* granulosa and theca cells proliferate and hypertrophy and invade the follicular lumen (Lehri, 1968; Tikare and Nadkarni, pers. comm.). Similar observations were made in *Channa gachua* (Sanwal and Khanna, 1972). In the goldfish *Carassius auratus* both theca and granulosa cells hypertrophy with a marked response in granulosa cells (Khoo, 1975). The granulosa cells also contain yellow luteal pigment. Khoo further reported that the granulosa lutein cells eventually differentiate into oogonial cysts.

In some teleosts, such as *Notopterus notopterus* (Srivastava, 1969), *Merluccius merluccius hubbsi* (Christiansen, 1971), *Oryzias latipes* (Kagawa and Takano, 1979) and *Perca fluviatilis* (Lang, 1981), the thecal cells form a distinct sheath surrounding the central mass of granulosa lutein cells. In other species, such as *Scomber scomber* (Bara, 1965b), *Cichlosoma nigrofasciatum* (Nicholls and Maple, 1972), *Cyprinus carpio* (Blanc-Livni, 1971; Guraya and Kaur, 1979), *Brachydanio rario* (Lambert and Van Oordt, 1974), *Carassius auratus* (Nagahama et al., 1976), *Oncorhynchus kisutch* and *O. gorbuscha* (Nagahama et al., 1978), *Salmo gairdneri* (Van den Hurk and Peute, 1979), and *Salvelinus leucomaenis* (Kagawa et al., 1981), the hypertrophied granulosa cells remain separated from the thecal cells.

Although there are numerous reports on the corpora lutea of fishes, descriptions of the histological changes are brief. The important steps occurring during the formation and regression of corpora lutea in *C. carpio*, as seen in histological sections described by Guraya and Kaur (1979), are as follows:

Stage 1: After expulsion of the egg, hypertrophy of the granulosa and theca layers which surround the lumen of the follicle occurs. The granulosa cells are columnar in shape with basal spherical nuclei while the theca is highly vascular and consists of fibrous elements and connective tissue. A few glandular cells are also seen in the theca.

Stage 2: The corpus luteum is reduced in size. The granulosa layer forms villi-like projections irregularly situated in the lumen. There is further hypertrophy of the granulosa cells and their cytoplasm is eosinophilic. Thecal vascularity increases.

Stage 3: The lumen of the corpus luteum reduces further. No change in the granulosa layer from that in the previous stage. No vascularization of the granulosa apparent but a few blood cells are seen. The theca gland cells appear vacuolated.

Stage 4: The hypertrophied granulosa lutein cell mass occupies the now much reduced lumen of the corpus luteum. Pyknosis of the granulosa cell nuclei increases. The corpus luteum now appears to be a multilayered structure. The thecal layer exhibits similar features as in the previous stage.

Stage 5: The corpus luteum is further reduced in size. The granulosa cells are arranged loosely and separated from each other due to dissolution of intercellular cohesion between them. All nuclei become pyknotic. Thecal vascularity is reduced and the theca becomes more fibrous.

Stage 6: The number of granulosa lutein cells decrease due to degeneration. The thecal layer invades the residual luteal cells. Thecal vascularity is greatly reduced.

In a lone study on trout, *S. gairdneri*, it was shown that thecal smooth muscle-like cells become phagocytotically active after ovulation and phagocytize adjacent collagen. The collagen bundles become indistinct 72 hours after ovulation indicating a possible hydrolysis (Szöllösi et al., 1978). These authors also showed the presence of a number of lytic enzymes in the corpora lutea of *S. gairdneri* in support of their morphological observations.

Ultrastructural studies on the corpus luteum are available for species *C. nigrofasciatum* (Nicholls and Maple, 1972); *C. auratus* (Nagahama et al., 1976), *O. kisutch* and *O. gorbuscha* (Nagahama et al., 1978), *B. rerio* (Lambert, 1978), and *S. gairdneri* (Van den Hurk and Peute, 1979). Following ovulation, the granulosa and/or theca of the corpora lutea in these species exhibit some features characteristic of steroidogenic cells.

In *C. nigrofasciatum* the granulosa cells of the corpus luteum possess mitochondria with lamellar cristae but have an abundant smooth endoplasmic reticulum while in *S. gairdneri*, the granulosa as well as the special thecal cells possess mitochondria with tubular cristae and some smooth endoplasmic reticulum.

The ultrastructure of the corpus luteum was studied in *C. auratus* soon after ovulation. Six hours following ovulation, granulosa cells of the corpus luteum contain numerous lipid droplets and mitochondria with lamellar cristae and a much less smooth endoplasmic reticulum. An extensive Golgi apparatus with lysosome-like bodies and a lesser number of mitochondria with tubular cristae and a smooth endoplasmic reticulum are seen in the theca cells. Lipid droplets are also rare. With advancement in age of the corpus luteum (30 h after ovulation), the mitochondrial cristae in the granulosa cells show an atypical arrangement. Lipid droplets are also irregular and vary in size and shape. Cellular organelles and other inclusions coalesce to form electron dense masses. Many lysosome-like bodies appear. Special theca cells show further signs of degeneration.

In salmonids, *O. kisutch* and *O. gorbuscha*, the granulosa cells of the corpus luteum are cuboidal with short cytoplasmic projections extending into the lumen. The apical cytoplasm contains numerous small spherical vacuoles and a few Golgi bodies. Cytoplasmic vacuoles of various sizes and lysosome-like bodies, indicative signs of degeneration of the granulosa cells, are seen. Also, the cellular organelles and other inclusions coalesce to form masses of electron-dense material. Cellular organelles of special theca cells are almost similar to those in the preovulatory follicles, but for a few differences. The endoplasmic reticulum becomes more abundant and some peripherally situated smooth endoplasmic reticulum forms concentric whorls. Lipid droplets appear in some cells. In general, the granulosa and/or theca cells of the corpora lutea of teleosts show the ultrastructural features of steroidogenic cells for some time following ovulation.

Steroidogenic potential of corpus luteum

Some histochemical studies report the steroidogenic ability of the corpora lutea in bony fishes. In *H. galli* the corpora lutea react negatively to the Schultz test, suggesting the absence of steroid precursors in them (Mackay, 1973). The presence of Δ^5-3β-HSDH activity is reported only in the granulosa cells in *O. latipes* (Iwasaki, 1973; Kagawa and Takano, 1979) and *B. rerio* (Lambert and Van Oordt, 1974) and in both granulosa and special theca cells or only theca cells in *S. scomber* (Bara, 1965a, b), *T. mediterraneus* (Bara, 1974), *C. auratus* (Khoo, 1975; Nagahama et al., 1976), *C. kisutch* and *O. gorbuscha* (Nagahama et al., 1978), and *S. gairdneri* (Van den Hurk and Peute, 1979). In *S. leucomaenis* Δ^5-3β-HSDH activity was observed only in the special theca cells although the granulosa cells exhibited some ultrastructural features generally associated with steroid-producing cells (Kagawa et al., 1981).

Some recent biochemical studies confirm the steroidogenic potential of the corpora lutea in teleosts. The high level of plasma progesterone following ovulation has been correlated to an intense steroidogenic activity in the special theca cells of corpora lutea in white spotted char (Kagawa et al., 1981). *In vitro* studies on isolated corpora lutea of amago salmon have provided direct evidence of steroidogenesis of young corpora lutea (Nagahama and Kagawa, 1982). Further, the high level of 17α-20β-diOH-progesterone after ovulation is associated with the hypertrophied granulosa lutein cells in these fishes (Young et al., 1983). Partially purified salmon gonadotropin (SG-G100) stimulated production of progesterone and to a lesser extent testosterone from young corpora lutea but not from the older ones. The possible physiological significance of these steroids in the postovulatory phase is not known.

Whether corpora lutea of fish produce any hormone other than steroids is still not known. The increased prostaglandin (PGF) production following ovulation in some species of teleosts has been attributed to the presence of ovulated oocytes within the ovarian lumen (Goetz, 1983).

Significance of corpora lutea

Opinions differ with regard to the role of corpora lutea in fishes. In elasmobranch fishes, Mathews (1955) suggested that luteal hormones may stimulate the hypertrophy of the uterine mucosa to produce "trophonemata" (uterine folds) in viviparous elasmobranchs during pregnancy. However, Chieffi (1961, 1962) found a correlation between lengthening of the trophonemata and an increase in number of atretic follicles rather than with corpora lutea in *T. marmorata*. Subsequently, Lupo di Prisco (1968) demonstrated *in vitro* production of steroids by corpora atretica of this species supporting, but not confirming, Chieffi's view. The only study on hypophysectomy performed so far, after the eggs had entered the uterus, demonstrated that in *M. canis* there was no effect on normal development of the embryo at least up to 3.5 months after surgery (Hisaw and Abramowitz, 1939). The study suggests that pituitary hormones may not be required in the formation of corpora lutea and the authors further suggested that these are not essential for maintenance of gestation in the fish. However, the yolk sac of *M. canis* does not fuse with the uterine wall until the fourth month of gestation and therefore the above conclusion is doubtful (Chieffi and Botte, 1970). Besides, to rule out the involvement of the corpus luteum in gestation the authors should have performed an ovariectomy or deluteinization. Hence comparative studies on other species are needed to understand the functional significance of the corpus luteum in gestation in elasmobranchs.

The corpora lutea are also known to have steroidogenic potential in some oviparous and all viviparous teleosts having intraovarian gestation (Saidapur, 1982; Nagahama, 1983; Xavier, 1987). However, the role of luteal hormones in oviparous forms and in the maintenance of gestation in viviparous teleosts is not known. Studies involving deluteinization have not been performed on viviparous teleosts; hence no data is available on the precise role of the corpora lutea in normal gestation in viviparous teleosts. The role of hypophyseal hormones in maintenance of corpora lutea in teleosts is also not clear. In *Mollienisia latipinna*, hypophysectomy did not affect gestation and embryonic mortality (Ball, 1962), while in *Gambusia* hypophysectomy increased embryonic mortality unless replacement therapy with salt solution and/or prolactin was done (Chambolle, 1964). Prolactin may not be a primary gestational hormone per se as this hormone is essential for maintenance of ionic balance in many freshwater fishes.

There is another line of evidence that the corpora lutea in some teleosts might have a role in courtship and nesting behavior. Such species are believed to have relatively long-lived corpora lutea (Lam et al., 1978a, b; 1979). These authors opine that the steroids secreted by corpora lutea have a role in maintenance of ovulated eggs within the ovarian cavity. The hormone(s) may do so directly and/or by stimulating the ovarian epithelium to secrete the cavity fluid in which the ovulated eggs are bathed. Their studies also indicate a relationship between the longevity of the corpora lutea and nature of reproductive behavior. In species, such as zebrafish and goldfish, which exhibit simple reproductive behavior and no mate selection and nest building, the corpus luteum is short lived. In Kongo cichlids, which display elaborate reproductive behavior involving mate selection and nest building, corpora lutea have a longer secretory life.

TERMINOLOGY

Various interchangeable terminologies have been used in research articles on the fish ovary to designate the degenerating follicles before ovulation and the structures that evolve from the follicular wall after ovulation in the fish ovary. Terms such as "corpora atretica," "preovulatory corpora lutea," "corpus luteum preovulationis," "corpus luteum atreticum" have been used to refer to the degenerating follicles. Because of their suspected endocrine role in some species, the terms "preovulatory corpora lutea" or "corpus luteum preovulationis" have been used to describe atretic follicles; these terms are erroneous as these follicles arise from follicular degeneration along with oocytes *in situ*. Therefore, a more precise term for any degenerating follicle *in situ*, be it small previtellogenic or large preovulatory vitellogenic, should be "corpus atreticum" or simply "atretic follicle".

The terms "ruptured follicle," "discharged follicle," "postovulatory follicle," "postovulatory corpus luteum," and "corpus luteum" have been used for the follicular remains following ovulation in the fish ovary. This has been due mainly to the fact that their role(s) in reproduction was not properly understood and whether or not they have a similar role as that of corpora lutea in mammals. In most oviparous fishes these structures of postovulation are frequently termed "postovulatory follicles" while in viviparous fishes they are designated "corpora lutea." Since some recent studies have pointed out that these structures show signs of luteinization in both oviparous and viviparous fishes (though short lived in the former), to avoid confusion and to maintain uniformity in terminology, it is suggested that all structures arising from follicular envelopes following ovulation be termed "corpora lutea" in fishes.

CONCLUSIONS AND FUTURE DIRECTIONS

The various histological, ultrastructural, histochemical, and biochemical changes occurring in the atretic follicles and corpora lutea in both cartilaginous and bony fishes have been reviewed.

Follicular atresia occurs at any stage of follicular development but the mode of atresia may differ in different follicles within the same ovary or in different species. In general, in most cases the granulosa cells hypertrophy, phagocytize the oocyte, and ultimately degenerate, leaving behind hypertrophied theca calls in the ovarian stroma. A few studies have pointed out that the interstitial gland cells in the ovary arise from the theca cells of large preovulatory atretic follicles in fish. Such studies are limited, however, to a few species and need to be extended before any generalization is possible.

Opinions differ with regard to the steroid synthesizing capacity of the atretic follicles in fish. Some authors hold that the atretic follicles are merely abortive follicles while others credit them with steroidogenic potential. It may be pointed out here that little progress has been made to resolve this aspect in the recent past and a careful correlative histochemical and ultrastructural study of both previtellogenic and vitellogenic atretic follicles and *in vitro* incubation studies with labeled steroid precursors should be carried out in a number of fish species to understand the endocrine potential and functional significance of atretic follicles in the ovarian physiology of fish.

Studies on the corpora lutea in fish have revealed that these structures degenerate very rapidly in most oviparous forms and are long lived in all viviparous fishes with ovarian gestation. Histological, histochemical, ultrastructural, and biochemical studies carried out on the corpora lutea of the fish ovary indicate that in all viviparous and some oviparous species the granulosa cells undergo luteinization and synthesize steroids. However, these studies are restricted to a handful of species. The role of the corpora lutea during gestation in viviparous forms, luteogenesis and luteolysis are still not properly understood due to lack of proper experimentation along these lines. Carefully controlled experiments involving hypophyseal hormones, ovariectomy or deluteinization (at least in such species where these techniques could be applied with ease) are needed to understand the precise role of the corpora lutea in gestation in fish. Lam et al. (1978a,b; 1979) proposed a highly innovative approach observing the role of corpora lutea in nesting and courtship behavior and maintenance of ovulated eggs to prevent overripening in the ovarian cavity in some fishes. Such type of studies should be extended to a larger number of species to confirm these views.

Future work should therefore be directed toward filling up the lacunae in our knowledge about the role of atretic follicles and corpora lutea. These findings would elucidate the status of both these structures in the fish ovary.

References

Agarwala, N. and R.K. Dixit. 1977. Pre- and Postovulatory corpora lutea in a freshwater fish, *Puntius sophore*. *Zool. Beitr.* 23:301-310.

Anand, T.C. and B.I. Sunderaraj. 1974. Ovarian maintenance in the hypophysectomized catfish, *Heteropneustes fossilis* (Bloch), with mammalian hypophyseal and placental hormones and gonadal and adrenocortical steroids. *Gen. Comp. Endocrinol.* 22:151-168.

Babu, N. and N.B. Nair. 1983. Follicular atresia in *Amblypharyngodon chakaiensis*. *Z. Mikrosk. Anat. Forsch.* 97:499-504.

Ball, N.J. 1962. Broad production after hypophysectomy in the viviparous teleost, *Millienisia latipinna* Leseur. *Nature* 194:787.

Bara, G. 1960. Histological and cytological changes in the ovaries of mackerel *Scomber scomber* L. during the annual cycle. *Instanb. Univ. Fen. Fak. Mecm. Ser.* B,25:4991.

Bara, G. 1965a. Histochemical localization of Δ^5-3β-hydroxysteroid dehydrogenase in the ovaries of a teleost fish, *Scomber scomber* L. *Gen. Comp. Endocrinol.* 3:197-204.

Bara, G. 1965b. Glucose-6-phosphate dehydrogenase activity in the ovaries of *Scomber scomber* L. *Experientia* 21:638-640.

Bara, G. 1974. Location of steroid hormone production in ovary of *Trachurus mediterraneus*. *Acta Histochem.* 51:91-101.

Barr, W.A. 1963. The endocrine control of the sexual cycle in the plaice, *Pleuronectes platessa* (L.). II. *Gen. Comp. Endocrinol.* 3:197-204.

Barr, W.A. 1968. Patterns of ovarian activity. In E.J.W. Barrington and C.B. Jorgensen (eds.), *Perspectives in Endocrinology*, pp. 164-238. Academic Press, New York.

Beach, A.W. 1959. Seasonal changes in the cytology of the ovary and the pituitary gland of the goldfish. *Can. J. Zool.* 37:615-625.

Belsare, D.K. 1962. Seasonal changes in the ovary of *Ophiocephalus punctatus* Bloch. *Indian J. Fish.* 9:140-156.

Bhat, G.K. and N.H. Dutt. 1989a. Histochemical localization of the steroidogenic enzymes in the ovary of *Puntius sarana* (Hamilton). *J. Reprod. Biol. Comp. Endocrinol.* 1:1-13.

Bhat, G.K. and N.H. Dutt. 1989b. Photoperiodic control on the ovary and pituitary gonadotrophs in *Puntius sarana* (Hamilton). *Arch. Anat. Hist. Embr. Norm. Et. Exp.* 72:113-124.

Bhat, G.K. and N.H. Dutt. 1989c. Histochemical localization of alkaline (EC3.1.3.1) and acid phosphatases (EC3.1.3.2) in the ovary of *Puntius sarana* (Hamilton) (Teleostei). *Zool. Anz.* 5/6:323-330.

Bhujle, B.V., V.B. Nadkarni, and M.A. Rao. 1979. Steroid synthesizing cellular sites in the ovary of the domestic pigeon, *Columba livia*. *Histochem. J.* 11:253-265.

Blanc-Livni, N. 1971. Ovarian histochemistry of the fishes *Cyprinus carpio*, *Mugil capito* and *Tilapia aurea* (Teleostei). *Histochem. J.* 3:405-415.

Borg, B. and T. Van Veen. 1982. Seasonal effects of photoperiod and temperature on the ovary of the three-spined stickleback, *Gasterosteus aculeatus* L. *Can. J. Zool*. 60:3387-3393.

Botte, V. 1963. Osservazione isotologiche ed istochiimiche sui follicoli post-ovulatori e atresici di *Raja* spp. *Acta Med. Romana* 1:117-125.

Brakevelt, C.R. and C.R. McMillan. 1967. Cyclic changes in the ovary of the brook stickleback, *Eucalia inconstans* (Kirtland). *J. Morphol*. 123:373-396.

Bretschneider, L.H. and J.J. Duvyene de Wit. 1947. Sexual endocrinology of nonmammalian vertebrates. *Monogr. Prog. Res. Holland* 11:146.

Browning, H.C. 1973. The evolutionary history of the corpus luteum. *Biol. Reprod*. 8:128-157.

Bye, V.J. 1984. The role of environmental factors in the timing of reproductive cycles. *In* G.W. Potts and R.J. Wooten (eds.), *Fish Reproduction: Strategies and Tactics*, pp. 187-203. Academic Press, New York.

Byskov, A.G. 1978. Follicular atresia. *In* R.E. Jones (ed.), *The Vertebrate Ovary*, pp. 533-562. Plenum Press, New York.

Chambolle, P. 1964. Influence de l'hypophysectomie sur la gestation de *Gambusia* sp. *C.R. Acad. Sci. (Paris)*, Ser. D, 259:3855-3857.

Chan, S.T.H., A. Wright, and J.G. Phillips. 1967. The atretic structures in the gonads of rice field eel (*Monopterus albus*) during natural sex reversal. *J. Zool*. 153:527-539.

Chieffi, G. 1961. La luteogenesi nei salacei ovovivipari: Richerche istologiche e istochimiche in *Torpedo marmorata* e *Torpeda ocellata*. *Publ. Staz-Zool. Napoli* 32:145.

Chieffi, G. 1962. Endocrine aspects of reproduction in elasmobranch fishes. *Gen. Comp. Endocrinol*. Suppl. 1:275-285.

Chieffi, G. and V. Botte. 1961. La luteogenesi in *Scylliorhinus stellaris* selacio oviparo. *Boll. Zool*. 28:203-209.

Chieffi, G. and V. Botte. 1970. The problem of luteogenesis in non-mammalian vertebrates. *Boll. Zool*. 21:16.

Christiansen, H.E. 1971. La reproduction de la merluzza en el mar argentino (merlucciidae, *Merluccius merluccius hubbsi*). *Biol. Inst. Biol. Mar*. 20:5-41.

Craig-Bennett, A. 1931. The reproductive cycle of the three-spined stickleback (*Gasterosteus aculeatus* Linn.). *Phil. Trans. R. Soc. London Ser*. B219:197-279.

Dodd, J.M. 1983. Reproduction in cartilaginous fishes (Chondrichthyes). *In* W.S. Hoar, D.J. Randall, and E.M. Donaldson (eds.), *Fish Physiology*, vol. IX, *Reproduction*, pt. A, pp. 31-95. Academic Press, New York.

Dodd, J.M. and J.P. Sumpter. 1984. Fishes. *In* G.E. Lamming, (ed.), *Marshall's Physiology of Reproduction*, vol. 1: *Reproductive Cycles of Vertebrates*, pp. 1-126. Churchill-Livingstone, New York.

Donaldson, E.M. 1973. Reproductive endocrinology of fishes. *Amer. Zool*. 13:909-927.

Eller, L.L. 1981. Histopathological lesions in cut-throat trout, *Salmo salmo clarki*, exposed chronically to the insecticide endrin. *Amer. J. Path*. 64:321-336.

Fostier, A., B. Jalabert, R. Billard, B. Breton, and Y. Zona. 1983. The gonadal steroids. *In* W.S. Hoar, D.J. Randall, and E.M. Donaldson (eds.), *Fish Physiology*, vol. IX. *Reproduction*, pt. A, pp. 227-372. Academic Press, New York.

Goetz, F.W. 1983. Hormonal control of oocyte final maturation and ovulation in fishes. *In* W.S. Hoar, D.J. Randall, and E.M. Donaldson (eds.), *Fish Physiology*, vol. IX. *Reproduction*, pt. B, pp. 117-170. Academic Press, New York.

Gokhale, S.V. 1957. Seasonal histological changes in the gonads of the whiting (*Gadus merlangus* L.) and the Norway Pout (*Gadus esmarkii* Nelson). *Indian J. Fish* 4:92-112.

Gouder, B.Y.M., V.B. Nadkarni, and M.A. Rao. 1979. Histological and histochemical studies on follicular atresia in the ovary of the lizard, *Calotes versicolor*. *J. Herpetol*. 13:451-465.

Guraya, S.S. 1972. Histochemical observations on the interstitial gland cells of dogfish ovary. *Gen. Comp. Endocrinol*. 18:409-412.

Guraya, S.S. 1973. Follicular atresia. *Proc. Indian Nat'l. Sci. Acad*. 39B:311-332.

Guraya, S.S. 1976. Recent advances in the morphology, histochemistry and biochemistry of steroid synthesizing cellular sites in nonmammalian vertebrate ovary. *Int. Rev. Cytol*. 44:365-409.

Guraya, S.S. 1979. Recent advances in the morphology and histochemistry of steroid synthesizing cellular sites in the gonads of fish. *Proc. Indian Nat'l. Sci. Acad*. 45B:452-461.

Guraya, S.S. 1985. *Biology of Ovarian Follicles in Mammals*. Springer, Heidelberg.

Guraya, S.S. 1986. The cell and mollecular biology of fish oogenesis. *In* H.E. Sauer (ed.), *Monographs in Developmental Biology*, vol. 18. S. Karger, Basel.

Guraya, S.S. and S. Kaur. 1979. Morphology of the postovulatory follicle (or corpus luteum) of teleost (*Cyprinus carpio* L.): Ovary. *Zool. Beitr.* 25:381-390.

Guraya, S.S. and S. Kaur. 1982. Cellular sites of steroid synthesis in the oviparous teleost fish (*Cyprinus carpio* L.): Histochemical study. *Proc. Indian Acad. Sci. (Anim. Sci.)*, 92:587-597.

Guraya, S.S., S. Kaur, and P.K. Saxena. 1975. Morphology of ovarian changes during reproductive cycle of fish, *Mystus tengara* (Ham). *Acta Anat.* 91:22-260.

Guraya, S.S., H.S. Toor, and S. Kumar. 1977. Morphology of ovarian changes during the reproductive cycle of the *Cyprinus carpio communis* (Linn). *Zool. Beitr.* 23:405-437.

Hisaw, F.L. and A.A. Abramowitz. 1939. Physiology of Reproduction in the Dogfish *Mustelus canis* and *Squalus acanthias*. *Rep. Woods Hole Oceanogr. Inst.* 22.

Hisaw, F.L. and A. Albert. 1947. Observations on the reproduction of the spiny dogfish (*Squalus acanthias*). *Biol. Bull. Woods Hole* 92:187-199.

Hisaw, F.L. Jr. and F.L. Hisaw. 1959. The corpora lutea of elasmobranch fishes. *Anat. Rec.* 135:269-277.

Hoar, W.S. 1965. Comparative physiology: Hormones and reproduction in fishes. *Amer. Rev. Physiol.* 27:51-70.

Hoar, W.S. and Y. Nagahama. 1978. The cellular sources of sex steroids in teleost gonads. *Ann. Biol. Anim. Biochem. Biophys.* 18:893-898.

Honma, Y. 1961. Studies on the endocrine glands of the salmonid fish, ayu, *Plecoglossus altivelis*. IV. The fate of unspawned eggs and the new crop of oocytes in spent ovary. *Bull. Jap. Soc. Sci. Fish.* 27:873-880.

Iwasaki, Y. 1973. Histochemical detection of Δ^5-3β-hydroxysteroid dehydrogenase in the ovary of medaka, *Oryzias latipes*, during annual reproductive cycle. *Bull. Fac. Fish. Hokkaida Univ.* 23:177-184.

Kagawa, H. and K. Takano. 1979. Ultrastructure and histochemistry of granulosa cells of pre- and postovulatory follicles in the ovary of the medaka, *Oryzias latipes*. *Bull. Fac. Fish. Hokkaido Univ.* 30:191-204.

Kagawa, H., K. Takano, and Y. Nagahama. 1981. Correlation of plasma estradiol-17β and progesterone levels with ultrastructure and histochemistry of ovarian follicles in the white spotted char, *Salvelinus leucomaenis*. *Cell Tissue Res.* 218:315-329.

Khoo, K.H. 1975. The corpus luteum of goldfish (*Carassius auratus* L.) and its function. *Can. J. Zool.* 53:1306-1323.

Kuo, C.M. 1982. Induced breeding of grey mullet, *Mugil cephalus*. In C.J.J. Richter and H.J.T. Goos (eds.), *Reproductive Physiology of Fish*, pp. 181-184. Centre for Agric. Publ. and Documentation, Wageningen.

Lam, T.J. 1983. Environmental influences on gonadal activity in fish. In W.S. Hoar, D.J. Randall, and E.M. Donaldson (eds.), *Fish Physiology*, vol. IX, *Reproduction*, pt. B, pp. 65-115. Academic Press, New York.

Lam, T.J., K. Chan, and W.S. Hoar. 1979. Effect of progesterone and estradiol-17β on ovarian fluid secretion in three-spined stickleback *Gasterosteus aculeatus* L. form *trachurus*. *Can. J. Zool.* 57:468-471.

Lam, T.J., Y. Nagahama, K. Chan, and W.S. Hoar. 1978a. Overripe eggs and postovulatory corpora lutea in the three-spined stickleback, *Gasterosteus aculeatus* L. form *trachurus*. *Can. J. Zool.* 56:2029-2039.

Lam, T.J., S. Pandey, Y. Nagahama, and W.S. Hoar. 1978b. Endocrine control of oogenesis, ovulation and oviposition in goldfish. In P.J. Gaillard and H.H. Boer (eds.), *Comparative Endocrinology*, pp. 55-64. Elsevier, Amsterdam.

Lambert, J.G.D. 1970. The ovary of the guppy *Poecilia reticulata*: the atretic follicle, a corpus atreticum or a corpus luteum preovulationis. *Z. Zellforsch.* 107:54-67.

Lambert, J.G.D. 1978. Steroidogenesis in the ovary of *Brachydanio rerio* (Teleostei). In P.J. Gaillard and H.H. Boer (eds.), *Comparative Endocrinology*, pp. 65-68. Elsevier, Amsterdam.

Lambert, J.G.D. and P.G.W.J. Van Oordt. 1974. Steroid transformation *in vitro* by the ovary of the zebrafish, *Brachydanio rerio*. *J. Endocrinol.* 64:73.

Lambert, J.G.D., J.A.J. Mattheij, and P.G.W.J. Van Oordt. 1972. The ovary and hypophysis of the zebrafish, *Brachydanio rerio* during the reproductive cycle. *Gen. Comp. Endocrinol.* 18:602 (abstract).

Lambert, P.G.D., G.I.G.G. Bosman, R. Van den Hurk, and P.G.W.J. Van Oordt. 1978. Annual cycle of plasma oestradiol-17β in the female trout *Salmo garidneri*. *Ann. Biol. Anim. Biochem. Biophys.* 18:923-927.

Lance, V.A. and I.P. Callard. 1969. A histochemical study of ovarian function in the ovoviviparous elasmobranch, *Squalus acanthias*. *Gen. Comp. Endocrinol.* 13:255-267.

Lang, I. 1981. Electron microscopic and histochemical study of the postovulatory follicles of *Perca fluviatilis* L. (Teleostei). *Gen. Comp. Endocrinol.* 45:219-233.

Lehri, G.K. 1968. Cyclical changes in the ovary of the catfish *Clarias batrachus* (Linn.). *Acta Anat.* 69:105-124.

Lofts, B. and H.A. Bern. 1972. Functional morphology of steroidogenic tissues. In D.R. Idler (ed.), *Steroids in Nonmammalian Vertebrates*, pp. 37-125. Academic Press, New York.

Lupo di Prisco, C. 1968. Biosintesi di ormoni steroidi nell'ovario di due specie di selacio: *Scylliorhinus tellaris* (oviparo) e *Torpedo marmorata* (Ovovivipora). *Riv. Biol.* 61:113-146.

Lupo di Prisco, C., V. Botte, and G. Chieffi. 1965. Differenze rella capacita di biosintesi degli ormoni steroidi tra folliculi post ovulatori ed atresici in *Torpedo marmorata* e *Scylliorhinus stellaris*. *Boll. Zool*. 32:185-191.

Mackay, N.J. 1973. The reproductive cycle of the firetail gudgeon, *Hypseliotris galii*. I. Seasonal histological changes in the ovary. *Aust. J. Zool*. 21:53-66.

Mani, K. and P.K. Saxena. 1985. Effect of safe concentrations of some pesticides on ovarian recrudescence in the freshwater murrel, *Channa punctatus* (B1): a quantitative study. *Ecotoxicol. Envir. Saf*. 92:241-249.

Mathews, L.H. 1950. Reproduction in the basking shark *Cetorhinus maximus* (Gunner). *Phil. Trans. R. Soc. London*, Ser. B234:247-316.

Mathews, L.H. 1955. The evolution of viviparity in vertebrates. *In* Chester Jones and P. Eckstein (eds.), *The Comparative Physiology of Reproduction and the Effects of Sex Hormones in Vertebrates*. Reproduced in *Mem. Soc. Endocrinol*. 4:129-144.

Mathews, S.A. 1938. The seasonal cycle in the gonads of *Fundulus*. *Biol. Bull*. (Woods Hole, Mass.) 75:66-74.

Moser, G.H. 1967. Seasonal histological changes in the gonads of *Sebastodes paucispinis* Ayres, an ovoviviparous teleost (family Scorpaenidae). *J. Morphol*. 123:329-354.

Nagahama, Y. 1983. The functional morphology of teleost gonads. *In* W.S. Hoar, D.J. Randall, and E.M. Donaldson (eds.), *Fish Physiology*, vol. IX: *Reproduction*, pt. A., pp. 223-276. Academic Press, New York.

Nagahama, Y. and H. Kagawa. 1982. *In vitro* steroid production in the postovulatory follicles of the amago salmon, *Oncorhynchus rhodurus* in response to salmon gonadotropin. *J. Exp. Zool*. 219:105-109.

Nagahama, Y., K. Chan, and W.S. Hoar. 1976. Histochemistry and ultrastructure of pre- and post-ovulatory follicles in the ovary of the goldfish *Carassius auratus*. *Can. J. Zool*. 54:1128-1139.

Nagahama, Y., W.C. Clark, and W.S. Hoar. 1978. Ultrastructure of putative steroid-producing cells in the gonads of coho, *Oncorhynchus kisutch* and pink salmon, *Oncorhynchus gorbischa*. *Can. J. Zool*. 56:2508-2519.

Nair, P.V. 1963. Ovular atresia and the formation of the so-called "corpus luteum" in the ovary of the Indian catfish, *Heteropneustes fossilis* (Bloch). *Proc. Zool. Soc. Calcutta* 16: 51-61.

Ng, T.B., C.M. Campbell, and D.R. Idler. 1980. Antibody inhibition of vitellogenesis and oocyte maturation in salmon and flounder. *Gen. Comp. Endocrinol*. 41:233-339.

Nicholls, T.J. and G. Maple. 1972. Ultrastructural observations on possible sites of steroid biosynthesis in the ovarian follicular epithelium of two species of cichlid fish, *Cichlasoma nigrofasciatum* and *Haplochromis multicolor*. *Z. Zellforsch*. 128:317-335.

Pant, M.C. 1968. The process of atresia and fate of the discharged follicle in the ovary of *Glyptothorax pectinopterus*. *Zool. Anz*. 18:153-160.

Patil, H.S. and S.K. Saidapur. 1989. Effect of pollution on reproductive cycles. *In* S.K. Saidapur (ed.), *Reproductive Cycles of Indian Vertebrates*, pp. 409-426. Allied Press, New Delhi.

Polder, J.J.W. 1964. On the occurrence and significance of atretic follicles (preovulatory corpora atretica) in ovaries of the bitterling, *Rhodeus amarus* Bloch. *Proc. K. Ned. Akad. Wet*., Ser. C67:218-222.

Rai, B.P. 1966. The corpora atretica and so-called corpora lutea in the ovary of *Tor* (Barbus) *tor* (Ham.). *Anat. Anz*. 119:459-465.

Rajalakshmi, M. 1966. Atresia of oocytes and ruptured follicles in *Gobius giuris* (Hamilton Buchanan). *Gen. Comp. Endocrinol*. 6:378-384.

Ram, R.N. and A.G. Sathyanesan. 1983. Effect of mercuric chloride on the reproductive cycle of the teleostean fish *Channa punctatus*. *Bull. Env. Contam. Toxicol*. 30:24-27.

Rastogi, R.K. 1966. A study of the follicular atresia and evacuated follicles in the viviparous Indian teleost, *Xenentodon cancila* (Ham.). *Acta Biol. Acad. Sci. Hung*. 17:52-63.

Rastogi, R.K. 1968. Occurrence and significance of follicular atresia in the catfish, *Mystus tengara* (Ham.). *Acta Zool*. 3:307-319.

Rastogi, R.K. 1969. The occurrence and significance of ovular atresia in the freshwater mudeel, *Amphipnous cuchia* (Ham.). *Acta Anat*. 73:148-160.

Saidapur, S.K. 1978. Follicular atresia in the ovaries of nonmammalian vertebrates. *Int. Rev. Cytol*. 54:225-244.

Saidapur, S.K. 1982. Structure and function of postovulatory follicles (corpora lutea) in the ovaries of nonmammalian vertebrates. *Int. Rev. Cytol*. 75:243-285.

Saidapur, S.K. and V.B. Nadkarni. 1976. Steroid synthesizing cellular sites in the ovary of catfish, *Mystus cavasius*: Histochemical study. *Gen. Comp. Endocrinol*. 30:457-461.

Saidapur, S.K., S. Pramoda, and M. Pancharatna. 1982. The occurrence of fibrous atretic follicles in the ovaries of *Rana cyanophylctis* and *Rana tigerina*. *Curr. Sci*. 51:1043-1044.

Saksena, D.N. and N.H. Bhargava. 1972. The corpora atretica, postovulatory follicles and spawning periodicity of Indian freshwater guppy, *Glossogobius giuris* (Ham.). *Zool. Jb. Anat*. 89:611-620.

Samuel, M. 1943. Studies on the corpus luteum in *Rhinobatus granulatus* Cuv. *Proc. Indian Acad. Sci*. B18:133-162.

Sanwal, R. and S.S. Khanna. 1972. Atretic and discharged follicles in a freshwater fish, *Channa gachua*. *Anat. Anz*. 130:297-303.

Sathyanesan, A.G. 1961. A histological study of ovular atresia in the catfish *Mystus seenghala* (Sykes). *Rec. Indian Mus*. 59:75-81.

Sathyanesan, A.G. 1962. The ovarian cycle in the catfish *Mystus seenghala*. *Proc. Nat'l. Acad. Sci. India* B28:497-506.

Saxena, P.K. and K. Anand. 1977. A comparison of ovarian recrudescence in the catfish *Mystus tengara* (Ham.) exposed to short photoperiods, to long photoperiods and to melatonin. *Gen. Comp. Endocrinol*. 33:506-511.

Saxena, P.K. and R. Bhatia. 1983. Effect of vegetable oil factory effluent on ovarian recrudescence in the freshwater teleost, *Channa punctatus* (B1). *Water Air Soil Pollut*. 20:55-61.

Schreibman, M.P., E.J. Berkawitz, and R. Van den Hurk. 1982. Histology and histochemistry of the testis and ovary of the platyfish, *Xiphophorus maculatus*, from birth to sexual maturity. *Cell Tissue Res*. 224:81-88.

Shanbhag, A.B. and V.B. Nadkarni. 1981. Identification of steroidogenic tissue in the ovary of a fish, *Channa gachua*. *J. Anim. Morph. Physiol*. 28:128-134.

Shanbhag, A.B. and V.B. Nadkarni. 1982. Occurrence of cystic atresia in the ovary of *Channa gachua*. *Curr. Sci*. 51:420-422.

Shrivastava, S.S. 1969. Formation of corpora atretica in *Notopterus notopterus* (Pallas). *Acta Zool*. 50:77-89.

Shukla, T.K. and K. Pandey. 1984a. Impaired ovarian functions in an arsenic-treated freshwater fish, *Colisa fasciatus*. *Toxicol. Lett. Amsterdam* 20:1-4.

Shukla, T.K. and K. Pandey. 1984b. Arsenic-induced structural changes during the ovarian cycle of a freshwater perch, *Colisa fasciatus*. *Bull. Inst. Zool. Acad. Sin. Taipei* 23:69-74.

Singh, H. and T.P. Singh. 1980a. Effect of two pesticides on ovarian ^{32}P uptake and gonadotropin concentrations during different phases of annual reproductive cycle in the freshwater catfish, *Heteropneustes fossilis* (Bloch). *Environ. Res*. 22:190-200.

Singh, H. and T.P. Singh. 1980b. Effect of two pesticides on total lipid and cholesterol contents of the annual reproductive cycle for the freshwater teleost, *Heteropneustes fossilis* (Bloch). *Environ. Pollut*. 23:9-17.

Singh, H. and T.P. Singh. 1980c. Short-term effect of two pesticides on the survival, ovarian ^{32}P uptake and gonadotropic potency in a freshwater catfish *Heteropneustes fossilis* (Bloch). *J. Endocrinol*. 85:193-199.

Stenger, A.H. 1959. A study of the structure and development of certain reproductive tissue of *Mugil cephalus* Linn. *Zoologica* (N.Y.) 44:53-70.

Stolk, A. 1951. Histo-endocrinological analysis of gestation in the cyprinodont *Lebistes reticulatus*, Peters II. The corpus luteum cycle during pregnancy. *Proc. K. Ned. Akad. Wet*., Ser. C. 54:558-565.

Sufi, G.B., K. Mori, and R. Soto. 1980. Histochemical changes in activities of dehydrogenases related to steroidogenesis in salmonid fishes (genus *Oncorhynchus*) during sexual maturation and spawning. *Tohoku J. Agric. Res*. 31:74-96.

Sunderaraj, B.I. and S.V. Goswami. 1968. Effect of short and long term hypophysectomy on the ovary and interrenal of catfish, *Heteropneustes fossilis* (B1). *J. Exp. Zool*. 168:85-104.

Sunderaraj, B.T. and P. Keshavnath. 1976. Effect of melatonin and prolactin treatment on the hypophyseal-ovarian system in the catfish *Heteropneustes fossilis* (Bloch.). *Gen. Comp. Endocrinol*. 29:84-96.

Szöllösi, D., B. Jalabert, and B. Breton. 1978. Postovulatory changes in the theca folliculi of the trout. *Ann. Biol. Anim. Biochem. Biophys*. 18:383-391.

Te Winkel, L.E. 1972. Histological and histochemical studies of post-ovulatory and preovulatory atretic follicles in *Mustelus canis*. *J. Morphol*. 136:433-457.

Tromp-Blom, N. 1959. The ovaries of *Gasterosteus aculeatus* (L.) (Teleostei) before, during and after the reproductive period. *Proc. K. Ned. Akad. Wet*., Ser. C62:225-237.

Upadhaya, S.N. 1977. *Morphologie des gonades immatures et étude expérimentale de l'induction de la gametogenese chez la truite arc-en-ciel juvenile* (*Salmo gairdneri* R). Ph.D. thesis, Univ. Paris.

Van den Hurk, R. and J. Peute. 1979. Cyclic changes in the ovary of the rainbow trout, *Salmo gairdneri*, with special reference to sites of steroidogenesis. *Cell. Tissue Res*. 199:289-306.

Van den Hurk, R. and C.J.J. Richter. 1980. Histochemical evidence for granulosa steroids in follicle maturation in the African catfish *Clarias lazera*. *Cell Tissue Res*. 211:345-348.

Van Ree, G.E. 1976. Effect of methallibure (ICI33,828) and ovine luteinizing hormone on the ovary and the pituitary of the female zebrafish, *Brachydanio rerio* I. *Proc. K. Ned. Acad. Wet.*, Ser. C79:150–159.

Vivien, J.H. 1939. Role de l'hypophyse dans le determinisme du cycle genitale female d'un teleosteen, *Gobius paganellus* L. *C.R. Hebd. Seances Acad. Sci.* 208:948–949.

Vlaming de, V.L. 1974. Environmental and endocrine control of teleost reproduction. *In* C. B. Schreck (ed.), *Control of Sex in Fishes*, pp. 13–83. Virginia Polytechnic Inst. State Univ., Blacksburg.

Wooten, R.J. 1982. Environmental factors in fish reproduction. *In* C.J.J. Richter and H.J.T. Goos (eds.), *Reproductive Physiology of Fish*, pp. 210–219. Centre for Agric. Publ. and Documentation, Wageningen.

Xavier, F. 1987. Functional morphology and regulation of the corpus luteum. *In* D.O. Norris and R.E. Jones (eds.), *Hormones and Reproduction in Fishes, Amphibians and Reptiles*, pp. 241–282. Academic Press, New York.

Yamamoto, K. and F. Yamazaki. 1961. Rhythm in the development of the oocytes in the goldfish *Carassius auratus*. *Bull. Fac. Fish. Hokkaido Univ.* 12:93–110.

Yamamoto, K. and H. Onozato. 1968. Steroid producing cells in the ovary of the zebrafish, *Brachydanio rerio*. *Annot. Zool. Jap.* 41:119–128.

Yaron, Z. 1971. Observations on the granulosa cells of *Acanthobrama terrae-sanctae* and *Tilapia nilotica* (Teleostei). *Gen. Comp. Endocrinol.* 17:247–252.

Young, G., L.W. Crim, H. Kagawa, A. Kambegawa, and Y. Nagahama. 1983. Plasma $17\alpha,20\beta$-dihydroxy-4-pregnen-3-one levels during sexual maturation of amago salmon (*Oncorhynchus rhodurus*): Correlation with plasma gonadotropin and *in vitro* production by ovarian follicles. *Gen. Comp. Endocrinol.* 51:96–105.

10

Structure of Fish Locomotory Muscle

Seth M. Kisia

This chapter will by no means deal exhaustively with the locomotory muscle in fish as only a small fraction of the several tens of thousands of living fish species have been studied. Fish have been in existence for more than 450,000,000 years and have undergone several evolutionary lines to occupy almost all the different aquatic habitats available on earth, some as warm as 40°C (such as the springs in the African Rift Valley) and others as cold as 0°C or less (Arctic seas). Fish also occupy freshwater, seawater and salty inland waters, some of which are alkaline. Adult fish range in weight from 1.5 g to over 4,000 kg. The locomotory muscle of fish is likely to vary structurally and physiologically due to the great diversity in form and habitat of the living fish. Studies carried out on some aspects of the gross and fine structure of living fish are discussed here.

GROSS STRUCTURE

The locomotory muscle constitutes a major proportion of the body musculature and body weight in most fishes. For example, it consists of 67% of total fish weight in the rainbow trout, *Salmo gairdneri* (Stevens, 1968), 68% in skipjack tuna, *Katsuwonus pelamis* (Fierstine and Walters, 1968), and 45% in the dogfish, *Scylliorhinus canicula* (Bone and Roberts, 1969). Such a proportion of muscle in relation to body weight is higher than in other vertebrates. The high proportion of muscle need not pose a problem to fish due to the support (buoyancy) received from an aqueous environment as opposed to terrestrial vertebrates. Fish also need a large amount of muscle to generate sufficient power for swimming due to the demands of the density of the medium. The muscle is mainly distributed along the axis of the body trunk with fin muscles usually being relatively small. The caudal fin (most powerful of fins) has the largest amount of direct musculature.

The fish body trunk muscle is mostly segmented. Each segment, known as a myomere, is a disk and constitutes a separate muscle. Myomeres increase phylogenetically in complexity of shape from amphioxus to the higher teleosts. Myomeres are the result of nonfusion of consecutive myotomes during myogenesis, as is the case in most muscles of higher vertebrates in which varying degrees of fusion of myomeres are seen. The number of myomeres is usually identical to the number of vertebrae in teleosts (Harder, 1975).

Myomeres originate from and insert on broad sheets of connective tissue (mainly collagen fibers) known as myosepts (myoseptae). The myosepts are attached to adjacent vertebrae and their vertebral processes (medially) and to the skin (laterally).

Myomeres either go through myosepts to attach to muscle fibers of neighboring myomeres, bend back into the same myomere and merge into neighboring fibers, or merge into abdominal fascia (forms the medial border of myomeres) and superficial fascia (forms the peripheral border of myomeres). The segmented structure permits any one or a number of the discrete segmental units to be contracted.

The axial musculature is divided into dorsal and ventral portions (at the lateral line region) by the horizonal ligament—found in the central part of the trunk, which divides the superficial musculature and part of the deep musculature.

Fibers in muscles of body extremities are arranged as in tetrapod muscles. Muscles are more or less spindle shaped in such areas; they are attached at their points of origin by means of a short tendon or aponeurosis and at their insertion by means of a longer narrow tendon. A guide to the nomenclature of fish muscles has been provided by Winterbottom (1974).

The axial and tail region muscles in fish produce the propulsive force whereas muscle in the body extremities serves to stabilize and steer the body. The role of swimming by fin muscles is important in some fish, such as rays, holocephali, labrids and trachyurids. The caudal fin and the paired pectoral fins are important in most fish for both cruising slowly and swimming rapidly for short bursts.

Postnatal growth of skeletal muscle in fish is by hypertrophic and hyperplastic mechanisms (Weatherly and Gill, 1981; Stickland, 1983; Romanello et al., 1987).

FINE STRUCTURE OF MUSCLE

In general, the fine structure of the fish locomotory muscle is similar to that of other higher vertebrates. As the structure of skeletal muscle in higher vertebrates is found in most textbooks of histology, the fine structure of the fish locomotory muscle will not be covered in this section.

The myomeres are mainly made up of fibers. The fibers insert on both ends of the myosepts and can be long (several centimeters), especially in large fish. The fibers tend to interdigitate at the myosept ends with connective tissue. In most fishes (mainly sharks and higher teleosts) most muscle fibers in myotomes do not run parallel to the long axis of the body, for example in gnathostome fishes, and the deeper muscle fibers are arranged in complex three-dimensional patterns in which some take angles of 30° or more to the long axis of the fish (Alexander, 1969). Although the superficial red muscle fibers can run more or less parallel to the long axis of the body in most fishes, nearly all the white muscle fibers are inclined at angles of 10° or more to the long axis in selachians (Alexander, 1969).

The size and shape of a muscle fiber depends on fiber type, age, activity, and environmental factors, such as nutrition, though fibers tend to be elliptical in cross section. An extreme case of long to short axis ratio of 12:1 has been observed in the red muscle fibers of the anchovy (Johnston, 1982b). This type of arrangement tends to shorten oxygen diffusion distances in fibers. Dimensions of fibers have been determined in several fish species (see section on fiber types). The figures vary—partly because measurements have been taken on fibers of fish of different sizes.

Muscle/fiber types

Three types of fibers, associated with three grossly distinct types of body trunk muscle, are recognized. This differs from most cases of vertebrates in which fibers of different types are mixed within each muscle. The red (slow) muscle is found beneath the lateral line region and is mainly located superficially and runs roughly parallel to the long axis of the body. In cross section the muscle has a flat triangular shape with the blunt tip pointing medially, though it can assume more complicated shapes (Fig. 1). Red muscle is also found in the superficial zone of fins, e.g. rays, and deep in the myotome, e.g. *Rhina*, tuna, and tilapia, or in fish in which the slow fibers are operated above ambient temperature (Carey and Teal, 1966). Red muscle works with high efficiency at the sort of speeds involved in slow cruising for which power requirements are very low. Its proportion varies along the length of the body, with the form of body movements, and with the mode of life of each species. Due to the slow continuous performance of red muscle, it is found in large proportions (relative to other fish) in pelagic species, such as tunas, herrings, and mackerels, and fish capable of continuous and sustained swimming, such as the carp (see Table 1). The proportion of red muscle in fish has been found to scale at a value close to unity in relation to body weight, for example in tilapia (*Oreochromis niloticus*). A scaling value of 1.16 has been observed (Kisia and Hughes, 1992). This value should also be investigated in other fish species.

White (fast) muscle forms the bulk of fish musculature. It is deeper than red muscle and is capable of fast contraction. It is inactive most of the time and is only recruited when rapid bursts of activity are required.

Table 1. Proportion of red muscle in various fish species

Species	% of red muscle in caudal region	% of red muscle overall
Cod (*Gadus morhua*)	17[a]	
Shark (*Galeus melastomus*)	8[a]	
Flounder (*Platichthys flesus*)	12[a]	
Mackerel (*Scomber scombrus*)	18[a]	
Sardina pilchardus	28.9[a]	
Dogfish (*Scylliorhinus stellaris*)	12.4[a]	
Dogfish (*Scylliorhinus canicula*)	18[a]	8[a]
Prionace glauca	22[a]	10–11[a]
Skipjack tuna (*Katsuwonus pelamis*)		10[b]
Rainbow trout (*Salmo gairdneri*)		1[c]
Tilapia (*Oreochromis niloticus*)		0.56–1.28[d]

Source: [a]Greer-Walker and Pull (1975)
[b]Mathieu-Costello et al. (1992)
[c]Stevens (1968)
[d]Kisia and Hughes (1992)

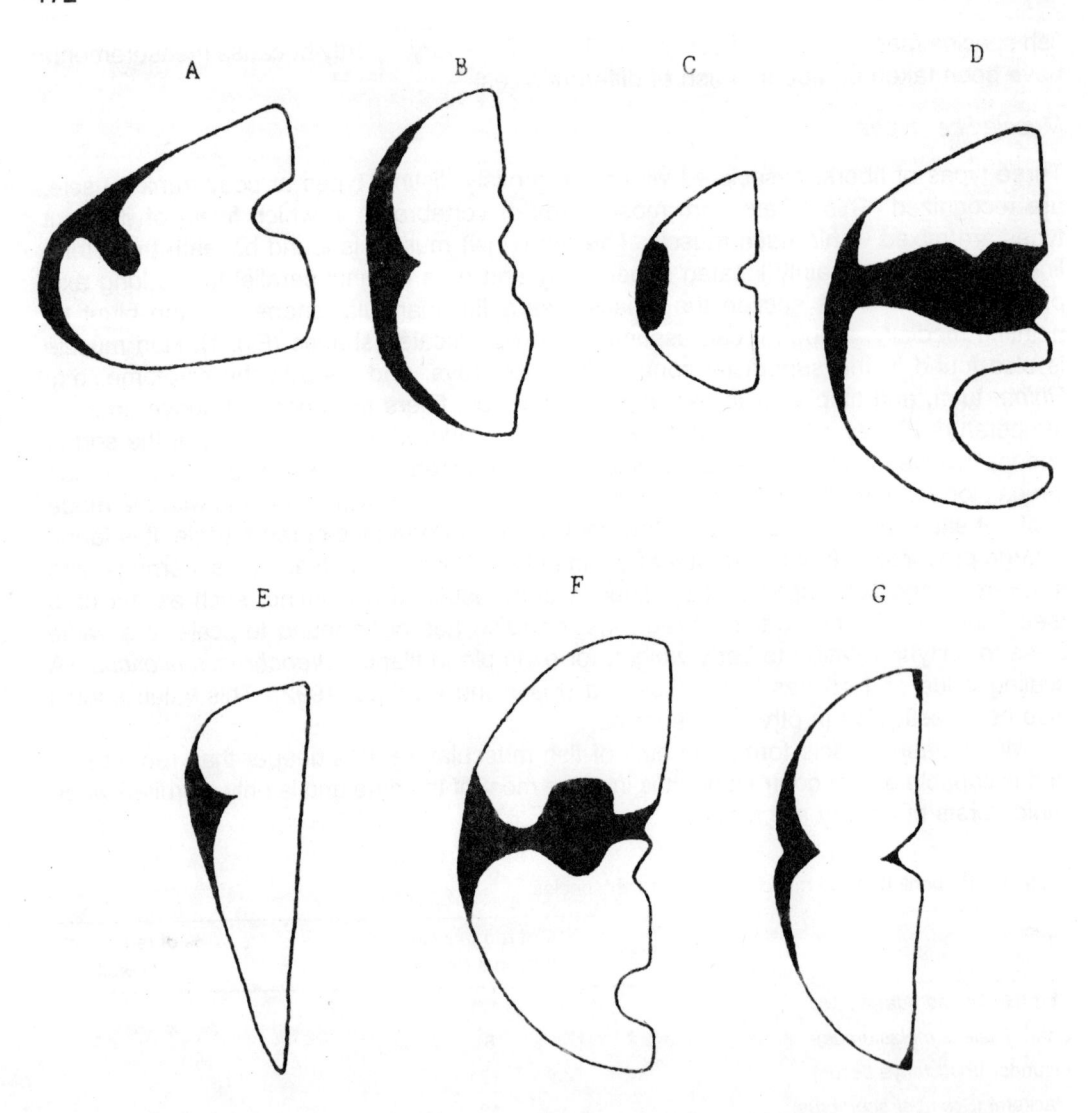

Fig. 1. Diagrams showing various shapes of red muscle (shaded) in cross sections of various fishes. Redrawn from: A to E—Bone (1978); F—Rayner and Keenan (1967), and G—Kisia (1989).

Pink (intermediate) muscle is found beneath the red muscle and is lateral to white muscle. Pink muscle has an intrinsic speed of shortening that is intermediate between that of the red and white muscle fibers.

The red, pink, and white fibers, with different optimum rates of shortening, enable the fish to swim at a range of speeds without much loss in efficiency. During swimming the red fibers are recruited first, the pink fibers next, and then the white fibers (if the speed of swimming has to be increased further). Recruitment is gradual. The presence of these different fiber types enables fish to perform speed ranges of many times over in individual species.

Fiber differences

Red muscle fibers are smaller and more uniform in diameter than white muscle fibers. They have greater quantities of mitochondria, myoglobin, fats, glycogen, and cytochromes, and have a more abundant vascular supply. The fibers respire aerobically and have higher activities of respiratory and citric acid cycle enzymes. Table 2 shows the dimensions of red muscle fibers in some fish. These fibers are 20–50% smaller in diameter than the white muscle fibers. The dimensions of red muscle fibers vary for different fish, depending on several factors, such as age/size of the fish. A scaling factor of 0.22 for red muscle fiber cross-sectional area in relation to body weight has been observed in tilapia, *Oreochromis niloticus* (Kisia and Hughes, 1992). The red fibers increase in size (cross-sectional area and diameter) with fish development.

White muscle fibers have a larger diameter (Table 2), are poorly vascularized, lack myoglobin, have fewer and smaller mitochondria with fewer cristae, have enzymes for anaerobic glycolysis, store glycogen and have little lipid. Buoyancy lipid is found in the white fibers of *Squalus*, *Cetorhinus*, and *Ruvettus* (Bone and Roberts, 1969; Bone, 1972). Lipid is stored in both white and red fibers and among the muscle fibers in herring and mackerel.

A variation to the normal fiber composition is seen in several families of mesopelagic teleosts in which there is great reduction in myofilaments with most of the rest of the fibers being occupied by water (Denton and Marshall, 1958; Blaxter et al., 1971). These fish lack a swimbladder and a reduction in the more dense myofilaments makes them more buoyant. Other mesopelagic teleosts (alepocephalids, stromateoids, and stomatoids) have reduced fiber diameters and large interfiber spaces for fish of similar sizes (Bone, 1978).

There may be several morphologically distinct red or white fibers in a particular fish, as seen in silver and yellow eels (Bostrom and Johansson, 1972), which shows that fibers of different fish in the same habitat may be more related than those of related fish in different habitats.

Mitochondria

These are the main sites for oxygen consumption and subsequent synthesis of ATP. The enzymes involved in ATP synthesis are located in the mitochondrial cristae. The function

Table 2. Comparison of diameters (μm) of muscle fibers in some fish

Species	Red fiber	White fiber	Reference
Skipjack tuna (*Katsuwonus pelamis*)	27	73	Mathieu-Costello et al. (1992)
Skipjack tuna	31	100	Bone (1978)
Kawakawa (*Euthynnus affinis*)	34.6	66.0	George and Stevens (1978)
Tilapia (*Oreochromis niloticus*)	37.6	—	Kisia and Hughes (1992)
Shark (*Galeus melastomus*)	47.5	—	Totland et al. (1981)
Anchovy (*Engraulis encrasicolus*)	37.7	—	Johnston (1982b)

of mitochondria depends to a large extent on the capillary density, which determines the volume of blood and oxygen supply to muscle to varying degrees. Other factors include blood flow rate, area over which diffusion of oxygen will occur, density of mitochondria, and diffusion distances.

A constant concentration of cristae has been observed in the mitochondria of various types of mammalian skeletal muscle and fully active mitochondria operate at a constant rate here (Hoppeler et al., 1987). Should this case apply to fish skeletal muscle, then the volume density of mitochondria should be a fair estimate of the concentration of respiratory chain enzymes in fish muscle fibers.

Mitochondrial volume densities have been determined in a number of fish species (for review see Dunn et al., 1981). The values vary considerably, for example from as low as 0.02 in white muscle of tuna (Hulbert et al., 1979) to as high as 0.455 for anchovy red muscle (Johnston, 1982b). Red muscle fibers have mitochondrial volume densities that are much higher than those of white muscle fibers. The variation seen in mitochondrial volume densities can be attributed to species of fish, temperature, fiber type, activity of fish, and environmental factors. Fish in colder regions of the world tend to have higher red muscle mitochondrial densities than those in warmer regions (Johnston, 1987) and acclimation to lower temperatures tends to increase mitochondrial volume density (Johnston and Maitland, 1980).

Most mitochondria are subsarcolemmal (in location) in fish muscle fibers, the rest being intermyofibrillar. 73.7–91.1% of mitochondria have been found to be subsarcolemmal in red muscle of tilapia, *Oreochromis niloticus* (Kisia, 1993) and are tightly packed in this region (Fig. 2). A slight decrease in volume density of mitochondria and a slight increase in absolute volume of mitochondria has been observed in *O. niloticus* with development (Kisia, 1993), which might suggest that the oxidative metabolism of red muscle does not change much with development. It would be interesting to carry out studies on enzyme activity and oxygen consumption of mitochondria and relate the findings to changes in mitochondrial volumes with development.

Oxygen diffusion distances

These distances are important in determining the rate at which oxygen will diffuse from red blood cells in capillaries to muscle fibers and metabolic wastes will diffuse back to the capillaries from the muscle fibers. Anatomical diffusion distances have been measured in red muscle of fish between capillaries and the radii of muscle fibers supplied by the capillaries as maximal hypothetical distances of 7.1 μm in crucian carp, *Carassius carassius* L. (Johnston, 1982a) and 25.7 μm, 47.5 μm, and 52.0 μm for the sharks, *Galeus*, *Etmopterus*, and *Chimaera* respectively (Totland et al., 1981). Harmonic mean diffusion distances have also been determined in relation to the distribution of mitochondria (from red blood cells in capillaries) in tilapia, *Oreochromis niloticus*, as 3.1 μm (Kisia and Hughes, 1994). The distances vary greatly due to different methods used in measurement and variation in diameter of fibers. Capillary volume densities and diameters and the position a red blood cell will occupy in a large diameter capillary also affect the oxygen diffusion distances. There is a slight increase in oxygen diffusion distances with development (Kisia and Hughes, 1994).

The measured anatomical diffusion distances might differ from the physiological diffusion distances due to blood flow in capillaries, interfiber and intrafiber convection, presence

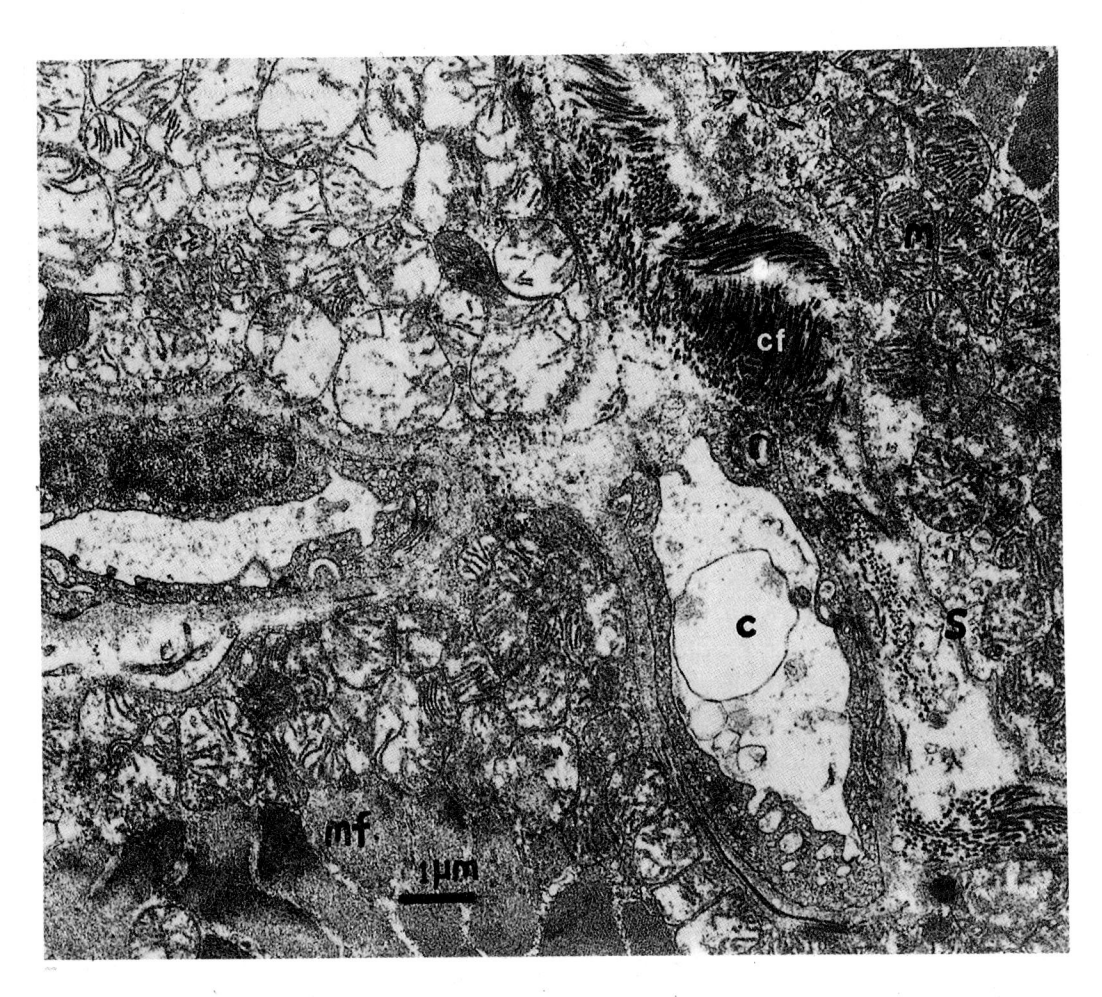

Fig. 2. Photomicrograph of red muscle mitochondria in an 82.7 g tilapia, *Oreochromis niloticus* (from Kisia, 1989). Magnification 7884 ×.

m—mitochondrion; cf—collagen fibers; c—capillary; s—sarcolemma; mf—myofibril.

of protein molecules in the anatomical diffusion path, and muscular contraction. These factors have to be considered before fully appreciating diffusion paths in different muscle types under different physiological conditions.

Blood supply

Muscle capillaries are important in supplying oxygen and nutrients to muscle, removal of different metabolites such as lactate from muscle, and reduction of heat produced mainly during muscle contraction. Blood capillaries are anisotropic (show a high degree of orientation) in relation to muscle fibers. An anisotropy coefficient ($c_2(K,O)$) of 1.016 has been seen in conger eel red muscle (Egginton and Johnston, 1983), and 1.44 in red muscle and 1.16 in white muscle of skipjack tuna, *Katsuwonus pelamis* (Mathieu-Costello et al., 1992). High anisotropy coefficients have been measured in red muscle of the antarctic fishes *Trematomus newnesi* and *Notothemia gibberifrons* as 1.73 and 1.45 respectively (Londraville and Sidell, 1990).

Quantitative data on capillary dimensions and volume densities in fish muscle, which describe the anatomical capillary bed, is scant. The number of capillaries around a muscle fiber varies greatly depending on fiber type and species of fish. Muscle fibers with a high aerobic metabolism have more capillaries (higher volume density) and a greater blood supply. For example, in rainbow trout, *Salmo gairdneri*, red muscle is 1% of body weight and contains 6% of the blood volume whereas white muscle is 66% of body weight and contains 15.8% of the blood volume (Stevens, 1968). The red muscle of the anchovy, *Engraulis encrasicolus*, shows a substantially higher capillary density with 12.9 capillaries in contact with a fiber and the number of capillaries per mm^2 of fiber cross-sectional area 6,000 (Johnston, 1982b). High values for number of capillaries per mm^2 of red muscle fiber area have been observed in tilapia (*Oreochromis niloticus*) as 3,170 (Kisia and Hughes, 1992), tench (*Tinca tinca*) 2,672 (Johnston and Bernard, 1982), and skipjack tuna (*Katsuwonus pelamis*) 2,880 (Mathieu-Costello, 1992). These values are several times higher than those for white muscle fibers. Red muscle capillary cross-sectional area values are available for crucian carp (20.3 μm^2; Johnston and Bernard, 1984), African catfish, *Clarias mossambicus* (20.8 μm^2; Johnston et al., 1983), and *Oreochromis niloticus* (25.1 μm^2; Kisia and Hughes, 1992).

There is a slight increase in number of capillaries in contact with a muscle fiber with increase in fiber diameter though the number of capillaries supplying a unit area of muscle decreases with development. This is due to the larger increase in fiber cross-sectional area with fish development (Kisia and Hughes, 1992).

INNERVATION OF MUSCLE

Muscle fibers are innervated by motor terminations derived from axons arising from the spinal medulla in two adjacent nerve roots. A single axon gives rise to a number of terminations to different muscle fibers.

The red muscle fibers are multiply innervated by small-diameter myelinated motor fibers that form several terminations along the length of a fiber. In teleosts the nerve terminals are usually embedded in the sarcolemma. The red fibers of the hagfish are innervated by two axons that pass onto the fiber at each of its myoseptal insertions (Anderson et al., 1963).

White muscle fibers, except those of most teleosts, are focally innervated at their myoseptal insertions (mainly at the end of the fiber). Most of the white fibers investigated in most teleosts are multiply innervated (Barets, 1961; Hudson, 1969) in a manner similar to that of the red fibers. In elasmobranchs and certain teleost groups, each white fiber is innervated by two separate axons which together contribute to the formation of a single motor end plate.

White muscle fibers of fin muscles in higher teleosts are multiply innervated whereas other groups are focally innervated. Focal innervation is not terminal and found in the midregion of the fibers. The motor end plates may be large and derived from several branches of the same axon which come together at their terminations.

Fish muscle does not contain muscle spindles, an important sensory organ, but has sensory nerve attachments as in higher vertebrates.

CONCLUSION

Fish muscle has received attention since the pioneering work of Lorenzini (1678) who first described the different muscle fiber types. Much study has been carried out on several aspects of fish muscle but our knowledge is still scant as only a few species have been studied from the several classes of living fish, which distinctly differ from each other as do classes of terrestrial vertebrates, amphibians, reptiles, birds, and mammals.

Fish display great diversity in body movement. Most have a streamlined body and vary in swimming behavior: surface swimmers such as needlefishes, halfbeaks, and tapminnows tend to be long and slender and are capable of dart movements; ocean flying fishes such as the South American freshwater flying fishes can swim above the water surface (with the lower lobe of the tail remaining in water); midwater swimmers such as trouts and tunas are fusiform and adapted for strong fast swimming; while those in quiet waters such as the sunfish and freshwater angelfish are not strong fast swimmers. Variation in demands placed on fish by the environment in relation to swimming behavior has brought about differences in the structure and physiology of the fish locomotory muscle. These differences are seen in the different fiber types and their dimensions and proportions in different fish, blood supply, innervation and densities of muscle fiber components, such as mitochondria. Phylogenetic changes are also evident in the gross and fine structure of the locomotory muscle in different orders of fish.

References

Alexander, R. McN. 1969. The orientation of muscle fibres in the myomeres of fishes. *J. Mar. Biol. Assoc. U.K.* 49:263–290.

Andersen, P., J.K.S. Jensen, and Y. Løyning. 1963. Slow and fast muscle fibres in the Atlantic hagfish *(Myxine glutinosa). Acta Physiol. Scand.* 57:167–179.

Barets, A. 1961. Contribution à l'étude des systémes moteurs lent et rapide du muscle latéral des téléostéens. *Arch. Anat. Morphol. Exp.*, Suppl. 50:91–187.

Blaxter, J.H.S., C.S. Wardle, and B.L. Roberts. 1971. Aspects of the circulatory physiology and muscle systems of deep-sea fish. *J. Mar. Biol. Assoc. U.K.* 51:991–1006.

Bone, Q. 1972. The dogfish neuromuscular junction. Dual innervation of vertebrate striated muscle fibres? *J. Cell Sci.* 10:657–665.

Bone, Q. 1978. Myotomal muscle fibre types in *Scomber* and *Katsuwonus*. In G.D. Sharp and A.E. Dizon (eds.), *The Physiological Ecology of Tunas*, pp. 183–205. Academic Press, New York.

Bone, Q. and B.L. Roberts. 1969. The density of elasmobranchs. *J. Mar. Biol. Assoc. U.K.* 49:913–937.

Boström, S.L. and Johansson, R.G. 1972. Enzyme activity patterns in white and red muscles of the eel (*Anguilla anguilla*) at different developmental stages. *Comp. Biochem. Physiol.* 42B:533–542.

Carey, F.G. and J.M. Teal. 1966. Heat conservation in tuna fish muscle. *Proc. Nat. Acad. Sci. USA* 56:1464–1469.

Denton, E.J. and N.B. Marshall. 1958. The buoyancy of bathypelagic fishes without a gas-filled swimbladder. *J. Mar. Biol. Assoc. U.K.* 37:753–767.

Dunn, J.F., W. Davison, G.M.O. Maloiy, P.W. Hochachka, and M. Guppy. 1981. An ultrastructural and histochemical study of the axial musculature in the African lungfish. *Cell Tissue Res.* 220:599–609.

Egginton, S. and I.A. Johnston. 1983. An estimate of capillary anisotropy and determination of surface and volume densities of capillaries in skeletal muscles of the conger eel (*Conger conger* L.). *Quart. J. Exper. Phys.* 68:603–617.

Fierstine, H.L. and V. Walters. 1968. Studies in locomotion and anatomy of scombroid fishes. *Mem. South Calif. Acad. Sci.* 6:1–31.

George, J.C. and E.D. Stevens. 1978. Fine structure and metabolic adaptation of red and white muscles in tuna. *Environ. Biol. Fishes* 3:185–191.

Greer-Walker, M. and G.A. Pull. 1975. A survey of red and white muscles in marine fish. *J. Fish Biol.* 7:295–300.

Harder, W. 1975. *Anatomy of Fishes*. E. Schweizerbart'sche Verlagsbuchhandlung (Nagele V. Obermiller), Stuttgart.

Hoppeler, H., S.R. Hayar, H. Classen, E. Uhlmann, and R.H. Karas. 1987. Adaptive variation in the mammalian respiratory system in relation to energetic demand. III. Skeletal muscles: Setting the demand for oxygen. *Respir. Physiol.* 69:27–46.

Hudson, R.C.L. 1969. Polyneural innervation of the fast muscles of the marine teleost *Cottus scorpius* L. *J. Exp. Biol.* 50:47–67.

Hulbert, W.C., M. Guppy, B. Murphy, and P.W. Hochachka. 1979. Metabolic sources of heat and power in tuna muscles. I. Muscle fine structure. *J. Exp. Biol.* 82:289–301.

Johnston, I.A. 1982a. Capillarisation, oxygen diffusion distances and mitochondrial content of carp muscles following acclimation to summer and winter temperatures. *Cell Tissue Res.* 222:325–337.

Johnston, I.A. 1982b. Quantitative analyses of ultrastructure and vascularization of the slow muscle fibres of the anchovy. *Tissue and Cell* 14:319–328.

Johnston, I.A. 1987. Respiratory characteristics of muscle fibres in a fish *(Chaenocephalus aceratus)* that lacks haem pigments. *J. Exp. Biol.* 133:415–428.

Johnston, I.A. and B. Maitland. 1980. Temperature acclimation in crucian carp, *Carassius carassius* L.: Morphometric analyses of muscle fibre ultrastructure. *J. Fish Biol.* 17:113–125.

Johnston, I.A. and L.M. Bernard. 1982. Routine oxygen consumption and characteristics of the myotomal muscle in tench: Effects of long-term acclimation to hypoxia. *Cell Tissue Res.* 227:161–177.

Johnston, I.A. and L.M. Bernard. 1984. Quantitative study of capillary supply to the skeletal muscles of crucian carp (*Carassius carassius* L.): Effects of hypoxic acclimation. *Physiol. Zool.* 57(1):9–18.

Johnston, I.A., L.M. Bernard, and G.M.O. Maloiy. 1983. Aquatic and aerial respiratory rates, muscle capillary supply and mitochondrial volume density in the air-breathing catfish (*Clarius mosambicus*) acclimated to either aerated or hypoxic water. *J. Exp. Biol.* 105:317–338.

Kisia, S.M. 1989. *Morphometry of gills and red muscle and oxygen consumption in different sizes of a tilapia, Oreochromis niloticus* (Trewavas). Ph.D. thesis, University of Bristol.

Kisia, S.M. 1993. Volume densities and absolute volumes of mitochondria in body trunk red muscle of a tilapia, *Oreochromis niloticus* (Trewavas). *Acta Biol. Hungarica* 44 (2-3):243–248.

Kisia, S.M. and G.M. Hughes. 1992. Red muscle fibre and capillary dimensions in different sizes of a tilapia, *Oreochromis niloticus* (Trewavas). *J. Fish Biol.* 40:97–106.

Kisia, S.M. and G.M. Hughes. 1994. Morphometry of some structural parameters affecting oxygen diffusion in body trunk red muscle at different sizes of tilapia, *Oreochromis niloticus*. *J. Fish Biol.* 44(2):233–239.

Londraville, R.L. and Sidell, B.D. 1990. Maximal diffusion distance within skeletal muscle can be estimated from mitochondrial distributions. *Respir. Physiol.* 81:291–302.

Lorenzini, S. 1678. *Osservazioni intorno alle Torpedini*. Onofri, Florence.

Mathieu-Costello, O., P.J. Agey, R.B. Logemann, R.W. Brill, and P.W. Hochachka. 1992. Capillary-fibre geometrical relationships in tuna red muscle. *Can. J. Zool.* 70:1218–1229.

Romanello, M.G., P.A. Scapolo, S. Luprano, and F. Mascarello. 1987. Post-larval growth in the lateral white muscle of the eel, *Anguilla anguilla*. *J. Fish Biol.* 30:161–172.

Stevens, E.D. 1968. The effect of exercise on the distribution of blood to various organs in rainbow trout. *Comp. Biochem. Physiol.* 25:615–625.

Stickland, N.C. 1983. Growth and development of muscle fibres in the rainbow trout (*Salmo gairdneri*). *J. Anat.* 137:323–333.

Totland, G.K., H. Kryvi, Q. Bone, and P.R. Flood. 1981. Vascularization of the lateral muscle of some elasmobranchiomorph fishes. *J. Fish Biol.* 18:223–234.

Weatherly, A.H. and H.S. Gill. 1981. Characteristics of mosaic muscle growth in rainbow trout *Salmo gairdneri*. *Experientia* 37:1102–1103.

Winterbottom, R. 1974. A descriptive synonymy of the striated muscles of the teleosts. *Proc. Acad. Nat. Sci. Philadelphia* 125:225–317.

11

Morphology of the Swim (Air) Bladder of a Cichlid Teleost: *Oreochromis alcalicus grahami* (Trewavas, 1983), A Fish Adapted to a Hyperosmotic, Alkaline, and Hypoxic Environment: A Brief Outline of the Structure and Function of the Swimbladder

J.N. Maina, C.M. Wood, A. Narahara, H.L. Bergman,
P. Laurent and P. Walsh

INTRODUCTION

Perhaps no other organ in fish has received as great and sustained scientific interest as the swim (air) bladder. The principal aspects which have elicited this interest have been those related to its evolutionary origin, structure, and function. For a review of early works, see Hall (1924), Marshall (1960), Alexander (1966), Fange (1966), Copeland (1969), and Steen (1970). The biophysics of buoyancy control through regulation of gas secretion (against high ambient pressures, particularly in the benthic deepwater fish) and resorption, has been the most intriguing attribute of the organ (Marshall, 1960; Scholander and van Dam, 1953; Scholander, 1954; Alexander, 1966). The current topics were recently reviewed by Pelster and Scheid (1992a, b). The variety of functions carried out by the airbladder in various species is perhaps greater than that of any other organ in fish (Hall, 1924). They include buoyancy control (Denton, 1961; Jones, 1952), respiration (particularly in physostomatous fish) (Jones, 1957; Dehadrai, 1962; Luling, 1964), hearing (Schwartzkopff, 1963; Tavolga, 1964), sound production (Green, 1924; Jones and Marshall, 1953; Schneider and Hasler, 1960; Tavolga, 1962), and pressure sensation (Vasilenko and Livanov, 1936; Koshtoyanz and Vassilenko, 1937; Qutob, 1962). In the exceptional case of Notopteridae (Greenwood, 1963), the airbladder seems to serve all the above functions. The airbladder, however, does not appear to be crucial for life since many fish, such as sharks and rays (elasmobranchs), do not have one and of about 20,000 extant teleostean species, approximately half lack an airbladder, particularly in the adult stage (Jones and Marshall, 1953; Marshall, 1962; Fange, 1966). Some of the well-known fish which lack an airbladder are the European mackerel (*Scomber*), the flounder (*Pleoronectes*), the angler (*Lophius*), and the bullhead (*Cottus*) (Jones and Marshall,

1953). In Scombroidei the airbladder is so variable in size that it is asserted by Jones and Marshall (1953) that its presence or absence has no selective advantage. It is highly probable that it conforms with the biological rule of necessity; whenever an airbladder has been retained, it must serve a definite adaptive role. Almost all species of fish are heavier than water when the airbladder lift is absent (Goolish, 1992), with the muscle density being about 1.05 g/cm^3 (Alexander, 1959) while the density of the vertebral axial skeleton ranges from 1.25 to 1.50 g/cm^3 (Webb, 1990).

Fange (1966) observed that "an intimate relationship between anatomy and function in the airbladder exists" and that "the varying shapes and sizes closely reflect functional diversities." Dobbin (1941) and Jones and Marshall (1953) observed that the structure of the airbladder can be related to the ecology of the teleosts. These general sentiments are supported by the following observations: for buoyancy reasons, adaptively, freshwater fish have larger airbladders than the marine ones (Marshall, 1960; Denton, 1962; Nielsen and Munk, 1963) and in some cases, the size of the airbladder varies remarkably within the same genus. Gee and Northcote (1963) observed that a species of *Rhinichthys* (Cyprinidae) which lives in rapid current has a much smaller airbladder than one which prefers slower currents. A similar observation was made by Hora (1922a, b) on a hillstream fish. Studying the Black Sea fish, Andriashev (1944) observed that the densities of fish with airbladders ranged from 1.012 to 1.021 compared to 1.061 to 1.085 for those without. Alexander (1959) noted that *Gobius flavescens*, the only pelagic British goby, has an airbladder of percentage volume 5.0 and a sinking factor of 1002, while the other British *Gobius* species have percentage volumes between 1.2 and 2.2 and a high negative buoyancy. Deepsea fishes have longer retia than shallow-water forms (Marshall, 1962) and the root effect is not found in fishes lacking an airbladder (Steen, 1970). Experimentally, fish restricted near the surface of the water were observed to be less buoyant (0.91 ml g^{-1}) than those similarly maintained at the bottom of the tank (0.98 ml g^{-1}) (Goolish, 1992).

Mainly due to the notable variability in the organization and function of the airbladder (Jones and Marshall, 1953; Marshall, 1962) and the relatively few species in which the organ has been studied in detail (Copeland, 1969; Morris and Albright, 1975), scant information is available to enable investigators to draw general conclusions regarding the form and function of the organ (Fange, 1966). Only in the cyprinids and salmonids (Dobbin, 1941; Fange and Mattison, 1956; Marshall, 1960), the eel (*Anguilla*) (Woodland, 1911; Krogh, 1936, 1959), and the euphysoclists in general (Scholander and van Dam, 1953, 1954; Scholander, 1954) has the airbladder been investigated to an appreciable detail. To date, electron (transmission) microscopic studies have been carried out only on those of *Fundulus* (Copeland, 1960, 1964, 1969), *Anguilla* (Dorn, 1961), *Opsanus* (Copeland 1964, 1969; Morris and Albright, 1975), *Lotta* and *Acernia* (Jacinski and Kilarski, 1964), *Coregonus* (Fahlen, 1967) and salmon (Fahlen, 1970, 1971). Detailed morphological studies are apparently entirely lacking in the important tilapiine group of fish.

We investigated the airbladder of *Oreochromis alcalicus grahami* (Trewavas, 1983), a tilapiine cichlid fish that lives in the volcanic lagoons of the Kenyan Lake Magadi, a severe habitat where the average temperature is 37°C (Coe, 1966; Wood et al., 1989), the osmolarity 525 mOsm·L^{-1} (Wood et al., 1989), and pH 10 (Johansen et al., 1975). The water is characterized by dramatic diurnal fluctuations in the partial pressure of oxygen (Narahara et al., unpub.). During the day the water is supersaturated with oxygen (due to the photosynthetic activity of the algae) with the PO_2 being in excess of 400 torr and at night it is virtually anoxic (PO_2 < 20 torr) mainly due to the respiratory activity of the abundant cyanobacteria. The airbladder was investigated by gross dissection and latex rubber casting to study its shape, size, and topographic anatomy and its basic structural components using transmission and scanning electron microscopy to establish its ultrastructural and spatial morphology. Our aim was to describe the structural design of the airbladder in this group of fish and to demonstrate the possible adaptive morphological changes the organ may have undergone to contribute to the survival of the fish in its unique habitat. The account also briefly reviews current knowledge of the structure of the airbladder while highlighting recent ideas on the plausible mode of function of this enigmatic organ.

MATERIALS AND METHODS

Specimens of *Oreochromis alcalicus grahami* (Trewavas, 1983) were captured by seine net from the Fish Spring Lagoons (FSL) of Lake Magadi (see Coe, 1966) in the months of January and February 1992. Additional fish were seined from the water-holding tank (WHT) of the Magadi Soda Company, which receives water directly pumped from the FSL, and in which a population of larger fish has developed, protected from avian predators. The fish were kept in aerated plastic containers filled with FSL water at 30-36°C before initiation of physiological and morphological studies. This research is part of an ongoing collaborative multidisciplinary research effort (see Walsh et al., 1993; Wood et al., 1994). The goal of the overall program is to study and understand the morphological and physiological adaptive devices that the fish has acquired during evaluation of the capacity for survival in an exceptionally severe habitat.

Gross dissection

For ease of identification and dissection, the largest of the specimens, especially those from the water-holding tank were selected. They ranged in body mass from 13 to 50 g. The fish were held out of water for about a minute to prompt them to inflate the airbladder and were subsequently euthanized. A cranicocaudal ventromedian incision was made through the gill arches and continued to the abdomen to expose the heart and the visceral organs. The topographic relationships between the gills, the airbladder, and the abdominal organs in particular the kidneys, were noted. A cannula was carefully inserted into the esophagus and subsequently pushed into the airbladder to establish the location of the connection between the esophagus and the bladder. The airbladder was removed for microscopic study after carefully isolating it from the kidneys and the ventrally located peritoneal membrane to which it is closely connected.

Latex casting of the gills and the airbladder. Stock solution of latex rubber (Latex White ZCP-652-OLOR, Griffin & George Ltd., U.K.) was diluted to 50% and colored red (for injection into the vasculature) and blue (for injection into the airbladder) to make the organs clearly visible. In vascular injection the heart was exposed and the ventricle cannulated. Physiological saline was infused from a pressure head of 30 cm (the perfusate running out through an incision made across the body just caudal to the anus) until the gills were seen to bleach. The airbladder was cannulated and the latex rubber injected simultaneously into the vasculature and the airbladder by applying gentle but continuous pressure onto the plunger of a 5-ml syringe. The injection was stopped as soon as the tissues were seen to tense with latex rubber. Signs of extravasations were continuously checked. After the injection, the cannulas were blocked in order to keep the latex in place and the specimen was kept at room temperature for 12 h to allow the latex to set.

The preparation was immersed in concentrated hydrochloric acid for maceration and finally flushed with distilled water before examination. Some of the preparations were cut into small pieces for study with the scanning electron microscope.

Transmission electron microscopy. The isolated airbladder (after gross dissection) was cut open, placed on a paraffin board, and pins placed at the corners to keep it stretched (to avoid undue folding and shrinkage). The preparation was immersed in 2.5% glutaraldehyde buffered in sodium phosphate (osmolarity 560 mOsmL^{-1}, pH 7.4) for 24 hrs. The airbladder was cut into small pieces which were processed for electron microscopy through standard techniques, entailing dehydration in ethanol, postfixation in 2% osmium tetroxide, infiltration and embedding in TAAB resin, counterstaining in uranyl acetate and examination on a JEOL 1200EX electron microscope.

Scanning electron microscopy

The latex rubber cast preparations of the airbladder and its vasculature as well as fixed tissues were dehydrated in absolute alcohol, critical-point dried in liquid carbon dioxide, sputtercoated with gold-palladium complex, and viewed on an ISI-DS 130 dual-stage scanning electron microscope at an accelerating voltage of 15 kV.

RESULTS

Oreochromis alcalicus grahami (Fig. 1) has a well-developed thin transparent airbladder (Figs. 2, 3, 4) which is connected to the esophagus through a short muscular pneumatic duct. Topographically, the bladder is located dorsal to the peritoneal cavity, ventral to the vertebral column, and dorsal to the alimentary tract. The airbladder is a rather conical membranous sac which starts just above the cardiac ventricle and decreases in size, tapering at a point close to the anus and the end of the peritoneal cavity. Ventrally, it is covered by the gonads (Fig. 2). A highly pigmented peritoneal membrane separates the airbladder from the kidneys and the peritoneal cavity (Figs. 2, 3).

Grossly, the airbladder has a silvery appearance except for reddish spots located near the middle of the organ on the cranioventral aspect; it is connected to the buccal cavity through a narrow sphincter. Examination under the electron microscope revealed that the red spots were the sites of the rete mirabile (Figs. 6,7), the color being attributable to the profuse degree of capillarization in such areas of the bladder. Also grossly discernible were transverse and longitudinal folds, observable both in the isolated preparations as well as in the *in situ* organs (Figs. 2, 4). This suggests that the folds may be normal structural features of the bladder which may facilitate its distension upon inflation. It was found in a collapsed state in almost every attempt at immediate removal the moment the fish was lifted out of water. Successful dissection required inducing the fish to gulp air. Once removed, the bladder did not collapse, a direct indication of the presence of a functional sphincter at the entrance.

Fig. 1. *Oreochromis alcalicus grahami*, a cichlid fish that lives in the alkaline Lake Magadi of Kenya. The fish grow to a maximum length of 12 cm but in protected surroundings may reach 16 cm. O—operculum.

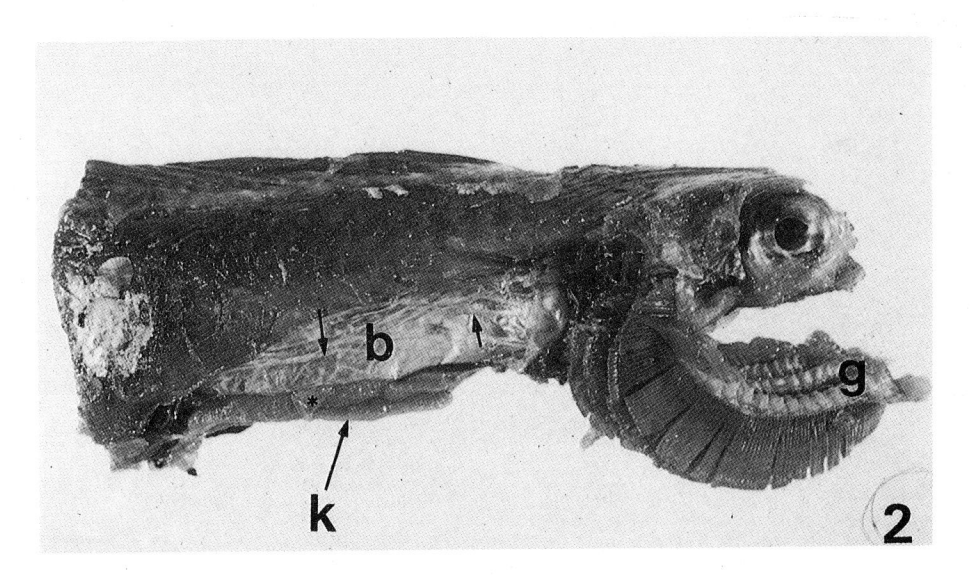

Fig. 2. A gross dissection of *O. a. grahami* showing the gill arches (g), swimbladder (b), and gonads (k). Arrows—folds on the wall of the swimbladder; star—peritoneal membrane separating the swimbladder from the gonads.

Fig. 3. A gross dissection of the side view of *O. a. grahami* showing the buccal cavity (asterisk) and a narrow sphincteric passage (arrow) running from the buccal cavity to the swimbladder. Star—peritoneal membrane.

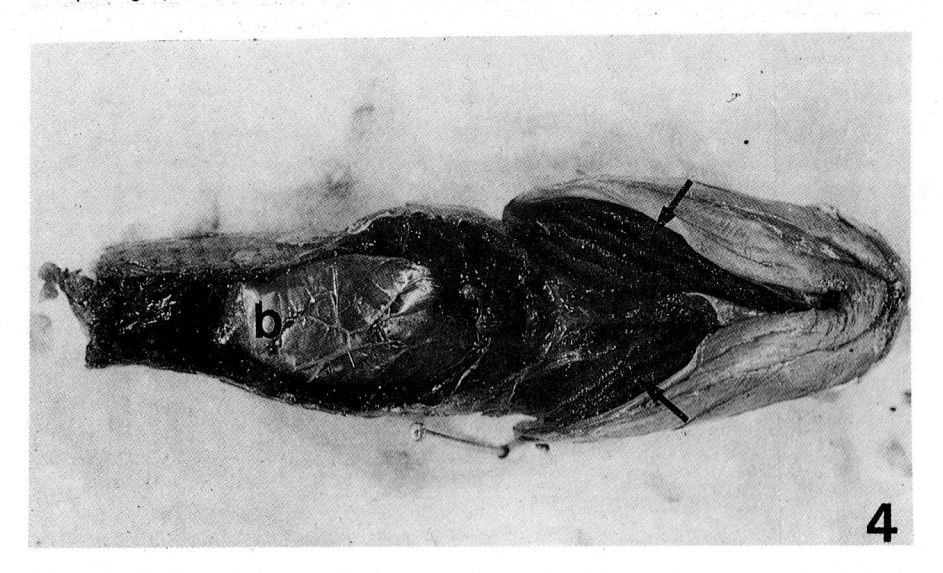

Fig. 4. A ventral view of a gross dissection of *O. a. grahami* showing the gills (arrows) and the swimbladder (b). The folds on the wall of the swimbladder are visible.

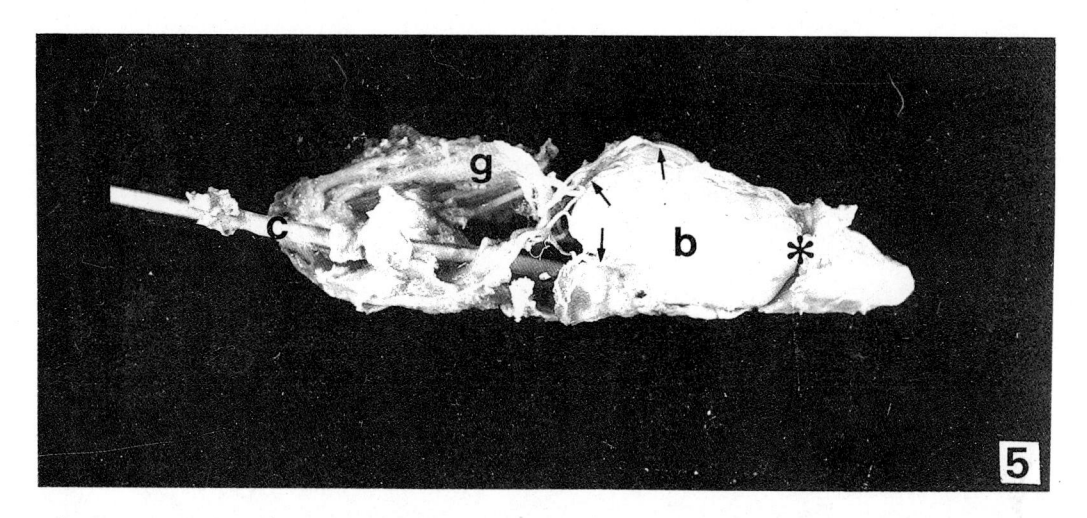

Fig. 5. Latex cast preparation of the gill arches (g) and the swimbladder (b) of *O. a. grahami*. The bladder is moderately internally divided (asterisk) into a larger cranial chamber and a small caudal one. c—cannula showing the connection between the buccal cavity and the swimbladder; arrows—venous blood drainage from the swimbladder; × 15.

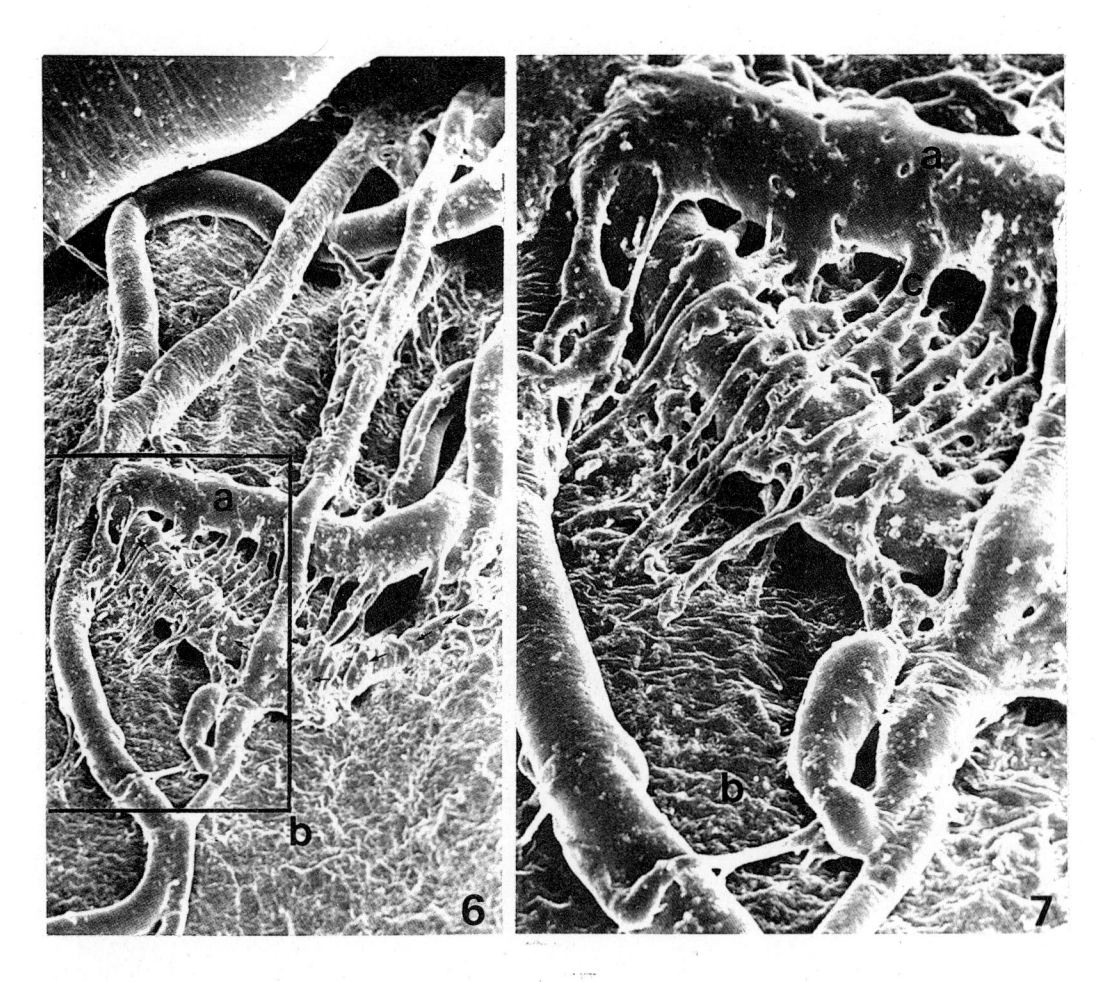

Figs. 6 and 7. Close-up SEM of the rete mirabile of the swimbladder of *O. a. grahami*. a—terminal arterioles; arrows—venules; b—surface of bladder; c—capillaries. Fig. 7 is an enlargement of the enclosed area in Fig. 6. Fig. 6 × 70; Fig. 7 × 210.

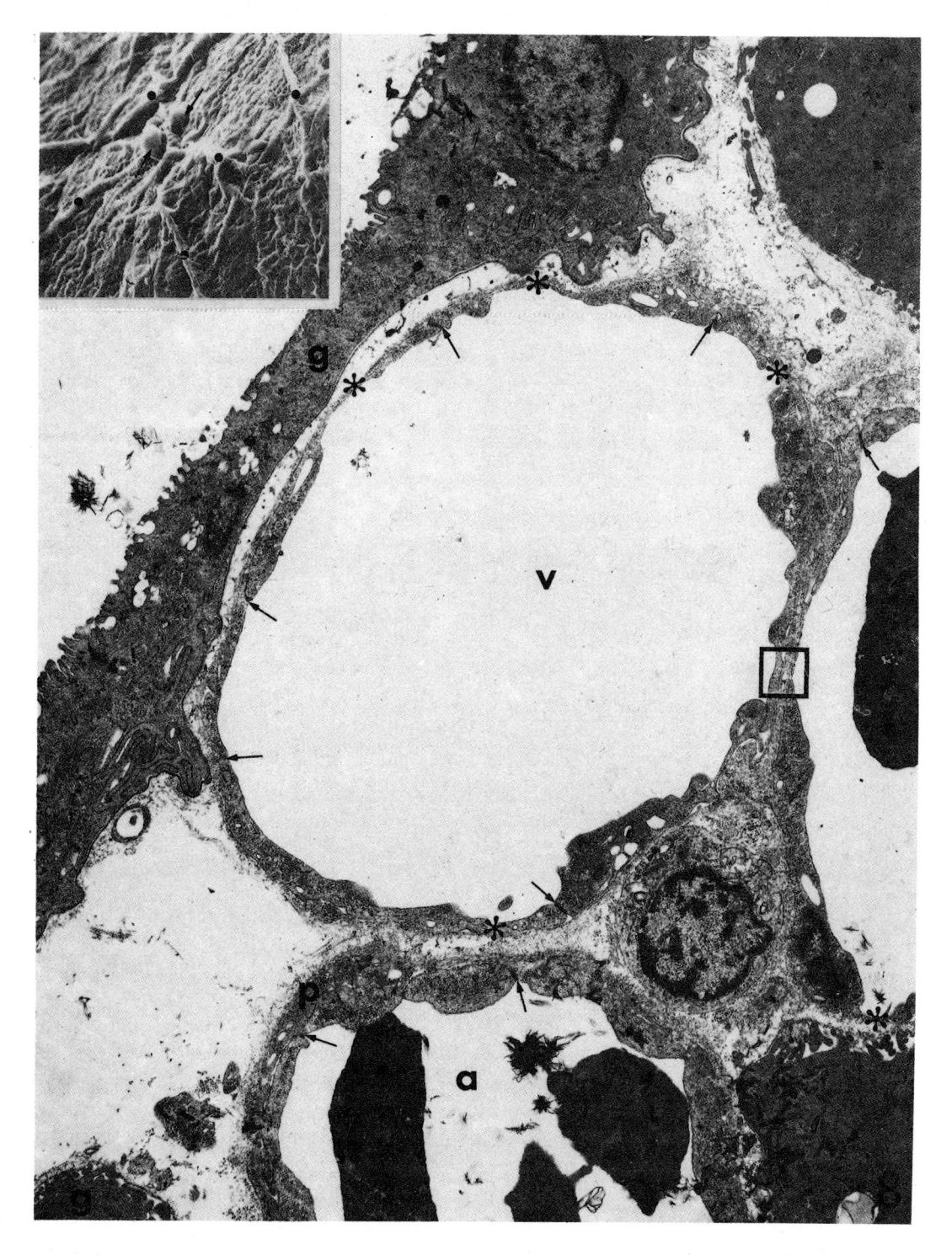

Fig. 8. Transmission electron micrograph of the vascular system of the rete mirabile of the swimbladder of *O. a. grahami*. The venular capillaries (v) are wider and more superficially located while the arteriolar ones (a) are more deeply situated and have a relatively thicker wall. g—epithelial cell with basal plasma membrane foldings; arrows—capillary endothelial cell tight junctions; asterisks—extremely attenuated areas of the endothelial cell; p—pericytes; square—site where venular capillaries lie in close proximity; × 18,000. Inset: view of the internal surface of the swimbladder of *O. a. grahami* showing intense vascularization (dots) and gas-secretory cells; × 250.

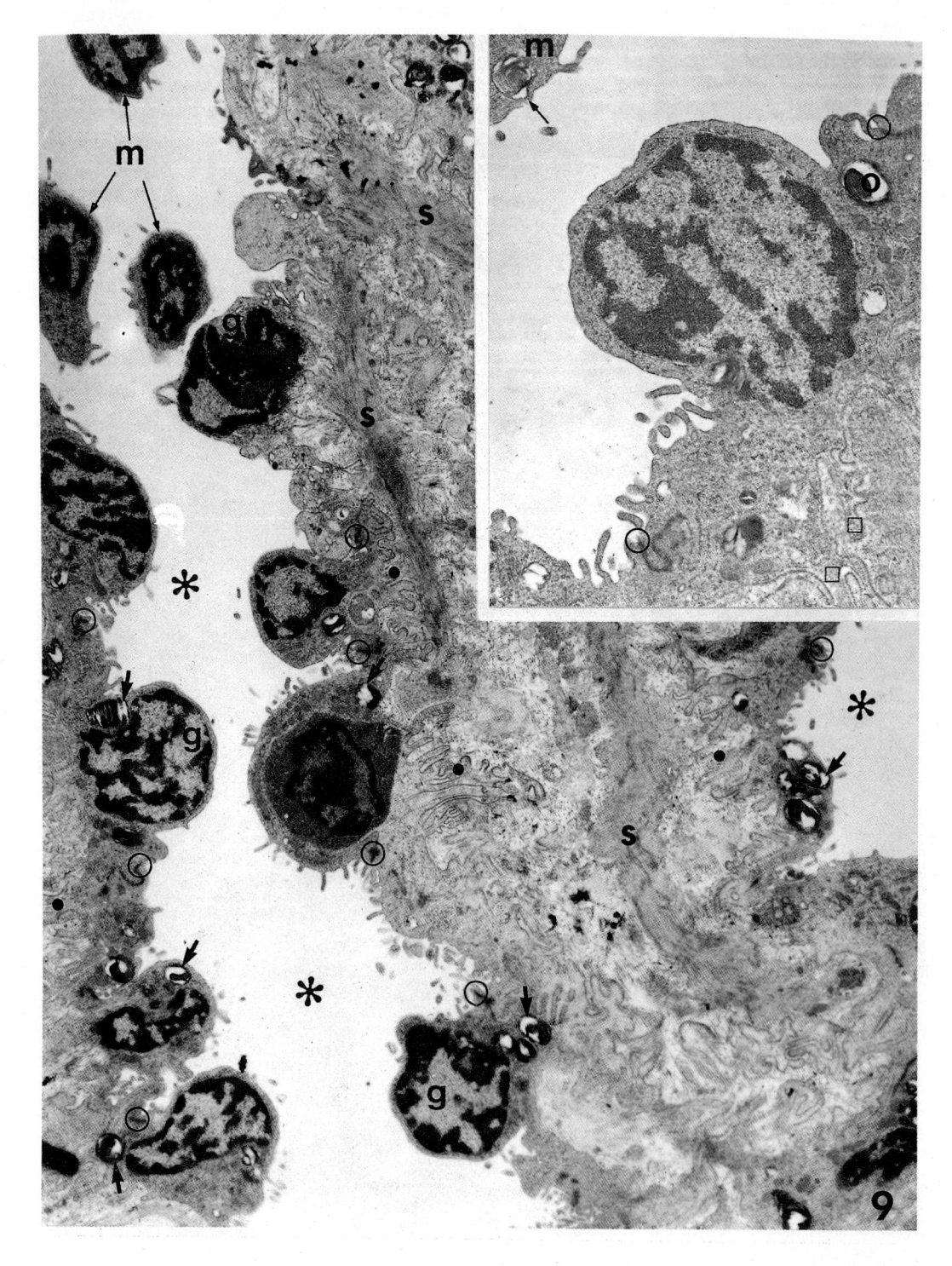

Fig. 9. Transmission electron micrograph of the gas-secretory cells (g) of *O. a. grahami*. Cells are well attached to the subepithelial tissue across amplifications of the basal plasma membranes (dots) and are tightly joined to each other (circles). s—smooth muscle; asterisk—air space; arrows—osmiophilic electron dense materials in the gas-secretory cells; m—macrophages; × 7,500. Inset: close-up of gas-secretory cell. o—osmiophilic electron dense body; circle—cell junction; square—basal infoldings; m—macrophage ingesting osmiophilic material (arrow); × 12,300.

Fig. 10. Transmission electron micrograph of the epithelial surface of the swimbladder of *O. a. grahami* showing a gas-secretory cell. Cell contains a centrally located nucleus with eccentrically placed mitochondria (m). Cells join each other through tight junctions (arrows) and are firmly attached to the basal aspect through well-developed plasma membrane foldings (asterisk). o—osmiophilic electron dense body; × 26,500. Inset: view of the gas-secretory cells with a somewhat rugged surface; × 2,500.

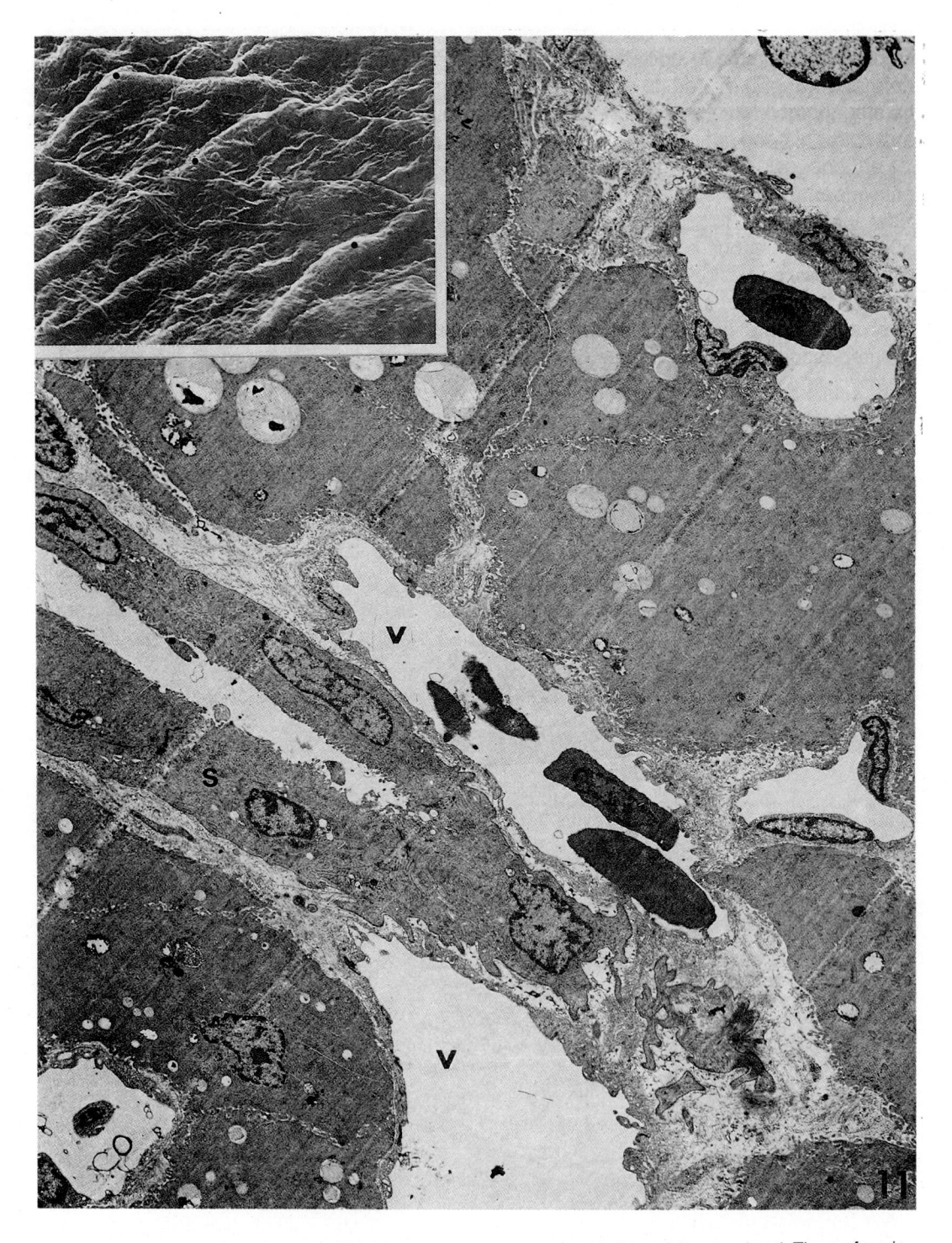

Fig. 11. Transmission electron micrograph of the surface of the swimbladder of *O. a. grahami*. The surface is very well vascularized, with some blood vessels (v) lying very close to the surface. s—smooth muscle; × 5,700. Inset: view of the vascular network (dots) on the surface of the swimbladder of *O. a. grahami*; × 1,500.

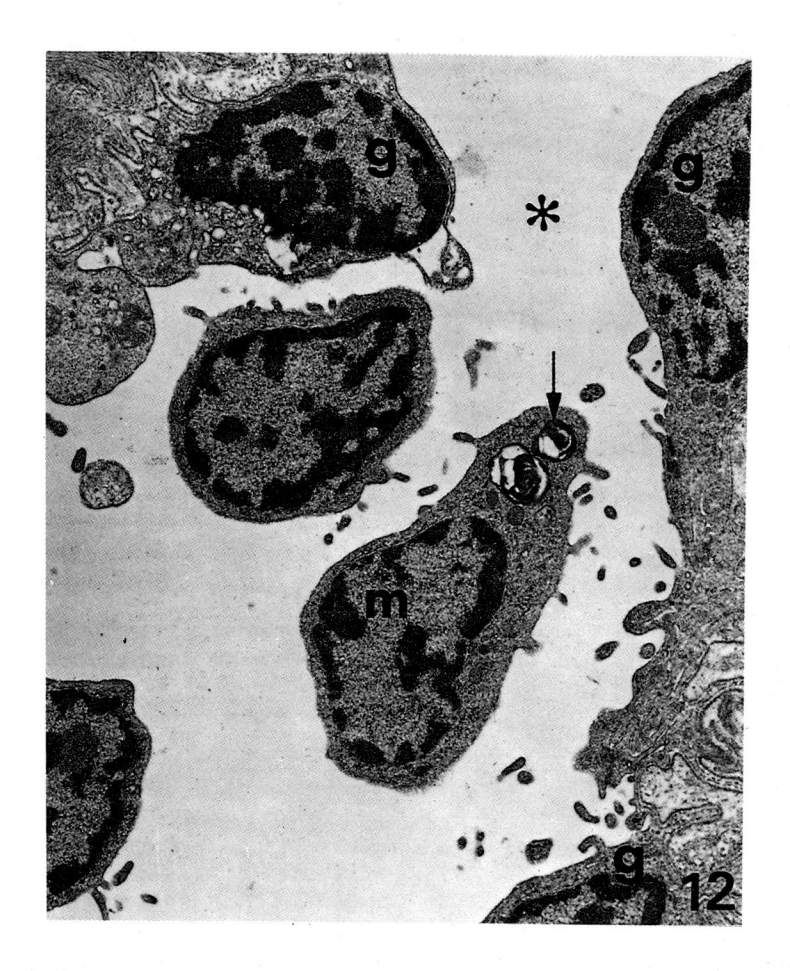

Fig. 12. View of the surface of the swimbladder of *O. a. grahami* showing gas-secretory cells (g) and macrophages (m). Arrow—osmiophilic electron dense bodies; asterisk—air space; × 19,300.

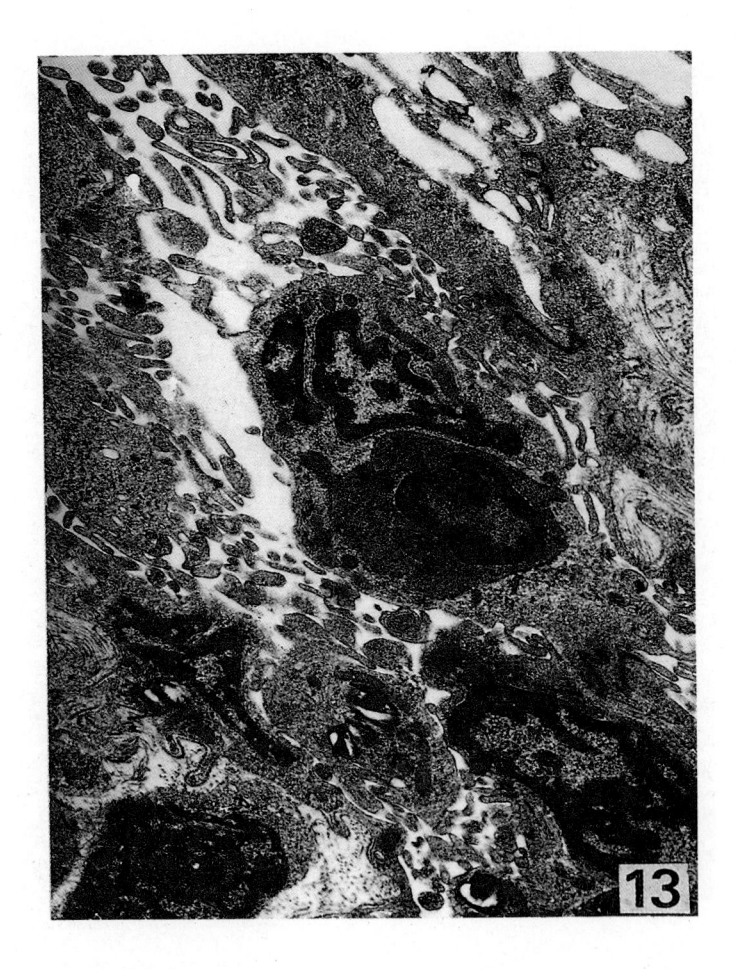

Fig. 13. Profusely interdigitating cells on the subepithelial space of the poorly vascularized area of the swimbladder of *O. a. grahami*; × 16,000.

Fig. 14. The tunica muscularis region of the swimbladder of *O. a. grahami* showing longitudinal bundles of smooth muscles separated by clusters of collagen (square) and elastic tissue (circle) fibers; × 15,200.

Fig. 15. Close-up of the basal aspect of the gas-secretory cell of the swimbladder of *O. a. grahami* showing amplifications (asterisk), osmiophilic electron dense bodies (o) and smooth muscle layer (s) onto which the cells are attached. Arrows—collagen fibers; × 31,500.

The outer wall of the airbladder, the tunica externa, mainly consists of dense collagenous connective tissue. This layer becomes continuous with the submuscularis layer proper, made up of circularly and longitudinally arranged muscle fibers interspersed with collagenous and elastic fibers (Figs. 11, 14). In this layer the blood vessels arising from the afferent arteries give rise to numerous arterioles (Fig. 11). Between the epithelial lining and the lumen of the airbladder is a distinct layer of loose, mainly collagenous connective tissue (Fig. 11). The collagen fibers in this layer attach to a thin smooth muscle layer to which the epithelial cells in the luminal aspect of the airbladder attach through profuse basal plasma membranous infoldings (Figs. 9, 10, 15). The inner surface of the airbladder is characterized by scattered macrophagic cells which have a large centrally located nucleus, pseudopodia and in some cells (Figs. 9, 12), intracytoplasmic electron-dense osmiophilic inclusions (Figs. 9, 15).

The swimbladder of *O. a. grahami* is supplied with blood through four vertebral branches of the dorsal aorta. On contacting the bladder, the blood vessels run along its surface, giving rise to branches which penetrate the tunica externa layer of the wall and then ramify into the submuscularis. The arterioles form into plexuses which constitute the arterial capillaries of the microretia mirabilia (Figs. 6, 7). These capillaries run somewhat perpendicular to the venular capillaries constituting the microretia mirabilia (Figs. 6, 7). Compared to the arterial retial capillaries, the venular capillaries are wider in diameter, have a thinner endothelial cell lining, manifest a greater degree of fenestration of the wall, and generally lie closer to the air space (Fig. 8). Frequently the venular retial capillaries lie directly beneath the attenuated part of the epithelial (gas-secreting) cells (Fig. 8). Spatially, the arterial capillaries surround the venular in a ratio of 3 to 1. In contrast to the venular capillaries, arterial capillaries are surrounded by pericytes (Fig. 8), a feature which suggests that their diameters may be variable, hence influencing the blood flow through the retial system. The endothelial cells of the retial capillaries contain numerous micropinocytotic vesicles (Fig. 8).

The retial vasculature topographically and structurally relates very closely to the columnar (gas-secreting) cells which are concentrated in the regions of the rete mirabile. In the luminal aspect the blood vessels converge onto the epithelial cells (Fig. 8, insert) and subepithelially attach directly onto the retial capillaries and in some cases onto the smooth muscle fibers through profuse basal cell wall amplifications (Figs. 8, 9, 10, 15). These features are highly indicative of functional interaction between the epithelial (gas-secreting) cells and the subepithelial structural components, principally the blood capillaries of the retial system.

In the anterior part of the airbladder the inner surface is more closely packed with columnar epithelial cells, the gas-secretory cells, which attach to each other through one or two elaborate desmosomes (Figs. 9, 10). These cells have large centrally located nuclei, microvilli on the apical surface, numerous osmiophilic bodies, and scattered filamentous mitochondria. The gas gland cells are firmly entrenched to the subepithelial surface through a highly amplified basal membrane (Figs. 9, 10). The shape of the outer cell (gas facing aspect) membrane is influenced by the large number of vacuoles which are intimately connected to the cell membrane (Fig. 10). The membrane appears discontinuous presumably due to the vacuoles physically passing through it. Vesicles of varying density are associated with the basal infoldings (Fig. 10). The outline of the outer cell (gas facing) membrane in some cells appears grossly interrupted (Fig. 10) while some of the cells contain numerous large perinuclear vacuolated spaces. Smooth muscles, collagen fibers, and amorphous ground substance lie in intimate contact with the gas-secreting cells (Figs. 9, 10). The luminal epithelium gradually changes to a rather squamous type towards the posterior aspect of the airbladder as the degree of vascularization increases. The squamous cells lack vacuolated bodies but are also well attached to the subepithelial tissue through basal folds.

DISCUSSION

Though basically a freshwater tropical and Near East fish, tilapia are ecologically a highly versatile group as evinced by the fact that they are now widely distributed across the tropical and subtropical regions of the world (Chervinski, 1982; Philippart and Ruwet, 1982). They can withstand and even breed in seawater while most species subsist in brackish water and moderately cool or even fairly high temperatures (Fryer and Iles, 1972; Matheus, 1984). A few species can endure very low environmental oxygen content and even withstand temporary anoxia (Welcomme, 1964; Verheyen et al., 1985). Among tilapines, and probably all extant fish, *O. a. grahami* lives in the harshest habitat of any known fish. It endures severe hypoxia, extremely high temperature (up to 42°C), and extreme ionoregulatory and acid base stress. *O. a. grahami* lives in the relatively shallow waters of the peripheral lagoons of Lake Magadi. In the water-holding tanks of the Magadi Soda Company, the water level can reach a maximum depth of 3 m, fluctuating with industrial use. The Fish Spring Lagoons are only about 1.5 m deep at most. This height has been brought about by construction of a barrier to raise the water level for ease of pumping to the factory for use in processing the trona. The specific gravity of the lake water, due to the high concentration of mainly sodium carbonate and bicarbonate, lies between 1.01 and 1.03 (Johansen et al., 1975), a value comparable to that of seawater. Our parallel investigations (Narahara et al., unpub.) revealed that when stressed by hypoxia, high temperature, exercise, or a combination of these factors, the fish skim the surface of the water. This behavior is probably a combination of pumping the better oxygenated top layer of water across the gills and possible use of the buccal cavity and/or the airbladder for gas exchange. The fish have been observed to later release air bubbles under the water, a feature which appears to support the latter interpretation. That the buccal cavity and/or the airbladder could be used as accessory respiratory organs in this fish would not be surprising, in view of the frequently very low oxygen concentration in the water, which ranges from 1.1 to 1.3 mgL^{-1} (Reite et al., 1974), and the dramatic fluctuations in its partial pressure between day (<400 torr) and night (<20 torr). The problem is further exacerbated by the high metabolic rate of the fish consequent to its generally small size and the high ambient temperature of the medium it thrives in. Preliminary investigations (Narahara et al., unpub.) revealed that in a relatively normoxic situation (PO$_2$, 109 torr), at an ambient temperature of 32°C, the gills met all the oxygen needs of the animal while in a hypoxic one (PO$_2$, 42 torr), the fish acquired 12.5% of its oxygen needs from the air. This proportion may increase dramatically at higher temperatures and also during the night when the water is hypoxic. The observation that the airbladder was in a collapsed state in unstressed animals suggests that use of the bladder may be discretionary, its significance being more notable during anoxic episodes or when there is specific need to maintain buoyancy, as when feeding on algal incrustations on the sides of the pond. On the other hand, it is plausible that the airbladder when in use may serve as an oxygen storage site in situations of hypoxia when the fish may not risk coming to the surface to "breathe". In such cases it may stay under water for as long as the oxygen in the airbladder is able to support its respiratory needs or until the partial pressure of oxygen in the water improves. Pearse and Achtenberg (1918) and Hall (1924) demonstrated that the perch could enter regions where conditions were unfavorable for respiration and remain in oxygen-free water as long as two hours without asphyxiation. Hall (1924) also showed that oxygen supply to the airbladder decreased with exposure to water of low oxygen tension. Perhaps this partly explains how some tilapiine species are able to survive hypoxia temporarily (Welcomme, 1964; Verheyen et al., 1985) by drawing oxygen from the airbladder.

The general structure of the airbladder of *O. a. grahami* is similar to that described in most other fish (Jones and Marshall, 1953; Copeland, 1970; Fahlen, 1970, 1971; Morris and Albright, 1975) but with some notable differences. The ciliated cells described by Fahlen (1967, 1968, 1971) in the salmonid airbladder were not observed in *O. a. grahami*. No specific function has been attributed to these cells. Further, the retial system and the gas-secretory cells are well developed and continuous with each other. The whole surface of the bladder is very well vascularized and the blood capillaries are well exposed to air by bulging out into the air space. The air-blood barrier in such areas is as thin as 0.5 μm in the more superficially located venular capillaries. The disposition and the ultrastructural differences between the venular and the arteriolar capillaries in *O. a. grahami* are similar to those observed in the swimbladder of the toadfish by Fawcett and Wittenberg (1959), and the well-developed basal (capillary facing) gas cells are similar to those described in *Fundulus* (Copeland, 1960), a feature which suggests possible active participation of these cells in secretion or transfer of materials (Pearse, 1956). Exposure of the blood capillaries on the gas facing side of the airbladder in *O. a. grahami* corresponds with that observed in the toadfish (Copeland, 1969). In *O. a. grahami* the venous blood flow from the bladder drains by a common trunk which joins with the branchial arches (Fig. 5). The swimbladder will hence additively increase the partial pressure of oxygen in the blood leaving the gills. This may be critical during hypoxic episodes when the bladder may play a significant respiratory role. Regulation of the blood flow to the airbladder may occur at the arteriolar and capillary level through constriction of these vessels by the action of pericytes. That type of capillary constriction was experimentally demonstrated by Krogh (1936). During the day, when oxygen is in abundance, the gills should be adequate for oxygen transfer and perfusion and correspondingly ventilation of the airbladder can be reduced. The observed collapse of the bladder during such times should minimize the energy expended by the fish to overcome flotation, especially during the day, a time when the fish would be most vulnerable to predators. It is probably not coincidental that it is during such times that the airbladder was observed to be in a nondilated state.

In the euphysoclistic bladders of fish such as the toadfish (Fange and Wittenberg, 1958), the bladder receives blood from a branch of the coeliac artery, while in salmonid *Argentina silus* (Fahlen, 1970) the bladder is supplied with blood by 4 to 6 arteries, which branch from the unpaired aorta distal to the coeliaco-mesenteric artery. In four other salmonids (*S. irideus*, *S. salar*, *S. fontinalis* and *S. trutta*) blood supply to the airbladder is effected through two branches of the coeliaco-mesenteric artery (Fahlen, 1971). In most physostomes and the majority of euphysoclists, the artery which supplies the airbladder arises from the coeliaco-mesenteric artery (Steen, 1970) and more distal branches of the dorsal aorta, the intercostal arteries (Jones and Marshall, 1953; Fange, 1966). Venous drainage from the airbladder occurs through the hepatoportal system and in fish such as pike through the cardinal veins as well (Corning, 1888; Saupe, 1939; Fange, 1966). Clearly, the airbladder circulation in *O. a. grahami* and perhaps in tilapia, as a group, deviates somewhat from that in most fish possessing an airbladder. Incorporation of the gills in this circuit may account for the respiratory capacity of the airbladder in fish, a radical innovation which may have been critical for survival in a stressful habitat. It may correspond with the observation made by Hall (1924) that the respiratory function of the airbladder is only important in a few species of fish. In such species, special structural adaptations may be essential, as appears to be the case in *O. a. grahami*.

The presence of a well-developed rete mirabile in the physostomatous *O. a. alcalicus* is most intriguing. It is nevertheless plausible that due to the very intense selective predation which occurs in its habitat, particularly from water birds such as flamingoes, ibises, pelicans, and plovers, small size despite its well-known disadvantages of high energetic demands, favors the survival of the fish. Indeed, this adaptive change has been well documented in stratigraphic studies (Copley, 1958). In some instances, especially during the day when the fish feed on algal incrustations on the walls of the lagoons, it may be expedient to regulate its buoyancy without exposing itself to predators by coming to the surface of the water to take in air. In such cases the retia mirabilia may play a significant part in the secretion of gas into the bladder and the vascularized areas in absorbing it. Fish without a swimbladder have to generate a continuous upward force equal to about 6% of body weight to stay in midwater (Goolish, 1992) and the hydrodynamic lift generated by swimming (either positive or negative) is an energetically costly solution, particularly when swimming at a high angle of attack (Alexander, 1966). The nature of the gas secreted into the airbladder appears to depend on the presence or absence of the rete mirabile; all fish which possess a rete accumulate O_2 and N_2 against a concentration gradient while those which lack a rete secrete only N_2 (Sundnes, 1963; Sundnes et al., 1958). Comparatively, the gas gland epithelium of *O. a. grahami* is more poorly developed than that of the euphysoclists such as toadfish (Fange and Wittenberg, 1958; Copeland, 1969) and *Fundulus heteroclitus* and *Gadus callarias* (Copeland, 1969; Morris and Albright, 1975) in which the epithelium is thicker, more vascularized and convoluted. The basal infoldings, features which indicate functional interaction with the subepithelial blood vessels, of the gas cells of *O. a. grahami* are as well developed as those of euphysoclistic fish (Dorn, 1961; Copeland, 1969). These features are characteristically absent in physostomous swim bladders (Fahlen, 1967, 1968). Similarly astounding is the well-developed euphysoclistic gas-secretory system (the gas gland and the rete mirabile) in the swimbladder of toadfish (Morris and Albright, 1975), a fish which lives in shallow waters (Perlmutter, 1961). It was observed by Hall (1924) that increased concentration of CO_2 in water altered the specific gravity of the perch, making the fish more buoyant. The remarkably high CO_2 concentration in Lake Magadi (164 mL^{-1}) (Wood et al., 1989) presumably makes *O. a. grahami* more buoyant, hence calling for it to regulate the size of its airbladder. An airbladder replaces part of the body of a fish (which is heavier than water) with air, hence lowering the overall specific weight and regulating it close to that of the water (Alexander,

1966). By maintaining neutral buoyancy at a definite depth, the fish can remain motionless in water and the energy required in horizontal motion is reduced. Despite the fact that energy is required to maintain a gas-filled bladder (Kanwisher and Ebeling, 1957), the savings achieved through hydrodynamic regulation to neutral buoyancy are remarkably high (Gray, 1953; Bainbridge, 1960; Alexander, 1966).

It is now generally accepted that production of the airbladder gas is largely caused by localized acidosis through the Bohr effect in the gas gland, produced by high glycolytic activity (D'aoust, 1970) and a concentration effect brought about by the rete mirabile (Fange, 1966). Wittenberg (1958) and Sundnes (1963) proposed that the high partial pressures of the inert gases found in such bladders may result from the interaction between a countercurrent mechanism of lower capacity than that found in physoclists, and a reabsorptive process of a high capacity. Fange (1958), Fahlen (1967), and Steen (1970) found it difficult to explain the high concentration of inert gases found in physostomes of which the rete is poorly developed. It is even more astounding that fish which lack a rete mirabile are able through unknown mechanisms to secrete gases into the airbladder (Sundnes et al., 1958) and regulate their buoyancy. The mode of gas deposition in the herring, which lacks a rete, has not been studied but may depend on acidification of the retial blood only (Fahlen, 1967), while that in salmonids which lack the retia (Sundnes et al., 1958) may be effected through an active cellular process. In the physostomous carp (*Cyprinus carpio*), in which oxygen in the blood and that in the airbladder are similar, physical diffusion may account for the passage of gases to and from the airbladder (Hall, 1924) while in the surface-dwelling bass (*Micropterus salmoides*), which does not experience depth changes, the process may occur by simple diffusion. This observation makes the presence of retia mirabilia in *O. a. grahami* even more enigmatic.

A number of fish, such as Megalops (Shlaifer and Breder, 1940), *Arapaima* (Luling, 1964), *Gymnarchus* (Bertin, 1938), *Notopterus* (Dehadrai, 1962), *Umbra* (Geyer and Mann, 1939), and some Erythrininae (Carter and Beadle, 1931; Wilmer, 1934), which live in poorly oxygenated waters are known to use the airbladder as an accessory respiratory organ and to a slight extent as an oxygen storage site (Black, 1940). In *Hoplerythrinus unitaeniatus*, the airbladder is very well vascularized with 60 to 80% of the cardiac output perfusing it (Blaxter and Tytler, 1978). In *Tetractenos glaber* (Green, 1984) the thickness at the resorptive area is 1 μm and at the gas gland region 20–30 μm. The capacity to secrete gas into the airbladder is correlated with the degree of development of the secretory epithelium and the presence and complexity of the retia mirabilia (Jones and Marshall, 1953). Activity of the gas gland is accompanied by an increased blood supply to it and the gland becomes larger and redder than normal (Hall, 1924; Akita, 1936) and the capillaries become dilated (Jacobs, 1930). In Salmonidae the glandular tissue is poorly developed and there are no retia mirabilia (Jones and Marshall, 1953). Jacobs (1934) found that salmonids such as *Hucho hucho, Salmo fario* and *S. irideus* were unable to replace gas lost from the airbladder when denied access to the surface. In fish such as the pike, carp, and minnows, which are able to secrete gas though at a slow pace (Evans and Damant, 1929; Jacobs, 1934), the secretory epithelium is better developed and is associated with capillary networks similar to the retia mirabilia of the physoclists. On the other hand, eels (Popta, 1910) and the physoclists (Jacobs, 1934), whose gas-secreting cells and retia mirabilia are well developed, can make remarkable adjustments to the airbladder within a short time. In freshwater the number of species of physostomous teleosts exceeds that of the physoclists but in the sea the reverse is seen (Jones and Marshall, 1953). This feature may correlate with the greater density of the marine environment. Certain benthic deep-sea dwelling fishes have entirely lost the airbladder (Jones and Marshall, 1953). That *O. a. grahami* has developed and retained a physostomous airbladder while surviving in a medium of equivalent density as that of seawater, strongly suggests that the organ in this species plays a more significant role in respiration than in buoyancy regulation.

CONCLUSION

The swim (air) bladder is widely found in fish. Whereas expectedly in deep water fish the organ may be of significant importance as an energy-saving device by buoyancy control, in some species the presence of the organ is, to say the least, enigmatic. Though the airbladder has been studied for a long time by various approaches, ranging from physiological, anatomical, and biophysical methods, the mode mechanism of secretion of gases, their resorption, and regulation of the internal pressures and composition is far from clear. The ontogeny of the airbladder and its possible role as the progenitor of the lung has long been debated yet the circumstances and factors which induced the change are far from clear. Two kinds of airbladders exist in fish: the physoclistous type is closed, i.e., it does not open to the outside, while the physostomous type does.

Oreochromis alcalicus grahami (Trewavas, 1983), a fish which lives in the shallow peripheral lagoons of the alkaline oxygen-deficient Lake Magadi, has a partly divided physostomous airbladder. The airbladder of *O. a. grahami* is internally remarkably well vascularized, a morphological feature which strongly suggests a possible role in gas exchange. It bears some structural features characteristic of a physoclistous airbladder in that the gas-secretory epithelium, which is concentrated to the cranial part of the bladder, is well developed: this attribute possibly affords a capacity of gas secretion. It is envisaged that the airbladder of *O. a. grahami* is designed to play dual roles of respiration (in extreme conditions of anoxia) and buoyancy control (whenever it is necessary for the fish to stay below the water surface for feeding or when threatened by predators). There is dire need to study the physiology of the airbladder of this fish which has adapted to subsist in a remarkably unique habitat by drastically refining its morphological features. The results would provide a classic paradigm of the extreme adaptive plasticity of animal life to ambient selective pressures.

Acknowledgments

This work was completed at the Department of Biology, McMaster University when the first author (J.N.M.) was a Visiting Hooker Professor under the sponsorship of CIDA and NSERC, Canada to whom he is indebted. The work would not have been possible without the kind permission and logistical support given by the following bodies: Office of the President, Nairobi, Kenya; Magadi Soda Company and the University of Nairobi. We acknowledge with thanks funding from the following bodies: NATO, NSERC Canada, International Collaboration Program and the U.S. National Geographic Society to the team, and from NSERC Research Program to C.M.W., U.S. NSF Program (IB-9118819) to P.J.W. and the University of Wyoming to H.L.B. and A.N.

References

Akita, Y.K. 1936. Studies on the physiology of the swimbladder. *J. Fac. Sci. Imp. Univ. Tokyo*, Sect. IV, 4:111-135.
Alexander, R. McN. 1959. The physical properties of the swimbladder in intact Cypriniformes. *J. Exp. Biol.* 36:315-332.
Alexander, R. McN. 1966. The physical aspects of swimbladder function. *Biol. Rev.* 41:141-176.
Andriashev, A.P. 1944. Determination of natural specific gravity in fish. *C.R. Acad. Sci., USSR* 43:80-102.
Bainbridge, R. 1960. Speed and stamina in three fish. *J. Exp. Biol.* 37:129-153.
Bertin, L. 1938. Formes nouvelles et formes larvaires de poissons apodes appartenant au sous-ordre des Lyomeres. *Dana Rep.* no. 15, pp. 1-26.
Black, E.C. 1940. The transport of oxygen by the blood of freshwater fish. *Biol. Bull.* (Woods Hole, Mass.), 79:215-224.
Blaxter, J.H.S. and P. Tytler. 1978. Physiology and function of the swimbladder. In O. Lowenstein (ed.), *Advances in Comparative Physiology and Biochemistry*, Vol. 7, pp. 311-367. Academic Press, London.
Carter, G.S. and L.C. Beadle. 1931. The fauna of swamps of the Paraguayan Chaco in relation to its environment. I. The physicochemical nature of the environment. *J. Linn. Soc. (Zool.)* 37:205-258.
Chervinski, J. 1982. Environmental physiology of tilapias. In R.S.V. Pullin and R.H. Lowe (eds.), *The Biology and Culture of Tilapia*, pp. 119-128. Internat. Ctr. Living Aquatic Resources, Proc. 7th ICLARM Conf., Manila (Philippines).
Coe, M.J. 1966. The biology of *Tilapia grahami* (Boulenger) in Lake Magadi, Kenya. *Acta Tropica* 23:146-177.
Copeland, D.C. 1960. Secretory epithelium of the swimbladder in *Fundulus*. *Biol. Bull.* 119:311-325.
Copeland, D.C. 1964. Gas secretion in teleost swimbladder, *Fundulus* and *Opsanus*. *Biol. Bull.* 127:367-374.
Copeland, D.E. 1969. Fine structural study of gas secretion in the physoclistous swimbladder of *Fundulus heteroclitus* and *Gadus callarias* and in the euphysoclistous swimbladder of *Opsanus tau*. *Z. Zellforsch.* 93:305-331.
Copeland, D.E. 1970. The gas bladder of *Argentisilus* L., with special reference to the ultrastructure of the gas gland cells and the counter-current vascular bundles. *Z. Zellforsch.* 95:45-76.
Copley, H. 1958. *Common Freshwater Fishes of East Africa*. Academic Press, London.
Corning, H.K. 1888. Beitrage zur Kenntnis der Wunernetzbilchungen in der Schwimmblase der Teleostier. *Morph. Jb.* 14:1-54.
D'aoust, B.G. 1970. The role of lactic acid in gas secretion in the teleost swimbladder. *Comp. Biochem. Physiol.* 32:637-668.
Dehadrai, P.V. 1962. Respiratory function of swimbladder of *Notopterus* (Lacapede). *Proc. Zool. Soc., Lond.* 139:341-357.
Denton, E.J. 1961. The buoyancy of fish and cephalopods. *Progr. Biophys. Chem.*, 11:178-234.
Denton, E.J. 1962. Buoyancy mechanisms of sea creatures. *Endeavour* 22:3-8.
Dobbin, C.N. 1941. A comparative study of the gross anatomy of the air-bladders of ten families of fishes of New York and other Eastern States. *J. Morphol.* 68:1-29.
Dorn, E. 1961. Über den Feinbau der Schwimmblase von *Anguilla vulgaris* L. Licht und electronenmikroscopische Untersuchungen. *Z. Zellforsch.* 55:849-912.
Evans, H.M. and G.C.C. Damant. 1929. Physiology of the swimbladder in Cyprinoid fishes. *J. Exp. Biol.* 6:42-57.
Fahlen, G. 1967. Morphology of the gas bladder of *Coregonus lavaretus* L. *Acta Univ. Lund* 2(28):1-37.
Fahlen, G. 1968. The gas bladder as a hydrostastic organ in *Thymallus thymallus* L., *Osmerus operlanus* L. and *Mallotus villosus* Mull. *Rep. Norweg. Fish Invest.* 14:199-228.
Fahlen, G. 1970. The gas bladder of *Argentina silus* L., with special reference to the ultrastructure of the gas gland cells and the counter-current vascular bundles. *Z. Zellforsch.* 11:350-372.
Fahlen, G. 1971. The functional morphology of the gas bladder of the genus *Salmo*. *Acta Anat.* 78:161-184.
Fange, R. 1958. The structure and function of the swimbladder in *Argentina silus*. *Quart. J. Microscop. Sci.* 43:195-102.
Fange, G. 1966. Physiology of the swimbladder. *Physiol. Rev.* 46:299-322.
Fange, G. and A. Mattison. 1956. The gas secretory structures and the smooth muscles of the swimbladder of Cyprinids. In K.G. Wingstrand (ed.), *B. Hanstrom, Zoological Papers in Honour of his 60th Birthday*. Zoological Institute, Lund, Sweden.
Fange, G. and J.B. Wittenberg. 1958. The swimbladder of the toadfish (*Opsanus tau* L.). *Biol. Bull.* 115:172-179.

Fawcett, D.W. and J. Wittenberg. 1959. The fine structure of capillaries in the rete mirabile of the swimbladder of *Opsanus tau*. *Anat. Rec.* 133:274.
Fryer, G. and T.D. Iles. 1972. *The Cichlid Fishes of the Great Lakes of East Africa*. Oliver and Boyd, London.
Gray, J. 1953. *The Locomotion of Fishes*. Oliver and Boyd, London.
Gee, J.H. and T.G. Northcote. 1963. Comparative ecology of two sympatric species of dace (*Rhinichthys*) in the Frazer River system, British Columbia. *J. Fish Res. Bd., Can.* 20:105–118.
Geyer, F. and M. Mann. 1939. Die Atmung des ungarischen Hundsfisches (*Umbra lacustris* Grossinger). Beitrange zur Atmung der Fisch I. *Zool. Anz.* 127:234–245.
Goolish, E.M. 1992. Swimbladder function and buoyancy regulation in the killifish *Fundulus heteroclitus*. *J. Exp. Biol.* 166:61–81.
Green, C.W. 1924. Analysis of the gases of the air-bladder of the California singing fish, *Porichthys notatus*. *J. Biol. Chem.* 59:615–621.
Green, S.L. 1984. Ultrastructure and enervation of the swimbladder of *Tetractenos glaber* (Tetraodontidae). *Cell Tissue Res.* 237:277–284.
Greenwood, P.H. 1963. The swimbladder in the African Notopteridae (Pisces) and its bearing on the taxonomy of the family. *Biol. Bull. Brit. Mus. (Nat. Hist.) Zool.* 11:377–412.
Hall, F.G. 1924. The functions of the swimbladder of fishes. *Biol. Bull.* (Woods Hole, Mass.) 47:79–126.
Hora, S.L. 1922a. The modification of the swimbladder in hillstream fishes. *J. Asiat. Soc. Beng.* 18:5–16.
Hora, S.L. 1922b. Notes on fishes in the Indian Museum. III. On fishes belonging to the family Cobitidae from high altitudes in Central Asia. *Rec. Indian Mus.* 24:63–75.
Jacinski, A. and W. Kilarski. 1964. The gas gland in the swimbladder of the burbot (*Lota lota* L.) and stone perch (*Acerina cernua* L.), its macro- and microscopic structure based on observations of electron microscopy. *Acta Biol., Cracov* VII:11–125.
Jacobs, W. 1930. Untersuchungen zur Physiologie der Schwimmblase. I. Über die Gassekretion in der Schwimmblase von Physoklisten. *Z. Vergl. Physiol.* 11:565–629.
Jacobs, W. 1934. Untersuchungen zur Physiologie der Schwimmblase Fische. III. Luftschlucjen und Gassekretion bei Physostomen. *Z. Vergl. Physiol.* 20:674–698.
Johansen, K., G.M.O. Maloiy and G. Lykkeboe. 1975. A fish in extreme alkalinity. *Respir. Physiol.* 24:159–167.
Jones, F.R.H. 1952. The swimbladder and the vertical movements of teleostean fishes. II. The restriction to rapid and slow movements. *J. Exp. Biol.* 29:94–109.
Jones, F.R.H. 1957. The swimbladder. *In* M.E. Brown (ed.), *The Physiology of Fishes*, Vol. 2, pp. 305–322. Academic Press, New York-London.
Jones, F.R.H. and N.B. Marshall. 1953. The structure and functions of the teleostean swimbladder. *Biol. Rev.* 28:16–83.
Kanwisher, J. and A. Ebeling. 1957. Composition of the swimbladder gas in bathypelagic fishes. *Deep-Sea Res.* 4:211–217.
Koshtoyanz, C.S. and P.D. Vassilenko. 1937. On the receptor function of the swimbladder of fishes. *J. Exp. Biol.* 14:16–19.
Krogh, A. 1936. *The Anatomy and Physiology of Capillaries*. Yale University, New Haven CT.
Krogh, A. 1959. *The Comparative Physiology of the Respiratory Mechanisms*. University of Pennsylvania Press, Philadelphia, PA.
Luling, K.H. 1964. Zur biologie und Okologie von *Arapaima gigas* (Pisces, Osteoglossidae). *Z. Morphol. Okol., Tierre* 54:436–530.
Marshall, N.B. 1960. Swimbladder structure of deep-sea fishes in relation to their systematics and biology. *Discovery Rep.* 31:1–22.
Marshall, N.B. 1962. The biology of sound producing fishes. *Symp. Zool. Soc., Lond.* 7:45–60.
Matheus, C.E. 1984. *Aspectos do crescimento e reproducao de Saratheradon niloticus (tilapia de Nilo) em lagoas de estabilizacaoe sua influencia no tratamento biologico*. Dessertacao de Mestrado, Fed. Univ., Sao Carlos.
Morris, S.M. and J.T. Albright. 1975. The ultrastructure of the swimbladder of the toadfish, *Opsanus tau* L. *Cell Tissue Res.* 164:85–104.
Nielsen, J.C. and O. Munk. 1963. A hadal fish (*Bassogigas profundissimus*) with a functional swimbladder. *Nature* 204:302–315.
Pearse, A.S. and H. Achtenberg. 1918. Habits of the yellow perch in Wisconsin Lakes. *Bull. U.S. Bur. Fish* 36:294–306.

Pearse, D.C. 1956. Infolded basal plasma membranes found in epithelia noted for water transport. *J. Biophys. Biochem. Cytol.* 2:203-208.

Pelster, B. and P. Scheid. 1992a. Countercurrent concentration and gas secretion in the fish swimbladder. *Physiol. Zool.* 65(1):1-16.

Pelster, B. and P. Scheid. 1992b. Metabolism of the swimbladder epithelium and the single concentrating effect. *Comp. Biochem. Physiol.* 105A(3):383-388.

Perlmutter, A. 1961. *Guide to Marine Fishes*. New York University Press, New York.

Philipart, R.C. and J.C. Ruwet. 1982. Ecology and distribution of tilapias. *In* R.S.V. Pullin and R.H.J. Lowe (eds.), The Biology and Culture of Tilapia, pp. 15-59. Int. Ctr. Living Aquatic Res. Manag. (Manila-Philippines), Proc. 7th ICLARM Conf.

Popta, C.M. 1910. Etude sur la vessie aerienne de poissons. Sa fonction. *Ann. Sci. Nat. Zool.*, ser. 9, 12:1-43.

Qutob, Z. 1962. The swimbladder of fish as pressure receptor. *Arch. Neerl. Zool.* 15:1-67.

Reite, O.B., G.M.O. Maloiy, and B. Aasenhaug. 1974. pH, salinity and temperature tolerance of Lake Magadi. *Nature, Lond.* 247:315.

Saupe, M. 1939. Anatomie und Histologie der Schwimblase des Flussbarsches (*Perca fluviatilis*) mit bederer Berucksichtigung des Ovals. *Z. Zellforsch Microskop. Anat.*, Abt. *Histochem.* 30:1-35.

Schneider, H. and A.D. Hasler. 1960. Laute und Lauterzeugung beim Susswassertrommler, *Aplodinotus grunniens* Rafinesque (Sciaenidae, Pisces). *Z. Vergl. Physiol.* 43:499-517.

Scholander, P.F. 1954. Secretion of gases against high pressures in the swimbladder of deep-sea fishes. II. The rete mirabile. *Biol. Bull.* 107:260-277.

Scholander, P.F. and L. van Dam. 1953. Composition of the swimbladder gas in deep-sea fishes. *Biol. Bull.* 104-75-86.

Scholander, P.F. and L. van Dam. 1954. Secretion of gases against high pressures in the swimbladder of deep-sea fishes. I. Oxygen dissociation in blood. *Biol. Bull.* 107:247-259.

Schwartzkopff, J. 1963. Vergeleichende Physiologie des Gehors und der Lautausserungen. *Fortschr. Zool.* 15:214-336.

Shlaifer, A. and C.M. Breder. 1940. Social and respiratory behaviour of small tarpon. *Zoologica* (NY) 25:493-512.

Steen, J.B. 1970. The swimbladder as a hydrostatic organ. *In* W.S. Hoar, D.J. Randall, and E.M. Donaldson (eds.), *Fish Physiology*, IV, pp. 414-443. Academic Press, New York-London.

Sundnes, G. 1963. Studies on the high nitrogen content in the physostome swimbladder. *Rept. Norweg. Fishery Invest.* 13:1-8.

Sundnes, G. T. Enns and P.F. Scholander. 1958. Gas secretion in fishes lacking rete mirabile. *J. Exp. Biol.* 35:671-676.

Tavolga, W.N. 1962. Mechanisms of sound production in the arid catfishes, Galeichthys and Bagre. *Bull. Am. Mus. Nat. Hist.* 126:1-30.

Tavolga, W.N. 1964. Psychophysics and hearing in fish. *Nat. Hist.* 73:34-41.

Trewavas, E. 1983. *Tilapiine Fishes of the Genera Sarotheradon, Oreochromis, and Danakilia*. Brit. Mus. Nat. Hist., Dorchester, England.

Vasilenko, T.D. and M.N. Livanov. 1936. Oscillographic studies of the reflex function of the swimbladder in fish. *Bull. Expt. Biol. Med., USSR* 2:264-266.

Verheyen, E., R. Blust and C. Doumen. 1985. The oxygen uptake of *Sarotheradon niloticus* L. and the oxygen binding properties of its blood and hemolysate (Pisces, Cichlidae). *Comp. Biochem. Physiol.* 81A:423-426.

Walsh, P.J., H.L. Bergman, A. Narahara, C.M. Wood, P.A. Wright, D.J. Randall, J.N. Maina, and P. Laurent. 1993. Effects of ammonia on survival, swimming and activities of enzymes of nitrogen metabolism in the Lake Magadi tilapia *O. a. grahami*. *J. Exp. Biol.* 180:323-387.

Webb, P.W. 1990. How does benthic living affect body volume, tissue composition, and density of fishes? *Can. J. Zool.* 68:1250-1255.

Welcomme, R.L. 1964. The habitats and habit preferences of the young of the Lake Victoria *Tilapia* (Pisces, Cichlidae). *Rev. Zool. Bot. Afr.* 70:1-28.

Willmer, E.M. 1934. Some observations on the respiration of certain tropical fishes. *J. Exp. Biol.* 11:283-306.

Wittenberg, J.B. 1958. The secretion of inert gas into the swimbladder of fishes. *J. Gen. Physiol.* 41:783-804.

Wood, C.M., S.F. Perry, P.A. Wright, H.L. Bergman, and D.J. Randall. 1989. Ammonia and urea dynamics in the Lake Magadi tilapia, a ureotelic fish adapted to extremely alkaline environment. *Respir. Physiol.* 77:1-20.

Wood, C.M., H.L. Bergman, P. Laurent, J.N. Maina, A. Narahara, and P.J. Walsh 1995. Urea production, acid base regulation and their interactions in the Lake Magadi tilapia, a unique teleost adapted to highly alkaline environment. *J. Exp. Biol.* 189:13–26

Woodland, W.N.F. 1911. On some experimental tests concerning the physiology of gas production in teleostean fishes. *Anat. Anz.* 40:225–242.

12

Cephalic Sensory Canal System of Some Cyprinodont Fishes in Relation to their Habitat

P.K. Ray and N.C. Datta

Fishes and aquatic amphibians are characterized by a unique surface receptor system, the lateral line system (McFarland et al., 1985). Receptors are actually the neuromasts, generally lodged within the canals but sometimes distributed nakedly over the head and body surface. The lateral line system occurs symmetrically on both sides of the head and trunk. In the head region of fishes the lateral line system is conspicuous and elaborate and referred to as the cephalic sensory canal system whereas in the trunk and tail regions it forms the lateral line proper. Neuromasts of this canal system are innervated by the seventh, ninth, and tenth cranial nerves. In teleosts the major canals belonging to the cephalic sensory canal system are the supraorbital, infraorbital, preopercular, mandibular, temporal, and posttemporal (Figs. 1A, B, C). However, the lateral line system is also known as the acoustico-lateralis system since the said system and ear are closely related. The organization of the sensory apparatus in fishes has been a subject of special interest to zoologists for many years because of its structural and functional specialization. The role played by the lateral line system in the life of fishes has been researched by many, viz., Sand (1937), Disler (1960), Dijkgraaf (1962), Van Bergeijk and Alexander (1962), Denton and Blaxter (1976) and Jorgenson (1985). Its taxonomic significance has been highlighted by Hubbs and Cannon (1935), Gosline (1949), Bailey (1951) and Nelson (1972).

A review of relevant literature revealed that almost no work worth mentioning has been done on Indian teleosts. The studies available are confined to the bony passage harboring the sensory canal (Khandelwal, 1963; Khandelwal and Rastogi, 1965; Khandelwal and Sharma, 1965; Kapoor, 1959, 1960, 1961, 1962, 1964; Sharma, 1964). Srivastava and Srivastava (1967) studied neuromast development in *Cirrhina mrigala* and *Ophicephalus punctatus*. The sensory canal system in relation to neuromast distribution and their functional morphology has been studied by Ray (1989) and Ray and Datta (1990).

Cyprinodontiformes is an interesting order of bony fishes, representatives of which are also known as tooth carps. These are predominantly carnivores and their jaws are provided with conical teeth. In this context it may be mentioned that Gosline (1949) and Rosen and Mendelson (1960) studied the sensory canal system of some poeciliid fishes but mostly from a taxonomic point of view. In the present study, four cyprinodont fishes

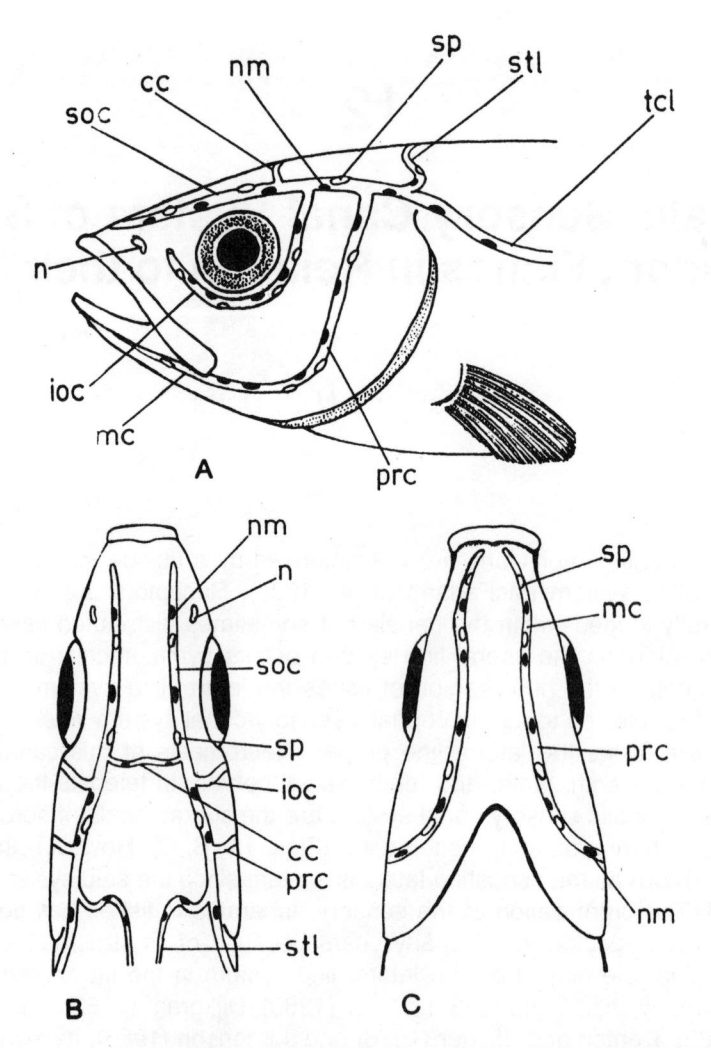

Fig. 1. Generalized view of the teleostean cephalic sensory canal system. A—lateral; B—Dorsal; C—Ventral views.

were selected, all of which inhabit surface to near-surface zones of the water body. The study was envisaged to evaluate the nature of the relationship between the stratum of water body (i.e., the habitat) and morphological specializations of the canal system.

MATERIAL AND METHODS

Material for the present study was collected from the water bodies of the Calcutta area as well as from aquarium fish traders. Some was preserved in 10% formalin and some unpreserved used for skull preparation. The course of the sensory canal in the prepared skull as well as in the preserved specimens was studied by injecting aqueous solution of methylene blue. Neuromasts in the canals were observed by carefully removing the outer wall under a high-power binocular microscope.

The taxonomic status of the study material, following Nelson (1984), is stated below:

Order	Cyprinodontiformes
Suborder	Adrianichthyoidei
Family	Oryziidae
Genus	*Oryzias*
Species	*Oryzias melastigma* (McClelland)
Suborder	Cyprinodontoidei
Family	Aplocheilidae
Genus	*Aplocheilus*
Species	*Aplocheilus panchax* (Hamilton)
Family	Poeciliidae
Subfamily	Poeciliinae
Genus	*Poecilia*
Species	*Poecilia reticulata* (Peters)
Genus	*Xiphophorus*
Species	*Xiphophorus helleri* Heckel

OBSERVATIONS

The sensory canals of these cyprinodont fishes are rudimentary in nature. Absence of bony encasement, nakedness of the neuromasts, and solitariness of the canals are the predominant features of the fishes studied.

In *A. panchax* the dorsal part of the head is devoid of any sensory canal, i.e., the supraorbital canal is absent (Fig. 3A). Some free neuromasts encapped with cupula are, of course, present on the dorsocranial surface. In *O. melastigma* also there is no supraorbital canal (Fig. 3C). In *P. reticulata* the supraorbital canal is represented by only two shallow depressions (Fig. 3E). There is no overlying membrane over these depressions. These elongated depressions are disposed roughly in an anteroposterior direction, one lying on the anterodorsal aspect and the other on the posterodorsal aspect of the eye. Both depressions are provided with two neuromasts each. In *X. helleri* the supraorbital canal is represented by three sections—anterior, middle and posterior (Fig. 3G). The anterior one is covered by a membrane and bears two terminal pores. It is disposed in an anteroposterior axis and bears a single neuromast. The middle and posterior sections of the supraorbital canal are open. The middle section is disposed in an anteroposterior direction and bears two neuromasts. The posterior section is oblique in the dorsoventral direction. This part of the supraorbital canal has a very short posterior extension which represents the temporal section. Three neuromasts are present in the posterior supraorbital section, of which two lie in the supraorbital proper and one in the temporal section.

The preopercular canal is retained in the outwardly facing bony trough of the preopercular bone. The face of the trough is guarded by scales except in the region of pores. In *A. panchax* the canal is perforated at seven places (Figs. 2A, B) for communication with the medium. The nature of the canal is identical in the rest of the three forms except

196   Fish Morphology

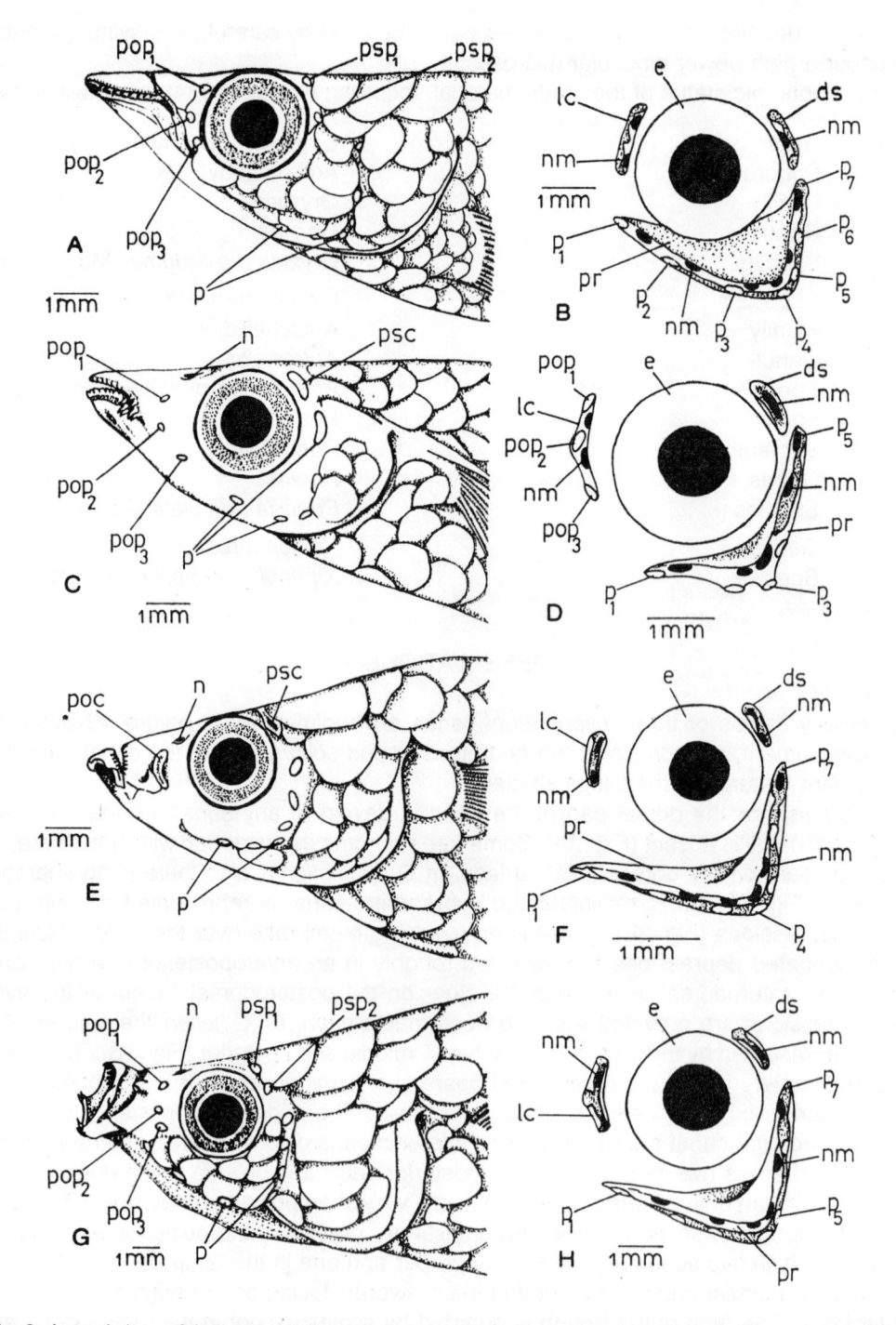

Fig. 2. Lateral views of the head showing distribution of sensory pores and enlarged views of the preopercular, preorbital and postorbital canals. *A. panchax* (A, B), *O. melastigma* (C, D), *P. reticulata* (E, F) and *X. helleri* (G, H).

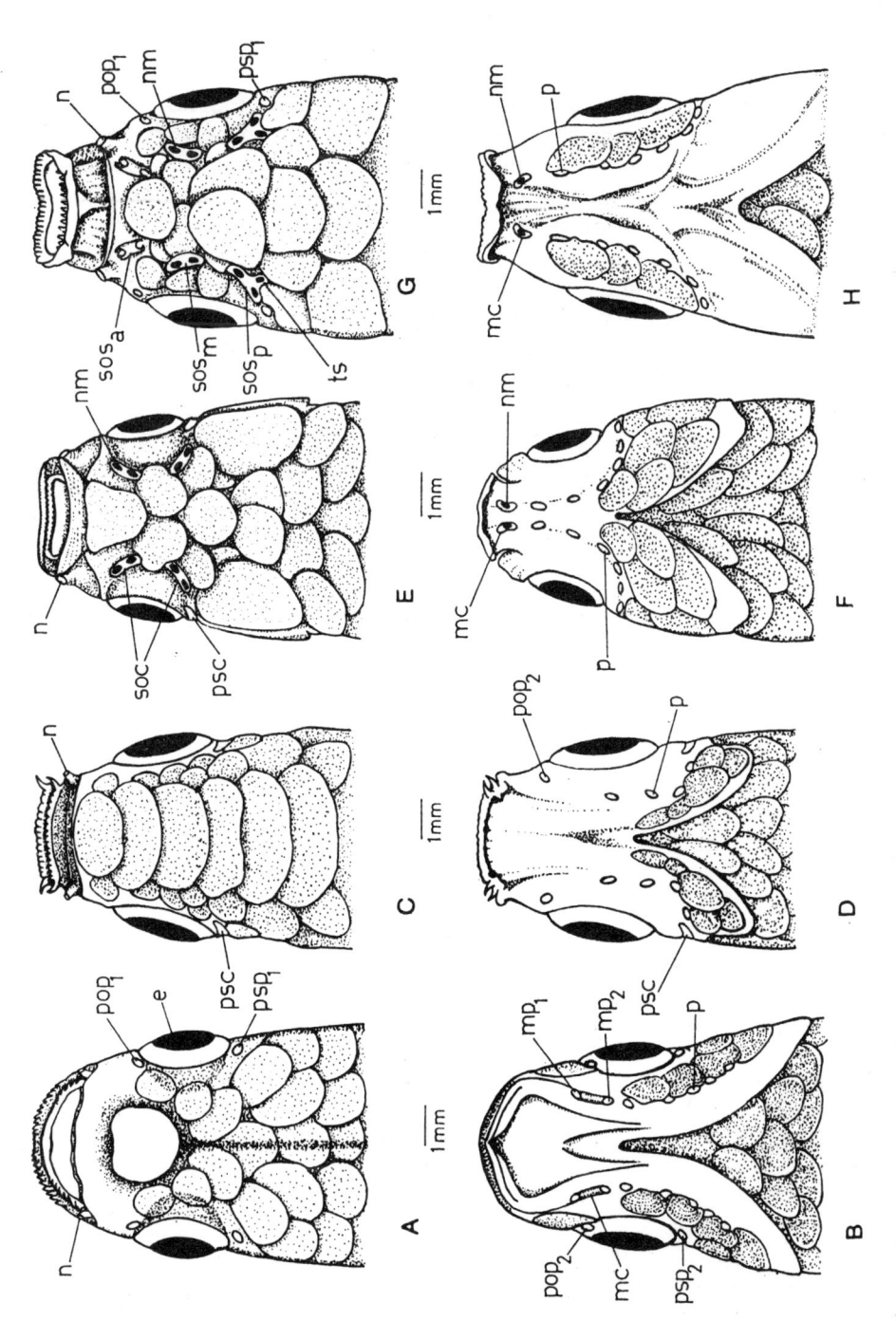

Fig. 3. Dorsal and ventral views of the head showing sensory pores and canal pattern. *A. panchax* (A, B), *O. melastigma* (C, D), *P. reticulata* (E, F) and *X. helleri* (G, H).

for the number of pores, which are five in *O. melastigma* (Figs. 2C, D), and seven in *P. reticulata* (Figs. 2E, F) as well as *X. helleri* (Figs. 2G, H). The number of neuromasts belonging to this canal are six in *A. panchax* (Fig. 2B), five in *O. melastigma* (Fig. 2D), and six in *P. reticulata* (Fig. 2F) and *X. helleri* (Fig. 2H).

A very short mandibular canal is present in *A. panchax* (Fig. 3B), which has only the terminal openings. The mandibular canal is absent in *O. melastigma* (Fig. 3D). It is represented by a very short groove, with no overlying membrane in *P. reticulata* (Fig. 3F) and *X. helleri* (Fig. 3H). The number of neuromasts is two in *A. panchax* (Fig. 3B) and one in *P. reticulata* and *X. helleri* (Figs. 3F, H).

A complete infraorbital canal was not found in any of the fishes studied. In all these forms it is represented by anterior (preorbital) and posterior (postorbital) sections only. The lacrymal and dermosphenotic bones form bony troughs for these pre- and postorbital sections respectively. The trough may be open or covered by membrane. In *A. panchax* the preorbital section (Figs. 2A, B) is covered by a complete membrane and disposed in a dorsoventral direction at the anterior border of the orbit. It is perforated at three places, of which two are terminal and one central. Two neuromasts are present in this canal. The postorbital section is also covered by a membrane and is much smaller in size

Table Summary of major cephalic canals along with distribution of pores and neuromasts

Species Habitat	*A. panchax* Upper surface	*O. melastigma* Upper surface	*P. reticulata* Mid-surface	*X. helleri* Lower surface
Supraorbital canal				
Anterior				
Pores	x	x	(0)	2
Neuromasts	x	x	2	1
Middle				
Pores	x	x	x	(0)
Neuromasts	x	x	x	2
Posterior				
Pores	x	x	(0)	(0)
Neuromasts	x	x	2	2 + 1
Infraorbital canal				
Preorbital				
Pores	3	3	(0)	3
Neuromasts	2	2	2	2
Postorbital				
Pores	2	(0)	(0)	2
Neuromasts	1	1	2	1
Preopercular canal				
Pores	7	5	7	7
Neuromasts	6	5	6	6
Mandibular canal				
Pores	2	x	(0)	(0)
Neuromasts	2	x	1	1

(0)—open canal, x—canal absent.
Note: Total number of neuromasts on one side (cephalic canal)
    A. panchax = 11, *O. melastigma* = 8, *P. reticulata* = 15, *X. helleri* = 16.

(Figs. 2A, B). It bears two terminal pores and a single neuromast. In *O. melastigma* the preorbital section is comparatively long and the lacrymal bone forms a complete bony tube for it (Figs. 2C, D). It also has three pores and two neuromasts. The postorbital section (Figs. 2C, D) is a large, oval, open groove adjoining the posterodorsal part of the orbit, containing a single large neuromast. In *P. reticulata* both pre- and postorbital sections (Figs. 2E, F) of the sensory canal are represented by short open grooves occupying the same position as in the former species. These bear two very small neuromasts each. In *X. helleri* the preorbital canal (Figs. 2G, H) is more or less identical to that of *A. panchax*. The number of neuromasts is two and both are considerably larger. The postorbital section (Figs. 2G, H) is also similar to that of *A. panchax* and bears a single neuromast.

DISCUSSION

Some interesting findings on the cephalic sensory canal system were recorded in this study. A bony encasement for the protection of delicate sensory canals in the head region, usually observed in most teleosts, is mostly lacking in these four forms. The only canal with an enveloping osseous wall in all the forms is the preopercular. Pending further study, it may provisionally be recognized as a common feature of cyprinodont fishes. Other canals are either open or bear a cutaneous covering. In open canals the neuromasts remain directly in contact with the aquatic medium, a condition observed earlier by Rosen and Mendelson (1960) in poeciliid fishes. Uncased canals were also observed in *H. nehereus* (Ray, 1989), in which the supraorbital canal is mostly naked except for some delicate bony arches extending specially over the neuromasts.

Cyprinodont fishes are small in size and essentially surface dwellers. Their thin and light skull probably cannot afford more ossification to cover the sensory canal. Therefore, absence of ossification over the cephalic sensory canals of cyprinodonts may be considered an adaptive feature.

The reduced nature of the cephalic canals is another important feature of cyprinodont fishes. Simplification of the cephalic canal system in poeciliid fishes has also been reported by Gosline (1949) and Rosen and Mendelson (1960). The lack of side branches or ramification of the major canals is an instance of simplification.

The cephalic canals were designated as nasal, supraorbital, postorbital, mandibular, preopercular, and preorbital by Denny (1937), Gosline (1949) and Van Bergeijk and Alexander (1962). The same nomenclature is followed here except that the nasal canal is considered under the supraorbital. The small isolated canal located posterior to the eye in *A. panchax* was described as supraorbital by Gosline (1949). But contrary to his observation, there is no true supraorbital discernible in *A. panchax*. The "supraorbital" he described is actually a postorbital part of the infraorbital since it occurs in the dermosphenotic. Thus its proper identification as postorbital is attested to by its morphological disposition. A supraorbital canal is completely absent in *O. melastigma* also. In *P. reticulata* two superficial grooves are noticed only on the dorsal plane of the head, bearing two neuromasts each, which probably represent the supraorbital system. In *X. helleri*, of the three sections of supraorbital canals, the first one is a complete tube and the other two are open.

The nature of the supraorbital system in particular in the fishes studied depicts a distinct relationship between habitat and structural specialization. *A. panchax* and

O. melastigma are true surface dwellers (Streba, 1962), especially *A. panchax*, whose dorsal cephalic plane usually remains above the water surface and therefore a supraorbital system is not necessary. In *O. melastigma* also the supraorbital canal is lacking. As for the other two forms, *P. reticulata* generally prefers a surface niche although not absolutely like *A. panchax* and *O. melastigma*. *P. reticulata*, despite generally taking food from the water surface through an upturned mouth, never exposes its dorsocranial plane. *P. reticulata* generally takes food from the water surface and since its dorsocranial part remains below water, the supraorbital canal is not eliminated. It is represented by shallow grooves with naked neuromasts. Although *X. helleri* is a surface-loving fish, it roams in the deeper layer of the surface water column. Hence its supraorbital canal is better developed, albeit simplified, following a poeciliid pattern.

The infraorbital canal was never a complete tube in any of the four forms studied. Gosline (1949) also reported the absence of a complete infraorbital canal in cyprinodonts. It is also interrupted in *Esox americanus, E. niger* (Nelson, 1972), and *T. lepturus* (Ray, 1989). The canal is represented only by pre- and postorbital sections, which may be either closed or open. These pre- and postorbital sections also show considerable variation: both sections are closed in *A. panchax* and *X. helleri*, only the preorbital remains closed in *O. melastigma*, while in *P. reticulata* both sections are open. The condition of the pre- and postorbital sections of the infraorbital canal is, without doubt, an indication of great reduction. In *P. reticulata* only small open grooves are present. The preorbital section is closed in the other three forms, however, particularly in *O. melastigma* where the lacrymal forms a complete bony tube around it. During movement friction of water is expected to affect the neuromasts, especially of the snout region, and hence encasement is necessary for protection of such delicate sensory structures. This is probably the reason why the antorbital section is provided with a cutaneous covering while some other canals, e.g. supraorbital, mandibular are without it. In this context it may be mentioned that Dijkgraaf (1962) observed well developed lateral line canals in a stream dwelling cobitid, whereas the neuromasts are exposed in the muddy water cobitid.

According to Gosline (1949), the mandibular canal is either a reduced canal or absent in cyprinodonts. But in the present study an intermediate condition was also noticed, i.e., the presence of an open groove. Very short mandibular grooves are present in *P. reticulata* and *X. helleri*.

The preopercular section is more or less identical in all the four fishes studied except for some variation in neuromast and pore number. The neuromasts are comparatively large in *O. melastigma*.

In summary, *A. panchax* and *O. melastigma* inhabiting the upper surface of the water body, have a much less developed cephalic canal system both in terms of canal pattern and number of neuromasts while *P. reticulata* and *X. helleri*, occupying the lower surface, have a better developed canal system. The difference is confined to the supraorbital canal system which is absent in the first two species. It may be concluded that the lateral line morphology of these cyprinodont fishes correlates with their ecological niche at the upper and lower strata of the water column.

LEGEND

cc	Coronal commissure	pop	Preorbital pore
ds	Dermosphenotic bone	pr	Preoperculum
e	Eye	prc	Preopercular canal
ioc	Infraorbital canal	psc	Postorbital canal
lc	Lacrymal bone	psp	Postorbital pore
mc	Mandibular canal	soc	Supraorbital canal
mp	Mandibular pore	sos	Supraorbital canal section (a—anterior, m—middle p—posterior)
n	Nostril	sp	Sensory pore
nm	Neuromast	stl	Supratemporal canal
p	Preopercular pore	tcl	Truncal canal
poc	Preorbital canal	ts	Temporal section

References

Bailey, R.M. 1951. A check list of fishes of Iowa with keys for identification. In J.R. Harlan and E.B. Speaker (eds.), *Iowa Fish and Fishing*, pp. 187-283. Iowa St. Conservn. Comm.

Denny, M. 1937. The lateral line system of the teleost, *Fundulus heteroclitus. J. Comp. Neurol.* 68:49-65.

Denton, E.J. and J.H.S. Blaxter. 1976. The mechanical relationship between the clupeid swimbladder, inner ear and lateral line. *J. Mar. Biol. Assn. U.K.* 56(3):787-807.

Dijkgraaf, S. 1962. The functioning and significance of the lateral line organs. *Biol. Rev.* 38:51-105.

Disler, N.N. 1960. The sense organs of the lateral line system and their significance in the behavior of fishes. In A.N. Severtsov (ed.), *Ecological-morphological Conformities in the Development of the Sense Organs of the Lateral Line System of Fishes*, pp. 264-298. Acad. Nauk., Moscow.

Gosline, W.A. 1949. The Sensory Canals of the Head in Some Cyprinodont Fishes, with Particular Reference to the Genus *Fundulus. Occ. Pap. Mus. Zool. Univ. Mich.* 519:21 pp.

Hubbs, C.L. and M.D. Cannon. 1935. The Darters of the Genera *Hololepis* and *Villora. Misc. Publ. Mus. Zool. Univ. Mich.* 30:1-93.

Jorgensen, J.M. 1985. On the fine structure of lateral line canal organs of the herrings (*Clupea harengus*). *J. Mar. Biol. Assn. U.K.* 65(3):751-758.

Kapoor, A.S. 1959. Sensory canals of the head in *Ophicephalus punctatus* (Bloch). *Zool. Anz.* 163:298-302.

Kapoor, A.S. 1960. Sensory canals and related dermal bones of the head in *Heteropneustes fossilis* Bloch. *Z. Wiss. Zool.* 164:315-323.

Kapoor, A.S. 1961. Sensory canals and related bones in the head of *Wallago attu. Trans. Amer. Micr. Soc.* 80(3):329-343.

Kapoor, A.S. 1962. Development of laterosensory canal bones in the fish *Ophicephalus punctatus. Copeia* 1962(4):854-855.

Kapoor, A.S. 1964. Functional morphology of laterosensory canals in the Notopteridae (Pisces). *Acta Zool., Stockh.* 45:77-91.

Khandelwal, O.P. 1963. Sensory canals of the head of a few cyprinids. *Curr. Sci.* 32:126-127.

Khandelwal, O.P. and M.G. Rastogi. 1965. Sensory canals and related bones of the head in *Rhinomugil corsula* (Ham.). *Agra Univ. J. Res. (Sci.)* 14(1):55-62.

Khandelwal, O.P. and R.S. Sharma. 1965. Sensory canals in the head of *Silonia silondia* (Ham.). *Agra Univ. J. Res. (Sci.)* 14(1):49-54.

McFarland, W.N., F.H. Pough, T.J. Cade, and J.B. Heiser. 1985. *Vertebrate Life*. MacMillan Publishing Company, New York, 636 pp.

Nelson, G.J. 1972. Cephalic sensory canals, pitlines, and the classification of escoid fishes with notes on the galaxiids and other teleosts. *Amer. Mus. Novit.* 2492:1-49.

Nelson, J.S. 1984. *Fishes of the World*. John Wiley and Sons, New York. 523 pp.
Ray, P.K. 1989. *Investigation on the cranial anatomy of some Indian Teleosts*. Ph.D. thesis, University of Calcutta, Calcutta.
Ray, P.K. and N.C. Datta. 1990. Cephalic sensory canal system of some Indian percoid fishes. *J. Freshwater Biol.* 2(3):221–231.
Rosen, D.E. and J.R. Mendelson. 1960. The sensory canals of the head in poeciliid fishes (Cyprinodontiformes), with special reference to dentitional types. *Copeia* 1960(3):203–210.
Sand, A. 1937. The mechanism of the lateral line sense organs of fishes. *Proc. Roy. Soc.* B123:472–495.
Sharma, M.S. 1964. The cephalic lateral line system in *Notopterus chitala* (Ham.). *Copeia* 1964:530–533.
Srivastava, M.D.L. and C.B.L. Srivastava. 1967. The development of neuromasts in *Cirrhina mrigala* Ham. Buch. (Cyprinidae) and *Ophicephalus (Channa) punctatus* Bloch (Channidae). *J. Morph.* 122(4):321–344.
Streba, G. 1962. *Freshwater Fishes of the World*. Vista Books, London, 877 pp.
Van Bergeijk, W.A. and S. Alexander. 1962. Lateral line canal organs on the head of *Fundulus heteroclitus*. *J. Morph.* 110(3):333–346.

13

Morphometrics of the Respiratory System of Air-Breathing Fishes of India

P.K. Roy and J.S. Datta Munshi

The term "morphometrics" means the measurement of dimension of any biological system. Today, morphometrics is used for a vast area of studies either for measurement of gross dimensional relationship of body weight, length, fin, or head size with respect to body length, or for minute dimensions of respiratory surface areas, blood vascular area and volume, diffusion thickness, etc. But it is appropriate to use this term in connection with morphological measurement of any structure along with statistical analysis to understand its functional relationships. Quantitative data on dimensions of a system in relation to growth and environmental conditions helps us in understanding its functional efficiency. The morphometrics of gills and accessory respiratory organs of air-breathing fishes in relation to growth gives a vivid picture of their relative importance in their life processes in relation to their natural environmental conditions.

The methodology for morphometrics varies in accordance with the structure and system to be measured. If the organ is of definite geometrical shape, its measurement is simple, but if the surface of the structure is complex and corrugated, measurement involves complex principles. The respiratory system of air-breathing fishes consists of gills and different types of accessory respiratory organs, which are very complex structures in configurations (Fig. 1) and as such, require special methods for measurement.

The morphometry of the gill and accessory respiratory organs of some air-breathing fishes of India, with particular reference to their diffusing capacity, is discussed below.

METHODOLOGY AND PRINCIPLES INVOLVED IN MORPHOMETRICS

Gills

Gills are provided with many long and parallel filaments which bear a number of secondary lamellae arranged parallel to each other and at a 90° angle to the filament axis. Due to this complexity several methods have been developed by different scientists considering some assumptions.

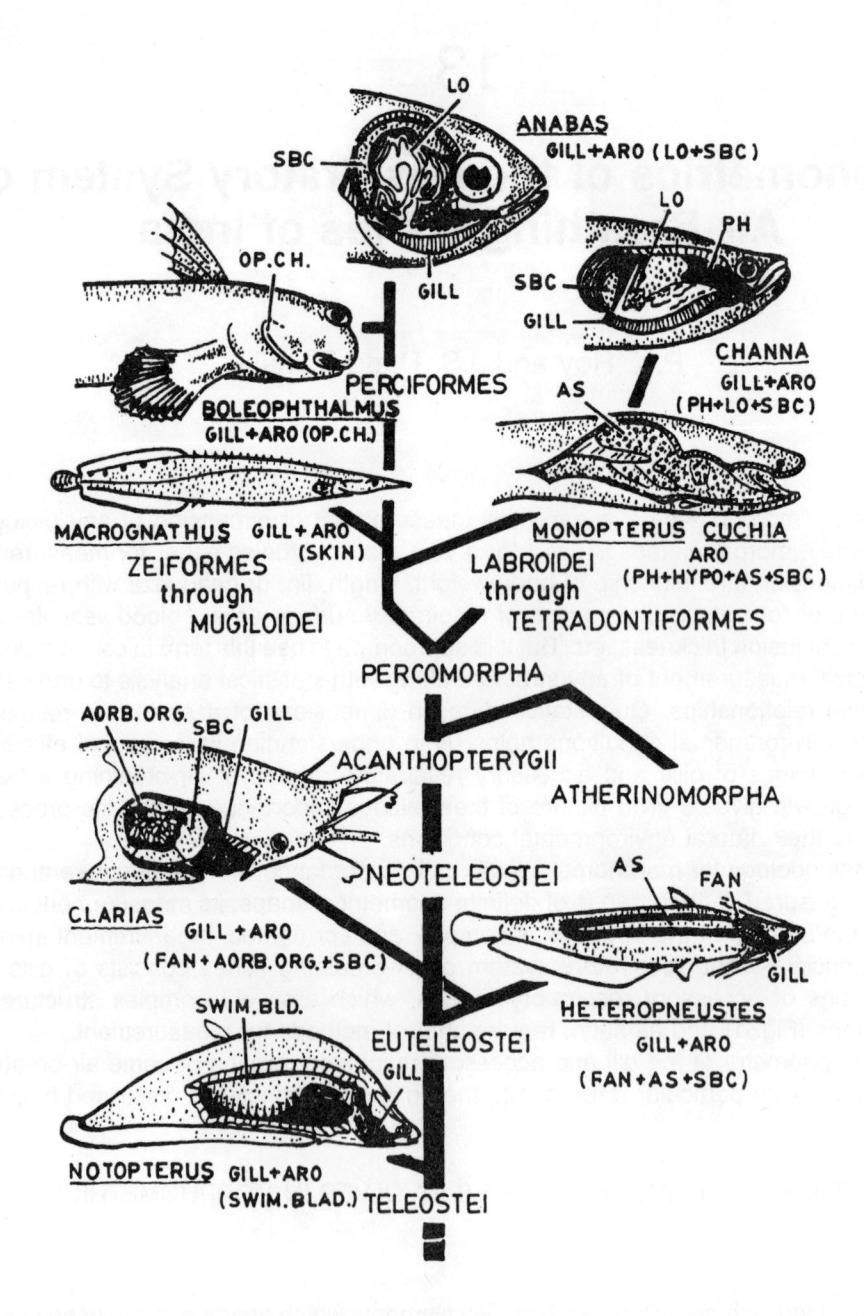

Fig. 1. Nature's innovation of varied designs of accessory respiratory organs (ARO) of Indian air-breathing fishes and their phylogenetic position.
AS—Air sac; AORB.ORG.—Arborescent organ; LO—Labyrinthine organ; OP.CHAM—Opercular chamber; PH—Pharynx; SBC—Suprabranchial chamber.

Price (1931) is credited as the first person who attempted to determine the total surface area of a gill (including both the filament and lamellae). He considered a filament as a triangle of which the two surfaces are increased by infoldings of rectangular secondary lamellae. Subsequently, emphasis was given only to the lamellar surface area. Most authors considered a single secondary lamella and measured its area using a camera lucida or projection microscope (Gray, 1954; Byczkowska-Smyk, 1957; Hughes, 1966; Munshi, 1980). The general formula used was $L \cdot n \cdot bl$ or $L \cdot 1/d \cdot bl$, Where L is the total filament length, n the number of secondary lamella/mm on both sides of filaments; d the distance between two consecutive secondary lamellae (l/d is equivalent to n); bl the bilateral surface area of a secondary lamella which can be taken as representative of all secondary lamellae.

This process is both laborious and time consuming, so only on a small sample of the millions of lamellae and hundreds of filaments present in the gill can be measured, for which different sampling techniques were developed.

It was formerly assumed that a gill is homogeneous except in filament length, so the number of secondary lamellae and their bilateral surface areas were measured irrespective of their position. This was labeled the unweighted method and was used in many early works (Gray, 1954; Byczkowska-Smyk, 1957; Hughes, 1966; Hughes et al., 1973; 1974).

Later, with advancement in knowledge of the structure of the gill sieve, it was established that the gills of most fishes are heterogeneous in nature, and so a different sampling method was developed, known as the weighted method (Muir and Hughes, 1969). This method considers all the differences and variations existing in their number, size of filaments and secondary lamella; sampling is done in a way to ensure proper representation of the whole organ. This method is based on the principle that "if samples represent portions of the total system that differ insignificantly from one another, in the average calculation more weight should be given to those samples which represent a lesser part in the whole". In fact, weighting of averages is done and the method termed the "weighted method" for gill morphometrics.

The gill system is viewed under a binocular microscope to determine its composition, i.e., how the different sectors vary in filament length, frequency, and lamellar areas. Each gill arch is sliced into small sectors containing 10–20 filaments (depending on their number). Generally, the middle (5th or 10th filament) of a sector/section is thought to be representative of that section. The average filament length (L_1) of the central filament is measured and multiplied with filament number (fn_1) to find the total filament length (L_1) of the section. Using a micrometer the lamellar frequency of this central filament length is counted from at least three regions, i.e., tip, middle, and base of the filament and doubled. The average of lamellar frequency (n_1) from the three regions is multiplied with L_1 to find the number of total secondary lamellae of the section. Thin hand-cut sections of the central filament are made perpendicular to the filament axis. The area of at least one lamella from each of the three respective regions is traced using a camera lucida or projection microscope.

The average area of three lamellae of the three regions of the filament is representative of the lamellae of that filament, which is representative of all the filaments of that section (sector) of the gill arch. The average area of a lamella is then doubled to ascertain the bilateral surface area (bl_1). The product of L_1, n_1, and bl_1 gives the lamellar area of that section. Similar measurements are made on all segments of that gill (Fig. 4).

Then, the total filament length (L_1, L_2, L_3 ...), the filament number (fn_1, fn_2, fn_3 ...), and the number of total secondary lamellae from all the sections is summed. By dividing the total secondary lamellae with total filament length, the weighted average for lamellar frequency (n) is calculated, which makes a consideration for all the filaments of the arch. The sum of the total lamellar area of all the sections after division by total number of secondary lamellae gives an estimate of the weighted average of the area of a secondary lamella (bl), since it had been weighted by lamellar number of all the sections. The product of the weighted average number of the secondary lamellae/mm filament length (n) and the weighted bilateral surface area of an average lamella (bl) together with the total filament length (L) gives the gill area. This value is considered to be more accurate in respect to gill area.

Accessory respiratory organs

Air-breathing fishes possess special accessory respiratory organs in the form of dendritic organ, labyrinthine organ, air sac or suprabranchial chamber, which are lined by respiratory membrane (Munshi, 1980). Formerly the whole of the respiratory membrane was thought to be a uniform surface suited for gaseous exchange. The area of this membrane was measured by planimetry. The membranous lining was removed, sliced, stretched flat, and the area traced on graph paper and estimated. This method was used by Hughes, Dube and Munshi (1973), and Hughes et al. (1974 a, b).

But later, after detailed use of scanning and transmission electron microscopes, it was observed that the surface of the accessory respiratory organs is beset with large numbers of respiratory islets (RI) and nonrespiratory narrow lanes (L). The surface is not smooth; it is corrugated. Respiratory and nonrespiratory areas are dissimilar even in histological details. When the surface is slightly folded, the area of gas-exchange surface is greater than that of projected area, i.e., the area obtained after tracing its outline shape. So, in recent studies this corrugation factor, by which the outer corrugated margin is more than the projected one, was taken into consideration. For proper measurement of the surface morphometric grids were superimposed on the surface and intersections of grid line both on the outer surface and a projected line were counted. Then, the final area was calculated as follows:

$$S = S_{pr} \times \frac{l_s}{l_{pr}},$$

where S is the surface area, S_{pr} the projected area, l_s is the intersections of grid with folded surface, l_{pr} is the intersections of grid with projected line (Fig. 5).

Types of grids. Different types of grids are available for stereological morphometric measurement, especially for densities in space, viz. volume density, surface density, length density, etc. These grids are simple-square lattice, double-square lattice, multipurpose hexagonal, triangular, and curvilinear (Merz) lattice test systems. For a sample which is isotropic in orientation, any type of test system can be used. But for a sample which shows a certain degree of anisotropy, i.e., preferred orientation of membranes or boundaries to a projected line, the curvilinear test line system is used. This test system is composed of semicircles disposed in a square lattice, so it becomes isotropic even for anisotropic samples, whereas other test systems composed of straight lines, themselves

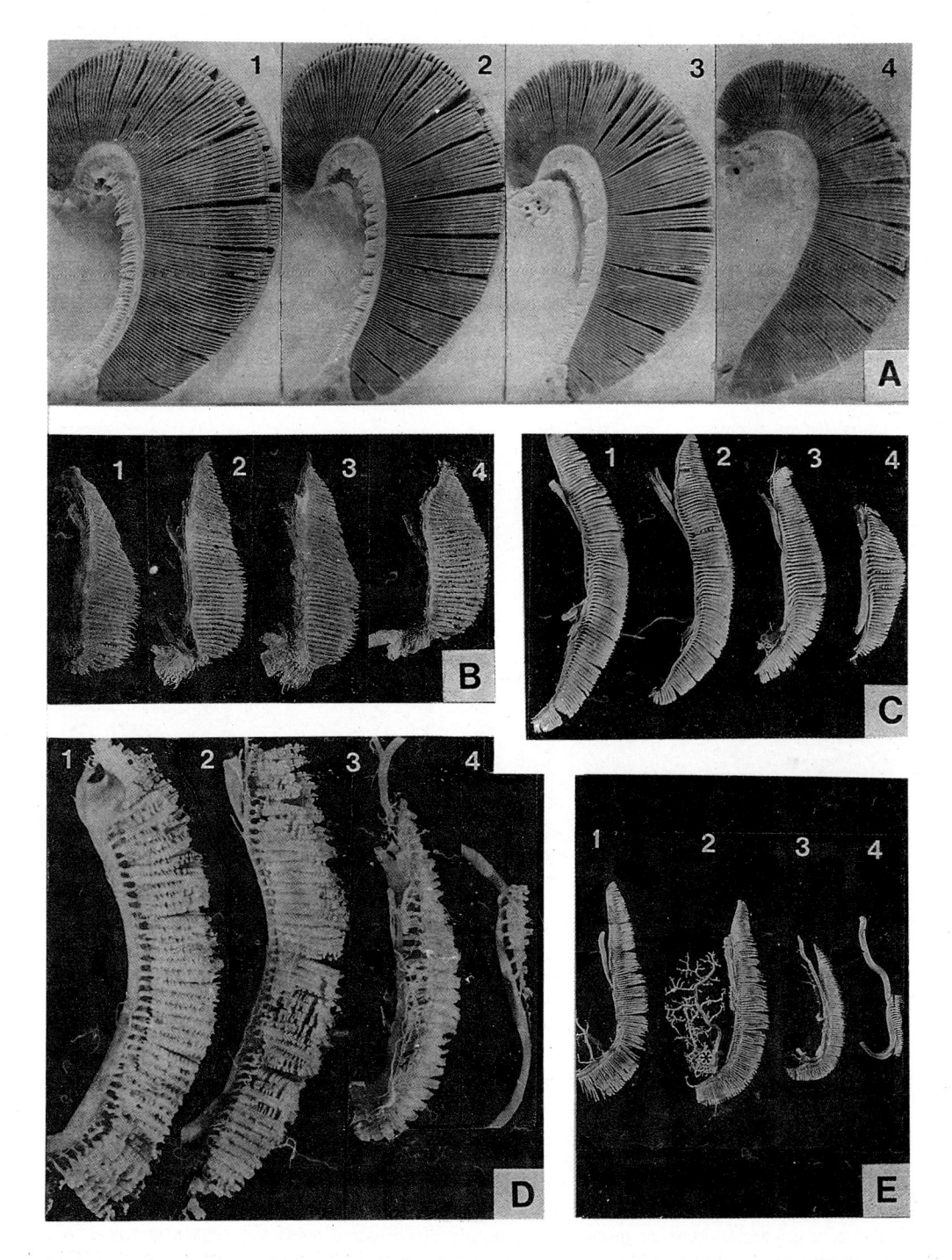

Fig. 2. Different gill arches of fishes to show their comparative size. A. *Cirrhinus mrigala* (water-breather; Roy, 1984) B. *Heteropneustes fossilis* (facultative air-breather; Olson et al., 1990); C. *Channa punctatus* (facultative air-breather; Olson et al., 1994); D. *Anabas testudineus* (obligate air-breather; Munshi et al., 1986); E. *Channa marulius* (obligate air-breather; Olson et al., 1994); N.B. B–E are corrosion cast replica of gills.

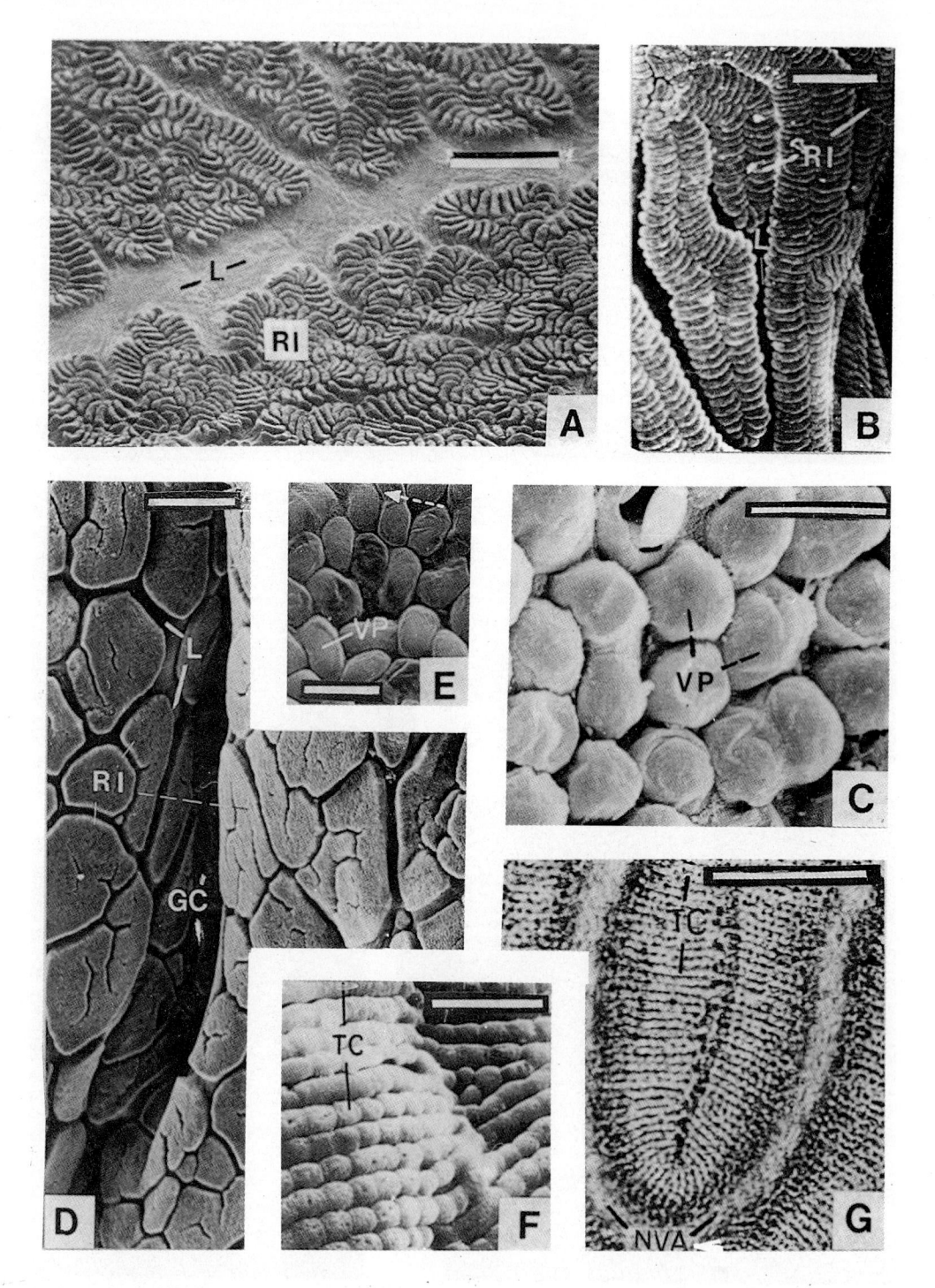

Fig. 3. Surface ultra structure of respiratory membrane of accessory respiratory organs of fishes. A. Air Sac of *Heteropneustes fossilis* (bar = 150 μm) B. Dendritic organ of *Clarias batrachus* (bar = 500 μm) C. Suprabranchial chamber of *Channa punctatus* (bar = 10 μm) D. Pharynx of *Monopterus cuchia* (bar = 200 μm). E. Higher magnification of D to show the vascular papillae (bar = 10 μm). F. Suprabranchial chamber of *Anabas testudineus* (bar = 50 μm). G. Light micrograph of respiratory membrane of *Anabas* (bar = 250 μm) GC —Gill cleft, L —lane (Nonvascular area, NVA), RI —Respiratory islets; TC —Transverse capillaries, VP —Vascular papillae.

show a high degree of anisotropy. This isotropic test system was proposed by Merz (1967) and is known as the "Merz Grid". These grids are available in different space length (d) and can be selected according to sample dimension.

Measurement of diffusing capacity. The oxygen flow mechanism into the tissues is maintained by two processes: (i) diffusion process in which movement of molecular O_2 through air or any fluid takes place. This movement is from water to blood in the case of gills, and from air through a film of water to blood in air-breathing organs; (ii) convection process in which mass transport of oxygen takes place by a carrier through the blood. The first process depends on the area, diffusion barrier and concentration gradient. Fick (1855) propounded a fundamental law governing the free diffusion of solute in a solution. The law states that diffusion takes place from higher concentration to lower concentration and the rate of linear diffusion (dn/dt) of quantity (n) in time (t) is proportional to the concentration gradient (dc/dx) and area (A) perpendicular to the direction of transmission (x).

$$\frac{dn}{dt} \ \alpha \ A \times \frac{dc}{dx}.$$

Or,
$$\frac{dn}{dt} = -D \cdot A \cdot \frac{dc}{dx}.$$

Here ($-$) sign denotes the downhill process from higher concentration to lower concentration and D is the diffusion coefficient. When the outflow of a solute is equal to its inflow, this is termed a steady-state condition. For a steady-state condition, dc is constant. This means that C_1 and C_2 will always be constant due to the balanced inflow and outflow. Thus, in case the diffusion is linear, the quantity of a substance (n) diffused in unit time is

$$n = -D \cdot A \cdot \frac{(C_1 - C_2)}{X} = \frac{-D \cdot A}{X} \cdot (C_1 - C_2),$$

(where D is diffusion coefficient, A the surface area, X diffusion barrier, and C_1 and C_2 concentrations of the solute at the two sides of the barrier (X) (Fig. 6)).

Concentration of a gas depends upon the partial pressure of the gas and its solubility in the medium. Thus

$$C_1 - C_2 = \beta \cdot (P_1 - P_2)$$
or
$$\beta \cdot \Delta P,$$

(where β is solubility or capacitance coefficient, P_1, P_2 are partial pressure of gas corresponding to two concentrations C_1 and C_2; and $\Delta P = (P_1 - P_2)$ is partial pressure difference).

So the quantity (n) of a gas diffused in unit time will be:

$$n = D \cdot \frac{A}{X} \cdot \beta \cdot \Delta P.$$

The product of two material properties, i.e., D and β of a gas is called Krogh's permeability coefficient (K) or diffusion constant. Since $K = D \cdot \beta$,

$$n = K \cdot \frac{A}{X} \cdot \Delta P.$$

This equation is called the modified form of Fick's equation and is used satisfactorily in many physiological measurements. Respiratory physiologists utilize this equation to measure the quantity of a gas diffusing into the blood from the medium in unit time. They take into consideration the air/water and blood diffusion barrier and respiratory surface area for gas exchange. For diffusion of O_2 from air/water into blood, the following equation holds:

$$\dot{V}o_2 = K_t \cdot \frac{A}{t} \cdot \Delta Po_2,$$

(where $\dot{V}o_2$ is oxygen uptake/unit time, A the area of gas exchange, K_t is Krogh's permeation coefficient of o_2 through tissue, t tissue barrier thickness, and ΔPo_2 difference in partial pressure of o_2 on both sides of the barrier, i.e., in water (Pwo_2) and blood (Pbo_2).) On rearranging, Fick's equation is

$$\frac{\dot{V}o_2}{\Delta Po_2} = \frac{K_t \cdot A}{t}$$

$K_t \cdot A/t$ is the diffusing capacity of the gas and is defined as the quantity of gas passing in unit time through a respiratory surface. This diffusing capacity provides valuable information on diffusion/exchange of a gas such as oxygen (O_2). For a morphometric estimate of the diffusing capacity, values of K_t, A, and t are needed. A is the total respiratory surface area of the gill/accessory respiratory organ in cm^2. Direct estimation of K_t is not available for fish tissue. The estimated value of K is available for a frog's connective tissue (0.00015), muscle (0.00019) and mammalian tissue (0.00033) in $mlO_2/min/mmHg/cm^2/\mu m$ at 20°C (Krogh, 1941). Hence it is customary to use the value of K of a frog's connective tissue as K for a fish gill. Recently, two values of K, i.e., for a frog's connective tissue and a mammalian tissue, have been applied for air-breathing surfaces of fish (Munshi et al., 1989; Hughes et al., 1992). "t" is the mean of water/blood barrier (μm). The harmonic mean of the tissue-blood barrier is more appropriate for measuring diffusing capacity. A gas can diffuse both at a right angle to the external surface as well as along longer pathways and, as such, a wide range of distances has to be taken into account. Further, the entire area and not merely the area just in front of the capillaries, is used in calculation. The use of arithmetic mean of the barrier under such circumstances is not proper; the harmonic mean should be used instead.

It is general practice to measure the length of a range of randomly placed lines across the barrier. For exact estimation, morphometric grids are superimposed on TEM electron micrographs or projection of semithin sections. When the plane of sectioning passes randomly through the gas exchange surface, a rectilinear (square lattice) grid may be used in place of a curvilinear grid, and the length of the line passing across the outer and inner surface measured. But when the sectioning plane is at 90° to the exchange surface (with a bias), a randomized grid, i.e., curvilinear (Merz) grid is essential. This grid is also used for randomly sectioned surfaces. It is necessary to observe the true intersections for the Merz grid. The true intersection is that intersection at which the grid line cuts the two surfaces. False intersections, on the other hand, are intersections at which a grid line crosses one surface only, and not the other, and returns again to the same surface. For barrier thickness, the shortest distance, or distance perpendicular to the surface at the intersecting point is measured, and the harmonic mean calculated.

Higher accuracy requires a large number of measurements (Hughes et al., 1992). For speedy measurements, the complete accuracy of each measurement is not essential. In harmonic mean calculation, the reciprocal of distance ($1/L$) is used; so longer distance need not be measured accurately as its contribution to mean calculation is negligible. On the other hand, short distances should be measured with care and accuracy since their contribution to calculations is vital.

For this purpose a ruler, based completely or partially on a logarithmic scale, and in which shorter intervals are more closely spaced is used (Weibel, 1979). This ruler enables grouping the measurements in many size ranges (Fig. 5). Each size range corresponds to its midvalue and the harmonic mean of the barrier is calculated as follows:

$$\tau_h = \frac{N \text{ (total number of measurements)}}{(n_1/L_1 + n_2/L_2 + n_3/L_3 + \ldots + n_{10}/L_{10})} \times \frac{2}{3}.$$

These calculations give the values for the diffusing capacity of the tissue barrier alone (D_t) and not for all the barriers across which the transfer of oxygen takes place in the course of its passage from water to hemoglobin molecules. The other barriers are water, mucus layer around the lamellar surface, the plasma, and the erythrocyte. Hence, the overall diffusing capacity may be derived as follows:

$$\frac{1}{D_g} = \frac{1}{D_w} + \frac{1}{D_m} + \frac{1}{D_t} + \frac{1}{D_p} + \frac{1}{D_e}.$$

The subscripts g, w, m, t, p and e denote gill, water, mucus, tissue, plasma, and erythrocytes respectively (Fig. 5).

To calculate D_w using the modified Fick's equation, the area used is the area of lamella and τ_w is one-fourth of the interlamellar distance because oxygen molecules can diffuse to the tissue both from the center and just near the lamellar surface. Krogh's permeation coefficient value for water (K_w) is 0.00045 mlO$_2$/cm^2/mmHg/min/μm (Krogh, 1941). However, in recent years another value of Krogh's permeation coefficient has been calculated for distilled water, namely, 0.00062 mlO$_2$/cm^2/mmHg/min/μm (Dejours, 1975).

To measure the mucous layer over the gills, the tissue fixed for electron microscopy is stained with ruthenium red to give a clear picture of the layer under TEM. This barrier is very thin in gill lamellae but can be measured and the harmonic mean of thickness (τ_m) can be calculated. The K_m value for mucus can be used as 0.00032 mlO$_2$/cm^2/min/mmHg/μm (Ultsch and Gros, 1979) or that of the connective tissue.

For calculation of the diffusing capacity of plasma (D_p), the surface area of the capillaries has to be estimated. By superimposition of the morphometric grid, the intersections of lamellae (I_l) and capillaries (I_c) can be counted. Then

$$\frac{I_c}{I_l} = \frac{\text{Surface area of capillaries } (S_c)}{\text{Surface area of lamellae } (S_l)}$$

or

$$S_c = \frac{I_c}{I_l} \times S_l.$$

Using the above equation and utilizing the number of both intersections (I_c and I_l), capillary surface area can be estimated. The method used for the measurement of τ_t is also used for the measurement of τ_p. The distance between the intersection of capillaries and nearby erythrocytes (L) is measured and the harmonic mean is calculated using the factor 3/4 instead of 2/3 to compensate the erythrocyte curvature.

$$\tau_p = \frac{N}{(n_1/L_1 + n_2/L_2 + n_3/L_3 + \ldots + n_{10}/L_{10})} \times \frac{3}{4}$$

The value of K_p used is the same for frog and mammalian connective tissue.
Calculation of the diffusing capacity of erythrocytes (D_e) involves a different formula:

$$D_e = \theta \cdot V_{ec},$$

where V_{ec} is volume of erythrocyte cytoplasm, θ the reaction velocity constant between hemoglobin and oxygen. The value for θ is not available for fish blood, so a modified equation is used due to the lower hematocrit and nucleated nature of the erythrocyte:

$$D_e = \frac{\theta_m \cdot V_{ec} \cdot H}{H_m},$$

where θ_m is the value of reaction velocity for mammalian blood (1.5 mlO_2/ml/min/mmHg) (Gehr et al., 1981), H = hematocrit value of the species and H_m the hematocrit value of mammalian blood (0.45).

At the time of superimposition of a grid on an electron micrograph of fish respiratory tissue, the points falling on capillaries (P_c), plasma (P_p), erythrocyte cytoplasm (P_{ec}), and erythrocyte nucleus (P_{en}) are also counted with the intersections (I_c and I_l). Volume of capillaries (V_c) and erythrocytes V_{ec} can be calculated as follows:

$$V_c = S_c \cdot \frac{P_c}{I_c} \cdot \pi \frac{d}{4} \quad \text{and} \quad V_{ec} = S_c \cdot \frac{P_{ec}}{I_c} \cdot \pi \cdot \frac{d}{4},$$

where d is the dimension of the grid during superimposition; $P_c = P_e + P_p$ whereas $P_e = P_{en} + P_{ec}$.

The hematocrit value can be calculated using the ratio of P_e/P_c. Thus all the different components for overall diffusing capacity can be measured. However, when the fish lives in normal water and ventilates water continuously, the thickness of the mucous layer is negligible and the total diffusing capacity is based on three barriers only, i.e., tissue, plasma, and erythrocyte.
Then,

$$\frac{1}{D_g} = \frac{1}{D_t} + \frac{1}{D_p} + \frac{1}{D_e}.$$

For air-breathing organs too the resistance to oxygen diffusion is similar and works in a series. So the diffusing capacity for the air-breathing organ is

$$\frac{1}{D_{abo}} = \frac{1}{D_m} + \frac{1}{D_t} + \frac{1}{D_p} + \frac{1}{D_e},$$

where abo = air-breathing organ.

The diffusing capacity of air-breathing organs of several species of air-breathing fishes has been calculated (Munshi et al., 1989; Hughes et al., 1992; Roy and Munshi, 1992).

MORPHOLOGICAL ANALYSIS

Gills

The gills are highly characteristic features of fishes which serve a multitude of vital functions, viz., gas exchange, ionic regulation, circulation of hormones, and detoxification (Hughes, 1984b; Munshi and Hughes, 1992; Olson, 1991). Their functional efficiency depends on their architectural plan, dimensions, nature of diffusion pathway, and also the ventilation and perfusion of the surface.

The basic architectural plan of the gills of all fishes is the same, but the four pairs of gills show a degree of modification according to breathing habits. In water-breathing fishes the gills look homogeneous and approximately equal in dimension (Fig. 2). However, the fourth pair of gills is slightly reduced (Roy and Munshi, 1986; Ojha, 1993). But in air-breathing fishes, the gills show a great degree of heterogeneity (Munshi, 1980). In obligate air-breathers such as *Anabas testudineus*, *Channa marulius*, and *Channa striatus*, the first pair of gills is comparatively well developed but the other three pairs poorly developed (Fig. 2). The fourth pair of gills is provided with short and stumpy filaments without functional lamellae (Munshi and Hughes, 1992). In another obligate air-breather, *Monopterus cuchia*, only the second pair of gills has short and stumpy gill filaments with no functional lamellae (Munshi, 1980; Munshi et al., 1989). In facultative air-breathers such as *Clarias batrachus*, *Heteropneustes fossilis*, *Channa punctatus* and amphibious air-breathers such as *Boleophthalmus boddaerti*, *Periophthalmus vulgaris*, *Macrognathus aculeatum* the four gills show gradual reduction from first to fourth arches, and they all have filaments with functional lamellae (Munshi and Hughes, 1992; Ojha, 1993). The shape and size of lamellae also show a great degree of heterogeneity. The lamellae differ from tip to base of filaments in shape and size, from triangular at the tip to almost rectangular at the base (Fig. 4), and are densely packed towards the tip of the filaments. New lamellae are added towards the tip region of filaments (Hughes, 1984b). The number is greatly reduced in air-breathers with the result that the interlamellar space is increased. In *Clarias* and *Heteropneustes* many of the lamellar capillaries remain buried in filament epithelium (Munshi, 1980; Hughes and Munshi, 1979), but in a water-breather the number of basal (buried) channels is fewer (Roy, 1984). The respiratory epithelium of secondary lamellae is heterogeneous in composition in air-breathers because of the presence of several types of cells, viz., mucous cells, pavement cells, mitochondria-rich cells, and acidophilic granular cells (Hughes and Munshi, 1979). The pavement cells (surface epithelial cells) are microridged in most fishes but are microvillous in *H. fossilis* and *C. batrachus* and smooth in the water-breathers *Cirrhinus mrigala* and *Labeo rohita*. The water/blood barrier is thin in water-breathers (0.6 μm–3.5 μm) and thick in air-breathers (3 μm–15 μm). It is thinner in amphibious fishes (Table 5). Thick barriers act as an oxygen conserving device in air-breathers when the surrounding water has low oxygen tension and concomitantly prevent collape of the lamellae when the fish is out of water (Hughes and Munshi, 1979). A great reduction in lamellar circulation of the fourth gill arch in obligate air-breathers and their modification as shunt vessels (Olson et al., 1986, 1994; Munshi et al., 1994) reduce the potential oxygen loss from blood to hypoxic environment. In *Boleophthalmus* the pillar cell flanges bear stiffening material which prevents collapse of blood vessels when the fish is out of water (Munshi and Hughes, 1992).

The heterogeneity and reduction of gill dimensions in air-breathing fishes are compensated by the development of some sort of accessory respiratory surfaces (Fig. 1). In fishes such as *H. fossilis*, *C. batrachus*, *M. aculeatum*, and *M. cuchia*, the skin is naked and vascularized, which helps in gaseous exchange. But the thickness of the tissue barrier (100 μm) is many times more than the gill's, and so, in spite of its higher surface area, its diffusing capacity is much lower. In these fishes the skin fulfills the oxygen demand to some extent for some time when the basic criteria for gaseous diffusion, i.e., moist surface, are available.

Accessory respiratory organs

In these air-breathing fishes specialized accessory organs have developed, either as extensions of the pharyngeal, branchial, or extrabranchial (opercular) chambers.

Modification of pharyngeal chambers. Monopterus cuchia has developed a pair of air sacs as lung-like extensions of the pharynx (Fig. 1), which are situated along the lateral sides of the head, partly covered over by the operculum (Munshi and Singh, 1968). The pharynx opens through an aperture into the air sac on each side, which is used both as an inhalant and an exhalant aperture. The mucosal lining of the respiratory pharynx, the hypopharynx, and the air sac consists of vascular and nonvascular areas (Hughes and Munshi, 1973; Munshi, 1985). Formerly, the whole air sac was thought to be covered with respiratory mucosa but Munshi et al. (1989) found that the posterior third of the air sac is covered by epithelium with microridges and is nonvascular and nonrespiratory. It probably serves for storage of residual air. The vascular areas of the respiratory mucosa are in the form of small and large respiratory islets formed of intraepithelial blood capillaries which run in spiral-like fashion (Munshi et al., 1990) (Fig. 3). The blood vessels penetrate the epithelium at many points to form vascular papillae. The endothelial cells with their characteristic round nuclei arise from the base or side wall of the capillaries to form the axis of the vascular papillae and may act as small valves to control blood flow. The air/blood pathway is three layers thick with a thin outer epithelium (0.30 μm), basement membrane (0.06 μm), and endothelial lining (0.40 μm).

The air-breathing organs of *Channa punctatus, C. gachua, C. striatus,* and *C. marulius* are in the form of a pair of suprapharyngeal chambers developed as dorsal extensions of the pharynx (Munshi, 1962a, b, 1985; Hughes and Munshi, 1973; Olson et al., 1994).

Each chamber is incompletely divided into two compartments by a shelf-like transverse outgrowth from the hyomandibula (Fig. 1). The pharynx opens through the inhalant aperture into an anterior compartment, which communicates freely with the posterior compartment, but the posterior compartment is also connected with the pharynx by another inhalant aperture. This inhalant aperture is guarded by a shutter of dendritic plate borne by the first branchial arch (Hughes and Munshi, 1986). The dorsal part of the first gill slit forms the exhalant aperture for the chamber. The suprabranchial chamber, buccopharynx, palate and even the tongue is lined with a respiratory epithelium comprised of respiratory islets studded with numerous vascular papillae (Fig. 3). These papillae are dome shaped, and in the buccopharynx, tongue, and palate are lodged in receptacles and are retractile (Hughes and Munshi, 1986). The dome is covered with smooth epithelium while the interdome area is covered with epithelial cells with microvilli. The domes are made up of infraepithelial blood capillaries running in spiral or wave-like fashion (Olson et al., 1994). The aperture of the dome is guarded by a valve originating from the base of the vascular papilla forming an axis of the structure (Munshi and Hughes, 1992). The valve is formed of unique endothelial cells bearing some sort of microvilli-like structure. The suprapharyngeal chambers take part mainly in air-breathing, though at times are also associated with water-breathing. Hughes and Munshi (1986) and Olson et al. (1994) observed collapsed papillae in the buccopharynx under SEM. Roy and Munshi (1992) found variations in number and tissue barrier thickness of papillae of suprabranchial chamber and bucco-pharynx

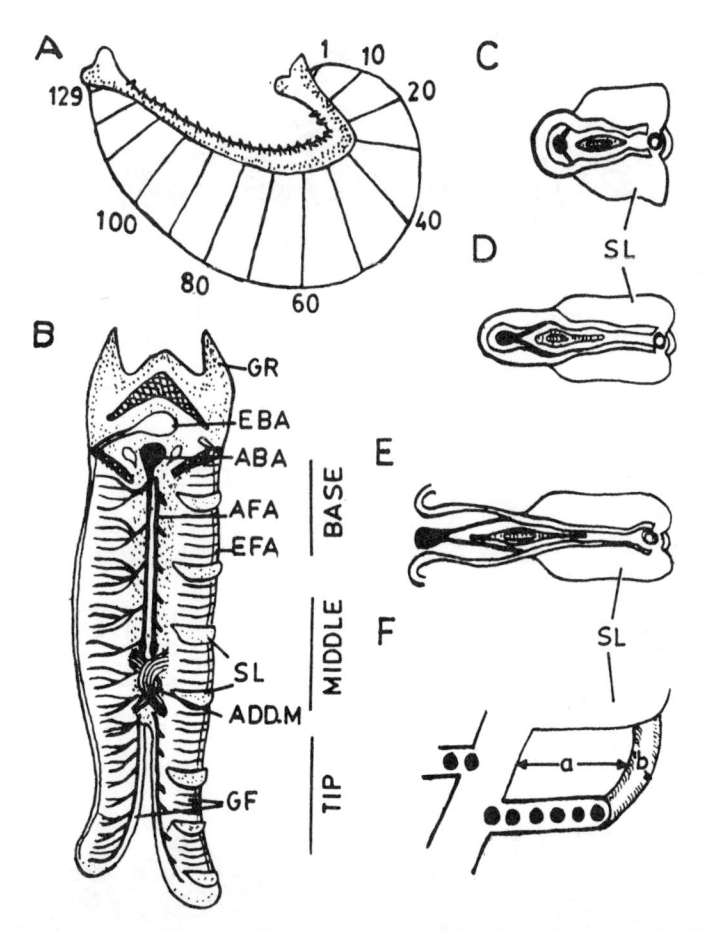

Fig. 4. Diagrammatic view of gill of carp *Cirrhinus mrigala*. A. A hemibranch to show the filaments, selected for measurement. B. A gill filament showing lamella for number estimation. C. Hand cut section from tip region. D. From middle region. E. From base region for lamellar area measurement F. To show that only area 'a' is measured and 'b' is left.

in *Channa marulius* and *C. striatus*. The barrier is relatively thinner in the suprabranchial chamber (0.1344 μm), compared to that of the buccopharynx (0.2458 μm). The number of papillae is denser in SBS (84%) than in the buccopharynx (69%). Though the thick barrier offers less diffusing capacity in the buccopharynx, it protects the vascular papillae from aberrations while feeding. The retractile nature of the vascular papillae is a functional adaptation.

The wave and spiral capillaries of *Channa* spp. are provided with sphincter muscles at the base of the papillae to control blood flow to the system (Olson et al., 1994). Other functional purposes of such arrangements seem to be pressure dampening, higher lodging time of RBC and their contact time with air, leading to more efficient gaseous exchange, and mixing of endothelial metabolites in blood (Olson et al., 1990, 1994).

Modification of branchial chamber. The accessory respiratory organs of *Heteropneustes fossilis* are in the form of four pairs of gill fans and one pair of air sacs or

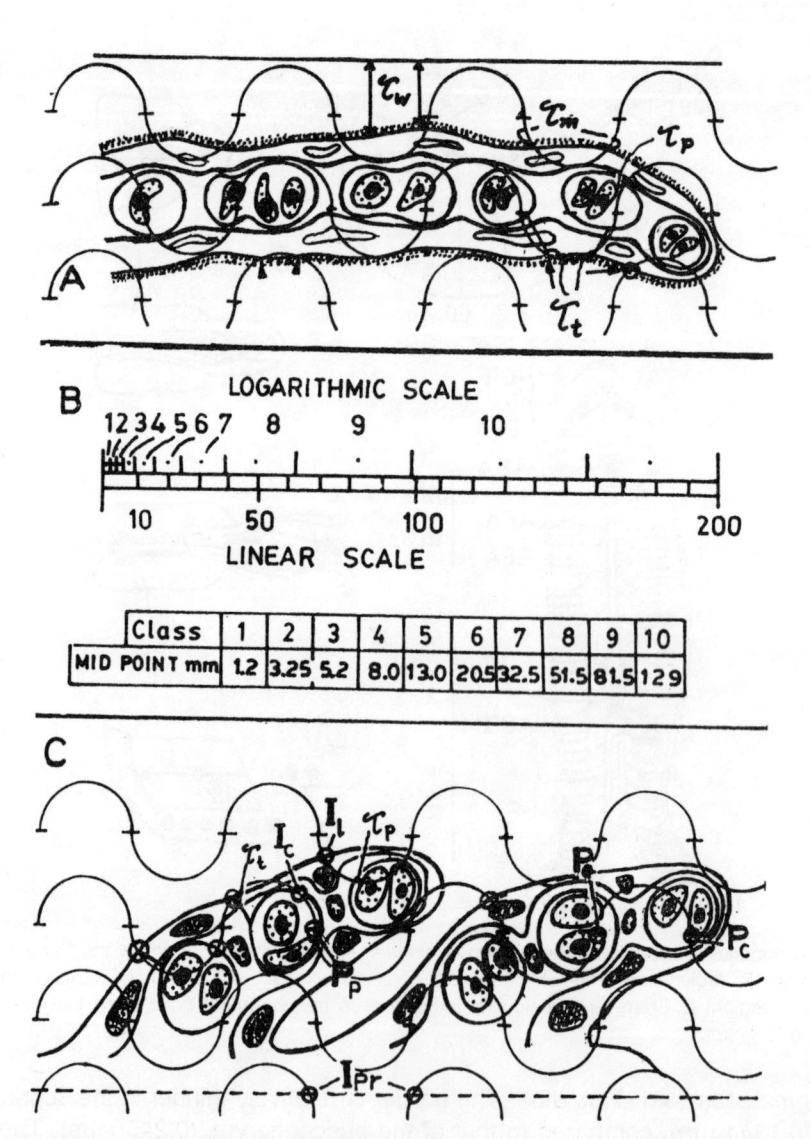

Fig. 5. A. A horizontal section of lamella with super-imposed Merz Grid to show the measurement of different diffusion barriers: water (T_w), Mucus (T_m), Tissue (T_t), Plasma (T_p). B. A new ruler based on logarithmic scale and compared with linear scale (after Weibel, 1979). C. A part of air sac lamella with super imposed Merz Grid to show the Intersections (I) & Points (P) (after Hughes et al., 1992). Note arrowheads are false intersections while arrows are true intersections.

air tubes (Munshi and Choudhary, 1994). The air tubes are backward extensions of the suprabranchial chambers into the trunk region embedded in myotomes (Fig. 1). Gill fans are formed by the fusion of gill filaments. The mucosal lining of the air tube exhibits many folds and ridges (Munshi, 1962a, b). Each ridge is covered by a respiratory epithelium which consists of vascular and nonvascular areas. The vascular area comprises many small and large islets and each islet is formed by a double row of modified secondary lamellae (Fig. 3). The lamellae are arranged at an angle of 75–90° in relation to the horizontal surface of the respiratory membrane (Munshi et al., 1986; Munshi and Choudhary, 1994). The gill fan also has a biserial arrangement of lamellae, covered by polygonal cells bearing microvilli, much like the lamellae of the air tube. In the air tube about three-fourths of each lamella remains buried in the epithelium with only one-fourth exposed as a flat free surface (Munshi et al., 1986a, b). The marginal channels and only two or three adjacent channels are broad and remain full of RBC, while the rest of the channels are discontinuous (Munshi et al., 1990) containing mainly amoebocytes and WBC in the plasma. TEM studies showed that marginal cells are formed by the pillar cell and endothelial cell containing Weibel-Pallade bodies (Munshi et al., 1986). On the basis of this structural feature, the whole air tube cannot be considered respiratory. Hughes et al. (1992) estimated the lamellar area to be 67% of the total air sac area. The inclined part of projected lamellae give the respiratory surface a corrugated appearance, increasing the total surface area severalfold. The total projected air sac thus gets increased by 20%. The barrier thickness of lamellar area is 0.342 μm. There is a slight interzonal difference in lamellar area and tissue barrier thickness. Further, a higher hematocrit value has been observed in the lamellar capillaries of the posterior portion (67.7%) than in the anterior portion (62.1%) of the air sac. In these lamellar capillaries the erythrocytes have a larger nuclear-cytoplasmic ratio in the anterior region (0.20) than other regions (0.15). All these factors indicate some sort of plasma skimming. The capillary loading was 2.122 cm^3/m^2.

The accessory respiratory organ of *Clarias batrachus* consists of four pairs of gill fans, two pairs of dendritic organs and one pair of suprabranchial chamber (Munshi, 1961). Gill fans are formed by the fusion of gill filaments at the posterior end of the arch. The first pair of dendritic organs are small tree-like structures borne by the second gill arches, while the second ones are gigantic and developed on the fourth gill arches. The suprabranchial chamber of each side consists of two recesses: (a) a dorsal cup-shaped recess formed by approximation of the second and third fan and (b) a ventroposterior recess behind the third fan. The first dendritic organ is lodged in the dorsal chamber, while the second dendritic organ is located in the ventroposterior chamber (Munshi, 1976). The surface of the suprabranchial chamber is formed by migration, modification, and fusion of the gill filaments of the four gill arches and consists of vascular and nonvascular areas. The vascular area comprises many respiratory islets, each of which has a biserial arrangement of secondary lamellae (Fig. 3) (Munshi, 1985). The tissue barrier is very thin, comprising a single epithelial layer (0.18 μm), very thin basement membrane (0.024 μm), and a thin lining of pillar cells (0.146 μm). A similar structure has been described in *Clarias mossambicus* by Maina and Maloiy (1986).

The accessory respiratory organs of *Anabas testudineus* and *Colisa fasciatus* are the labyrinthine organs borne on the first pair of gill arches lodged in the suprabranchial chambers. The labyrinthine organ is complex in *Anabas* with many plates (Figs. 1, 3) but simple in *Colisa fasciatus* with only two plates. The labyrinthine organs and the inner surface of the suprabranchial chambers bear numerous respiratory islets surrounding the nonvascular lanes. Each islet consists of a double row of parallel channels. The contiguous blood channels of the islet are separated from each other by a series of epithelial cells, which give off long flanges to cover the blood capillaries (Hughes and Munshi, 1968; Munshi and Hughes, 1991). These cells were thought to be the pillar cells, but it has now been established that the pillars between these structures of contiguous capillaries

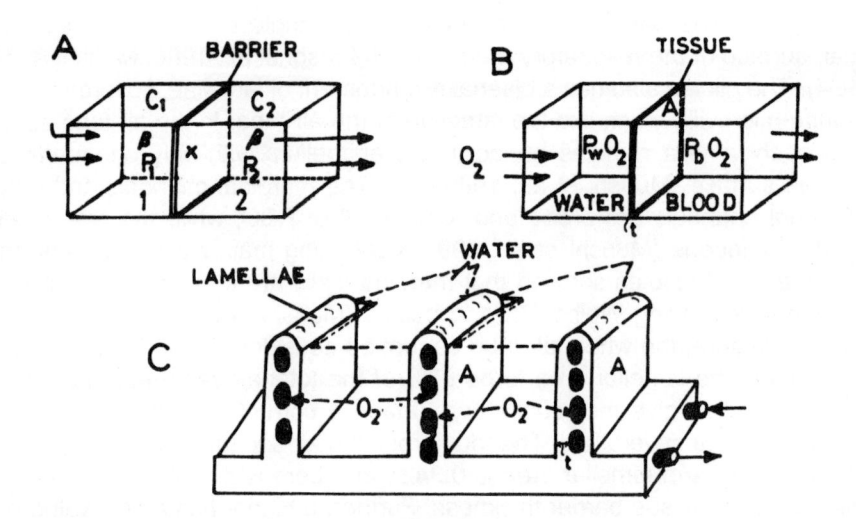

Fig. 6. A diagram explaining the Fick's equation for diffusion of a substance from chamber 1 to chamber 2 with the same medium but with different concentration (C_1 & C_2) through a barrier 'X'. B. Explaining modified Fick's equation for diffusion of oxygen from water to blood through a tissue barrier of area 'A' and thickness 't'. C. Showing the flow of water and blood and diffusion of O_2 to capillaries of lamellae of area 'A'.

are formed of the epithelial cells which are analogous to pillar cells (Hughes and Munshi, 1968). The blood capillaries are formed from a series of endothelial cells having large spherical nuclei and a very characteristic tongue-like process, which may act as small valves controlling the flow of blood (Munshi, 1985). The transverse capillaries (Fig. 3) of the respiratory islets are the modification of only marginal channels of the lamellae (Munshi and Hughes, 1991). The cells covering the transverse capillaries show many similarities with the lung epithelial cell type II, in having a crown of microvilli and a central bald region (Weibel, 1984). The epithelial cell type I of the human lung, which contains a lamellated body, is not found in the suprabranchial and labyrinthine epithelium.

Modification of opercular chambers. Certain estuarine fishes of the family Gobiidae utilize their opercular chambers for air-breathing. Among the Indian gobies, *Periophthalmus vulgaris, P. pearsi, Boleophthalmus boddaerti, B. viridis*, and *Apocryptes lanceolatus* extend their opercular chambers with fresh air and oxygen diffuses into the blood via subepithelial and epithelial capillaries of the operculum. The opercular epithelium is thin, richly vascular and consists of elastic opercular bones (Munshi, 1985). Singh and Munshi (1969) found a safety valve on the branchiostegal apparatus in *Periophthalmus vulgaris* for ventilation of air. Detailed studies of these chambers are lacking.

Modification of gastrointestinal tract. Some Indian cobitids such as *Lepidocephalichthys guntea* inhale air and pass it into the alimentary canal. The intestine works as a gas-exchange surface and the used air is passed out to the exterior either through the mouth or through the anus. A detailed study of intestinal morphometrics has not been made while that of gill morphometrics has been done in *L. guntea* (Singh et al., 1981).

Modification of swimbladder. In India, *Notopterus chitala* and *N. notopterus* have evolved their swimbladder as an accessory respiratory organ (Dehadrai, 1962). The

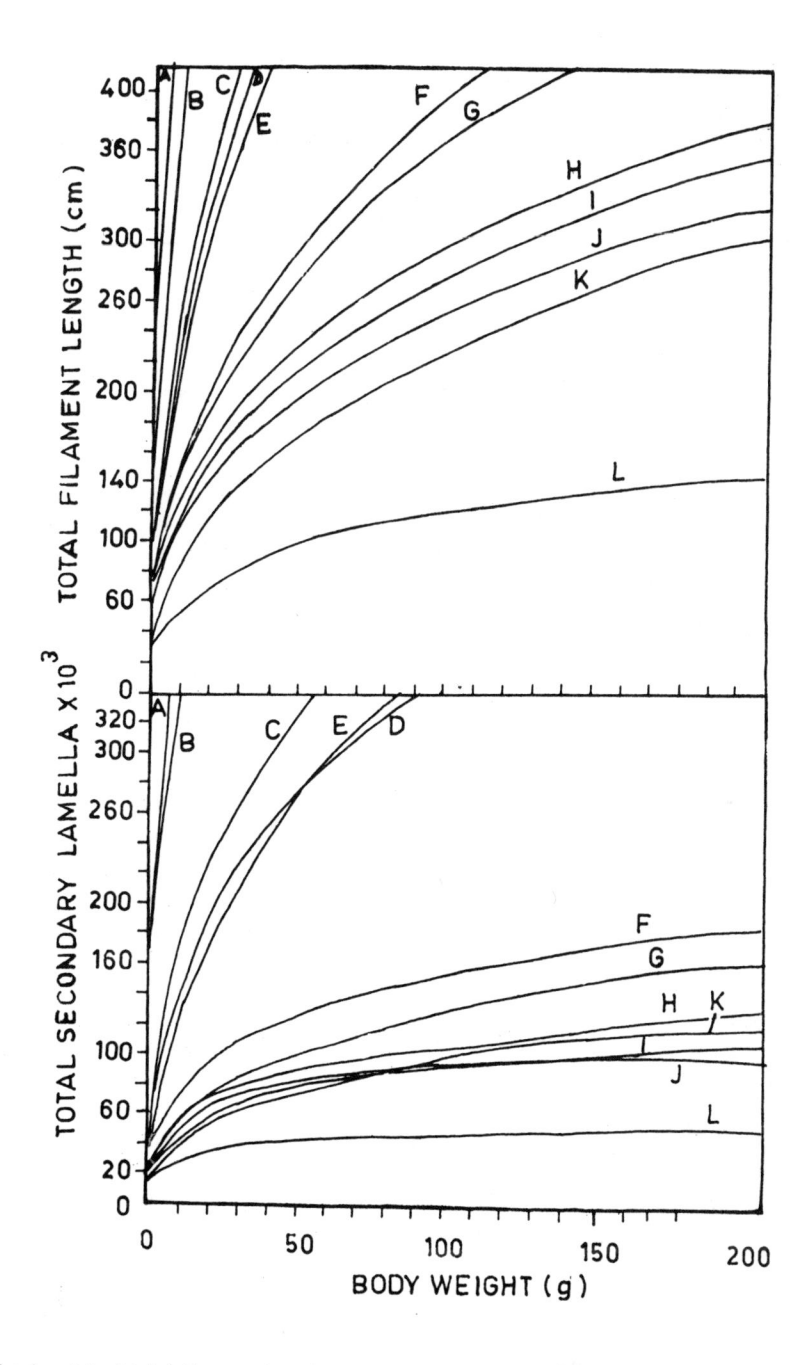

Fig. 7. Simple plot of total filament length and total secondary lamellae in different fishes showing their growth pattern. A. *Catla catla*. B. *Cirrhinus mrigala*. C. *Mystus cavasius*. D. *Mystus vittatus*. E. *Rhinomugil corsula*. F. *Channa punctatus*. G. *Clarias batrachus*. H. *Channa striatus*. I. *Boleophthalmus boddaerti*. J. *Channa gachua*. K. *Heteropneustes fossilis*. L. *Anabas testudineus*. (A–E—Water-breathers. F–L—Air-breathers).

swimbladder is large and has a pneumatic duct connecting it with the esophagus. The wall of the bladder is very thin and well vascularized and is supported by elastic ribs of the body. The epithelium is a single layer and has many crevices on the surfaces which facilitate easy diffusion of gases. A thorough morphometric study is in progress. This kind of adaptation is probably a secondary one in these fishes.

DIMENSIONAL ANALYSIS

The gill area and its other parameters and accessory respiratory organ area have been measured and diffusing capacities estimated for different weight groups of fishes. Using the least-square minimization of variance after double logarithmic transformations of the data, allometric regression equations have been established. The slope value or regression coefficient states how one parameter grows with an increase in body weight, this coefficient is known as the scaling factor or scaling coefficient. The general equation is $Y = aW^b$, where Y is the parameter analyzed, W the body weight, a the intercept, and b the slope (regression coefficient or scaling coefficient).

Scaling of respiratory organs is also carried out in order to study the effect of life stages by dividing the data of morphometrics of fishes into two groups: adult and juvenile; two scaling coefficients are thus observed, giving an inflection in regression lines.

Scaling of Gill and its Parameters

Allometric equations for different gill parameters are given in Table 1. Filaments are heterogeneous in size in almost all fish species depending on the space available in branchial cavities. In air-breathing fishes, the filaments are shorter and decrease in size from gill arches I to IV (Munshi, 1980). But water-breathers have long filaments, though the fourth gill arches in general bear a smaller number of filaments. The total filament length of the fourth gill arch does not differ significantly from the first gill arch in water- and facultative air-breathers but in obligate air-breathers the difference is significant (< 0.05). Ojha (1993) estimated the average exponent value (=regression coefficients) for total filament length in obligate water-breathers, air-breathers, and amphibious fishes, and concluded that coefficients are higher for water-breathers ($b = 0.48 \pm 0.06$), moderate in air-breathers ($b = 0.43 \pm 0.09$), and low in amphibious fishes ($b = 0.345 \pm 0.06$). Such variations in the coefficients are anatomical adaptations of particular fish species depending on their breathing behavior.

Lamellae are the actual sites for gaseous exchange, and so their distribution, dimension, and number of gill filaments regulate the efficiency of gills. Lamellar frequency is low in air-breathers but high in obligate water-breathers (Table 1). It decreases more rapidly in amphibious fishes ($b = -0.14 \pm 0.06$) than in air-breathers ($b = -0.106 \pm 0.06$) and in obligate water-breathers ($b = -0.05 \pm 0.05$) (Ojha, 1993). This pattern is an adaptation to a particular mode of life in fish. The compactness of secondary lamellae reduces the physiological dead space and increases the efficiency of the gills of water-breathers. Hughes and Gray (1972) remarked that air-breathing, amphibious and sluggish fishes have fewer but larger lamellae, but obligate water-breathers have densely packed smaller lamellae (Fig. 7). Intercepts for lamellar area and numbers in Table 1 support this conjecture of Hughes and Gray (1972). Scaling coefficients for bilateral surface area of a lamella

Table 1. Allometric equation ($Y = aW^b$) to show the relationship between body weight and gill parameters of various fishes

Fish species	Mode of breathing	Total length of gill filament (mm)	No. of secondary lamellae/mm filament length	Average bilateral surface area (mm²)	References
Air-breathing:					
Anabas testudineus	Obligate	$258\ W^{0.335}$	$73\ W^{-0.152}$	$0.0149\ W^{0.426}$	Hughes et al., 1973
Heteropneustes fossilis	Facultative	$306\ W^{0.435}$	$63\ W^{-0.095}$	$0.0096\ W^{0.408}$	Hughes et al., 1974a
Clarias batrachus	Facultative	$537\ W^{0.415}$	$51\ W^{-0.083}$	$0.0084\ W^{0.450}$	Munshi et al., 1980
Channa punctatus	Facultative	$558\ W^{0.425}$	$72\ W^{-0.138}$	$0.0117\ W^{0.304}$	Hakim et al., 1978
Channa gachua	Facultative	$519\ W^{0.344}$	$65\ W^{-0.127}$	$0.0039\ W^{0.568}$	Dandotia, 1978
Channa striatus	Obligate	$549\ W^{0.364}$	$51\ W^{-0.075}$	$0.0068\ W^{0.431}$	Choudhary, 1992
Macrognathus aculeatum	Facultative	$457\ W^{0.467}$	$84\ W^{-0.069}$	$0.0054\ W^{0.347}$	Ojha & Munshi, 1974
Lepidocephalichthys guntea	Facultative	$265\ W^{0.639}$	$90\ W^{-0.221}$	$0.0207\ W^{0.328}$	Singh et al., 1981
Amphibious:					
Boleophthalmus boddaerti	Facultative	$518\ W^{0.362}$	$49\ W^{-0.083}$	$0.0111\ W^{0.430}$	Niva et al., 1981
Boleophthalmus boddaerti	Facultative	$675\ W^{0.427}$	$53\ W^{-0.229}$	$0.0026\ W^{0.851}$	Hughes & Alkodhomiy, 1986
Periophthalmus lanceolatus	Facultative	$785\ W^{0.270}$	$55\ W^{-0.110}$	$0.0180\ W^{0.410}$	Yadav & Singh, 1989
Periophthalmus schlosseri	Facultative	$653\ W^{0.330}$	$55\ W^{-0.130}$	$0.0164\ W^{0.628}$	Yadav et al., 1990
Water-breathing:					
Cirrhinus mrigala	Obligate	$1,414\ W^{0.444}$	$126\ W^{-0.129}$	$0.0066\ W^{0.501}$	Roy & Munshi, 1986
Mystus cavasius	Obligate	$774\ W^{0.520}$	$80\ W^{-0.097}$	$0.0099\ W^{0.490}$	Ojha et al., 1985
Mystus vittatus	Obligate	$811\ W^{0.490}$	$71\ W^{-0.101}$	$0.0079\ W^{0.400}$	Ojha & Singh, 1987
Catla catla	Obligate	$2,359\ W^{0.430}$	$78\ W^{-0.050}$	$0.0081\ W^{0.380}$	Kunwar, 1984
Rhinomugil corsula	Obligate	$750\ W^{0.468}$	$63\ W^{-0.028}$	$0.0250\ W^{0.344}$	Ojha & Munshi, 1993
Sicamugil cascasia	Obligate	$569\ W^{0.497}$	$82\ W^{-0.027}$	$0.0250\ W^{0.297}$	Ojha & Mishra, 1993
Botia dario	Obligate	$634\ W^{0.420}$	$82\ W^{-0.050}$	$0.0190\ W^{0.340}$	Singh et al., 1988
Garra lamta	Obligate	$773\ W^{0.380}$	$71\ W^{-0.006}$	$0.0075\ W^{0.497}$	Rooj, 1984
Naemachilus rupicola	Obligate	$309\ W^{0.570}$	$74\ W^{-0.100}$	$0.0083\ W^{0.870}$	Rooj, 1984
Glossogobius giuris	Obligate	$642\ W^{0.308}$	$62\ W^{-0.139}$	$0.0340\ W^{0.315}$	Singh, 1982
Labeo bata	Obligate	$997\ W^{0.500}$	$112\ W^{-0.087}$	$0.0070\ W^{0.354}$	Sinha, 1983

approximate 0.5 in most fishes but in a few species may be 0.8 and even more. No significant differences were observed in these values. Ojha (1993) estimated the average values of b to be (0.58 ± 0.21) in amphibious fishes, (0.42 ± 0.17) in air-breathers, and (0.42 ± 0.07) in water-breathers.

The gill area is directly proportional to the efficacy of the gill sieve and shows the adaptive features for a particular mode of life. Intercept and regression coefficients along with allometric equations for various fishes are given in Table 2. For air-breathing fishes the gill-area scale to body weight is depicted by coefficients of 0.59 to 0.80, ranging from about 0.7 to 1.3 in various species. Table 2 shows greater variation in the scaling coefficients but for water-breathers the values approximate 0.75. Ursin (1967) reported the mean value for scaling coefficients for Gray's intermediates as 0.82 while Jager and Dekkers (1975) calculated the mean value for some species as 0.811. Variation in the gill

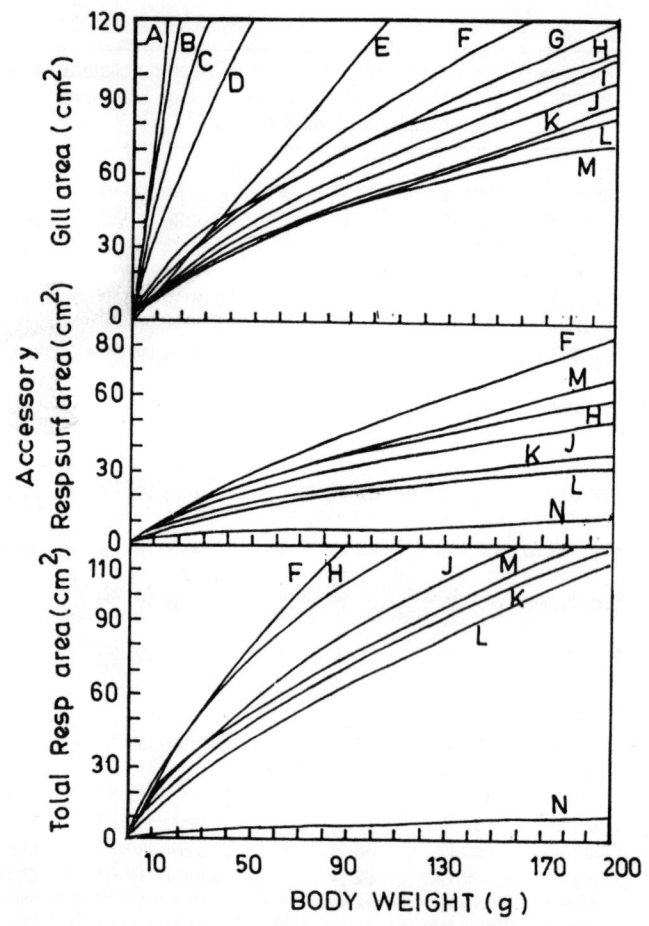

Fig. 8. Simple plot of gill area, accessory respiratory surface area and total respiratory surface area of different fishes. A. Gray's Intermediates. B. *Cirrhinus mrigala*. C. *Cyprinus carpio*. D. *Opsanus tau*. E, G. *Boleophthalmus boddaerti*. F. *Clarias batrachus*. H. *Channa punctatus*. I. *Macrognathus aculeatum*. J. *Heteropneustes fossilis*. K. *Channa striatus*. L. *Channa gachua*. M. *Anabas testudineus*. N. *Amphipnous cuchia*.

area of different fishes has been observed (Table 2). The gill area for 1 g of Gray's Intermediates was calculated to be 1,392 mm^2 (Ursin, 1967) but for active tunny 3,151 mm^2. The area for air-breathing fishes is much smaller compared to that for water-breathers (Fig. 8).

Juvenile and adult groups (divided on the basis of natural growth of fishes). In all the air-breathing fishes a diphasic allometry was observed for the gill area and body weight relationship (Fig. 9). Filament number in adults of *Anabas, Heteropneustes, Clarias*, and *Channa punctatus* increased with a greater slope value than in juveniles except for *Channa striatus*. An opposite trend was seen in water-breathers (Table 3). In air-breathers *Anabas* and *Clarias* the filament length increased in adults with a higher and significantly different slope value than in juveniles. But the slope of lamellar frequency differs insignificantly in the two groups (Table 3). The average bilateral surface area of

[Figure: Log-log plot with y-axis "GILL AREA (mm²)" ranging from 10² to 10⁶, and x-axis "BODY WEIGHT (g)" ranging from 1 to 1000. Legend lists: Cirrhinus mrigala, Anabas testudineus, Channa striatus, Heteropneustes fossilis, Clarias batrachus, Channa punctatus, Channa gachua.]

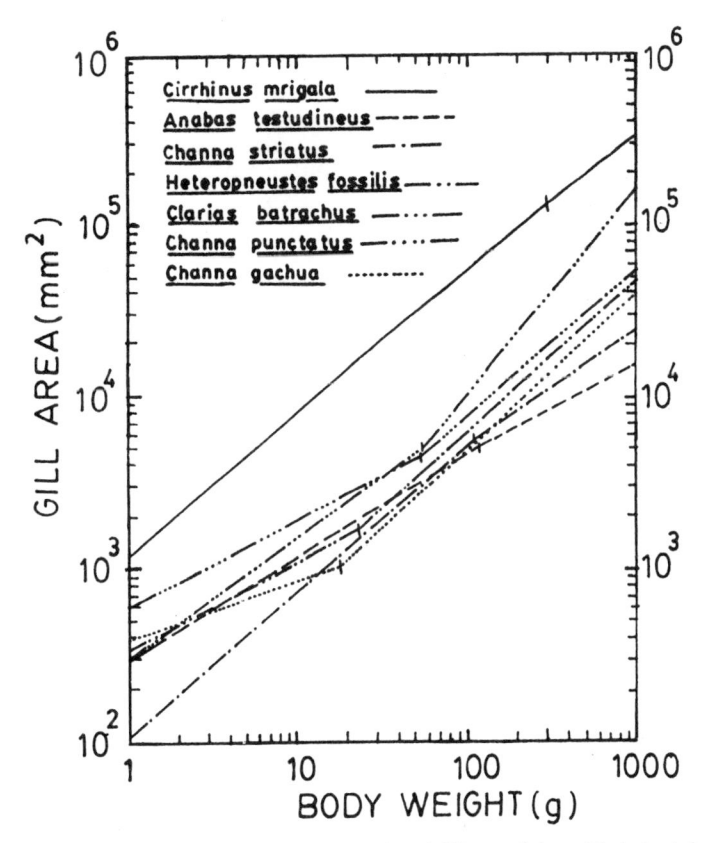

Fig. 9. Log-log plot of gill area of juveniles and adults of different fishes. Mark the inflexion points.

secondary lamella in adult fishes increased with a greater slope value than in juveniles except for *Anabas* (Roy, 1994). Gill area of all the facultative air-breathers increased with a higher exponent value in adults than in juveniles. But in obligate air-breathers such as *Anabas* and *Channa striatus* an opposite trend was seen. Water-breathers *Cirrhinus mrigala* showed no change in slope values of gill area in either the adult or juvenile groups. In aquatic breathers such as *Labeo rohita*, the juveniles had a lower exponent value than the adults. The bilogarithmic plot of regression lines shows inflection in the lines at different body weights in different species (Fig. 9).

Price (1931) measured the gill area of *Micropterus dolomieu* in the weight range of 0.02–850 g, but gave a single slope of 0.785. De Silva (1974) divided herring and plaice into premetamorphic and postmorphic stages and observed slopes of 3.36 and 0.78 for herring and 1.59 and 0.85 for plaice. Al-Kadhomiy (1985) also gave respective slopes of 2.213 and 0.824 for two stages of flounder. Both authors discussed these changes in slope values with respect to the metamorphosis of fishes. Oikawa and Itazawa (1985) gave three slope values of 7.066, 1.222, and 0.794 respectively for prelarvae (0.0016–0.003 g), postlarvae (0.003–0.2 g) and juvenile and later stages (0.2 to 2250 g) of *Cyprinus carpio*. On this basis the authors drew a general conclusion that the slope for gill area decreases in later stages of life compared to the earlier stage. However, a weight-related

Table 2. Intercept *(a)* and slope *(b)* values for regression analysis of various fish species

Fish species	Intercept *(a)*	Slope *(b)*	References
Micropterus dolomeiu	865.00	0.780	Price, 1931
Gray's Intermediates	1,392.00	0.820	Ursin, 1967
Katsuwonus pelamis	5,218.00	0.850	Muir & Hughes, 1969
Thunnus thynnus	3,151.00	0.875	
Thunnus albacares			
Coryphaena hippurus	5,208.00	0.713	Hughes, 1970a
Scomber scombrus	424.10	0.997	
Tinca tinca	867.20	0.698	Hughes, 1970b
Scyliorhinus canicula	262.30	0.961	
Blennius pholis	1,156.10	0.850	Milton, 1971
Opsanus tau	560.70	0.790	Hughes & Gray, 1972
Salmo gairdneri	314.80	0.932	Hughes, 1980
Stizostedian vitreum	224.60	1.070	Nimi & Morgan, 1980
Salmo gairdneri	156.20	1.058	
Hoplias malabaricus	73.12	1.250	Fernandes & Rantin, 1987
Oreochromis niloticus	515.16	0.750	Fernandes & Rantin, 1986
Scyliorhinus stellaris	621.00	0.779	Hughes et al., 1986
Rhinomugil corsula	1,199.00	0.784	Ojha & Mishra, 1993
Sicamugil cascasia	1,177.00	0.767	
Mystus cavasius	617.00	0.915	Ojha et al., 1985
Mystus vittatus	513.00	0.780	Ojha & Singh, 1987
Garra lamta	414.00	0.800	Rooj, 1984
Botia dario	1,047.00	0.720	Singh et al., 1988
Botia lohachata	913.30	0.699	Sharma et al., 1982
Naemachilus rupicola	191.20	1.300	Rooj, 1984
Catla catla	1,506.00	0.760	Kunwar, 1984
Cirrhinus mrigala	1,183.40	0.816	Roy & Munshi, 1986
Cyprinus carpio	846.00	0.794	Oikawa & Itazawa, 1985
Wallago attu	736.00	0.700	Singh, 1990
Glossogobius giuris	1,264.20	0.538	Singh, 1982
Labeo bata	766.80	0.767	Sinha, 1983
Anabas testudineus	278.00	0.615	Hughes et al., 1973
Heteropneustes fossilis	186.10	0.746	Hughes et al., 1974a
Macrognathus aculeatum	217.30	0.733	Ojha & Munshi, 1974
Anguilla anguilla	1,219.60	0.715	Bennett, 1988
Colisa fasciatus	288.00	0.800	Singh, 1986
Lepidocephalichthys guntea	493.60	0.745	Singh et al., 1981
Channa punctatus	470.40	0.592	Hakim et al., 1978
Channa gachua	148.80	0.757	Dandotia, 1978
Channa striatus	191.90	0.718	Choudhary, 1992
Clarias batrachus	227.50	0.781	Munshi et al., 1980
Clarias mossambicus		0.971	Maina & Maloiy, 1986
Boleophthalmus boddaerti	281.30	0.709	Niva et al., 1981
Boleophthalmus boddaerti	92.70	1.050	Hughes & AlKadhomiy, 1986
Periophthalmus lanceolatus	779.00	0.560	Yadav & Singh, 1989
Periophthalmus schlosseri	607.00	0.827	Yadav et al., 1990

decrease in slope value, which could be related to stages of maturity, was found only in obligate air-breathers and not in facultative ones.

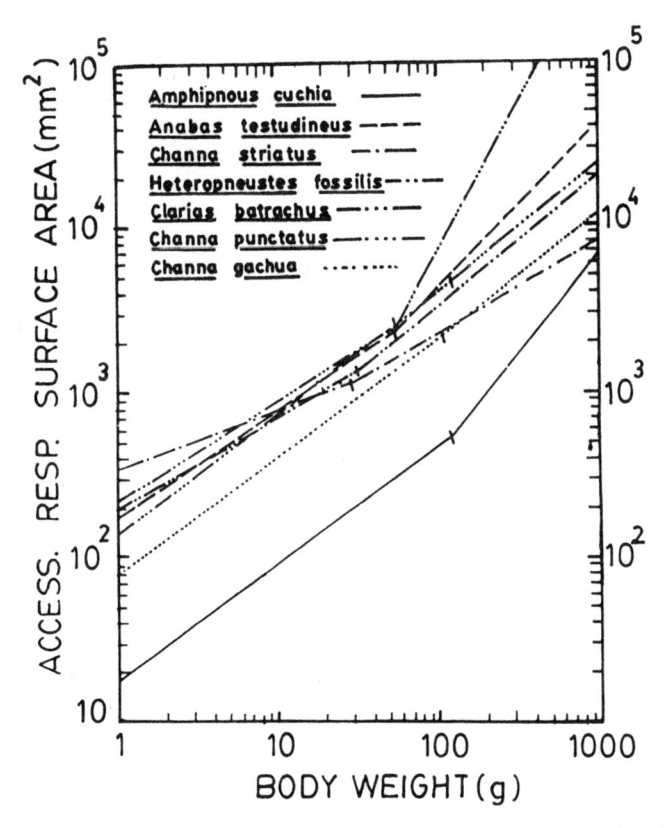

Fig. 10. Log-log plot of accessory respiratory surface area of juveniles and adults of different fishes.

Scaling of air-breathing organs

Considering entire fish group as one. The results of morphometric analysis for the air-breathing organs are set out in Table 4. The labyrinthine organ of *Anabas testudineus* increases with a higher slope value (0.799; Hughes et al., 1973) compared to the suprabranchial chamber ($b = 0.574$). Similarly, the dendritic organ of *Clarias batrachus* grows faster ($b = 0.840$) than the gills and other organs (Munshi et al., 1980). In *Anabas* the total accessory respiratory surface area increases by a slope of 0.713, while in *Clarias batrachus*, it increases by a power of 0.790 (Table 4). In *Channa* spp. the suprabranchial chamber grows with nearly the same value as in facultative breathers (*C. punctatus*: $b = 0.696$; Hakim et al., 1978 and *C. gachua*: $b = 0.678$; Dandotia, 1978) but shows a lesser slope value in an obligate air-breather (*C. striatus*: $b = 0.543$; Choudhary, 1992). The intercept value is the highest for *Channa striatus* (189.4 mm^2) and lowest for *C. gachua* (86.0 mm^2). For *Monopterus cuchia* the air sacs grow by a power of 0.797 with body weight (Hughes et al., 1974b), whereas in *Heteropneustes fossilis*, the air sacs grow with a slope of 0.662 (Hughes et al., 1974a). All these values are based on planimetry of the respiratory membrane. In a 40 g fish *H. fossilis*, the air-sac area was calculated to be 0.421 cm^2/g (Hughes et al., 1974a). But by utilizing modern stereological methods, the air-sac area was estimated to be as high as 0.889 cm^2/g, whereas the lamellar area

Table 3. Allometric equations showing relationship between gill area and its parameters in certain fishes in two different life stages (J = Juvenile; A = Adult).

Fish species	Weight range (g)	Filament number	Average filament length (mm)	Sl/mm filament length	Average bilateral surface area (mm²)	Gill area (mm²)
Air-breathing:						
Anabas testudineus[1]	J 1–45	428 $W^{0.084}$	0.665 $W^{0.282}$	68 $W^{-0.126}$	0.0179 $W^{0.351}$	294.7 $W^{0.591}$
	A 45–120	384 $W^{0.092}$	0.324 $W^{0.406}$	90 $W^{-0.202}$	0.0361 $W^{0.236}$	405.9 $W^{0.532}$
Channa striatus[2]	J 1–50	773 $W^{0.156}$	0.365 $W^{0.381}$	51 $W^{-0.081}$	0.0077 $W^{0.371}$	111.4 $W^{0.826}$
	A 5–1,000	1,227 $W^{0.053}$	0.552 $W^{0.272}$	52 $W^{-0.080}$	0.0072 $W^{0.421}$	260.1 $W^{0.661}$
Heteropneustes fossilis[3]	J 1–25	570 $W^{0.031}$	0.530 $W^{0.404}$	61 $W^{-0.084}$	0.0172 $W^{0.185}$	319.6 $W^{0.536}$
	A 25–120	381 $W^{0.145}$	0.977 $W^{0.244}$	64 $W^{-0.099}$	0.0059 $W^{0.522}$	110.4 $W^{0.871}$
Clarias batrachus[4]	J 1–30	682 $W^{0.035}$	1.366 $W^{0.184}$	48 $W^{-0.056}$	0.0067 $W^{0.540}$	296.1 $W^{0.704}$
	A 30–60	402 $W^{0.177}$	0.517 $W^{0.473}$	44 $W^{-0.049}$	0.0035 $W^{0.660}$	41.1 $W^{1.194}$
Channa punctatus[5]	J 1–45	829 $W^{0.130}$	0.709 $W^{0.276}$	70 $W^{-0.125}$	0.0142 $W^{0.232}$	585.5 $W^{0.512}$
	A 45–100	594 $W^{0.211}$	0.661 $W^{0.296}$	72 $W^{-0.138}$	0.0055 $W^{0.483}$	154.4 $W^{0.852}$
Channa gachua[6]	J 1–25	697 $W^{0.109}$	0.760 $W^{0.230}$	59 $W^{-0.089}$	0.0121 $W^{0.103}$	379.2 $W^{0.354}$
	A 25–100	761 $W^{0.070}$	0.742 $W^{0.253}$	55 $W^{-0.087}$	0.0026 $W^{0.672}$	80.3 $W^{0.905}$
Water-breathing:						
Cirrhinus mrigala[7]	J 5–200	1,040 $W^{0.136}$	1.245 $W^{0.331}$	129 $W^{-0.134}$	0.0073 $W^{0.475}$	1217.4 $W^{0.808}$
	A 200–2,000	1,389 $W^{0.079}$	1.644 $W^{0.288}$	101 $W^{-0.095}$	0.0056 $W^{0.533}$	1290.3 $W^{0.805}$
Labeo rohita[8]	J 1–294	1,170 $W^{0.180}$	1.828 $W^{0.246}$	83 $W^{-0.047}$	0.0155 $W^{0.264}$	2765.4 $W^{0.640}$
	A 400–5476	1,730 $W^{0.105}$	1.962 $W^{0.267}$	104 $W^{-0.094}$	0.0031 $W^{0.533}$	873.7 $W^{0.810}$

Based on 1. Hughes et al., 1973; 2. Choudhary, 1992; 3. Hughes et al., 1974a; 4. Munshi et al., 1980; 5. Hakim et al., 1978; 6. Dandotia, 1978; 7. Roy and Munshi, 1986; 8. Pandey, 1988.

was only 0.598 cm²/g (Hughes, Roy and Munshi, 1992). This higher value was due to the uneven corrugated nature of the respiratory membrane. Similarly, for a 200 g *Monopterus cuchia*, the calculated air-sac area was 9.50 cm² (Hughes et al., 1974b), but in recent studies by Munshi et al. (1989) the estimated respiratory surface came to 20.0 cm² in a total air-sac area of 71.36 cm².

Roy and Munshi (1992) found 49.04 cm² of vascular papillary area alone in a 325 g *Channa striatus*; Choudhary (1992) estimated the total respiratory membrane area to be 44.91 cm². This was mainly due to the occurrence of vascular papillae on the palate, roof, and floor of the buccopharynx. Vascular papillae are found even on the tongue. The skin also acts as an accessory respiratory organ in *Monopterus cuchia*, *H. fossilis*, and *Clarias batrachus* and shows the respective slope values of 0.707, 0.684, and 0.743.

Juvenile and adult groups. When air-breathing organs are scaled with body weight, their surfaces show higher scaling coefficients in adults of each air-breathing species than in juveniles (Table 5). Regression lines depicted diphasic allometry and the coefficient was positive (*b* more than one) in the case of adults of *Monopterus* and *Clarias* (Fig. 10). The differences in their coefficient values were significant. In *Channa punctatus* and *Channa gachua* the intercepts for both juveniles and adults gave lower values than when they were pooled together into one group. Inflection in regression lines was evident in different weight groups in most of the species (at 55 g in *Anabas* and *Clarias*; at 30 g in *Heteropneustes* and *Channa striatus*, and at 110–120 g in *C. punctatus* and *C. gachua*). The faster rate of development of air-breathing organs was reflected in their slope values. These species

Table 4. Allometric equations showing relationship between accessory respiratory surface area and total respiratory surface (Gill + ARS) in various air-breathing fishes

Fish species	Body wt. life stages (g)	Accessory respiratory organ area (mm²)	Total respiratory surface area (gill + ARS) (mm²)	Skin area (mm²)
Anabas testudineus [1]	Pooled (1–120g)	147.2 $W^{0.713}$	419.2 $W^{0.658}$	
	Juvenile (1–45g)	172.6 $W^{0.659}$	448.6 $W^{0.632}$	
	Adult (45–120g)	46.3 $W^{0.975}$	316.8 $W^{0.723}$	
Channa striatus [2]	Pooled (1–1,000g)	189.4 $W^{0.543}$	360.5 $W^{0.661}$	
	Juvenile (1–50g)	340.4 $W^{0.354}$	328.1 $W^{0.651}$	
	Adult (50–1,000g)	175.0 $W^{0.557}$	421.4 $W^{0.632}$	
Monopterus cuchia [3]	Pooled (1–250g)	12.5 $W^{0.797}$	12.5 $W^{0.797}$	877.4 $W^{0.706}$
	Juvenile (5–60g)	17.8 $W^{0.714}$	17.8 $W^{0.714}$	
	Adult (60–250g)	1.7 $W^{1.190}$	1.7 $W^{1.190}$	
Heteropneustes fossilis [4]	Pooled (1–120g)	145.9 $W^{0.662}$	328.2 $W^{0.715}$	851.1 $W^{0.684}$
	Juvenile (1–25g)	187.2 $W^{0.569}$	506.8 $W^{0.549}$	
	Adult (25–120g)	95.1 $W^{0.765}$	200.0 $W^{0.834}$	
Clarias batrachus [5]	Pooled (1–80g)	124.7 $W^{0.790}$	351.2 $W^{0.787}$	564.9 $W^{0.743}$
	Juvenile (1–30g)	214.1 $W^{0.608}$	505.3 $W^{0.671}$	
	Adult (30–80g)	2.0 $W^{1.785}$	27.4 $W^{1.400}$	
Channa punctatus [6]	Pooled (1–100g)	159.1 $W^{0.696}$	629.8 $W^{0.623}$	
	Juvenile (1–45g)	149.5 $W^{0.720}$	719.8 $W^{0.574}$	
	Adult (45–100g)	121.6 $W^{0.755}$	270.9 $W^{0.818}$	
Channa gachua [7]	Pooled (1–100g)	86.0 $W^{0.678}$	213.5 $W^{0.749}$	
	Juvenile (1–25g)	80.0 $W^{0.701}$	430.4 $W^{0.469}$	
	Adult (25–100g)	70.8 $W^{0.727}$	143.3 $W^{0.850}$	

Based on 1. Hughes et al., 1973; 2. Choudhary, 1992; 3. Hughes et al., 1974b; 4. Hughes et al., 1974a; 5. Munshi et al., 1980; 6. Hakim et al., 1978; 7. Dandotia, 1978.

depend more on an aerial mode of respiration as they grow in size. When the total respiratory area (gill + ARS) is scaled (Fig. 11), the coefficients are always higher in adults than in juveniles (Table 4).

Diffusing capacity. Diffusing capacity depends on two variables, namely, area and air/water/blood barrier. Very few studies have been conducted to ascertain variation in barrier thickness with growth of fishes. In practice, the average value of barrier thickness has been used for all weight groups. Thus the diffusing capacity scales with body weight have been determined by the same coefficient as that of the respiratory area. The diffusing capacities for a 100 g specimen of various fish species are given in Table 5. Obligate water-breathers have greater gill area with smaller diffusion distance, so have greater diffusing capacities; air-breathers have larger diffusion distance and smaller gill area, and hence smaller diffusing capacities. Larger diffusion distance is an adaptation to check the transfer of O_2 from oxygen-rich lamellar blood to the hypoxic water of swamps. The smaller gill-diffusing capacities of air-breathers are compensated by higher diffusing capacities of air-breathing organs. These organs have smaller diffusion distances, which increases their capacity for oxygen uptake. Use of modern stereological morphometric methods has enabled biologists to make more accurate measurements. Hughes et al. (1974b) measured the air/blood diffusion barrier as 0.44 μm in *Monopterus cuchia* and calculated the diffusing capacity as 0.0165 mlO_2/min/mm Hg/kg. Recently Munshi et al. (1989) estimated the air-sac area to be double the earlier estimates in a 200 g

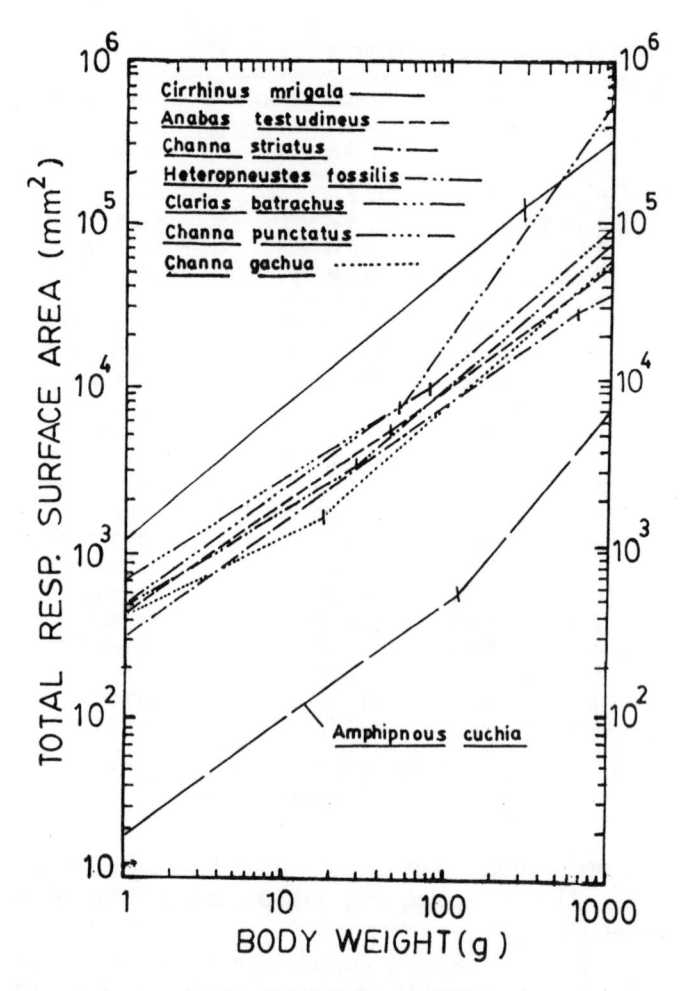

Fig. 11. Log-log plot of total respiratory surface area of juveniles and adults of different fishes.

Monopterus and the barrier as 0.72 μm. Applying these values, they calculated the diffusing capacity (D_t) as 0.02085 ml0$_2$/min/mm Hg/kg. Similarly, Hughes et al. (1974a) used harmonic mean thickness of 1.6 μm in *Heteropneustes fossilis* and the area based on planimetry and calculated diffusing capacity as 0.0288 ml0$_2$/min/mm Hg/kg (Table 5). But recently, Hughes et al. (1992) measured the area of the air-sac lamellae and the diffusion barrier (0.342 μm), and thus calculated D_t as 0.2631 ml0$_2$/min/mm Hg/kg for a 40 g *H. fossilis*. Considering all the components of barrier, i.e., mucus, tissue, plasma, and erythrocyte, the overall diffusing capacity of *Monopterus* and *Heteropneustes* became 0.0008 and 0.0532 ml0$_2$/min/mm Hg/kg respectively (Table 5). Due to a very thin diffusion barrier (0.21 μm) of the air-breathing organs in *Anabas testudineus*, the diffusing capacity becomes the highest, i.e., 0.2825 ml0$_2$/min/mm Hg/kg (Hughes et al., 1973). Using the new model, Roy and Munshi (1992) estimated the diffusing capacity of the accessory respiratory organ of a 325 g *Channa striatus* as 0.2006 ml0$_2$/min/mm Hg/kg. Vascular

Table 5. Respiratory area, diffusion barriers and diffusing capacity of various fishes. Respective body weights are also given

Fish species	Weight	Gill Area (mm²/g)	Gill Diffusion barrier (μm)	Gill Diffusing capacity mlO₂/min/mmHg/Kg	Accessory respiratory surface Area (mm²/g)	Accessory Diffusion barrier (μm)	Accessory Diffusing capacity mlO₂/min/mmHg/Kg	References
Water-breathing:								
Cirrhinus mrigala	100	507.14	1.290	0.5891				Roy & Munshi, 1986
Catla catla	100	498.68	0.885	0.8452				Kunwar, 1984
Labeo rohita	100	526.94	1.320	0.5988				Pandey et al., 1989
Mystus vittatus	100	186.26	1.380	0.2111				Ojha & Singh, 1987
Mystus cavasius	100	417.14	2.150	0.2910				Ojha et al., 1985
Rhinomugil corsula	100	443.42	3.630	0.1832				Mishra, 1984
Wallago attu	100	184.87	1.250	0.2218				Singh, 1990
Botia lohachatta	1	913.30	1.710	0.8000				Sharma et al., 1982
Naemachilus rupicola	3.5	280.06	2.250	0.1867				Rooj, 1984
Garra lamta	4.1	321.36	1.750	0.2759				Rooj, 1984
Salmo gairdneri	100	240.00	6.000	0.0600				Hughes, 1972
Tinca tinca	100	250.00	3.000	0.1250				Hughes, 1972
Scyliorhinus canicula	100	260.00	11.000	0.0272				Hughes, 1972
Tuna	100	2000.00	0.500	6.0000				Hughes & Gray, 1972
Opsanus tau	100	215.70	5.000	0.0630				Hughes & Gray, 1972
Air-breathing:								
Anabas testudineus	100	47.20	10.000	0.0071	39.60	0.210	0.2825	Hughes et al., 1973
Heteropneustes fossilis	100	57.70	3.580	0.0242	30.70	1.600	0.0288	Hughes et al., 1974a
	40				59.80	0.342	0.2631	Hughes et al., 1992
Clarias batrachus	100	83.00	7.670	0.0162	3.42 GF	7.900	0.0007	Munshi et al., 1980
					44.20 LO	0.450	0.1473	
Channa punctatus	100	71.86	2.030	0.0530	39.17	0.780	0.0753	Hakim et al., 1978
Channa gachua	100	48.60	2.400	0.0382	39.23	0.800	0.0366	Dandotia, 1978
Channa striatus	100	52.36	6.980	0.0115	23.04	1.359	0.0254	Choudhary, 1992
	325				15.09 SBC	0.134	0.0547	Roy & Munshi, 1992
					5.27 BP	0.246	0.0046	

(Contd.)

Table 5. (Continued)

Fish species	Weight	Gill Area (mm²/g)	Gill Diffusion barrier (μm)	Gill Diffusing capacity mlO₂/min/mmHg/Kg	Accessory respiratory surface Area (mm²/g)	Accessory respiratory surface Diffusion barrier (μm)	Accessory respiratory surface Diffusing capacity mlO₂/min/mmHg/Kg	References
Monopterus cuchia	100				4.84	0.440	0.0165	Hughes et al., 1974b
	200				10.00	0.720	0.0208	Munshi et al., 1989
Boleophthalmus boddaerti	100	73.72	1.439	0.0770				Niva et al., 1981
		116.43	4–12	0.2160				Hughes & Al-Kadhomiy, 1986
Macrognathus aculeatum	33	178.61	1.150	0.4660				
Clarias mossambicus	100	17.30	1.970	0.0213	7.79 SBC	0.313	0.0500	Ojha & Munshi, 1976
					4.65 LO	0.287	0.0700	Maina & Maloiy, 1986

BP—Buccopharynx; GF—Gill fan; LO—Labyrinthine organ; SBC—Suprabranchial chamber papillae are not only present in the suprabranchial chambers, but also on the palate and tongue of the fish; the barrier thickness of papillae varies according to their location, however. The tissue barrier was 0.3769 μm thick in the buccopharynx and 0.2538 μm thick in the respiratory membrane. The difference in thickness was due to the nature and location of vascular papillae to suit their specific functional needs.

Morphometric estimation of oxygen uptake: Using the modified Fick's equation, the oxygen uptake of a fish can be estimated:

$$\dot{V}_{O_2} = \frac{K \cdot A \cdot \Delta P_{O_2}}{t} \quad \text{or} \quad D_t \times \Delta P_{O_2},$$

where $\dot{V}_{O_2}$ is oxygen uptake, D_t diffusing capacity, and ΔP_{O_2} difference in partial pressure of oxygen between inspiring water and blood. ΔP_{O_2} varies in fishes inhabiting water with different oxygen content and its measured value is not available for all fish species. So for comparison in various fishes, ΔP_{O_2} was assumed to be 100 mm Hg and the morphometric oxygen uptake then calculated (Table 6). This estimate is many times higher than the actual oxygen uptake and shows only the maximum limit up to which oxygen can diffuse through the respiratory membrane of air-breathing organs. The actual oxygen uptake depends on many other factors. In calculating diffusing capacities and morphometric oxygen uptake, resistance offered by the water layer, mucous layer, plasma, and erythrocyte are factors to be taken into account. Further, ΔP_{O_2} does not normally approximate about 100 mm Hg. Hughes (1984a) suggested that the area used for gaseous exchange is not actually the entire measured area, but varies due to change in blood flow in channels during breathing. The ventilation rate in fishes also depends on several variable factors and changes the steady state condition. The thickness of barrier also varies according to environmental oxygen concentrations.

FUTURE PERSPECTIVES IN MORPHOMETRICS

In spite of our sincere effort to obtain morphometric measurements of the respiratory area and diffusing capacity, these studies lack some valuable information and do not consider certain special features of the structure, for example: i) The lamellar margins are semicircular in shape and these are the places with the thinnest diffusion barrier. While measuring the lamellar area, normally the flat surface of the lamella is measured, and the curved surface area of lamellar margins are ignored. ii) The lamellar epithelium comprise many lymphatic spaces which vary in dimensions under different environmental conditions. Oxygen may diffuse directly to the lymphatic fluid and come into contact with blood without facing any further barrier. iii) Air-breathers live in swamps and breathe in water with varying oxygen tension. So in different seasons the barrier thickness may change with variation in oxygen content either to facilitate oxygen diffusion or to check back diffusion. iv) The change in scaling of respiratory surface in juveniles and adults has been established and the breathing behavior of the fish also changes and should be taken into account. v) In many air-breathers parts of lamellae remain buried in filament epithelium and this particular feature of the structure should be taken into account. The environmental conditions, viz., O_2, CO_2, pH of water, and temperature affect gas exchange. In the light of these facts, the morphometric work on the respiratory organs in fishes needs careful planning.

Table 6. Morphometric oxygen uptake through the tissue barrier of respiratory organs of various fishes. Values for gills and accessory respiratory organs given

Fish species	Body weight (g)	Aquatic $\dot{V}o_2$ through gills $mlO_2/Kg/h$	Aerial $\dot{V}o_2$ through ABO $mlO_2/Kg/h$	References
Water breathing:				
Cirrhinus mrigala	100	3,534.6		Roy & Munshi, 1986
Catla catla	100	5,071.2		Kunwar, 1984
Labeo rohita	100	3,592.8		Pandey et al., 1989
Mystus vittatus	100	1,266.6		Ojha & Singh, 1987
Mystus cavasius	100	1,746.0		Ojha et al., 1985
Rhinomugil corsula	100	1,099.2		Mishra, 1984
Wallago attu	100	1,330.8		Singh, 1990
Botia lohachatta	1	4,800.0		Sharma et al., 1982
Garra lamta	4.1	1,655.4		Rooj, 1984
Salmo gairdneri	100	360.0		Hughes, 1972
Tinca tinca	100	750.0		Hughes, 1972
Tuna	100	36,000.0		Hughes & Gray, 1972
Opsanus tau	100	378.0		Hughes & Gray, 1972
Air breathing:				
Anabas testudineus	100	42.6	1,695.0	Hughes et al., 1973
Heteropneustes fossilis	100	145.2	172.8	Hughes et al., 1974a
	40		1,578.6	Hughes et al., 1992
Clarias batrachus	100	97.2	887.7	Munshi et al., 1980
Channa punctatus	100	318.0	451.8	Hakim et al., 1978
Channa gachua	100	229.2	219.6	Dandotia, 1978
Channa striatus	100	69.0	152.5	Choudhary, 1992
	325		356.0	Roy & Munshi, 1992
Monopterus cuchia	100		99.0	Hughes et al., 1974b
	200		125.1	Munshi et al., 1989
Boleophthalmus boddaerti	100	462.0		Niva et al., 1981
	100	129.6		Hughes & A-Kadhomiy, 1986
Macrognathus aculeatum	33	1,435.2		Ojha & Munshi, 1976
Clarias mossambicus	100	127.8	720.0	Maina & Maloiy, 1986

SUMMARY

Morphological and morphometric studies contribute much to our understanding of respiratory organ functions with respect to growth, development, habit, and habitat of fishes. This article presents an account of commonly accepted methods for measurement of the gill area of fishes, giving more emphasis to weightage of different parameters. The structural complexity of air-breathing organs of Indian fishes necessitates the application of stereological techniques.

The relationship between respiratory area and body mass was examined during development of water- and air-breathing organs of the fishes through the method of logarithmic transformations. The regression coefficients for filament numbers, filament length, and secondary lamellae are always higher for a water-breather than for an air-breather. The coefficients of gill area vary from 0.5 to 1.0 in different fish species but is always lower in air-breathers. In obligate air-breathers the gills are much reduced from the first to fourth arches. The morphometrics of some fishes was analyzed after grouping them into juveniles and adults. Great variation in the slope of the two groups of air-breathers was observed. The gill area of obligate air-breathers, such as *Anabas* and *Channa striatus*, scale with higher coefficients in juveniles than in adults, but the situation is reversed in the case of facultative air-breathers, *Clarias* and *Heteropneustes*. The air-breathing organs scale with higher coefficients in adults of almost all the air-breathing species. These studies clearly indicate that obligate air-breathers show a tendency to have a relatively lower gill area even in the juvenile stage, but all the air-breathers require and utilize more and more accessory respiratory organs as they grow in size and attain maturity.

The resistance to gas transfer from water/air to blood involves many aspects, viz., the water layer itself, mucous film around the lamella, tissue barrier, plasma barrier, and the erythrocyte reaction rate. Measurement methods of barrier thickness were discussed and more emphasis was given to harmonic mean thickness. The gill diffusing capacities of water-breathers are greater due to their larger surface area and the thinner water/blood barrier, but the diffusing capacities of accessory respiratory organs of air-breathers are higher due to their thinner diffusion barrier and characteristic blood capillaries.

Acknowledgment

The authors acknowledge their gratitude to the various scientists whose works are incorporated in this article. Special thanks are due to our collaborators, Professor G.M. Hughes and Professor E.R. Weibel for their contributions in morphometrics of the fish respiratory system. This review chapter was prepared during the tenure of the Government of India's Council of Scientific and Industrial Research (CSIR) Emeritus Scientist Fellowship of J.S. Datta Munshi.

References

Al-Kadhomiy, N.K. 1985. *Gill development, growth and respiration of the flounder, Platichthys flesus I*. Ph.D. thesis, Bristol University, U.K.
Bennett, M.B. 1988. Morphometric analysis of the gills of the European eel, *Anguilla anguilla*. *J. Zool., London*, 215: 549–560.
Byczkowska-Smyk, W. 1957. The respiratory surface of the gills in teleosts. 1. The respiratory surface of the gills in the flounder (*Pleuronectes platessa*) and perch (*Perca fluviatilis*). *Zool. Pol.*, 8:91–111.
Choudhary, D.P. 1992. Morphometrics of the respiratory organs of the Indian snake headed fish, *Channa striata* (Bloch) (Ophicephaliformes, Channidae). *J. Freshwater Biol.* 4(2):81–98.
Dandotia, O.P. 1978. *Studies on the functional capacity of the respiratory organs of a freshwater amphibious fish, Channa (= Ophiocephalus) gachua*. Ph.D. thesis, Bhagalpur University, India.
Dehadrai, P.V. 1962. Respiratory function of the swimbladder of *Notopterus* (Lacapede). *Proc. Zool. Soc. Lond.*, 139:341–357.
Dejours, P. 1975. *Principles of Comparative Respiratory Physiology*. North Holland Publishing Company, Amsterdam, 253 pp.
De Silva, C. 1974. Development of the respiratory system in herring and plaice larvae. *In* J.H.S. Blaxter (ed.) *The Early Life-History of Fish*. Springer-Verlag, Berlin.
Fernandes, M.R. and F.T. Rantin. 1986. Gill morphometry of cichlid fish, *Oreochromis (Sarotherodon) niloticus* (Pisces, Teleostei). *Ciencia E. Cultura* 38(1):192–198.
Fernandes, M.R. and F.T. Rantin. 1987. Gill morphometry of the teleost, *Hoplias malabaricus* (Bloch). *Fisiol. Anim. Univ. S. Paulo*. 9:57–65.
Fick, A. 1855. *Ann. Physik.*, 94:59. See Samson Wright's *Applied Physiology* (revised by C.A. Keele and Erick Neil). Oxford University Press, Oxford (1971).

Gehr, P., D.K. Mwangi, A. Ammann, G.M. Maloiy, C.P. Taylor, and E.R. Weibel. 1981. Design of the mammalian respiratory system. V. Scaling morphometric diffusing capacity to body mass: wild and domestic animals. *Respir. Physiol.* 44:61–86.

Gray, I.E. 1954. Comparative study of the gill area of marine fishes. *Biol. Bull.* (Woods Hole, Mass.) 107:219–225.

Hakim, A., J.S.D. Munshi, and G.M. Hughes. 1978. Morphometrics of the respiratory organs of the Indian green snake-headed fish, *Channa punctata*. *J. Zool., London* 184:519–543.

Hughes, G.M. 1966. The dimensions of fish gills in relation to their function. *J. Exp. Biol.* 45:177–195.

Hughes, G.M. 1970a. Morphological measurements of the gills of fishes in relation to their respiratory function. *Folia Morphol. (Prague)* 18:78–95.

Hughes, G.M. 1970b. A comparative approach to fish respiration. *Experientia* 26:113–122.

Hughes, G.M. 1972. Morphometrics of fish gills. *Respir. Physiol.* 14:1–25.

Hughes, G.M. 1980. Functional morphology of fish gills. In B. Laholu (ed.), *Epithelial Transport in the Lower Vertebrates*, pp. 15–36. Cambridge University Press, London-New York.

Hughes, G.M. 1984a. Scaling of respiratory areas in relation to oxygen consumption of vertebrates. *Experientia* 40:519–524.

Hughes, G.M. 1984b. General anatomy of the gills. In W.S. Hoar, D.J. Randall, and E.M. Donaldson (eds.), *Fish Physiology*, vol. XA, pp. 1–72. Academic Press, New York.

Hughes, G.M. and J.S.D. Munshi. 1968. Fine structure of respiratory surfaces of an air-breathing fish, the climbing perch, *Anabas testdineus* (Bloch). *Nature (London)* 219:1382–1384.

Hughes, G.M. and I.E. Gray. 1972. Dimensions and ultra structure of toadfish gills. *Biol. Bull.* (Woods Hole, Mass.) 143:150–161.

Hughes, G.M. and J.S.D. Munshi. 1973. Nature of the air-breathing organs of the Indian fishes, *Channa, Amphipnous, Clarias,* and *Saccobranchus* as shown by electron microscopy. *J. Zool., London* 170:245–270.

Hughes, G.M. and J.S.D. Munshi. 1979. Fine structure of the gills of some Indian air-breathing fishes. *J. Morphol.* 160:169–194.

Hughes, G.M. and N.K. Al-Kadhomiy. 1986. Gill morphometry of the mudskipper, *Boleophthalmus boddaerti*. *J. Mar. Biol. Assoc. U.K.* 66:671–682.

Hughes, G.M. and J.S.D. Munshi. 1986. Scanning electron microscopy of the accessory respiratory organs of the snake-headed fish, *Channa striata* (Bloch). (Channidae, Channiformes). *J. Zool. London*, 209:305–317.

Hughes, G.M., S.C. Dube, and J.S.D. Munshi. 1973. Surface area of the respiratory organs of the climbing perch, *Anabas testudineus*. *J. Zool., London* 170:227–243.

Hughes, G.M., S.F. Perry, and J. Piper. 1986. Morphometry of the gills of the elasmobranch, *Scylliorhinus stellaris* in relation to body size. *J. Exp. Biol.* 121:27–42.

Hughes, G.M., P.K. Roy, and J.S.D. Munshi. 1992. Morphometric estimation of oxygen diffusing capacity for the air sac in *Heteropneustes fossilis*. *J. Zool., London* 227:193–209.

Hughes, G.M., B.R. Singh, G. Guha, S.C. Dube, and J.S.D. Munshi. 1974a. Respiratory surface areas of an air-breathing siluroid fish *Saccobranchus* (=*Heteropneustes*) *fossilis* in relation to body size. *J. Zool., London* 172:215–232.

Hughes, G.M., B.R. Singh, R.N. Thakur, and J.S.D. Munshi. 1974b. Areas of the air-breathing surfaces of *Amphipnous cuchia*. *Proc. Indian Nat'l. Sci. Acad.* 40B(4) 379–392.

Jager, S. de and W.J. Dekkers. 1975. Relations between gill structure and activity in fish. *Netherlands J. Zool.* 25:276–308.

Krogh, A. 1941. *The Comparative Physiology of Respiratory Mechanisms*. Univ. of Pennsylvania Press, Philadelphia, PA.

Kunwar, G.K. 1984. *The structure and function of respiratory organs of a major carp, Catla catla*. Ph.D. thesis, Bhagalpur University, India.

Maina, J.N. and G.M.O. Maloiy. 1986. The morphology of the respiratory organs of the African air-breathing catfish (*Clarias mossambicus*): A light, electron, and scanning microscopic study with morphometric observations. *J. Zool., London* 209:421–445.

Merz, W.A. 1967. Die Streckenmessung an gerichteten Strukturen im Mikroskop und ihre Anwendung zur Bestimmung von Oberflächen-Volumen-Relationen im Knochengewebe. *Mikroskopie* 22:132.

Milton, P. 1971. Oxygen consumption and the osmoregulation in the shanny, *Blennius pholis*. *J. Mar. Biol. Assoc. U.K.*, 51:247–265.

Mishra, A.K. 1984. *Studies on the structure and function of the respiratory organs of certain freshwater mullets*. Ph.D. thesis, Bhagalpur University, India.

Muir, B.S. and G.M. Hughes. 1969. Gill dimension for three species of tunny. *J. Exp. Biol.* 551:271–285.

Munshi, J.S.D. 1960. The structure of the gills of certain freshwater teleosts. *Indian J. Zool. Memoir* 4:1-40.
Munshi, J.S.D. 1961. The accessory respiratory organs of *Clarias batrachus* (Linn.) *J. Morph.* 109 (2):115-139.
Munshi, J.S.D. 1962a. On the accessory respiratory organs of *Heteropneustes fossilis*. *Proc. Roy. Soc. Edin.* 68:128-146.
Munshi, J.S.D. 1962b. On the accessory respiratory organs of *Ophiocephalus punctatus* and *O. striatus* (Bloch). *J. Linn. Soc. Zool.* 44:616-626.
Munshi, J.S.D. 1976. Gross and fine structure of the respiratory organs of air-breathing fishes. *In* G.M. Hughes (ed.), *Respiration of Amphibious Vertebrates*, pp. 73-104. Academic Press, London.
Munshi, J.S.D. 1980. *The structure and function of the respiratory organs of air-breathing fishes of India.* pp. 32-70. Presidential Address. Section of Zoology, Entomology and Fisheries. 67th Session, Indian Science Congress Association.
Munshi, J.S.D. 1985. The structure, function and evolution of the accessory respiratory organs of air-breathing fishes of India. *In* H.R. Duncker and G. Fleischer (ed.), *Fortschritte der Zoologie: Functional Morphology in Vertebrates*, vol. 30. pp. 353-366. Gustav Fischer-Verlag, Stuttgart-New York.
Munshi, J.S.D. and G.M. Hughes. 1991. Structure of the respiratory islets of accessory respiratory organs and their relationship with the gills in the climbing perch, *Anabas testudineus*. *J. Morphol.* 209:241-256.
Munshi, J.S.D. and G.M. Hughes. 1992. *Air-breathing Fishes: Their Structure, Function and Life History.* Oxford & IBH Publishing Co. Ltd., New Delhi.
Munshi, J.S.D. and B.N. Singh. 1968. On the respiratory organs of *Amphipnous cuchia* (Ham.) *J. Morph.* 124: 423-444.
Munshi, J.S.D. and S. Choudhary. 1994. *Ecology of Heteropneustes fossilis (Bloch): An Air-breathing Catfish of Southeast Asia.* Monograph Series. Freshwater Biological Association of India, Bhagalpur, 174 pp.
Munshi, J.S.D., J. Ojha, and A.L. Sinha. 1980. Morphometrics of the respiratory organs of an air-breathing catfish, *Clarias batrachus*, in relation to body weight. *Proc. Indian Nat'l. Sci. Acad.* B46:621-635.
Munshi, J.S.D., E.R. Weibel, P. Gehr, and G.M. Hughes. 1986a. Structure of the respiratory air-sac of *Heteropneustes fossilis* (Bloch) (Heteropneustidae, Pisces)—An electron microscope study. *Proc. Indian Nat'l. Sci. Acad.* B52(6):703-713.
Munshi, J.S.D., K.R. Olson, J. Ojha, and T.K. Ghosh. 1986b. Morphology and vascular anatomy of the accessory respiratory organs of the air-breathing climbing perch, *Anabas testudineus* (Bloch). *Amer. J. Anat.* 176:321-331.
Munshi, J.S.D., G.M. Hughes, P. Gehr, and E.R. Weibel. 1989. Structure of the air-breathing organs of a swamp mud-eel, *Monopterus cuchia* (Ham.). *Japan. J. Ichthyol.* 35(4):453-465.
Munshi, J.S.D., K.R. Olson, T.K. Ghosh, and J. Ojha. 1990. Vasculature of the head and respiratory organs in an obligate air-breathing fish, the swamp eel, *Monopterus (=Amphipnous) cuchia*. *J. Morph.* 203:181-201.
Munshi, J.S.D., P.K. Roy, T.K. Ghosh, and K.R. Olson. 1994. Cephalic circulation in the air-breathing snakehead fish, *Channa punctata, C. gachua* and *C. marulius* (Ophiocephalidae, Ophiocephaliformes). *Anat. Rec.* 238:77-91.
Nimi, A.Z. and S.L. Morgan. 1980. Morphometric examination of the gills of walleye, *Stizostedian vitreum vitreum* (Mitchell) and rainbow trout, *Salmo gairdneri* (Richardson). *J. Fish Biol.* 16:685-692.
Niva Biswas, J. Ojha, and J.S.D. Munshi. 1981. Morphometrics of the respiratory organs of an estuarine goby *Boleophthalmus boddaerti*. *Japan J. Ichthyol.* 27:316-326.
Oikawa, S. and Y. Itazawa. 1985. Gill and body surface areas of the carp in relation to body mass, with special reference to the metabolism size relationship. *J. Exp. Biol.* 117:1-14.
Ojha, J. 1993. Functional organization of teleostean gills. *In* B.R. Singh (ed.), *Advances in Fish Research*, vol. 1. pp. 71-98. Narendra Publishing House, New Delhi.
Ojha, J. and J.S.D. Munshi. 1974a. Morphometric studies on the gill and skin dimensions in relation to body weight in a fresh-water mud-eel, *Macrognathus aculeatum* (Bloch). *Zool. Anz.* 193 (5/6):364-381.
Ojha, J. and J.S.D. Munshi. 1974b. Morphometric estimation of gill diffusing capacity of a fresh water mud-eel, *Macrognathus aculeatum* (Bloch), in relation to body weight. *Zool. Beitrage* 22 (1):87-98.
Ojha, J. and R. Singh. 1987. Effect of body size on the dimensions of the respiratory organs of a freshwater catfish, *Mystus vittatus*. *Japan J. Ichthyol.* 34:59-65.
Ojha, J. and A.K. Mishra. 1993. Interspecific variations in the functional organization, dimensions and scaling of gills in relation to body weight of Gangetic mullets, *Rhinomugil corsula* (Ham.) and *Sicamugil cascasia* (Ham.) (Mugilidae, Mugiliformes). *Proc. Indian Nat'l. Sci. Acad.* B59:576-580.
Ojha, J., R. Singh, and N.K. Singh. 1985. Structure and dimensions of the respiratory organs of a freshwater catfish *Mystus cavasius*. *Proc. Indian Nat'l. Sci. Acad.* B51:202-210.

Olson, K.R. 1991. Vasculature of the fish gill: Anatomical correlates of physiological functions. *J. Electron Micros. Technique* 19:389-405.

Olson, K.R., J.S.D. Munshi, T.K. Ghosh, and J. Ojha. 1986. Gill microcirculation of the air-breathing climbing perch, *Anabas testudineus* (Bloch): Relationships with the accessory respiratory organs and systemic circulation. *Amer. J. Anat.* 176:305-320.

Olson, K.R., J.S.D. Munshi, T.K. Ghosh, and J. Ojha. 1990. Vascular organization of the head and respiratory organs of the air-breathing catfish, *Heteropneustes fossilis. J. Morph.* 203:165-179.

Olson, K.R., P.K. Roy, T.K. Ghosh, and J.S.D. Munshi. 1994. Microcirculation of gills and accessory respiratory organs from the air-breathing snakehead fish, *Channa punctata, C. gachua* and *C. marulius. Anat. Rec.* 238:92-107.

Pandey, A. 1988. *The structure and function of the respiratory organs of a major carp, Labeo rohita (Ham)*. Ph.D. thesis, Bhagalpur University, India.

Pandey, A., G.K. Kunwar, and J.S.D. Munshi. 1989. Gill diffusing capacity of a major carp, *Labeo rohita* (Ham), in relation to body weight. *Proc. Indian Acad. Sci. (Anim. Sci)*. 98(6):385-389.

Price, J.W. 1931. Growth and gill development in the small-mouthed black bass, *Micropterus dolomieu* Lacapede. *Contrib. Theodore Stone Lab. Ohio Univ.* 4:1-46.

Rooj, N.C. 1984. *Structure and function of the respiratory organs of certain hill stream fishes of Chhotanagpur division.* Ph.D. thesis, Bhagalpur University, India.

Roy, P.K. 1984. *Morphometrics of the respiratory organs of certain major carp.* Ph.D. thesis. Bhagalpur University, India.

Roy, P.K. 1994. Effect of body size on scaling of respiratory organs of Indian air-breathing fishes. 4th International Congress of Vertebrate Morphology, Chicago. *J. Morph.* 220(3):389.

Roy, P.K. and J.S.D. Munshi. 1986. Morphometrics of the respiratory organs of a freshwater major carp, *Cirrhinus mrigala* (Ham.), in relation to body weight. *Japan J. Ichthyol.* 33(3):269-279.

Roy, P.K. and J.S.D. Munshi. 1992. Morphometric assessment of accessory respiratory surface area and their diffusing capacity in an adult snakeheaded fish, *Channa striata* (Bloch) *J. Freshwater Biol.* 4(2):99-107.

Sharma, S.N., G. Guha, and B.R. Singh. 1982. Gill dimensions of a hill stream fish, *Botia lohachata* (Pisces, Cobitidae). *Proc. Indian Nat'l. Sci. Acad.* B 48:81-91.

Singh, B. 1990. *Light microscopy, SEM, morphometrics and histochemistry of the gills of some commercially important fishes of India.* Ph.D. thesis, Bhagalpur University, India.

Singh, B.N. and J.S.D. Munshi. 1969. On the respiratory organs and mechanics of breathing in *Periophthalmus vulgaris* (Eggert.). *Zool. Anz.* 183 (1/2):92-110.

Singh, B.R., A.M. Yadava, J. Ojha, and J.S.D. Munshi. 1981. Gross structure and dimensions of the gills of an intestinal air-breathing fish, *Lepidocephalichthys guntea. Copeia* 1: 224-229.

Singh, N.K. 1986. *The structure and function of the respiratory organs of a freshwater air-breathing fish, Colisa fasciatus.* Ph.D. thesis, Bhagalpur University, India.

Singh, O.N. 1982. *Biology of certain gobiid fish of India.* Ph.D. thesis, Bhagalpur University, India.

Singh, O.N., P.K. Roy, and J.S.D. Munshi. 1988. Gill dimensions of an Indian hill-stream cyprinid fish, *Botia dario* (Ham.). *Arch. Biol.* 99:169-182.

Sinha, S.K. 1983. *Biology of Labeo bata (Ham.).* Ph.D. thesis, Bhaglapur University, India.

Ultsch, G.R. and G. Gros. 1979. Mucus as a diffusion barrier to oxygen: Possible role in O_2 uptake at low pH in carp (*Cyprinus carpio*) gills. *Comp. Biochem. Physiol.* 62(A):685-689.

Ursin, E. 1967. A mathematical model of some aspects of fish growth, respiration and mortality. *J. Fish. Res. Bd. Can.* 24:2365-2453.

Weibel, E.R. 1979. *Stereological Methods* vol.1. *Practical Methods for Biological Morphometry.* Academic Press Inc., (London) Ltd.

Weibel, E.R. 1984. *The Pathway for Oxygen: Structure and Function in the Mammalian Respiratory System.* Harvard University Press, Harvard, Mass.

Yadav, A.N. and B.R. Singh. 1989. Gross structure and dimensions of the gill in an estuarine goby, *Pseudapocryptes lanceloatus. Japan J. Ichthyol.* 36:252-259.

Yadav, A.N., M.S. Prasad, and B.R. Singh. 1990. Gross structure of the respiratory organs and dimensions of the gills in the mud skipper, *Periophthalmodon schlosseri. J. Fish Biol.* 37:383-392.

14

Effects of Gill Dimensions on Respiration

Brian A. Hills

The dimensions of many creatures have been studied extensively, especially with respect to age, and none more so than fish, and that includes their gills. Since the primary function of the gill is the exchange of both respiratory gases and other solutes with the water inhabited by the fish, the purpose of this article is to outline how those dimensions determine the rate of solute exchange without invoking the very complex mathematics which tend to accompany texts on fluid dynamics.

RESPIRATION

The overall transport of respiratory gases with respect to metabolic demand of fish has been particularly well outlined by Hughes (1965), who also makes interesting comparisons between water and air as the respiratory medium for a wide range of vertebrates. Fundamental to these considerations is the trade-off between extraction of oxygen and energy expenditure in ventilating the gas exchange interface with the surrounding medium. Whereas mammals extract about 25% of the oxygen from inspired air, fish can utilize up to 80% of the O_2 dissolved in water.

This higher extraction rate is determined by the relatively low solubility of oxygen in water (0.031 mlO_2/mlH$_2$O) and the much higher work load in pumping a much denser and more viscous respiratory medium. In ventilating a human on hyperoxygenated saline, it was found necessary to use a mechanical ventilator at three breaths per minute (Schoenfisch et al., 1976), while deepsea divers need to substitute helium for nitrogen at depths in excess of 150 feet in order to reduce the work of breathing (Lanphier, 1969).

Another way of viewing this trade-off is that in man an estimated 1–2% of the O_2 uptake is used to liberate the energy requisite for breathing, whereas this figure ranges from 10 to 25% for fish (Hughes, 1965). Thus, in considering the effect of gill dimensions upon respiration, we must not only address their impact upon gas exchange, but also their effect upon the resistance to fluid flow in the gill.

Gas exchange

In all transport processes the rate of transfer is determined by two basic entities—driving force and resistance. In a purely passive system, such as respiratory gas exchange, the driving force is oxygen tension (P_{O_2}), whether providing a gradient to force O_2 into solution or to displace the position of chemical equilibrium as occurs in the conversion of hemoglobin to oxyhemoglobin. These processes lie in series, and one can trace the fall in P_{O_2} from an inspired value of 150 mmHg or so in air or water down to almost zero in the vicinity of the mitochondria of the peripheral tissues (Fig. 1). This is known as the "oxygen conduction line" (Holmgren, 1966) and applies as much to lower vertebrates as it does to man, in whom it has been quantified in great detail (Hills, 1974). Although the various transport processes lie essentially in series, O_2 molecules travel in parallel in some parts of the lung and the gradients in them may differ. This applies in particular to the mismatch which may occur at the local level between ventilation and blood perfusion, otherwise known in the lung as $\dot{V}_A/\dot{Q}$ inequality (West, 1965)—a phenomenon much studied in humans as a major cause of hypoxia resulting from disease. The same applies to fish. A similar conduction line for CO_2 can be traced in reverse.

Having briefly discussed gas tension as the driving force, we can now consider resistance to gas exchange and the particular point in the overall oxygen conduction line at which O_2 molecules are transferred from the aqueous respiratory medium to the blood. This is the step directly influenced by gill dimensions.

Gill dimensions

The gill system has been well described by Hughes (1965) who also discusses the variations between bony fish, dogfish and sharks. Teleosts have four branchial arches, two on either side of the body, each having a double row of plates, the filaments displayed as a V (see Fig. 2a). These filaments are stacked one upon the other but separated by a space through which water can pass. The surface of each filament is folded rather than flat, which increases the epithelial gas exchange surface. These secondary folds are termed secondary lamellae, being separated from each other to form water channels of the order of 300 μm in width and 10 mm in length for a surface area per lamella of 0.5 mm².

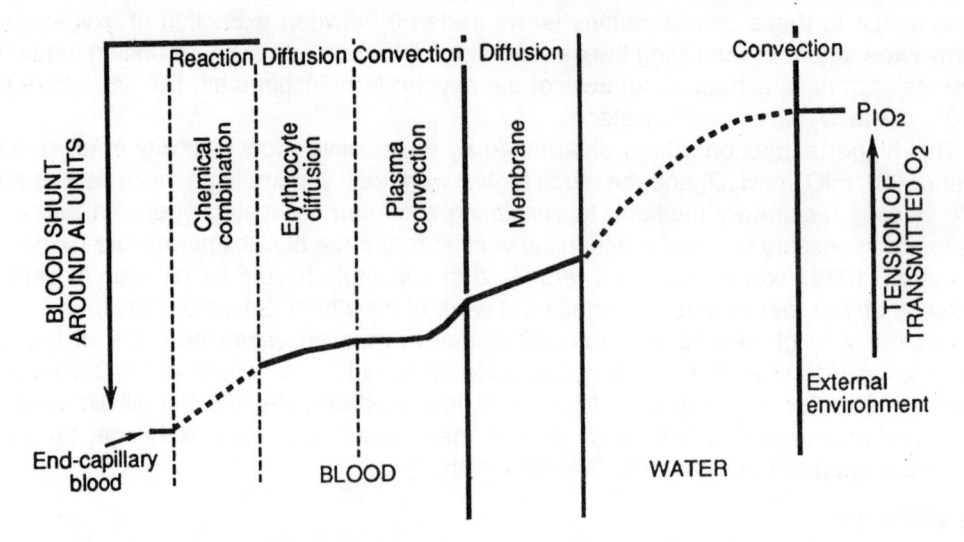

Fig. 1. Diagrammatic representation of the "oxygen conduction line" (Holmgren, 1966) depicting the additive nature of the successive steps in P_{O_2}, each step "driving" oxygen in the direction of its eventual metabolic assimilation.

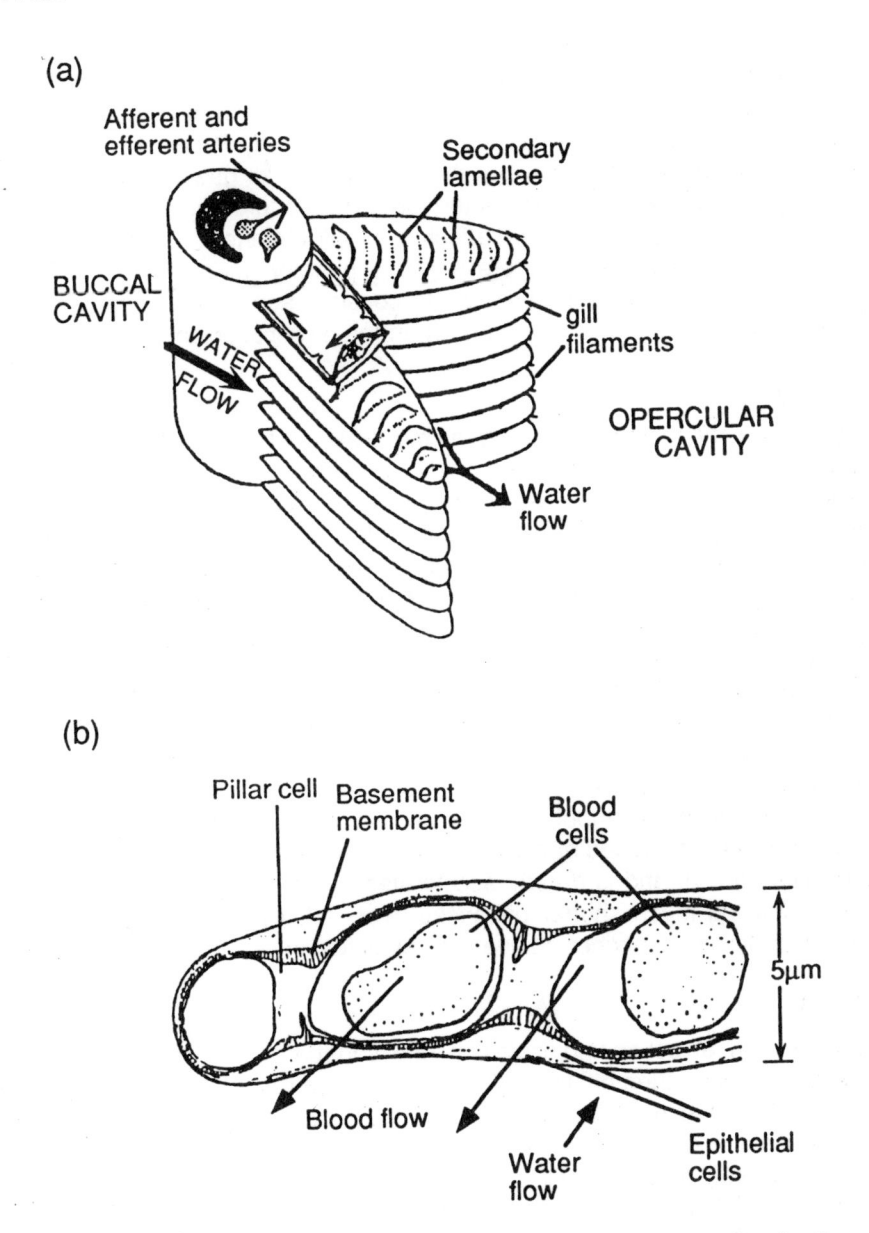

Fig. 2. General configuration of gills of teleost fishes, showing (a) filaments and water flow direction within each branchial arch, (b) a section through a secondary lamella. Redrawn from Hughes (1965).

Capillaries run lengthwise in each secondary lamella, forming a sheet separating the two epithelial surfaces. Thus each capillary is bordered on opposite sides by epithelial surfaces, separated only by the basement membrane to give a mean blood-to-water distance which can range from less than 1 μm up to 10 μm, depending on the species. This is depicted in Fig. 2b. Fish invoke the countercurrent principle (Hughes, 1965) much exploited by chemical engineers in enhancing the concentration gradient to improve mass transfer (Coulson and Richardson, 1965), blood flowing in the opposite direction in the capillaries to water in the adjacent channels. The dimensions of the channels vary widely among the 93 species of fish for which data were collated by Piiper (1970).

Having outlined the flow systems and the driving force for gas exchange, we can now consider the resistance posed by the interface between gill epithelium and ventilating water—the transport process primarily affected by gill dimensions. Before pursuing the question of which gill dimensions are relevant to respiratory gas exchange, or other transport properties of that organ, it is desirable to consider some fundamental principles of how solutes move in solutions—a topic known in the physical sciences as mass transfer.

Basics of mass transfer

Within any medium the molecules and atoms move at very high velocities, whether vibrating about a fixed position—as occurs in solids—or allowed to move anywhere within the container in the case of gases, or anywhere within its boundary for a liquid. Any foreign molecule will experience collisions with these molecules with the net result that, at the microlevel, any accumulation of these foreign molecules will tend to be dispersed.

An appropriate analogy in two dimensions is a snooker table. The red balls make perfectly elastic collisions with each other and with the cushions; in the case of gases this is manifest as a pressure in accordance with the Kinetic Theory of Gases. When the group of colored balls is placed on the table, these "foreign molecules" are soon randomly dispersed by collisions with the red balls. At the macroscopic level, the redistribution of solute molecules is very slow but they do eventually reach a uniform distribution, i.e., a state of equilibrium, as witnessed by adding a concentrated salt solution to water without stirring. This process of redistribution at the molecular level is known as diffusion. The equilibration process is hastened by raising the temperature, which increases the velocities of all molecules present and is greatly accelerated by stirring to effect macrodispersion of the solute, leaving much shorter distances over which to effect equilibration by the intermolecular collisions described above, i.e., by diffusion.

Steady supply

If the concentration of solute molecules at one surface of the medium is maintained from an external source, then diffusion continues until the rest of the medium reaches the same concentration, assuming the other boundaries to be impermeable. The mathematics describing the attainment of this state are horrendous, taking the form of transcendental equations. These can only be applied to simple geometric systems whose boundaries can be described by simple equations (Crank, 1956) and, even then, only for simple changes in these boundary conditions. Since the conduction of heat obeys physical laws exactly analogous to those governing diffusion, further mathematical solutions can often be obtained from the heat transfer literature (Carslaw and Jaeger, 1959). The advent of computers has enabled certain finite difference models to be used for less regular shapes and changes in gas tension but computation is still difficult. If the other boundaries are no longer impermeable and one or more lose solute molecules to the surroundings, then a continuous transport of molecules from the source to this sink will arise. A very common example in physiology is the passive transport of any substance across a membrane, where the solute is prevented from accumulating by removal in flowing fluid such as blood.

Steady state

The transport system can reach a state where the number of solute molecules reaching a particular point in the system equals the number leaving and so, at the macroscopic level, the concentration remains constant with respect to time. When this occurs at all points, the system is said to have reached a steady state. This may apply whether the medium is static or flowing but the simplest case, and the one to consider first, is that when it is static.

Diffusion

In a solid or a gas or liquid which is static at the macrolevel, the process is known as diffusion; the rate at which the solute moves from a higher to a lower concentration is proportional to the concentration gradient. This is Fick's law of diffusion, which is readily demonstrated for a steady state by considering a flat parallel-faced slab of the static medium as shown in Fig. 3. If one face is the source and the other is the sink, then the rate of transport of solute molecules ($\dot{Q}$) is found to be proportional to the area (A) across which the solute is diffusing from the higher concentration of the source (C_1) to the lower concentration of the sink (C_2). $\dot{Q}$ is also inversely proportional to the path length (x), i.e., thickness of the slab to be traversed:

$$\dot{Q} = \frac{DA(C_1 - C_2)}{x}, \qquad \ldots(1)$$

where the proportionality constant (D) is termed the diffusion coefficient or diffusivity.

Sometimes the rate of transfer is quoted as the flux ($\dot{Q}/A$), while the concentration gradient ($C_1 - C_2$)/x may be replaced by the tension gradient ($P_1 - P_2$)/x. This term is preferable when describing gases since these always diffuse *down* the tension gradient whereas

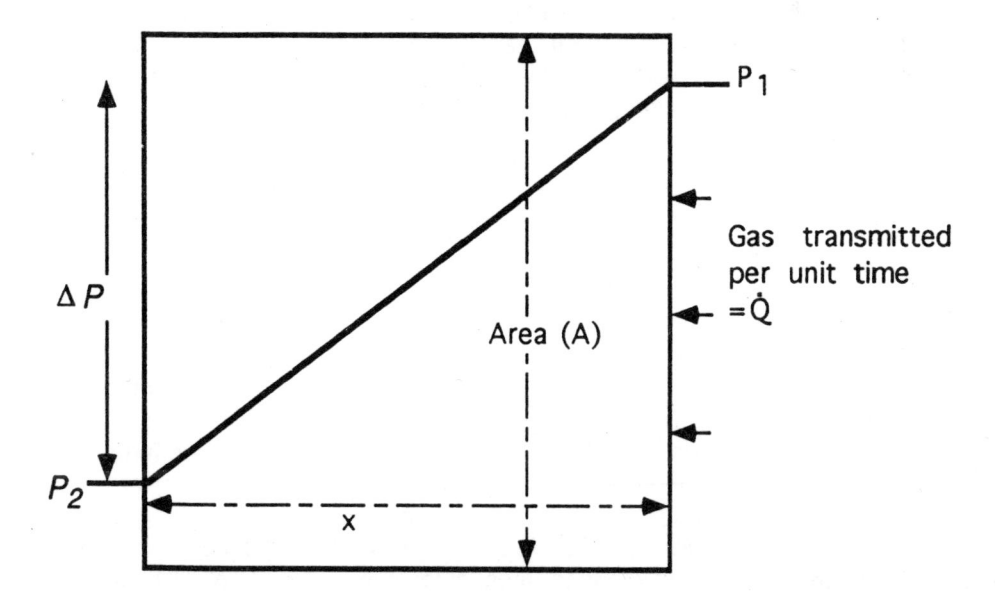

Fig. 3. Fick's law of diffusion (Equation 1) illustrated by the uniform tension gradient obtained *under steady-state conditions* across a flat, parallel-faced slab of *uniform* tissue when opposite faces are maintained at constant yet different tensions (P_1 & P_2).

they may move from a lower to a higher concentration if the gas is much more soluble in the second medium ($S_2 > S_1$). The solubility (S) relates tension (P) to concentration (C) by Henry's law ($C = SP$), wherein Fick's law (equation 1) can be reexpressed as:

$$\dot{Q} = \frac{DSA(P_1 - P_2)}{x}. \qquad \ldots (2)$$

This is particularly useful in lung respiratory systems wherein P_1 may refer to a gaseous phase and P_2 to a liquid.

In some texts DS is merged as the permeability, while A or x or both may also be incorporated (Hills, 1970) to define a membrane diffusion capacity (ADS/x); this can lead to some confusion. The only sure method of avoiding misunderstanding in using any value quoted in the literature is to look for the dimensions which accompany the quoted value. Thus a true diffusion coefficient (D) has the dimension of L^2/T, where L is length, T is time, translating into typical units of cm^2/s, although it would still be technically correct to use miles^2/h.

Diffusivities

The rate of diffusion of a gas such as oxygen in water is very slow with values of D quoted as 3.3×10^{-5} cm^2/s for water and 2.3×10^{-5} cm^2/s for whole lung tissue (Grote, 1967) compared with 0.19 cm^2/s for O_2 diffusing across nitrogen at 25°C (Reid and Sherwood, 1966).

Since the diffusion coefficient is so low, the biological creature circumvents this problem by compensating with a large surface area—as much as 90 m^2 in the human lung (Weibel, 1967)—and a very thin membrane corresponding to a harmonic mean thickness of only 0.7 µm (Weibel, 1971). For a typical oxygen uptake of 200 ml/min, equation (2) gives a tension gradient of only 0.43 mm Hg (Hills, 1974) which is negligible by comparison with an inspired-arterial difference $(I - a)DO_2$ of about 50 mm Hg in the mammalian lung and so the major resistance, i.e., bottleneck to oxygen exchange, has been sought elsewhere in the overall transport process.

GAS EXCHANGE IN THE GILL

The exchange of oxygen and carbon dioxide in the secondary lamellae approximates to a steady state condition with a continuous transfer of O_2 from the water to the blood. However, when Fick's Law is applied to the gill membrane, the actual flux would appear to be only about one-quarter of that predicted by equation (2). This applies after allowing for countercurrent flow and applying the appropriate Bohr integration to allow for the nonlinear relationship between PO_2 and oxygen content of blood as hemoglobin is converted to oxyhemoglobin (Hughes and Hills, 1971) and, hence, the change in O_2 driving force $(P_1 - P_2)$. Without this correction the kinetics of oxygenation would appear an order of magnitude faster than deoxygenation, as proposed in the chorioallantoic capillary (M), but this is not the case (Tazawa and Mochizuki, 1976). In the gill physical as opposed to chemical limitations also determine oxygen exchange rates, at least in carp and eel (Hills et al., 1982). The fourfold apparent deviation from Fick's law in attempting to describe oxygen exchange in terms of the gill membrane only has led to a search for other physical resistances to gas transfer, such as the water boundary layer in particular (Hills and Hughes, 1970).

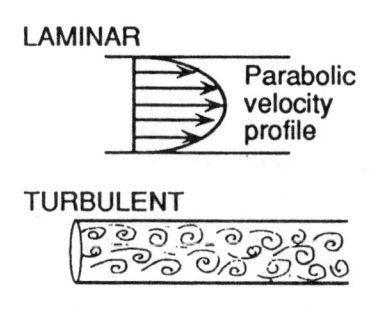

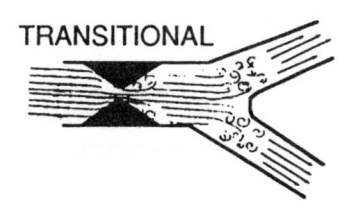

Fig. 4. Demonstrating three different types of fluid flow (a) streamline or laminar as characterized by a parabolic velocity profile, (b) turbulent flow and (c) transitional flow where turbulence has been introduced by the bifurcation or a constriction.

Laminar flow

Earlier, I mentioned the similarity between mass transfer and heat transfer; the simple household radiator provides a very appropriate analogy. The rate of transfer of heat from the hot water to the room is not limited to any appreciable extent by the thickness of the metal but, rather, by the water layer on one side and the air layer on the other. Taking the hot water side, the water immediately adjacent to the metal surface is static. If the surface is flat and the flow is slow, the next layer or lamina will be moving parallel to the surface and the one farther away will be moving faster but still parallel to the surface. This type of flow is known as laminar or streamline flow as all of these layers move parallel to each other as depicted in Fig. 4. For a long flat surface the velocity increases at a uniform rate as the distance from the surface increases. At the microscopic level, wherein molecules are moving at the high velocities described above, an analogy can be drawn between adjacent laminae and two trains on parallel tracks as one overtakes the other. If people—representing molecules—keep jumping back and forth between the trains, the slower will speed up and the faster will slow down as this causes an exchange of momentum in the direction of travel. To maintain the difference in speed, a greater force needs to be applied by the engine of the faster train. In streamline flow this situation is envisaged as one of shear, with one lamellation sliding over the other, the force per unit area of contact being known as the shear stress. The velocity gradient is the rate of shear and for ideal or "Newtonian" liquids, such as water, the ratio is a constant which is known as the viscosity (η) of the particular fluid. Thus a greater force is needed to maintain the same velocity gradient in treacle as in water. If the liquid is "non-Newtonian", flow can still be laminar and velocity will still be lower closer to the surface, although the velocity gradient is no longer uniform.

Boundary layer

If we return to the analogy of the hot water radiator, the net result of flow is that whereas we may know the temperature of the bulk of the water well away from the surface and the temperature of the wall, it is very difficult to calculate the distribution of temperature in between (Fig. 5). Most of the temperature drop occurs in the region adjacent to the surface where most of the velocity gradient arises; this is termed the boundary layer. Sometimes it is termed an "unstirred layer" but this is not strictly true. When fresh water enters a gill channel, one can envisage contours for the uptake of CO_2 similar to those shown in Fig. 6.

Film coefficient

Even if we cannot determine the velocity distribution within the boundary layer, we still find that the rate of transfer of heat is proportional to the surface area and the temperature gradient between the bulk of the hot water and the radiator wall. In the same way, we find that the rate of transfer of a dissolved solute between the bulk water flowing in a channel and the membrane wall is proportional to the concentration difference ($C_1 - C_2$) and the surface area (A). In that way chemical engineers define a film coefficient for mass transfer (k) as

$$\dot{Q} = kA(C_1 - C_2). \qquad \ldots (3)$$

Referring back to equation (2), we can equate k to D/x' and sometimes the effective thickness of the boundary layer (x') is cited in preference to k. However k and hence x' are functions of the dynamics of the fluid in the vicinity of the surface.

Types of flow

We have already discussed the parallel nature of the movement of different laminations of fluid elements which characterizes laminar flow (Fig. 4). If the surface is still flat but the velocity is greatly increased, the parallel, well-ordered flow will break up as some elements start moving in other directions with some even moving perpendicular to the surface and to the direction of bulk flow. This turbulence may not reach the surface itself but will certainly tend to disrupt the boundary layer, reducing its inherent resistance to

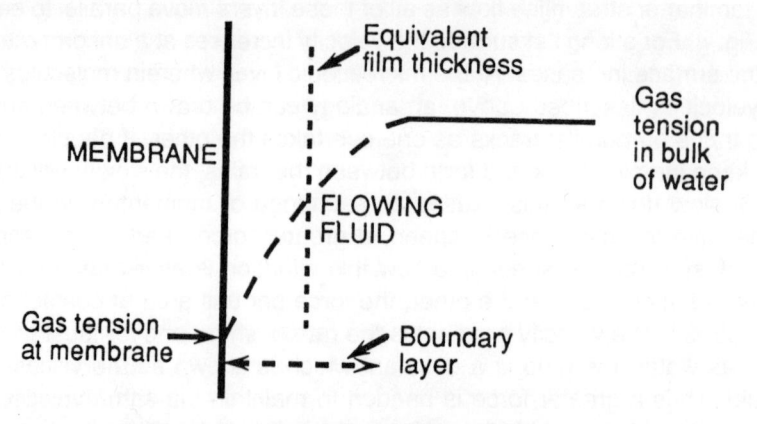

Fig. 5. Depicting the boundary layer of fluid flowing parallel to the surface across which the concentration gradient varies but is equivalent to an unstirred (static) layer of thickness x'.

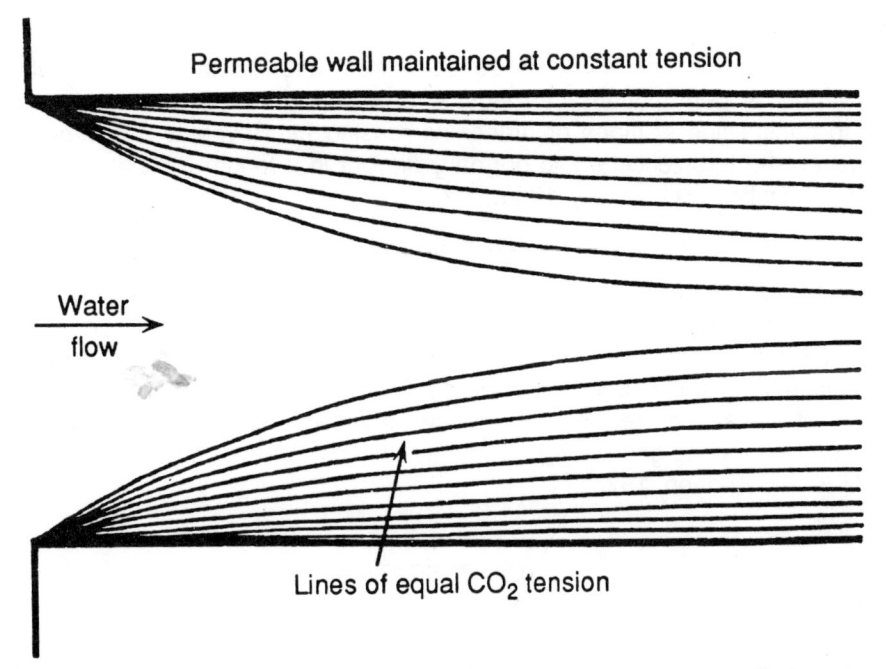

Fig. 6. Diagrammatic representation of lines of equal CO_2 tension as water enters a gill channel, assuming P_{CO_2} at the wall to remain uniform.

both mass transfer and heat transfer. Thus modern hot water radiators use much smaller pipes with far higher water velocities in order to reduce the boundary layer and improve heat transfer. The price to be paid for this apparent increase in efficiency is the additional work of pumping needed to attain higher velocities, which is particularly relevant to gas exchange in the gill as we shall see later.

Reynolds number

Turbulence is much easier to induce in air than in water, thus invoking viscosity (η) and density (ρ) in addition to velocity (u). Engineers use the Reynolds number (Re) as the best inherent index of flow:

$$\mathrm{Re} = \frac{ud\rho}{\eta}. \qquad \ldots (4)$$

This is a dimensionless number of the type much favored by engineers because it is a pure number with no units. The dimensions: LT^{-1} for velocity (u); ML^{-3} for density (ρ) and $ML^{-1}T^{-1}$ for viscosity (η) and L for diameter (d) actually cancel out when applied to the right side of equation (5).

One of the statements most misquoted, or quoted out of context, is that turbulent flow occurs for Re > 2,100 and laminar flow for Re < 2,100. This applies only to the end of a very long and perfectly smooth tube, as used by Sir Osborne Reynolds when he performed his original experiment. For Re < 2,100, turbulence will eventually die out in progressing along a perfectly smooth tube.

Transitional flow

In most practical situations flow is in transition from turbulent to laminar or vice versa. A wide variety of perturbations can induce turbulence ranging from roughness of the surface to constriction in pipes or branching, arising for Reynolds numbers as low as 45. However, the turbulence would normally die out in a length (l) related to Re by the Schlichting equation (Schlichting, 1960), viz.,

$$l/r = 0.1(\text{Re}) \qquad \ldots (5)$$

where r is the radius of the pipe. Thus more turbulent flow (Re↑) takes longer to die out.

Ridges in the surface are often employed to break up the boundary layer and the drag associated with it, a common example being the use of spoilers on cars. Constrictions can induce turbulence and the sounds associated with it, such as the wheezing of asthmatics; bifurcation of airways in the lung can induce transitional flow, which has the functional advantage of promoting gas mixing and hence minimizing the effect of anatomical dead space (Hills and Kuonen, 1973).

QUANTIFICATION

Having outlined the different types of flow, it is now desirable to try to quantify mass transfer in each case. There are two basic approaches: the first involves the most horrendous mathematical analysis for which the whole field of fluid dynamics is notorious and, even then, the equations derived seem difficult to apply to real situations. Most engineers faced with the need to come up with practical solutions to real problems opt for the much simpler approach of relating film coefficients for heat transfer (h) or for mass transfer (k) to dimensionless groups, of which the Reynolds number is one and the one which represents the state of flow. Another dimensionless group is needed to describe the properties of the solute in solution in the case of mass transfer or the thermal properties of the system in the case of heat transfer. The relevant group for mass transfer is the Schmidt number (Sc) which is defined as:

$$\text{Sc} = \eta/\rho D. \qquad \ldots (6)$$

In keeping with the tradition of using dimensionless groups, the mass transfer coefficient (k) is given by:

$$\frac{(kd)}{D} = \lambda(\text{Re})^\alpha (\text{Sc})^\beta (d/l)^\omega, \qquad \ldots (7)$$

where λ, α, β and ω are all dimensionless constants, d is the width and l is the length of the channel.

This simple empirical equation has been very widely used by engineers in which values of λ, α, β and ω are obtainable for standard cases (see Table).

Flow within tubes

For fully developed turbulent flow within tubes, Sherwood (1937) gives: $\lambda = 0.023$, $\alpha = 0.83$, $\beta = 0.44$, and $\omega = 0$, i.e., the film coefficient is independent of tube length. It is also independent of diameter. It is also independent of d and l ($\omega = 0$) for laminar flow which has developed a full parabolic profile (Fig. 4) when $\lambda = 4.1, \alpha = 0$ and $\beta = 0$

(Sherwood, 1937). In transitional flow, by contrast, k is dependent on length ($\omega = 0.33$) since turbulence is dying out. Values of the various constants for the major flow types are given in the Table.

Constants for mass transfer equations

Type of flow	λ	α	β	ω
Fully developed turbulent	0.023	0.83	0.44	0
Fully developed laminar	4.1	0	0	0
Transitional	1.62	0.33	0.33	0.33

Reynolds analogy

Many more cases have been studied for heat transfer which may be adapted to mass transfer by using the Reynolds analogy. Thus the exponent of the Prandtl number (Pr) can be applied to the Schmidt number (Sc) for the corresponding flow conditions as β in equation (7). This makes sense intuitively since the same type of flow (same Re and exponent α) should promote heat transfer to the same degree as mass transfer.

This analogy can be taken one step further to include frictional factors in flow. In the physical sciences this is regarded as momentum transfer in keeping with the analogy of the passing trains described earlier in discussing viscosity. In classical texts (Rohsenow and Choi, 1961) expressions for heat, mass, and momentum transfer are derived simultaneously. Those analyses are far beyond the scope of this paper but qualitatively serve to illustrate the point that whatever the fish might do to enhance gas exchange, it is also likely to increase frictional drag on the water and hence the work of ventilation.

THE GILL

Reynolds analogy with heat transfer has been applied to "gilled systems" in engineering by Brown (1965) to derive the flux ($\dot{Q}/A$) of respiratory gases across the boundary layer of water adjacent to the membrane as:

$$\frac{\dot{Q}}{A} = DS(\Delta P) \left(\frac{3\mu}{Dld}\right)^{0.33}, \qquad \ldots (8)$$

where d is the width of the water channel between adjacent secondary lamellae and l is the length. A similar expression was derived quite independently by Hills and Hughes (1973) by performing a dimensional analysis of growing fish of the same species. A small difference of 0.4 rather than 0.33 for the exponent of the Schmidt number (β) indicates rather more turbulence than indicated by Brown's expression (equation 8) when compared with values in the Table.

In their analysis Hills and Hughes (1973) showed how the water boundary layer provides about 80% of the resistance to oxygen exchange in the secondary lamellae of the dogfish. Similar proportions have been indicated by those who have attempted what mathematicians term "exact solutions" (Kylstra et al., 1966; Scheid and Piiper, 1971), although it is difficult to define a functional unit in $\dot{Q}$ in the same way that one can with reciprocating flow in the lung (Power, 1969; Hills, 1971).

Boundary layer analysis

In adopting the form of analysis employed by design engineers rather than academic physicists or mathematicians in quantifying mass transfer, it is still necessary to confirm that the expression derived, viz. equation (1), makes sense qualitatively. Firstly this expression emphasizes the solubility (S) of O_2 in water and the tension gradient (ΔP) as determined by the oxygen conduction line discussed above. Gas exchange will be moderately reduced by wider channels ($d\uparrow$) giving less intimate contact between blood and water and long channels ($l\uparrow$) in which the turbulence of water in the buccal cavity has more distance in which to diminish in each channel before exiting. It is interesting to note a lesser dependence upon diffusivity ($\dot{Q} \alpha D^{0.67}$) than would occur with no convection, i.e., diffusion alone ($\dot{Q} \alpha D$).

For water channels of various configurations there are many combinations of the empirical constants λ, α, β, and ω. These have been compiled as charts of flux *versus* velocity by Stoever (1941) and cover some 23 combinations involving flow *within or outside* tubes, *parallel* or *perpendicular* to those tubes for flow which is either *turbulent, laminar* or in *transition*.

Water velocity

The expression derived by Brown (equation 8) has a shortcoming insofar as it does not relate the oxygen flux directly to water velocity but only indirectly via the viscosity (μ). In the dimensional analysis by Hills and Hughes (1970) the data for bass and dogfish is far more compatible with turbulent flow than laminar when:

$$\dot{Q} \alpha (\text{Re})^{0.83} \alpha \mu^{0.83}. \qquad \ldots (9)$$

Thus mass transfer is appreciably more effective at higher velocities as originally predicted qualitatively by Steen and Kruysse (1964). However the induction of turbulence raises the work load.

Energy considerations

For laminar flow wherein a pressure drop (Δp) is required to achieve a mean water velocity ($\overline{u}$):

$$\Delta P = \frac{12 \overline{u} \mu l}{d^2}. \qquad \ldots (10)$$

Thus the pressure gradient and hence the energy expenditure, is particularly dependent on the width of the channel.

References

Brown, C.E. 1965. *Gilled Systems for Hydrospace*. Amer. Soc. Mech. Engrs. publ. 65-WA/UNT-6.
Carslaw, H.S. and J.C. Jaeger. 1959. *Conduction of Heat in Solids*, p. 101. Univ. Press, Oxford.
Coulson, J.M. and J.F. Richardson. 1965. *Chemical Engineering*. Pergamon, Oxford.
Crank, J. 1956. *The Mathematics of Diffusion*. Univ. Press, Oxford.
Grote, J. 1967. Die Sauerstoffdiffusionkonstanten in Lungengewebe und Wasser und ihre Temperaturabhangigkeit. *Pflügers Arch. Ges. Physiol.* 295:245-254.
Hills, B.A. 1970. An assessment of the expression $C = \dot{Q}[1-\exp(-PS/\dot{Q})]$ for estimating capillary permeabilities. *Phys. Med. Biol.* 15:705-713.
Hills, B.A. 1971. Analysis of relative contributions to the alveolar-arterial oxygen gradient. *Bull. Math. Biophys.* 33:259-260.

Hills, B.A. 1974. *Gas Transfer in the Lung*. Univ. Press, Cambridge.

Hills, B.A. and G.M. Hughes. 1970. A dimensional analysis of oxygen transfer in the fish gill. *Respir. Physiol.* 9:126–140.

Hills, B.A. and E. Kuonen, E. 1973. Longitudinal dispersion of composition differences in the airways of the lung. *Math. Biosciences* 18:351-364.

Hills, B.A., G.M. Hughes and T. Koyama. 1982. Oxygenation and deoxygenation kinetics of red cells in isolated lamellae of fish gills. *J. Exp. Biol.* 98:269-275.

Holmgren, A. 1966. The "oxygen conduction line" of the human body. *In* D. Hatcher and J.B. Jennings (eds.), *Proc. Int. Symp. Cardiovasc. Respir. Effects of Hypoxia*, pp. 391–400. Hafner, New York.

Hughes, G.M. 1965. *Comparative Physiology of Vertebrate Respiration*. Heinemann: London.

Hughes, G.M. and B.A. Hills. 1971. Oxygen tension distribution in water and blood at the secondary lamella of the dogfish gill. *J. Exp. Biol.* 55:399–408.

Kylstra, J.A., C.V. Paganelli, and E.H. Lanphier. 1966. Pulmonary gas exchange in dogs ventilated with hyperbarically oxygenated liquid. *J. Appl. Physiol.* 27:177-184.

Lanphier, E.J. 1969. Pulmonary function. *In* P.B. Bennett and D.H..Elliott (eds.), *The Physiology and Medicine of Diving and Compressed-air Work*, pp. 58–112. Bailère, Tindall and Cassell, London.

Mochizuki, M, H. Tazawa, and T. Ono. 1973. Microphometry for determining the reaction rate of O_2 and CO_2 with red blood cells in the chorioallantoic capillary. *In* D.F. Bruley and I.H. Bicher (eds.), *Oxygen Transport to Tissue*, pp. 997–1006. Plenum, New York.

Piiper, J. 1970. Gill surface area: fishes. *In* P.L. Altman and D.S. Dittmer (eds.), *Respiration and Circulation*, pp. 120–1. FASEB, Bethesda, MD.

Power, G.G. 1969. Gaseous diffusion between airways and alveoli in the human lung. *J. Appl. Physiol.* 27:701–9.

Reid, R.C. and T.K. Sherwood. 1966. *The Properties of Gases and Liquids*. McGraw-Hill, New York (2nd ed.).

Rohsenow, W.M. and H.Y. Choi. 1961. *Heat, Mass and Momentum Transfer*. Prentice-Hall, Englewood Cliffs, N.J.

Scheid, P. and J. Piiper. 1971. Theoretical analysis of respiratory gas equilibration in water passing through fish gills. *Respir. Physiol.* 13:305–318.

Schlichting, H. 1960. *Boundary Layer Theory*. McGraw-Hill, New York (4th ed.).

Schoenfisch, W.H., G.D. Blenkarn, B.A. Hills, and J.A. Kylstra. 1976. Liquid breathing: expiratory flow and CO_2 elimination using fluorocarbon and aqueous solutions. *In* C.J. Lambertsen (ed.), *Underwater Physiology* (V). FASEB, Bethesda, MD.

Sherwood, T.K. 1937. *Absorption and Extraction*. McGraw-Hill, New York.

Steen, J.B. and A. Kruysse. 1964. The respiratory function of teleostean gills. *Comp. Biochem. Physiol.* 12:127–142.

Stoever, H.J. 1941. *Applied Heat Transmission*. McGraw-Hill, New York.

Tazawa, H.T. and M. Mochizuki. 1976. Oxygenation and deoxygenation velocity factors of chorioallantoic capillary blood. *J. Appl. Physiol.* 40:399–403.

Weibel, E.R. 1967. Airways and the respiratory surface. *In* (eds.): *The Lung*, ch.1. Williams and Wilkins, Baltimore.

Weibel, E.R. 1971. Morphometric estimation of pulmonary diffusion capacity. *Respir. Physiol.* 11:54–75.

West, J.B. 1965. *Ventilation/Blood Flow and Gas Exchange*. Blackwell, Oxford.

15

A Composite Approach for Evaluation of the Effects of Pesticides on Fish

Hiran M. Dutta

OVERVIEW

Indiscriminate use of pesticides, careless handling, accidental spillage, or discharge of untreated effluents into natural waterways have harmful effects on the fish population. According to Haider and Inbaraj (1988), insecticides used for pest control are also toxic to nontarget organisms in the aquatic ecosystem. The indiscriminate use of pesticides can be considered one of the factors which changes the environment, causing several imbalances in the ecosystem, especially the denizens of the aquatic environment (Reddy and Rao, 1990).

According to EPA News (October 30, 1989), nearly half of the acreage in 34 states (USA) is badly polluted or likely to become that way soon. The EPA survey shows that pollution has impaired about one-fourth of more than 12 million acres of lakes examined in those states. Pollution threatens another 20%. Three-fourths of the pollution was attributed to agricultural and urban runoffs. Another 11% was blamed on industrial or sewage releases.

According to Tandon and Dubey (1983), overuse of pesticides and other chemicals in some agricultural and public health operations may not include disturbance and disequilibrium in aquatic and terrestrial environment, but it may have adverse effects on nontarget organisms such as fish. They researched the toxic effects of different concentrations of two organophosphorus pesticides, malathion, and dimecron on the aldolase enzyme of the liver, brain, and gill tissues in *Clarias batrachus*, exposed for periods of 24 to 96 h and aldolase activity became elevated. They also indicated that the toxic condition produced by the pesticides finally leads to death of the fish, owing to failure of the respiratory center of the brain. The effect of diazinon, an organophosphorus pesticide, on the activities of brain enzymes of *Ophiocephalus punctatus* has been studied by Sastry and Sharma (1980).

The effect of diazinon toxicity on protein and nucleic acid metabolism in the liver of zebrafish, *Brachydanio rerio* (Cyprinidae) was researched by Ansari and Kumar (1988). Haider and Inbaraj (1988) studied *in vitro* the effect of malathion and endosulfan on the LH-induced oocyte maturation in the common carp, *Cyprinus carpio*. Effects of aldrin and malathion on blood chloride in the Indian catfish *Heteropneustes fossilis* were studied by

Srivastava and Srivastava (1988). The deleterious effects of malathion on survivability and growth of fingerlings of Channa punctatus were investigated by Shukla et al. (1987). Effects of other pollutants, such as pentachlorophenol, on growth of young-of-year largemouth bass, Micropterus salmoides were studied by Johansen et al. (1987). The effect of acute and sublethal concentrations of monocrotophos on the kidney of Puntius conchonius was investigated by Kumar and Pant (1985). Orsatti and Colgan (1987) studied the effects of sulfuric acid on the behavior of largemouth bass, Micropterus salmoides.

Brown et al. (1987) investigated the impairment of early feeding behavior of largemouth bass by pentachlorophenol (PCP) exposure. Pentachlorophenol is a broad-spectrum biocide frequently used to preserve wood and causes contamination in many water bodies. Wester (1991) studied the histopathological effects of β-hexachlorocyclohexane (β-HCH) and methyl mercury on reproductive organs in Poecilia reticulata (guppy) and Oryzias latipes (medaka), and found that β-HCH induced vitellogenesis and hermaphrodism and methyl mercury impaired spermatogenesis.

The use of organophosphorus insecticides has been increased because of their rapid biodegradability, replacing the more persistent organochlorines. They are now produced in larger quantities than the chlorinated hydrocarbon (Lawless et al., 1972). Organophosphorus pesticides are being added to the global environment at an annual rate of tens of thousands of tons (Kozlovskaya and Mayer, 1984). Organophosphorus pesticides such as malathion are used in large quantities as an insecticide on a wide range of fruits and vegetables in agriculture and horticulture, as well as against ectoparasites in poultry, cattle, and pigs. Malathion and diazinon are also used as insecticides in paddy fields. They pollute the aquatic environment by direct application, spray drift, aerial spraying, washing from the atmosphere by precipitation, erosion and runoff from agricultural land, and by discharge of effluent from factories and in sewage (Frank et al., 1987). These two important pesticides are discussed here.

Malathion has a low toxicity for mammals and a relatively high toxicity for fish (Mount and Stephan, 1967). This happens due to the lack of hydrolytic enzymes in insects and fish (Arecchon and Plumb, 1990). Oxygen analogues of malathion (malaoxon) appear to be the active part that binds vigorously to acetylcholinesterase (O'Brien et al., 1974). This malaoxon is hydrolyzed rapidly in mammals thus becoming inactive, but such hydrolysis does not occur in insects and proceeds very slowly in fish (Arecchon and Plumb, 1990).

Diazinon is a broad-spectrum insecticide. Despite its several advantages, the possibility of water contamination by diazinon due to degradation of pH etc. cannot be overlooked. Diazinon has been found in several rivers in the USA at concentrations of slightly below $1.0\,\mu g/L^{-1}$ (Ansari and Kumar, 1988). The major victims of this contamination are nontarget aquatic animals, especially fish.

In almost every instance of diazinon poisoning, there has been a general reduction in cholinesterase activity levels, especially in the brain and blood (Mantz, 1983). Diazinon is not a potent inhibitor of cholinesterase and must be converted into its oxygen analogues (axons), especially diazoxon (diethyl 2-isopropyl-6-methylpyrimidin-4yl phosphate) *in vivo* before it can inhibit cholinesterase. It has been found that diazoxon is about 10,000 times more effective in reducing cholinesterase activity levels than diazinon (Fujii and Asaka, 1982). It is generally agreed that diazinon is metabolized to diazoxon through the action of the liver (McLean et al., 1984). Very few reports are available on the effects of diazinon on nontarget aquatic organisms (Doggell and Rhodes, 1991).

Although compounds such as malathion and diazinon have been shown to be eliminated from the body of the fish in a few days, they are detrimental to fish and other organisms during their short presence inside the body of these animals. According to Sastry and Sharma (1980), alkaline phosphatase activity of the brain was reduced in *Ophiocephalus punctatus* exposed for 96 h to diazinon. Both short- and long-term exposures of malathion and diazinon have been shown to be detrimental for fish (Kanazawa, 1978; Sastry and Sharma, 1980; Dutta et al., 1992a–c; Richmonds and Dutta, 1992a).

Chemically, malathion is known as 0,0-Dimethyl S-(1,2-dicarbethoxy-ethyl) phosphorodithioate and diazinon is 0,0-Diethyl 0-(2-isopropyl-6-methyl-4-pyrimidinyl) phosphorothioate. Commercial grade malathion was found to be more toxic than technical grade malathion to the fish *Tilapia mossambica* (Sailatha et al., 1981). Normally, the commercial grade malathion consists of 50% malathion (active ingredient), 33% organic solvent, and 17% inert ingredients; the commercial grade diazinon contains 25% active diazinon, 57% aromatic petroleum derivative solvent, and 18% inert ingredients.

In the United States the generally established tolerance residue for malathion is 8 ppm for a wide range of fruits and vegetables and for postharvest application to grains in storage (Spiller, 1961). For diazinon the established tolerance residue ranges from 0.1 to 0.75 ppm.

Cook et al. (1976) reported malathion concentrations ranging from 0.08 to 500 $\mu g/l$ in some surface waters. Previous researchers have made an assessment of the patterns of effects of organophosphate pesticides using one or two measures. This limited number of measures cannot provide an indepth and comprehensive picture of the effects of pesticides on fish. Therefore, a composite measure from different disciplines should be taken into account when we formulate the tolerance limit which induces ecologically significant changes in structures, physiology, and behavior. These changes decrease the ability of an animal (fish) to adapt or survive in its environment (Doving, 1991). The composite measures might include: histopathology, changes in ultrastructures, analyses of blood serum proteins, acetylcholinesterase, and optomotor behavior. Physiology is the closest relative of toxicology. The purpose of toxicology is to study the disturbances that toxic chemicals may cause in physiological functions (Koeman, 1991).

Six major aspects of the effects of pesticides on fish are discussed here. The first concerns changes in the serum proteins due to the action of these pesticides. Blood is the most important and abundant body fluid. Its composition often reflects the total physiological condition. The main route of entry for any pesticide is through the gills. From the gills it is transported to various parts of the body via the blood stream. Blood provides an ideal medium for toxicity studies (Lee, 1969). Electrophoresis of serum proteins and cholesterol has proven to be very helpful in the diagnosis of chemical pollution (Jarvinen, 1971; Dutta et al., 1983; Dutta and Haghighi, 1986; Dutta et al. 1992b; Richmonds and Dutta, 1992a).

Fish blood can serve as a valuable tool in detecting physiological changes taking place in the animal. Polyacrylamide gel electrophoresis is the technique employed for this aspect of study (Richmonds and Dutta, 1992a). Changes in the amount of different serum proteins provide necessary information with regard to changes induced by the toxicant (Menezes and Qasim, 1984). These quantitative changes can be analyzed using a densitometer. Noteworthy research with regard to electrophoretic analyses of various fish sera has been done by Ohkawa et al. (1987), Wester (1991), Dutta et al. (1992b), and Richmonds and Dutta (1992a).

While addressing the impact of pesticides on the behavior of fish we shall consider the effects on the enzymes (acetylcholinesterase) and the related chemical (acetylcholine) which controls motor behavior. This second aspect involves a discussion of the changes in acetylcholinesterase (AChE) activity in fishes exposed to pesticides.

The studies of inputs and stability of organophosphate pesticides in aquatic ecosystems have demonstrated that aquatic organisms are frequently subjected to either periodic or prolonged exposures of low concentrations of the toxicants (Kozlovskaya and Flerov, 1980). In such periodic exposures to sublethal concentrations, a cumulative increase in brain acetylcholinesterase inhibition was noticed. Thus, because of the cumulative effect, prolonged or periodic exposures to organophosphate compounds may have consequences as detrimental to the fish population as acute exposures (Kozlovskaya and Mayer, 1984). Antwi (1987), Galgani and Bocquene (1990), Dutta et al. (1992a), Richmonds and Dutta (1992b), Reddy and Philip (1994), and Martinez-Tabche et al. (1994) have studied various anticholinesterase agents and their effects. Studies of Cook et al. (1976), Reddy and Rao (1988), Galgani and Bocquene (1990), and Reddy et al. (1992) have indicated that AChE measurements are probably the best and most sensitive indicators of organophosphorus poisons. The toxic effects of the organophosphorus compounds result from their ability to severely inhibit the enzyme acetylcholinesterase (Kozlovskaya and Mayer, 1984; Dutta et al., 1992a; Richmonds and Dutta, 1992b).

Abiola et al. (1991) studied the blood cholinesterase activity of field pesticide applicators. Their results showed an inhibition of cholinesterase activity ranging from 5% to 28%. Galgani and Bocquene (1990) studied the inhibition of AChE by organophosphate and carbamates for four marine species and found sensitivity to be highest for fish, compared to shrimp and mussel. Kozlovskaya and Mayer (1984) likewise observed that fish mortality is related to the inhibition of the acetylcholinesterase activity caused by durations and concentrations of organophosphates.

Acetylcholine (ACh) is a neurotransmitter substance which carries the nerve impulse at the synapse, neuromuscular, and neuroglandular junctions. This acetylcholine is hydrolyzed and removed by the enzyme acetylcholinesterase after it has completed its function (Inesterosa and Perelman, 1990; Moya et al., 1991; Richmonds and Dutta, 1992b). The presence of acetylcholine without removal will keep the acetylcholine receptors in continuous stimulation; this produces severe physiological disturbances, eventually leading to tetanic paralysis and death. Bardach and Life (1983) explained the selective toxicity of malathion to certain cells as follows. Selective toxicity of chemicals to cells involves the presence of specific targets or receptor systems in exposed cells. In this case the concentration of the toxicant is the same for all cells but only certain cells are affected. This is due to the specificity of toxicant action on receptors that are normally occupied by endogenous hormonal or neurohormonal substances. In this example acetylcholinesterase is considered to be the receptor. There are several methods of measuring acetylcholinesterase activity (Ellman et al., 1961; Murphy, 1968). The Ellman et al. (1961) method is considered one of the straightforward methods developed for acetylcholinesterase analysis.

In the Ellman method acetylthiocholine (ASCh) is used as the substrate and dithiobisnitrobenzoic acid (DTNB) as the reagent. Thiocholine, formed upon enzymatic hydrolysis of ASCh, reduces DTNB to the yellow anion of thionitrobenzoic acid, whose absorbance is measured with a colorimeter at 412 millimicrons. Ellman's (1961) method has more relative merits than others.

The third part of the discussion below includes behavioral changes of the fish due to the action of these pesticides. Changes in behavior have been suggested for use as a sensitive indicator of chronic sublethal toxicant exposure (Orsati and Colgan, 1987). Some fish behaviors (e.g., locomotor activity and avoidance) are extremely sensitive to pollutant chemicals whereas others (e.g., aggression) seem to be rather refractory (Heath, 1987; Doving, 1991). Rand (1985) published an extensive essay on behavioral methods for fish experiments in toxicological studies and Dantzer (1980) compiled animal models in behavioral toxicology. Warner et al. (1976) declared: "The behavior [or activities] of an organism represents the final integrated result of a diversity of biochemical and physiological processes. Thus, a single behavioral parameter is generally more comprehensive than a physiological or biochemical parameter." Borlakoglu and Kickuth (1990) indicated that very low levels of environmental pollutants could alter the behavioral pattern of an animal.

The observation of significant behavioral changes in bluegills, such as cough, yawn, fin flick and threat, with copper concentrations as low as 0.034 mg/L demonstrated that behavior can be a sensitive indicator of copper stress (Atchison et al., 1987). Other behavior-related studies on fish due to pesticides or chemicals include: Kennedy et al. (1987), Naqvi and Hawkins (1988), Hartwell and Doving (1989), Faber et al. (1989), Ysargil and Sandri (1990), Steele et al. (1990), Baatrup (1991), Dutta et al. (1992c; 1994a), Richmonds and Dutta (1992c), and Campbell and Bettoli (1992).

The fourth, fifth, and sixth parts of the discussion below deal with microscopic, scanning, and transmission electron microscopic examination of the gill, liver, and ovary tissues from fish exposed to these pesticides. Histopathology is regarded to be an indispensable and powerful technique in assessing toxic effects of pesticides on fish. Its value lies not only in sensitivity in terms of toxic levels, but particularly in revealing target organs and mechanisms of actions (Wester et al., 1990; Wester and Canton, 1987, 1991). Further, Wester (1991) indicated that data which he obtained on toxic mechanisms of fish was largely comparable with that obtained in mammals, and therefore fish could serve as an alternative species for mechanism studies.

Histopathological changes in the gills of fishes due to pesticides and chemicals have been reported by Munshi and Singh (1971), Mallatt (1985), Roy et al. (1986), Richmonds and Dutta (1989), Roy and Munshi (1987, 1991), Powell et al. (1992), and Dutta et al. (1993a). Since the gills are the primary route of entry for the pesticide and the liver the organ of detoxification (Mclean et al., 1984; Dutta et al., 1993b), they are preferentially discussed here. Dubale and Shah (1979) reported that malathion was hepatopathic and induced hepatopathy in *Channa punctatus*. Arecchon and Plumb (1990) observed the damage in gill lamellae and the liver by sublethal effects of malathion in *Ictalurus punctatus*. Terrant et al. (1992) reviewed the histopathological effects of pesticides on the liver of fishes. Histopathology of ovaries has been studied by Mani and Saxena (1985), Sukumar and Karpagaganapathy (1992), and Dutta et al. (1994c).

The scanning electron microscope is an important modern tool for revealing the surface ultrastructure of various organs including the gill. It also elucidates modifications of the various gill units in relation to food and feeding habits and the chemical nature of the aquatic environment of the fish with special reference to the architectural plan of the microridges, density and distribution of mucous glands and chloride cells (Munshi and Singh, 1992).

The importance of electron microscopy was observed by Hughes and Munshi (1973) when they revealed that the "pillars" separating the capillaries in the air-breathing organs of *Anabas* are modified epithelial cells. Unfortunately, very few researchers have investigated the effects of organophosphates on the ultrastructures of gills, liver, or ovaries.

Roy et al. (1986) investigated the effect of saponin extract (a biocidal plant) on *Anabas testudineus* by using scanning electron microscopy and found a progressive loss of microridges of the epithelial cells, dissociation of the epithelium, and disappearance of chloride cells. The gill filaments of control fish showed well-defined microridged epithelial cells and the mucous glands were discernible. The overall effect was the reduction of interlamellar space, restricting the flow of water through the gill sieve for respiration. Alteration in gill morphology of *Heteropneustes fossilis* caused by sublethal doses of saponin extract of plants *Mollugo pentaphylla* and *Acacia auriculaeformis* were studied under light and scanning electron microscopy by Roy et al. (1990). Maina (1990) studied the ultrastructure of the chloride cells and their modifications after exposing the extremely hyperosmatic and alkaline-adapted fish *Oreochromis alcalicus grahami* to diluted water. He found that the mitochondria of the chloride cells became widely dispersed, increased in size, and subsequently underwent progressive autolytic changes.

Prasad (1991) conducted an SEM study on the effects of crude oil on the gills and air-breathing organs of climbing perch, *Anabas testudineus*. Roy and Munshi (1991) studied the structural and morphometric changes in gills of a freshwater carp, *Cirrhinus mrigala* (Ham.) after 48-hour exposure to a sublethal dose of malathion. They found 1.25 times increase in the gill area of *Cirrhinus mrigala* but a fivefold increase in water/blood diffusion thickness, leading to four times lesser diffusing capacity. They also observed inflammatory alterations in the lamellar epithelium and hyperplasia caused by short-term exposure under light and SEM.

Avella et al. (1993) studied morphological changes in the gill epithelium of two tropical fishes, *Oreochromis aureus* and *O. niloticus*, exposed to three types of salinity. The study revealed an increased number of chloride cells (CC). The ultrastructure of CC was not altered after transfer to 20% salinity. SEM of gills revealed epithelial cells with denser and more numerous ridges when external salinity was raised. Dutta et al. (1994b) studied changes in the microridges of the gill lamellae, gill filament, and gill arch of diazinon-exposed bluegills by scanning electron microscope. Dutta et al. (1994d) used transmission electron microscopy to reveal a drastic structural change in the secondary lamellae of *H. fossilis* exposed to malathion, especially in the chloride cells and the lymphatic system.

As the liver is involved in detoxification of pollutants, it merits analysis. Casillas et al. (1983) observed cellular coagulation, necrosis, increased cytoplasmic eosinophilia, and subcapsular necrotic foci in English sole *Parophrys vetulus* injected with 3.0 ml of carbon tetrachloride. Dutta et al. (1993b) found changes in the diameter, cellular coagulation, and necrosis in the hepatocytes of *H. fossilis* exposed to malathion. Wajsbrot et al. (1993) investigated the chronic toxicity of ammonia in the gills, liver, and kidney of juvenile gillhead seabream, *Sparus aurata* and found at 13 mg/L^{-1} TAN (0.74 mg/L^{-1} NH$_3$-H) most of the hepatocytes showed some degree of atrophy, a heterogeneous cytoplasm, and fatty vacuolation.

Fish reproduction is greatly dependent on healthy gonads. Therefore a study of the effects of pesticides on the structures of the ovary is highly warranted.

The effects of organophosphorus and organochlorine insecticides on ovarian growth, steroidogenesis, ovulation, and other aspects of reproduction in various species of teleosts have been investigated (Mani and Saxena, 1985; Haider and Inbaraj, 1988). However, a few researchers have dealt with the ultrastructural abnormalities and histopathology of the ovary caused by organophosphorus pesticides. Wester (1991) found B-HCH-induced excessive vitellogenesis and yolk formation on premature oocytes in juvenile guppies. Wester et al. (1988) observed the effects of hormonal imbalances and stress on the ovaries and testis as a result of exposure to methyl bromide and sodium bromide.

Murugesan and Haniffa (1992) observed histopathological and histochemical changes in oocytes of the air-breathing fish *Heteropneustes fossilis* (Bloch) after exposure to textile-mill effluent. They observed complete karyolysis, disappearance of the chromatin reticulum of the nucleoli, and vacuolation of the cytoplasm. Sukumar and Karpagaganapathy (1992) found fewer mature oocytes, most of which had become atretic in the ovaries of fish exposed to sublethal concentrations of carbofuran.

AUTHOR'S RESEARCH

The results obtained from the multiple measures (six aspects) I used for assessing the effects of organophosphorus pesticides (malathion and diazinon) on freshwater fish such as bluegill *Lepomis macrochirus* and *Heteropneustes fossilis* (an Indian air-breathing fish) are discussed in this section. It is followed by a composite model in which the effects of the aforementioned pesticides on these six aspects and their interrelationship in fish are analyzed. This section comprises six parts: the first covers electrophoretic studies, the second acetylcholinesterase activity, the third discusses optomotor behavioral changes, the fourth histopathology with a light microscope, the fifth magnified structures with scanning, and the sixth ultrastructures with transmission electron microscopy.

Electrophoresis

The electrophoretic technique remains a promising procedure for identifying stressful but sublethal levels of pollution (Bouck and Ball, 1966; Dutta et al., 1983). The methods for electrophoresis have been adopted from Richmonds and Dutta (1992a).

The serum showed five different protein fractions after SDS-gel electrophoresis. The protein fractions of the bluegills were labeled in the order of increasing mobility as fractions 1, 2, 3, 4, and 5 respectively. After exposure to malathion there were significant quantitative changes in fractions 1 and 2, which included globulins and albumin respectively. The increase in quantity of serum protein fraction 1 was significant only at the two highest exposure concentrations, namely, 0.032 and 0.048 ppm (Fig. 1). Linear regression analysis showed a decrease in the second fraction after exposure to different concentrations of malathion. This fraction started to decrease at 0.018 ppm and continued to decline to 0.048 ppm (Fig. 2). Fraction 3 decreased at 0.048 ppm (Fig. 3) but there was no change in fractions 4 and 5 (Figs. 4 and 5).

The five fractions of bluegill serum were identified as globulin, albumin, transferrin, prealbumin 1, and prealbumin 2. This identification was based on the Ney and Smith

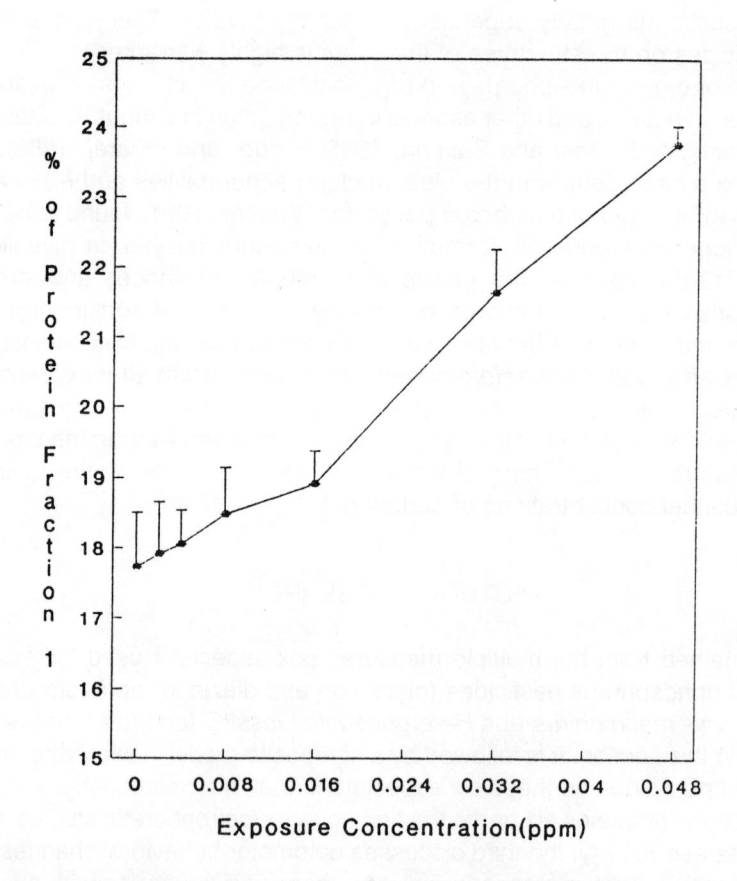

Fig. 1. Percentage of serum protein fraction 1 from the blood of fish—control and exposed to different concentrations of malathion (from Richmonds, 1989).

(1976) findings on bluegill serum. In bluegill transferrin is located anodal to albumin. Fraction 1 is a low mobility serum protein and, according to Menzel (1970), the slowest migrating fraction is a globulin fraction. One reason for the increase in fraction 1 may be the following. Pesticides after entering the circulatory system quickly bind to the blood proteins (Plack et al., 1979). This may result in the recognition of these pesticides by the immune system as foreign. This type of response by the immune system may be manifested as an increase in the serum protein fraction 1. Damage to the gill epithelium was identified upon exposure to malathion (Richmonds and Dutta, 1989; Dutta et al., 1993a). This damage to the epithelium may result in some infection, which in turn may induce an immune response. This immune response will increase the quantity of fraction 1, that includes the globulin.

Ney and Smith (1976) suggested that the second fraction in the bluegill serum contained albumin. In this study fraction 2 showed a general trend of decrease in quantity. This decrease can be explained on a functional basis. Albumin is supposed to have three main functions: osmotic regulation of blood volume, an easily available protein reserve, and a transport protein (Anderson, 1979).

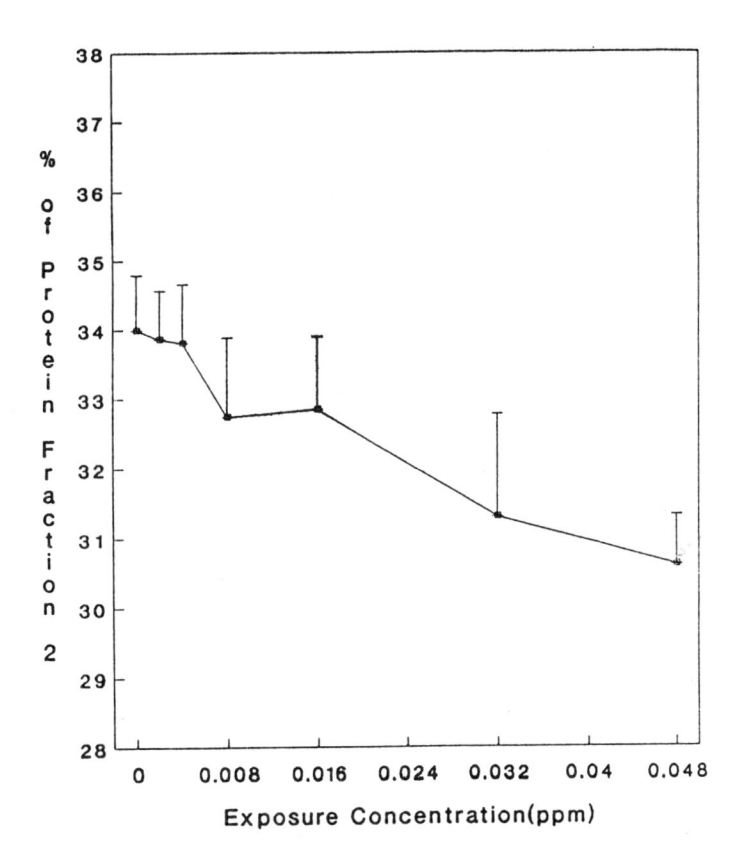

Fig. 2. Percentage of serum protein fraction 2 from the blood of fish—control and exposed to different concentrations of malathion (from Richmonds, 1989).

Hyperactivity was observed around an exposure concentration of 0.016 ppm. This hyperactivity may have caused a utilization of the easily available protein reserve fraction containing albumin resulting in a decrease in its quantity. Shrinkage of the nuclei and vacuolization in the cells of the liver (Dutta et al., 1993b) indicate damage to the hepatocytes which may lead to a possible decreased albumin synthesis. This might be another reason for the lower amount of albumin observed in this study. Transferrin (fraction 3) is mainly responsible for the transport of iron. This fraction showed no significant difference in quantity due to malathion exposure.

There was no significant difference in the amount of fractions 4 and 5 after exposure to various concentrations of malathion. Fractions 4 and 5 (prealbumin) have been shown to be involved in the binding and transport of the thyroid hormones, thyroxine (T_4) and triiodothyronine (T_3) (Cederblad, 1979).

Researchers, such as Bouck and Ball (1966), have shown that changes in the serum proteins may be stress-related due to methods of capture and also by subsequent procedures for holding fish. It is suspected that changes in the serum proteins may also be due to the direct effects of chemicals or pesticides. However, it is very difficult to differentiate

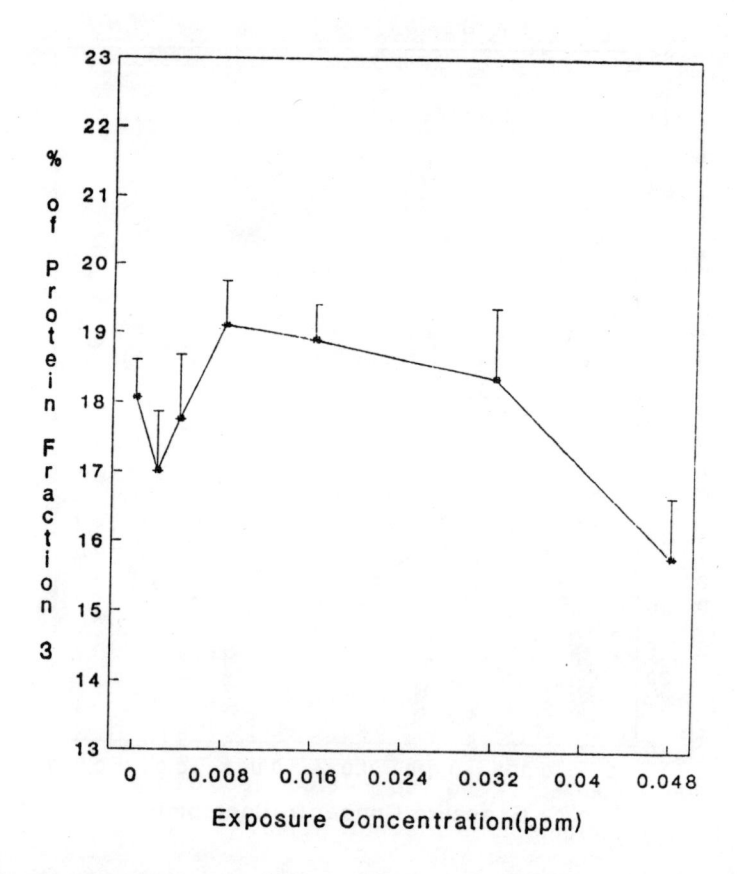

Fig. 3. Percentage of serum protein fraction 3 from the blood of fish—control and exposed to different concentrations of malathion (from Richmonds, 1989).

whether the changes are in fact due to stress or a direct effect of chemicals or both. In this study, however, one should plan to maintain similar environmental conditions for control as well as pesticide-exposed animals and thus stress other than pesticide exposure will be equitable in both situations. Electrophoretic analysis of serum protein fractions revealing an increase in serum protein fraction 1 containing globulin and decrease in serum protein fraction 2 containing albumin can be used to detect environmental contamination caused by toxic substances such as malathion.

Acetylcholinesterase Activity

Anticholinesterase insecticides are widely used in agricultural pest control and pose potential hazards to both wildlife and mankind. Measurement of acetylcholinesterase (AChE) activity is useful in establishing the degree of pollution caused by anticholinesterase agents used as pesticides. Organophosphorus compounds severely inhibit the enzyme acetylcholinesterase (O'Brien et al., 1974; Richmonds and Dutta, 1992b; Martinez-Tabche et al., 1994). Intermittent or extended exposure of fish to sublethal concentrations of anticholinesterase compounds is more likely to cause a cumulative toxicity due to the relative

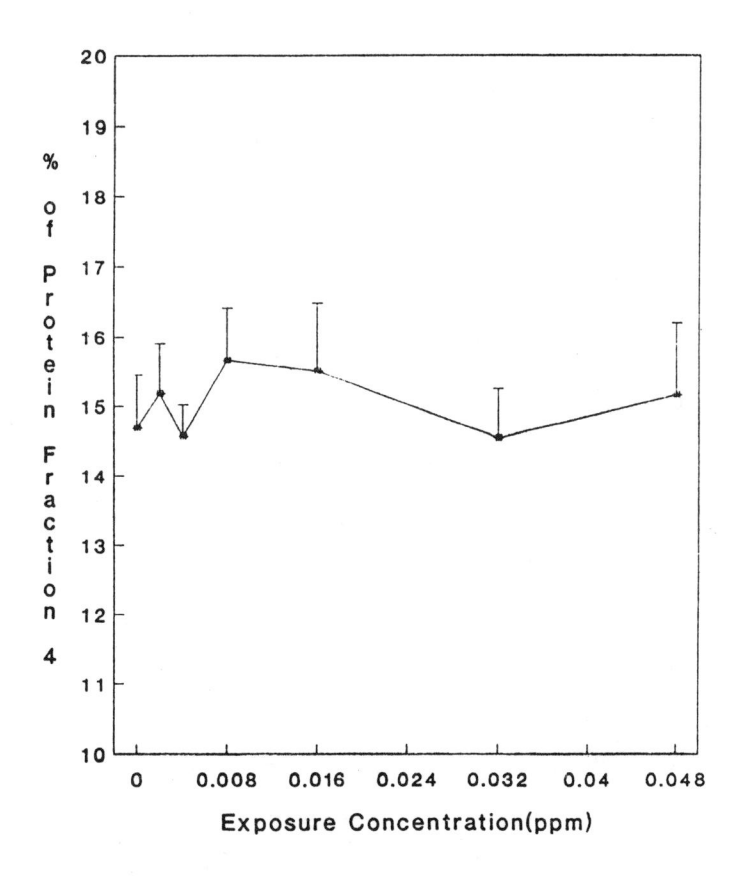

Fig. 4. Percentage of serum protein fraction 4 from the blood of fish—control and exposed to different concentrations of malathion (from Richmonds, 1989).

irreversibility of AChE inhibition. In almost every instance of malathion and diazinon poisoning there has been a general reduction of cholinesterase activity levels, especially in the brain and blood (Mantz, 1983; Dutta et al., 1992a).

In the present study, effects of both malathion and diazinon pesticides on brain acetylcholinesterase activity were studied. The method for this study was that followed by Dutta et al. (1992a) and Richmonds and Dutta (1992b). Acetylcholinesterase activity showed a significant decrease above the exposure concentration of 0.016 ppm malathion (Fig. 6). A reduction in the mean acetylcholinesterase activity was seen in the juvenile *H. fossilis* at all exposure durations (24, 48, 72, and 96 h) of malathion. But in the case of adults a reduction was seen only at the 72-h exposure duration (Dutta et al., 1995). The studies on diazinon yielded the following results. The test concentrations used were 0.015, 0.030, 0.045, 0.060, and 0.075 ppm. Acetylcholinesterase activity showed a highly significant decrease above 0.045 ppm of diazinon (Fig. 7). There was a significant decrease in the "forward" optomotor activity in concentrations above 0.030 ppm (Fig. 7). Changes in acetylcholinesterase activity induce behavioral changes which are detrimental to the existence of the species. Reddy et al. (1992) found a maximum inhibition of AChE activity in the

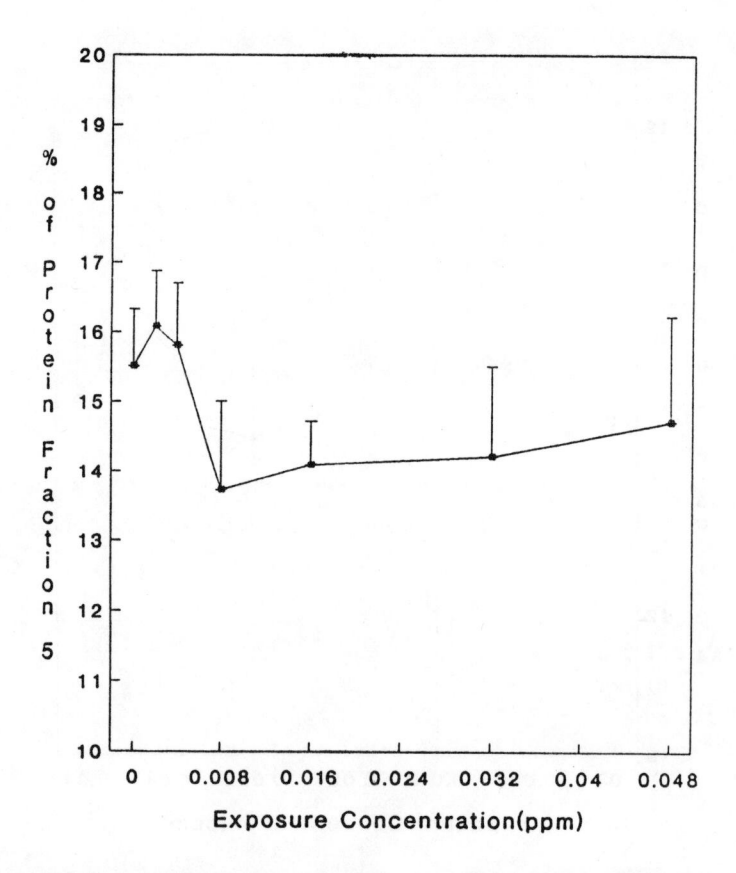

Fig. 5. Percentage of serum protein fraction 5 from the blood of fish—control and exposed to different concentrations of malathion (from Richmonds, 1989).

tissues, such as, gill, brain, liver, and muscle at 48-h exposure to fenvalerate (0.01 ppm, a sublethal dose). The inhibition of AChE and elevation of ACh content may be due to a decrease in the ionic composition of the tissue of *C. carpio* under fenvalerate stress (Malla Reddy, 1988). The greater decrease in AChE with a simultaneous increase in ACh content in the brain tissue is an implication of greater inhibition in the integrative activity of the central nervous system. ACh accumulation in the brain and other tissues may cause uncontrolled hormonal release, which may cause degeneration of many biochemical and physiological functions (Carbett, 1974). Inhibition of AChE and accumulation of ACh at the synaptic junction in fish may lead to behavioral changes creating widespread disturbance in the normal physiology, ultimately causing death of the organism. The physiological and ecological significance of malathion- and diazinon-induced disturbances in AChE of fish may be used as valuable indices for determining environmental pollution by these two or any other pesticides. Pavlo et al. (1992) found that exposure of *Abramis brema* to the organophosphorus pesticide DDVP resulted in decreased food consumption and inhibited AChE activity. Intraperitoneal injection of the cholinergic drug TMB-4 recovered the AChE activity and feeding efficiency. These authors concluded that the cholinergic system in the brain constitutes the biochemical mechanism controlling feeding behavior in fish. Therefore, an examination of changes in acetylcholinesterase and behavior is crucial and should be a part of any toxicological study.

AChE is not only essential for normal behavior, it has an important role in embryonic development. The AChE activity role in somitogenesis was assessed by using diisopropyl-fluorophosphate (DFP), an inhibitor of AChE activity, which led to an abnormal development of somitic mesoderm. This supports the hypothesis that AChE activity has a role in somitogenesis in zebrafish (Hanneman, 1992) and any type of inhibitor can disrupt the normal embryonic development.

Optomotor Behavior

The optomotor response is widespread throughout the animal kingdom and is considered to be extremely important for maintaining position within the habitat and for schooling in fish. Optomotor response, a component of rheotaxis, is a fish's reaction to a current of water, responding to visual tactile and inertial stimuli, resulting from displacement of the

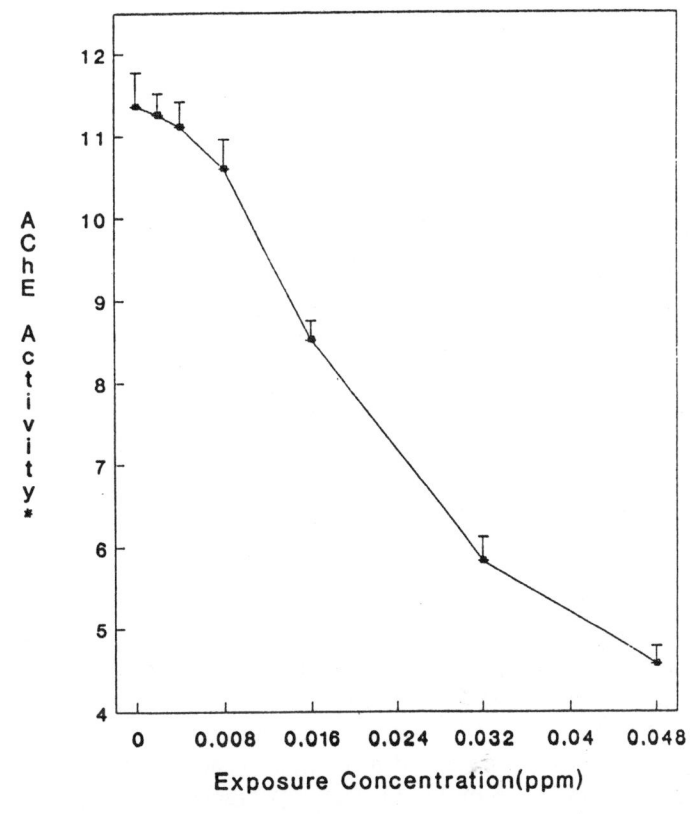

* µM ASCh hydrolyzed/min/gm brain tissue

Fig. 6. Mean brain acetylcholinesterase activity in fish—control and exposed to different concentrations of malathion (from Richmonds, 1989).

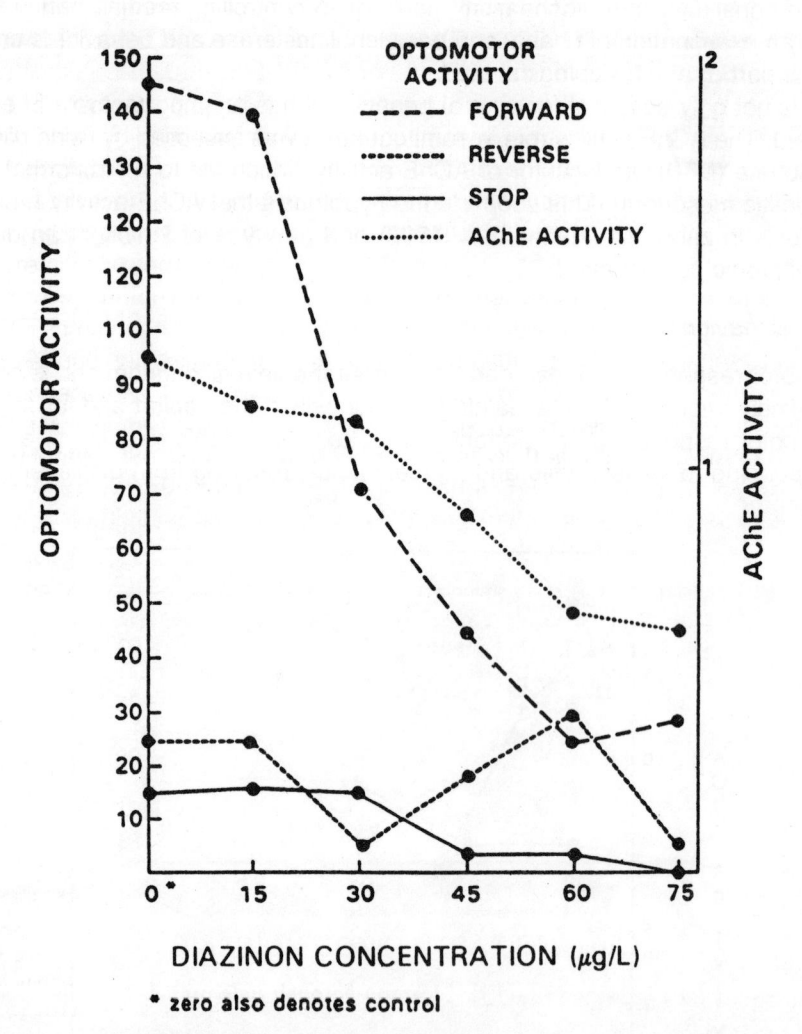

Fig. 7. Mean brain acetylcholinesterase activity and optomotor behavior of bluegill fish—control and exposed to different concentrations of diazinon (from Dutta et al., 1992a).

fish relative to the position of natural landmarks along the streambed and to water flowing over its body (DePeyster and Long, 1993). Clearly, the rheotropic response requires integration of numerous visual, tactile and kinetic cues and responses used by fishes to feed, mate, and avoid predators (Smith, 1984). Any type of disbalance in the rheotropic response brought on by pollution will certainly be detrimental to the existence of the fish. Scherer and Harrison (1979) described a method for detecting impairment of visual orientation. The method of Richmonds and Dutta (1992c) was followed in conducting the experiments on optomotor behavior reported below.

In the case of malathion exposure, the mean score for "following" (clockwise movement) was significantly different at 0.016, 0.032, and 0.048 ppm compared to control, 0.002, 0.004, and 0.008 ppm levels. At 0.016 ppm, the fish became hyperactive, followed by a decline in activity. At the highest concentration (0.048 ppm) the fish became lethargic (Fig. 8). Mean "reversal" (anticlockwise movement) scores during "stripes on" showed more or less the same pattern as in the case of "following" (Fig. 9) but the number of quarter turns was very low compared to "following" during "stripes on." There was not much variation in scores during the "stripes off" periods (Fig. 8). "Stripes on" refers to the period when the turntable is on and "stripes off" when the turntable is stopped (Richmonds, 1989).

In the case of diazinon exposure, a significant difference was exhibited during the "following/stripes on" period at concentrations of 30 μg/L–75 μg/L compared to control. During "reversal/stripes on," except for 15 μg/L and 45 μg/L concentrations, the fish exhibited significant differences in all other concentrations of diazinon. These results differ from those with malathion, wherein all the concentrations during "stripes on/reversal" exhibited the same pattern as in the case of "stripes on/following." In the case of off condition the "following" scores showed significant differences at all concentrations except 15 μg/L and 30 μg/L exposures. These results also differ from those of malathion pesticide exposure in which the "stripes off" period did not show much variation in scores (Figs. 7, 8).

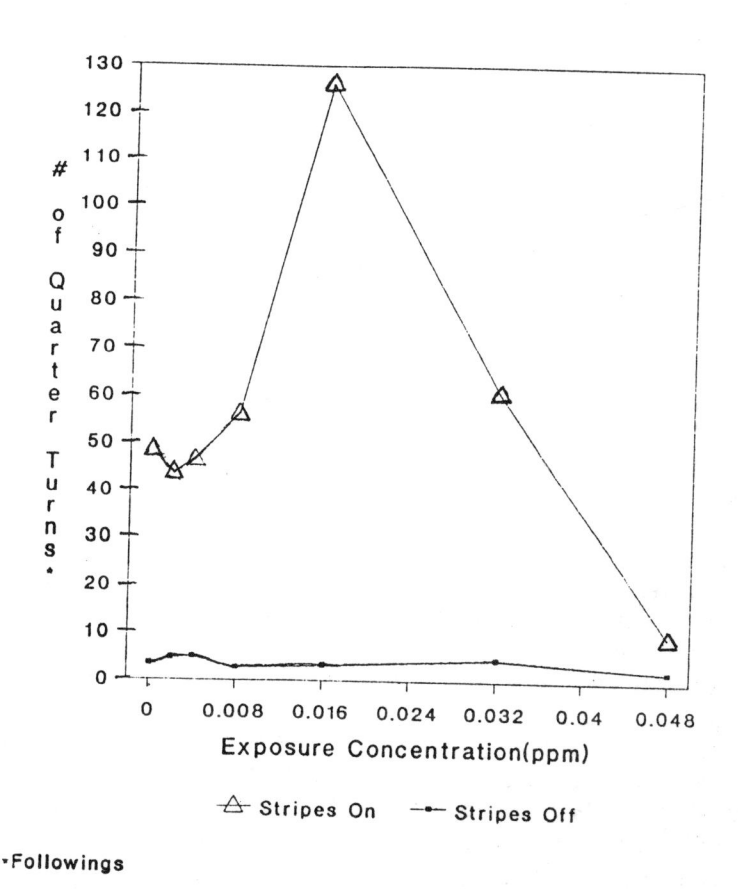

Fig. 8. Effect of malathion on the "following" response of fish (from Richmonds, 1989).

[Figure: Plot with x-axis "Exposure Concentration(ppm)" from 0 to 0.048, y-axis "# of Quarter Turns", showing two curves labeled "Stripes On" (triangles) and "Stripes Off" (circles). Below: •Reversals]

Fig. 9. Effect of malathion on the "reverse" response of fish (from Richmonds, 1989).

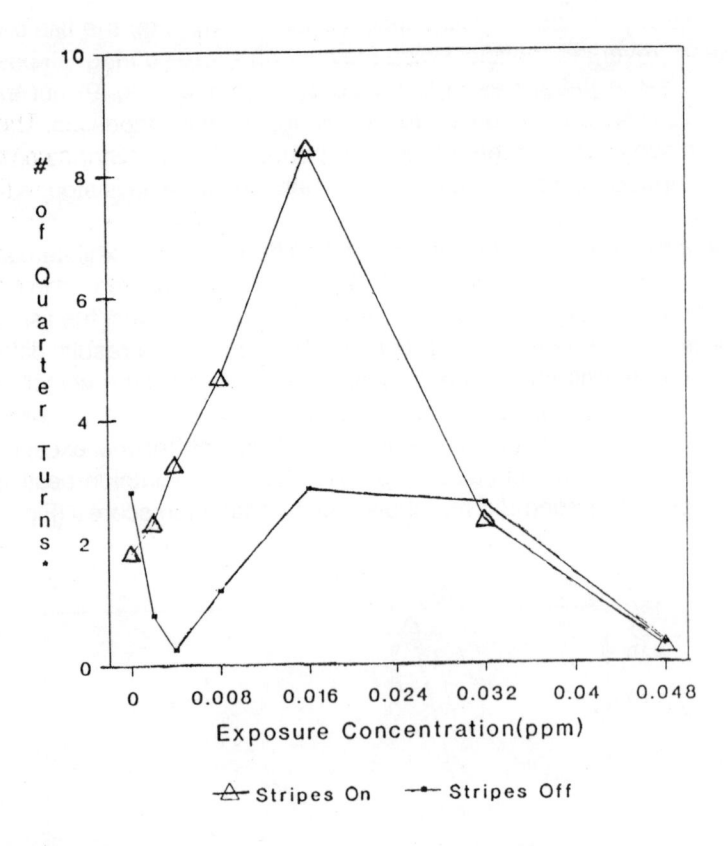

The mean scores of the "following/stripes on" response to different concentrations of diazinon are depicted in Figure 7 and the significant decrease at a concentration of 30 µg/L is readily apparent. This sharp decline in the mean scores of "following" continued with an increase in concentration of diazinon up to an exposure concentration of 60 µg/L. Thereafter the score for "following" showed a small increase.

The mean "reversal" scores during "stripes on" showed a significantly inconsistent pattern. No change was exhibited at a concentration of 15 µg/L. A significant decline in the "reversal" scores was observed at a concentration of 30 µg/L. A significant increase was observed at a concentration of 60 µg/L. This was followed by a decline at a concentration of 75 µg/L (Fig. 7).

After the turntable was turned off, that is "stripes off," the combined mean values of "following" and "reversal" scores exhibited no significant change in activity up to a concentration of 30 µg/L. A sudden decline in activity was observed at 45 µg/L. There was no change in activity at a 60 µg/L concentration, followed by very little activity at an exposure concentration of 75 µg/L.

The optomotor function is essential for behaviors such as searching for food, orienting toward food odor, locating a mate, and avoiding predators. The optomotor responses quantify changes in some of the movement pattern of fish. The optomotor study allowed a quantification of these changes in some of the locomotory pattern of bluegills due to the effects of malathion and diazinon exposure. In the case of malathion 0.016 ppm concentration exposure and in the case of diazinon 45 µg/L concentration, the fish exhibited hyperactivity. Hyperactivity was also noticed by Henry (1984) in bluegills exposed to methylparathion. Around an exposure level of 0.048 ppm malathion and 75 µg/l of diazinon fatigue was observed and the fish became lethargic. This would make them easy prey for any predator.

The results of this study show that exposure of fish to malathion and diazinon in an aquatic environment may not cause their immediate death but may bring about certain behavioral changes or an inability to continue the normal activity pattern essential to maintaining their position in the aquatic environment, to search for food, and to escape from predators. A behavioral bioassay is easily performed. It is quick and sensitive. Changes in behavior are sensitive indicators of pollution by any toxicant.

Histopathology

Histopathology appears to be a very sensitive parameter and is crucial in determining cellular changes that may occur in the target organs such as gills, liver, and ovary. The procedure for histopathology given by Dutta et al. (1993a) was followed.

Gills. Compared to the gill structure of control fish, exposure to 15 µg/L of diazinon produced some mild degenerative changes. Other changes included dilation of blood sinusoids. Gills from fish exposed to 30 µg/L were afflicted with mild forms of hyperplasia. This was accompanied by epithelial necrosis. Epithelial rupture was frequently seen when the dosage was increased to 45 µg/L. Secondary lamellae were shortened and in some places fusion occurred. Extensive lamellar fusion, severe hyperplasia, and mucous cell hypertrophy were found in the gills of fish exposed to 60 µg/L of diazinon. Clavate lamellae and extensive fusion were observed with 75 µg/L exposure. Exposure to 15 µg/L and 30 µg/L produced some mild degenerative changes. The most common of these was lifting of the epithelial layer from the secondary lamellae. This lifting and swelling of the epithelium could serve as a defense mechanism in protecting the internal structures from the contaminated water.

The bulbing of secondary lamellae might be the result of an inflammatory reaction induced by diazinon. In mammals, inflammatory reaction begins with dilation of the blood vessels. As a result of this dilation, there is an increased blood flow to the injured site. During this phase migration of blood plasma and cellular components across the vessel wall and into the surrounding tissue occurs. This leads to the accumulation of blood cells within the vessels.

Mucous cell hypertrophy was also observed as diazinon concentrations increased. Increased mucus secretion and mucous cell hypertrophy can have adverse effects on the respiration of fish. Adverse effects can occur by decreasing microridge surface area. There are three possible functions of epithelial microridges. These include (a) increasing surface area, (b) creating microturbulence to increase gas exchange, and (c) anchoring layers of mucus to the surface of the gill. Therefore, decreasing the microridge surface area, in conjunction with other lesions, may decrease surface oxygen uptake, which could result in hypoxia. Coagulation of mucus on the gills has also been reported to result in disruption of other bodily functions, for example gas exchange, nitrogenous excretion, salt balance, and circulation of blood. Gill alterations such as observed in this study may result in severe physiological problems, ultimately leading to death of the fish (Richmonds and Dutta, 1989; Dutta et al., 1993a).

Liver. The liver is an important organ performing vital functions such as detoxification, synthesis of several components of blood plasma, storage of glucose in the form of glycogen, and release of glucose. The morphological, histological, and histopathological alterations in the liver of pollutant-exposed fish have been studied by various scholars (Mandal and Kulshrestha, 1980; Kulshrestha and Jauhar, 1984; Ahmad and Srivastava, 1985). Their studies showed that these pesticides can cause severe damage to the liver cells. The present study was undertaken to evaluate the histopathological changes caused by a sublethal dose of malathion in the liver of a common catfish, *Heteropneustes fossilis*, with special emphasis on changes in the size of hepatocytes after exposure to malathion.

The normal liver of *H. fossilis* is a large bilobed, orange-colored organ with a homogeneous mass of polygonal hepatic cells with centrally located nuclei and granular cytoplasm. The hepatocytes line the bile canaliculi which open into the hepatic ducts. The hepatic cells are supplied by a fine reticular network of connective tissue (Fig. 10A).

After a 24-h exposure the hepatic cell diameters decreased (Fig. 10B) and were still decreasing after 48 h of exposure. The decrease in diameter was due to shrinkage of the cell. The nuclei became pyknotic and eccentric. There was some degeneration of the cell membrane and vacuolation in the cytoplasm (Fig. 10C). After 72 h of exposure the diameter began to increase. There were vacuoles in the cytoplasm and the nuclei continued to be pyknotic and eccentric. The cell membrane continued to disintegrate (Fig. 10D). After 96 h of exposure the cellular organization was damaged to a greater extent. The hepatic cells lost their normal polygonal shape with eccentric nuclei. The cell membrane showed ruptures and fusion between two or more cells took place, exhibiting binucleate or multinucleate cells at several areas. Some cells became necrotic and complete extrusion of nuclei occurred. Hemorrhage and dilation of blood sinusoids were also noticed although the size of the hepatocyte returned to normal (Figs. 10E, 10F). The histopathological alterations resulting from an exposure to malathion may affect the functional efficiency of the liver, leading to malfunctioning of several organ systems of the fish. This in turn may cause death of the fish, which would eventually cause a change in the population structure (Dutta et al., 1993b).

Ovary. The action of pesticides on gonadal tissue exemplifies sublethal effects which may often remain unrecognized. This study examined the effect of malathion on the ovary of an air-breathing catfish, *Heteropneustes fossilis*, with specific reference to ovigerous lamellae, oocytes at different stages of development, and the nucleus of an immature oocyte. The effect on the level of estrogen in the serum was also noted.

This work was done on fish in the prespawning phase. Therefore, the immature ovary consisted of the ovarian wall (OW), comprising an outer peritoneum, middle connective tissue and an inner germinal epithelium which projects into the ovocoel in the form of ovigerous lamellae. The lamellae has oogonia and a number of oocytes (OC) representing different stages of development (Fig. 11A). The oogonium has a highly stained smooth cytoplasm and a centrally located large nucleus with one nucleolus (NC). The immature oocytes could be classified into different stages of development. Stage I oocytes have a deeply stained homogenous cytoplasm with a large nucleus containing 1–2 nucleoli. Stage III oocytes have a thin layer of follicular epithelium (FE) around the cytoplasm and

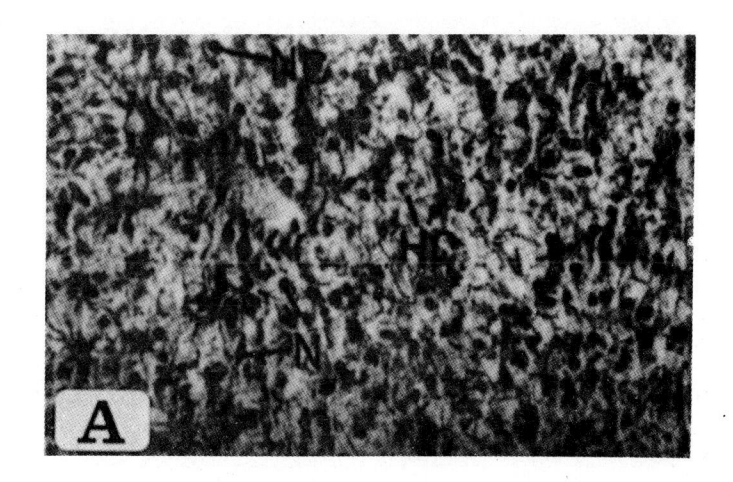

Fig. 10A. T.S. of liver from control *Heteropneustes fossilis*. HC—hepatocyte; N—nucleus; 4 × 40.

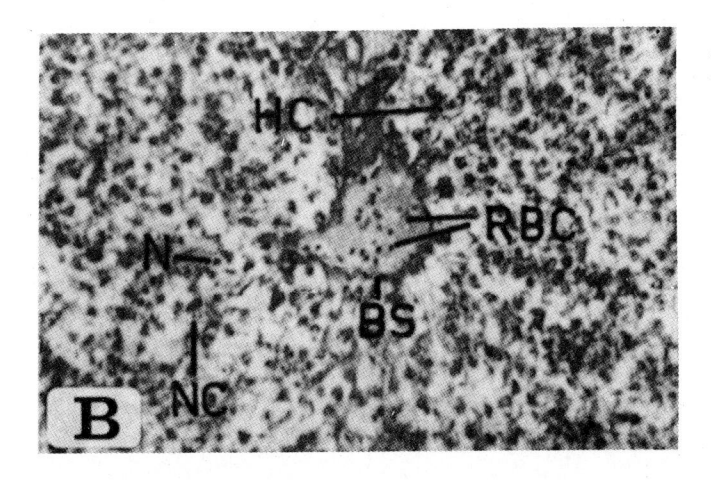

Fig. 10B. T.S. of liver from *H. fossilis* exposed to 1.2 mg/L malathion for 24 h. BS—blood sinusoid; HC—hepatocyte; N—nucleus; NC—nucleolus; RBC—red blood cell; 4 × 40.

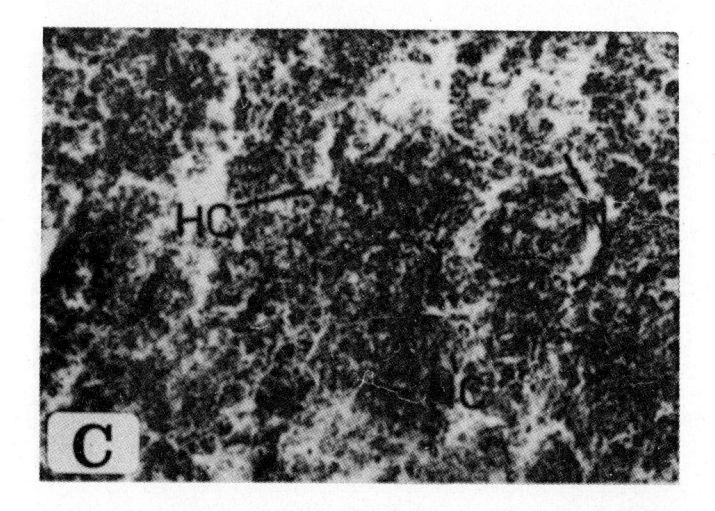

Fig. 10C. T.S. of liver of *H. fossilis* exposed to 1.2 mg/L malathion for 48 h. HC—hepatocyte.

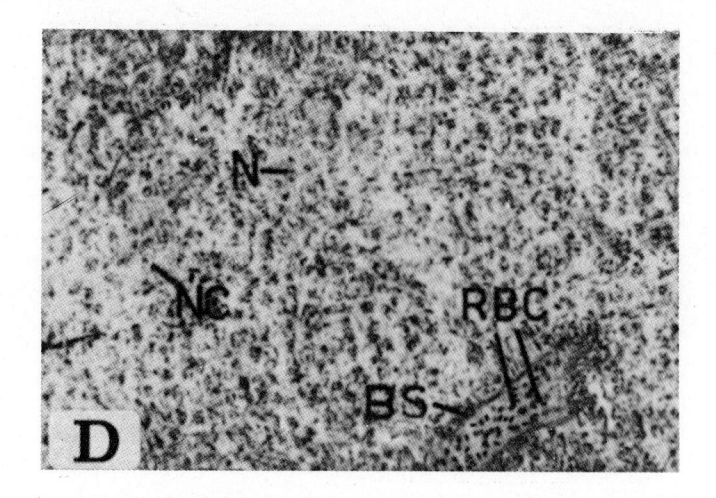

Fig. 10D. T.S. of liver of *H. fossilis* exposed to 1.2 mg/L malathion for 72 h. BS—blood sinusoid; N—nucleus; NC—nucleolus; RBC—red blood cell; 4 × 40.

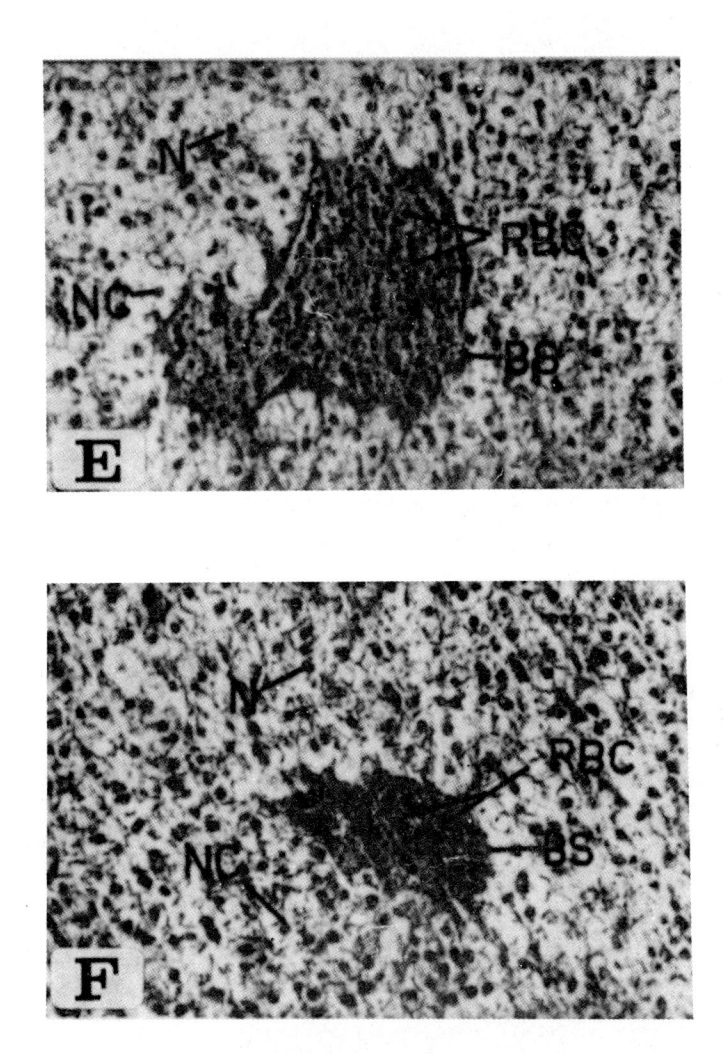

Figs. 10E and 10F. T.S. of liver of *H. fossilis* exposed to 1.2 mg/L malathion for 96 h (from Dutta et al., 1993b). BS—blood sinusoid; N—nucleus; NC—nucleolus; RBC—red blood cell; 4 × 40.

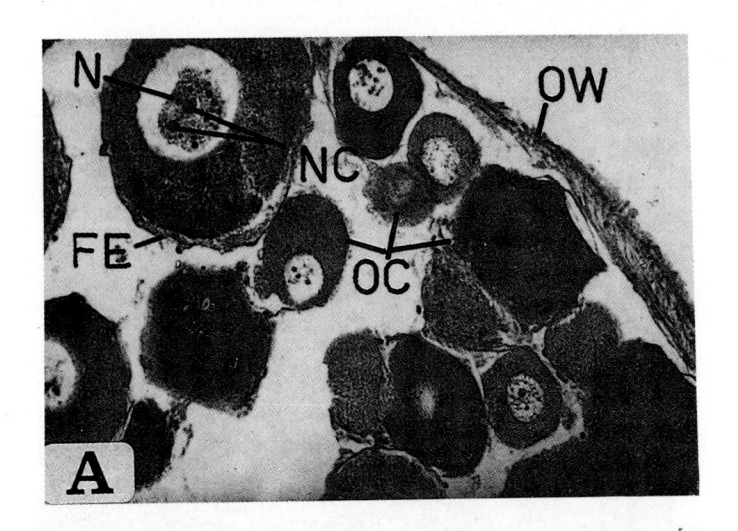

Fig. 11A. Section of an ovary from control *Heteropneustes fossilis* fish.
FE—follicular epithelium; N—nucleus; NC—nucleolus; OC—oocyte; OW—ovarian wall; 4 × 40.

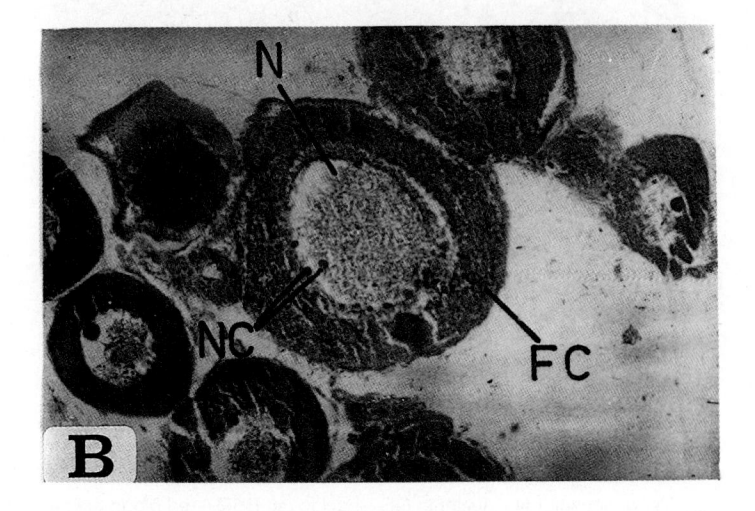

Fig. 11B. Section of an ovary from *Heteropneustes fossilis* exposed to 1.2 mg/L malathion for 24 h.
FC—follicular cell of oocyte; N—nucleus; NC—nucleolus; 4 × 40.
(Source for Fig. 11A, 11B, 11C, 11D and 11E: Dutta et. al., 1994c)

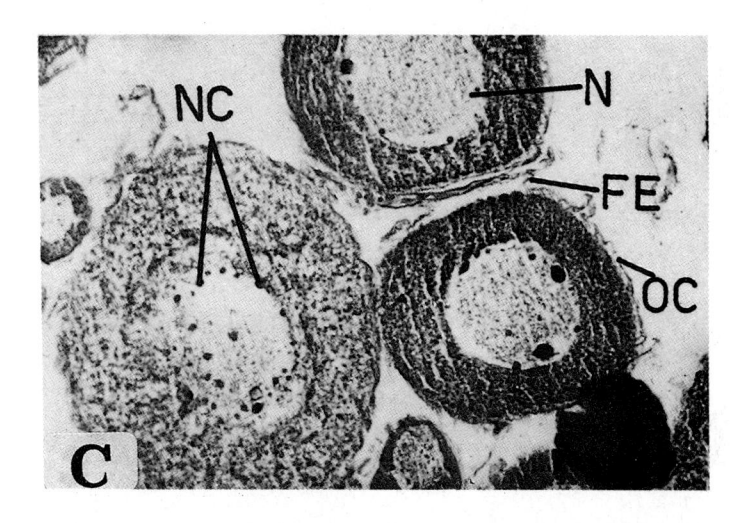

Fig. 11C. Section of an ovary from *H. fossilis* exposed to 1.2 mg/L malathion for 48 h. FE—follicular epithelium; N—nucleus; NC—nucleolus; OC—oocyte; 4 × 40.

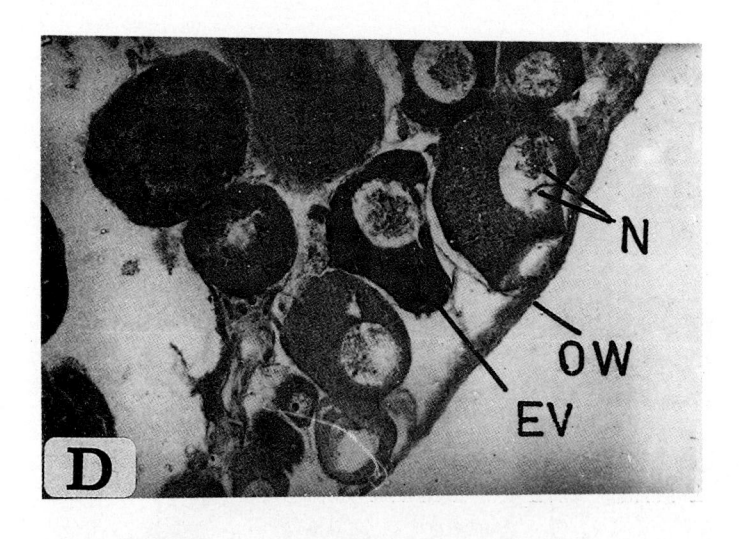

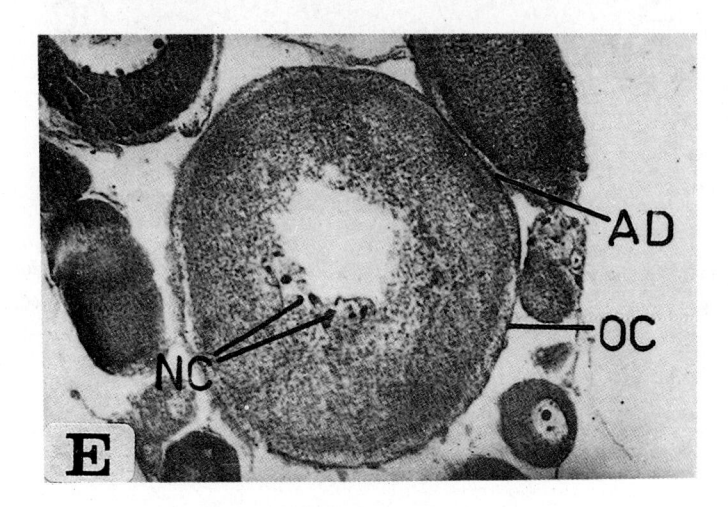

Figs. 11D and 11E. Section of an ovary from *H. fossilis* exposed to 1.2 mg/L malathion for 72 h. AD—adhesion of oocyte; N—nucleus; NC—nucleolus; OC—oocyte; OW—ovarian wall; 4 × 20 and 4 × 40.

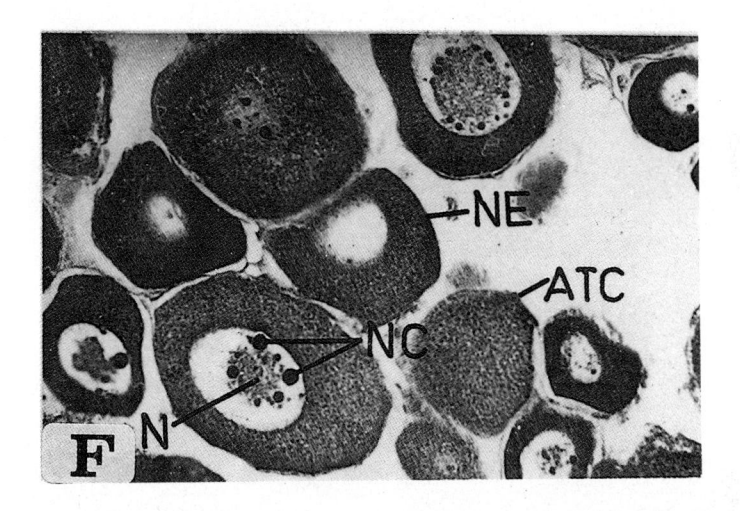

Fig. 11F. Section of an ovary from *H. fossilis* exposed to 1.2 mg/L malathion for 96 h. ATC—atretic cell; N—nucleus; NC—nucleolus; NE—necrotic cell; 4 × 20.

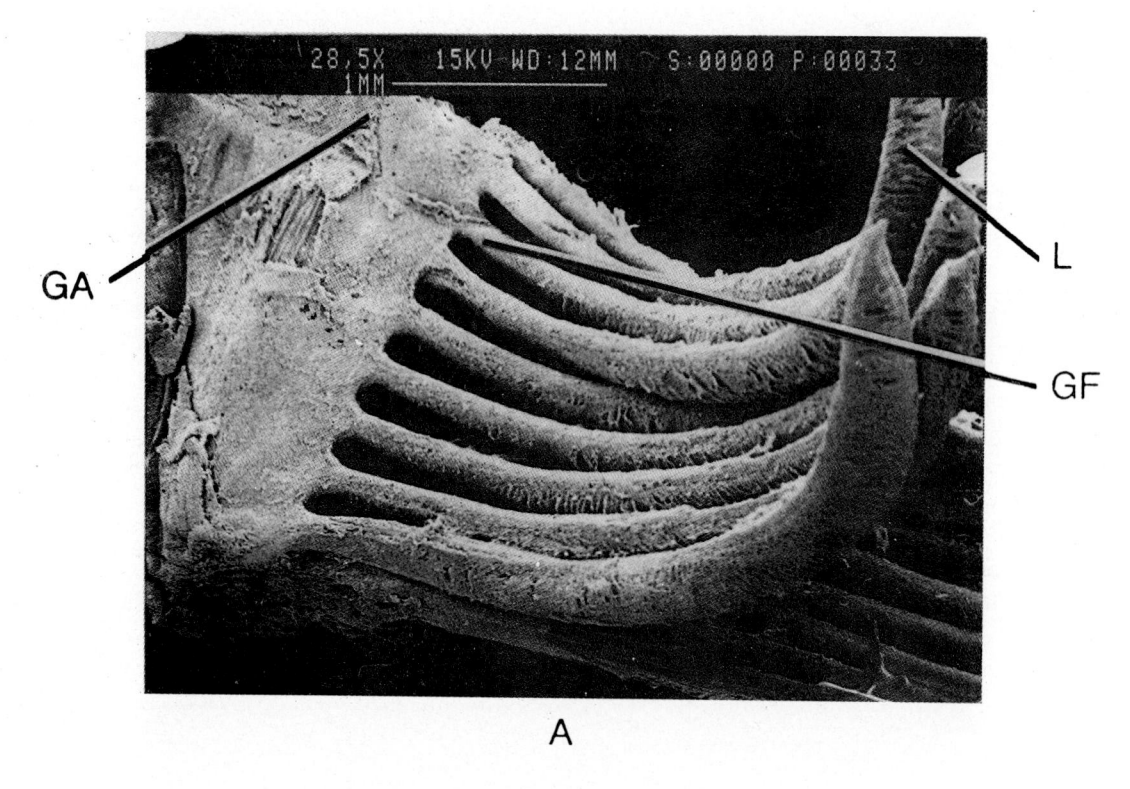

Fig. 12A. Scanning electron micrograph of a gill of control bluegill fish, *Lepomis macrochirus*. GA—gill arch; GF—gill filament; L—lamella.

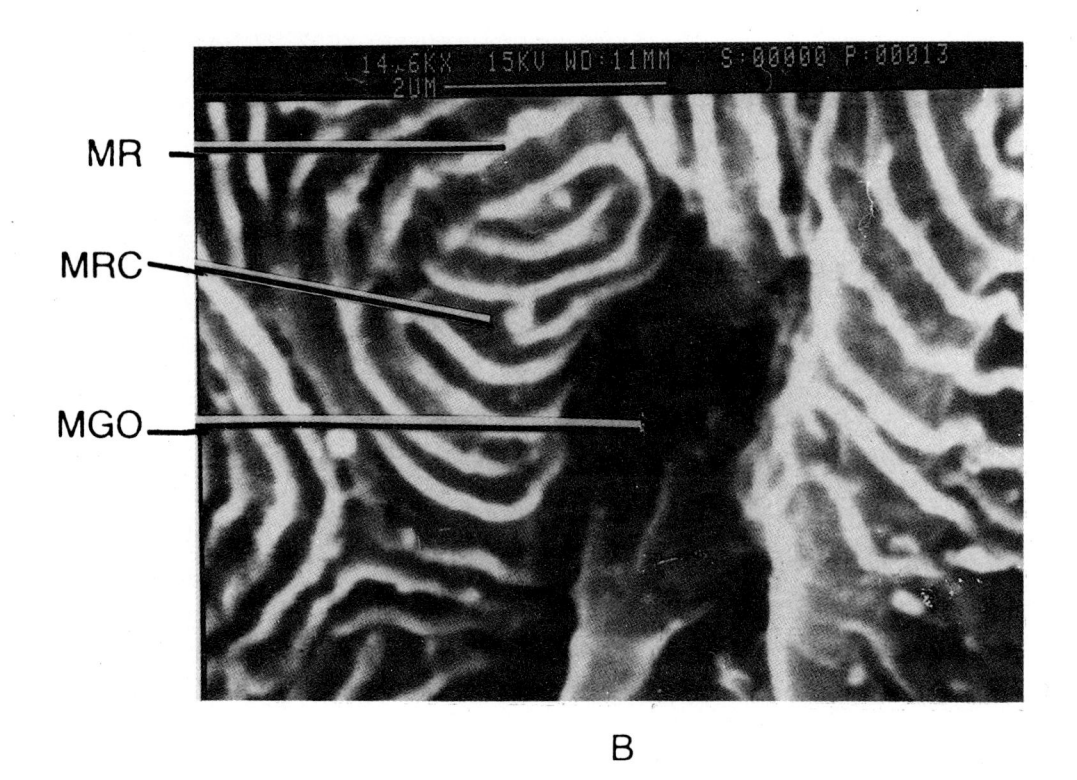

Fig. 12B. Surface of a gill arch of a control bluegill showing microridged epithelial cells and mucous gland opening.

MGO—mucous gland opening; MR—microridge; MRC—microridge cell; ×14,600.

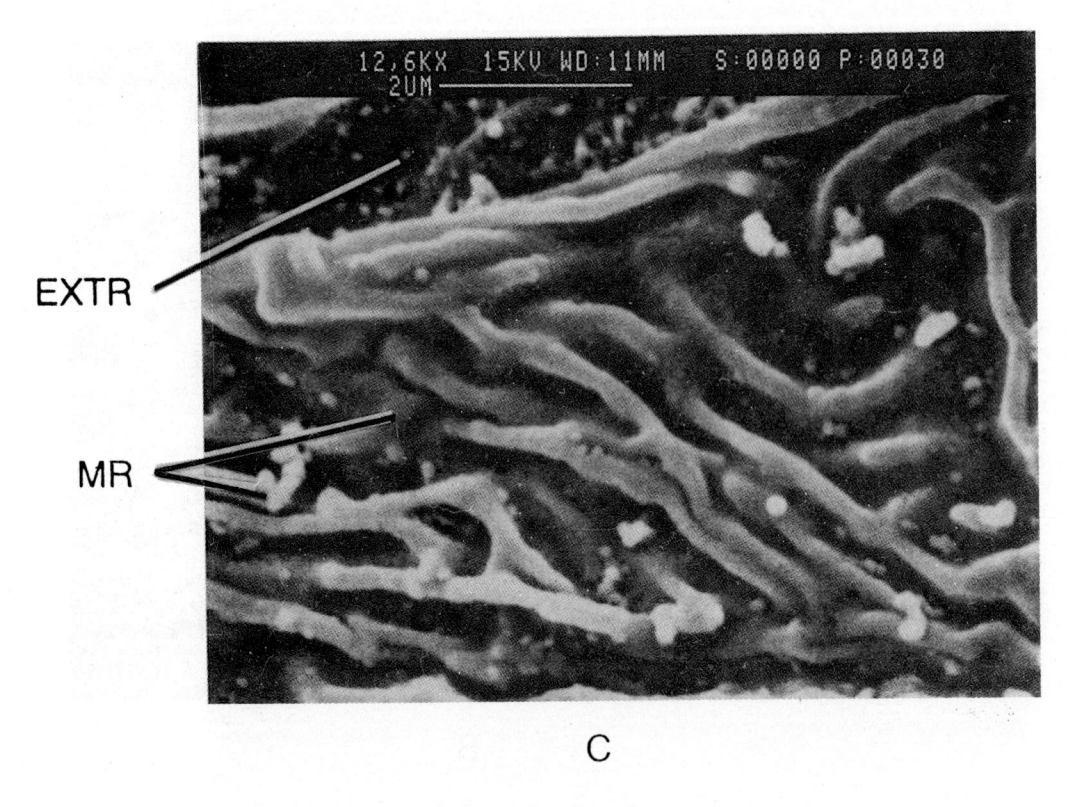

Fig. 12C. Surface of a gill arch of bluegill fish exposed to 15 µg/L diazinon. EXTR—extrusion; MR—microridge; ×12,600.

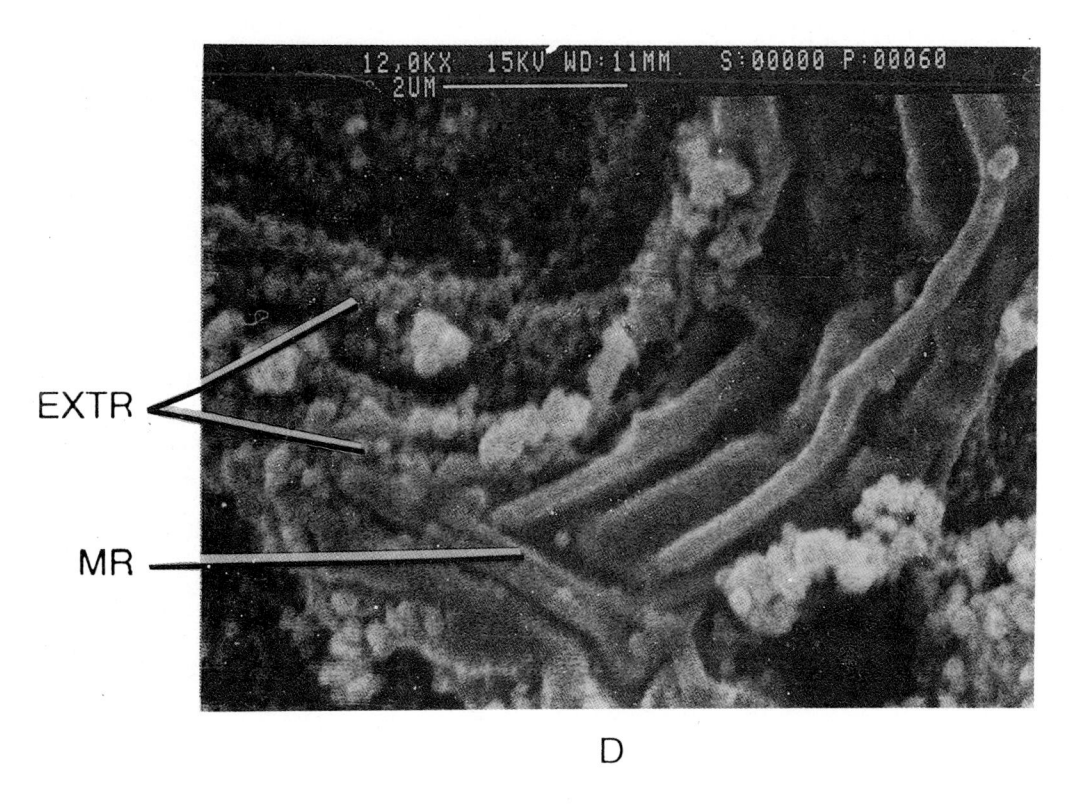

Fig. 12D. Surface of a gill arch of bluegill fish exposed to 30 µg/L diazinon. EXTR—extrusion; MR—microridge; ×12,000.

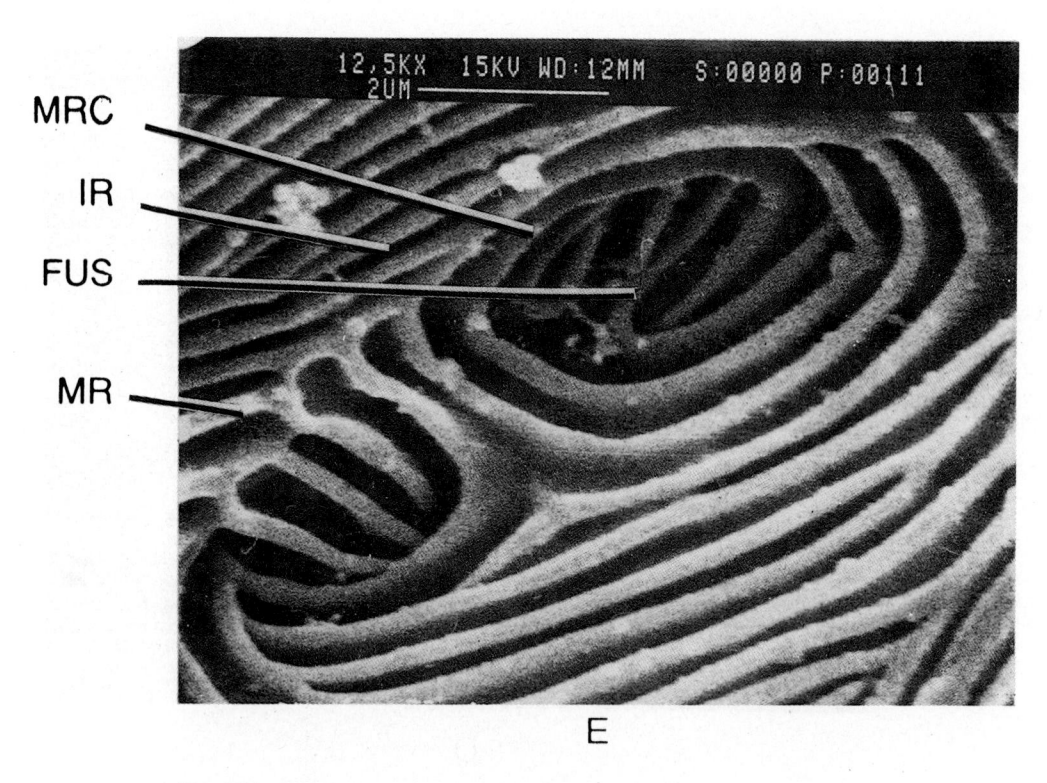

Fig. 12E. Surface of a gill arch of bluegill fish exposed to 60 µg/L diazinon. FUS—fusion; IR—intermicroridge space; MR—microridge; MRC—microridge cell; ×12,500.

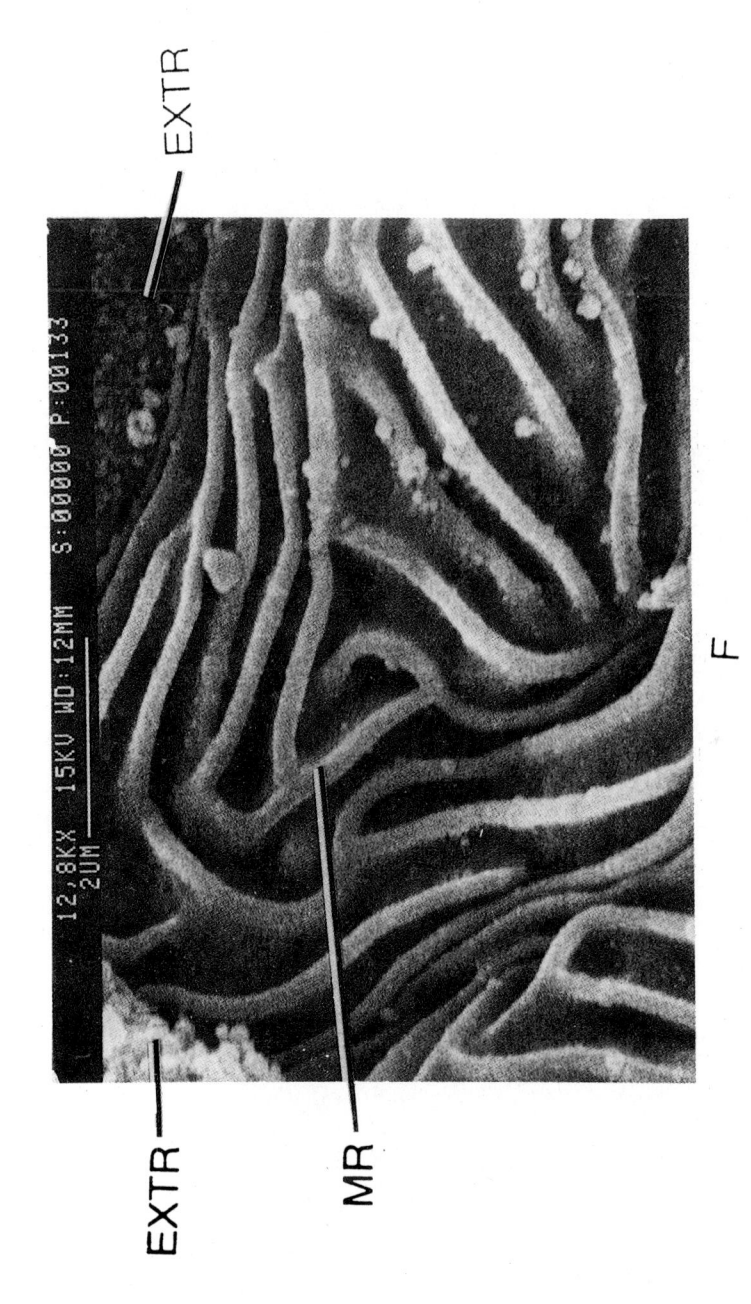

Fig. 12F. Surface of a gill arch of bluegill fish exposed to 75 µg/L diazinon. EXTR—extrusion; MR—microridge; ×12,800.

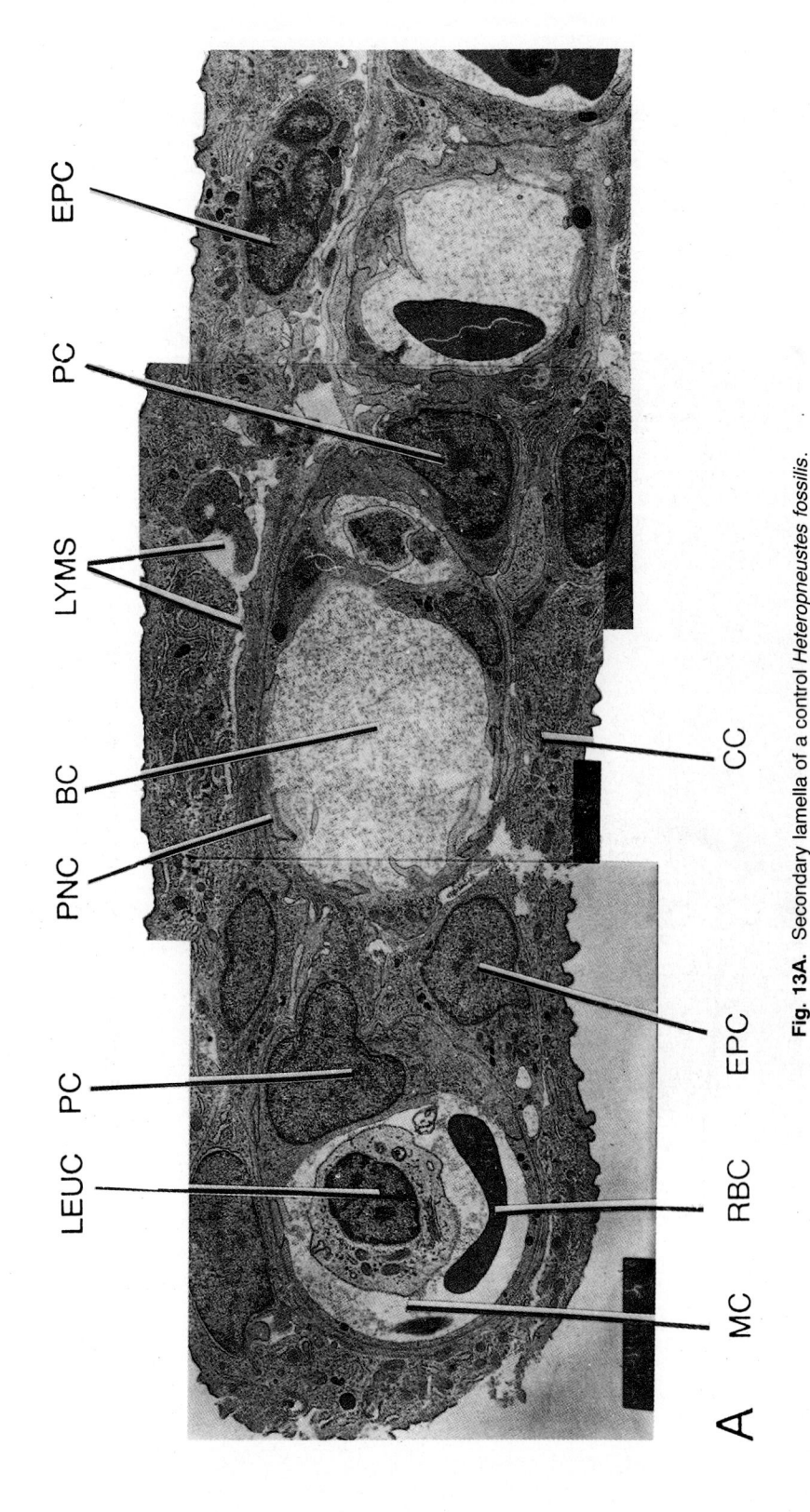

Fig. 13A. Secondary lamella of a control *Heteropneustes fossilis*.

BC—blood capillary; CC—chloride cell; EPC—epithelial cell; LEUC—leukocyte; LYMS—lymphatic space; MC—marginal capillary; PC—pillar cell; PNC—pinocytosis; RBC—red blood cell; × 6,000.

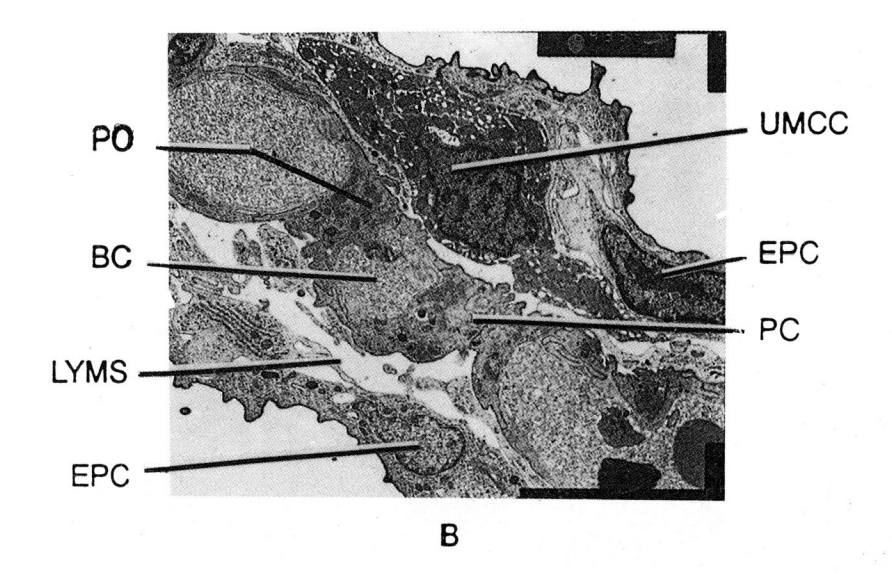

Fig. 13B. Secondary lamella of 24 h in 6 mg/L malathion-exposed fish.
BC—blood channel; EPC—epithelial cell; LYMS—lymphatic space; PC—pillar cell; UMCC—upward movement of chloride cell.

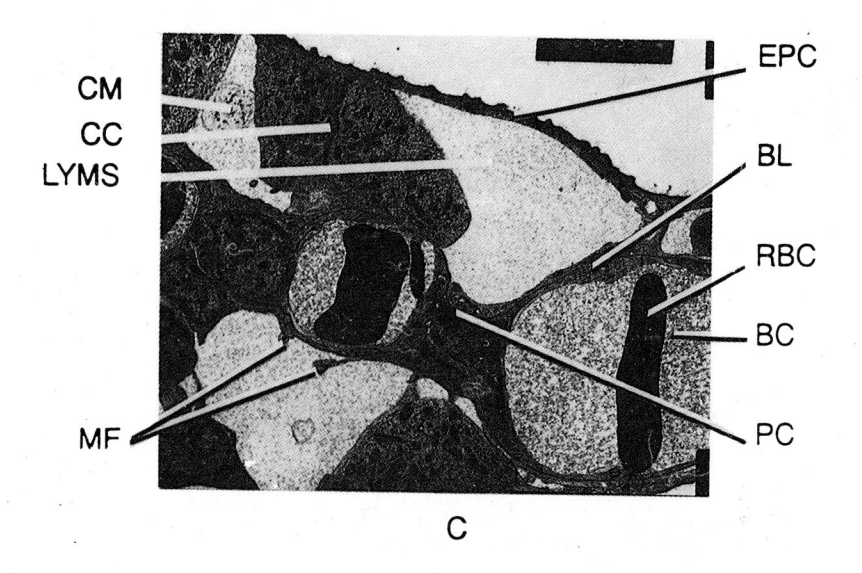

Fig. 13C. Secondary lamella of 48 h in 6 mg/L malathion-exposed fish.
BC—blood channel; BL—basal lamina; CC—chloride cell; CM—cytoplasmic masses; LYMS—lymphatic space; MF—marginal fold; PC—pillar cell; × 6,000.

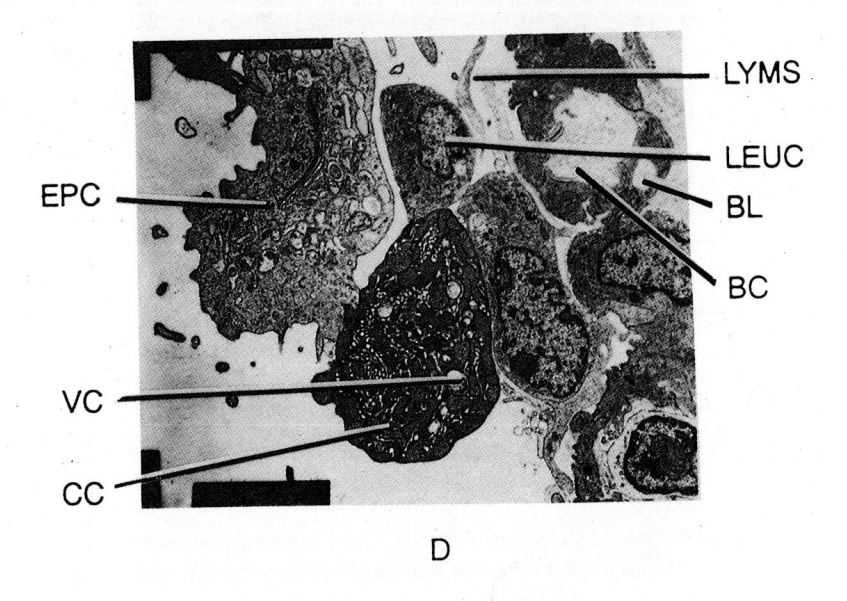

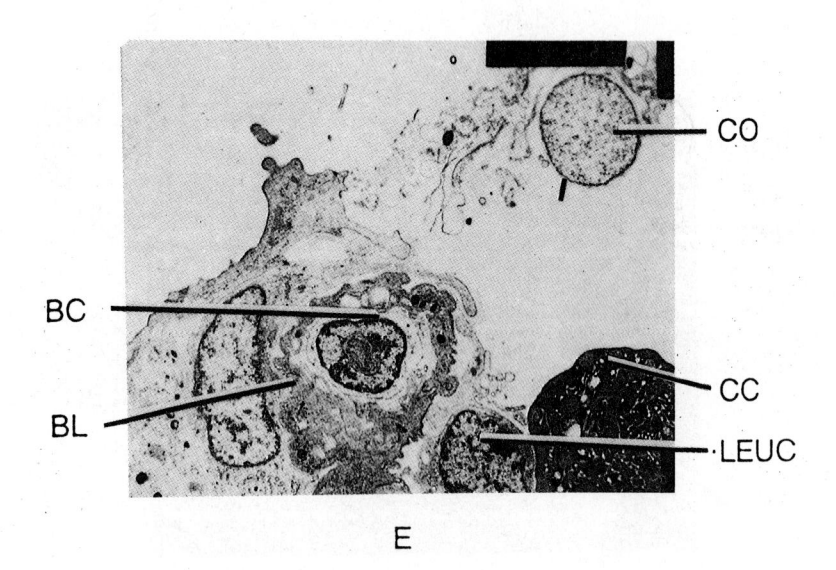

Figs. 13D and 13E. Secondary lamella of 72 h in 6 mg/L malathion-exposed fish.
BC—blood channel; BL—basal lamina; CC—chloride cell; CO—cell organelle; EPC—epithelial cell; LEUC—leukocyte; LYMS—lymphatic space; VC—vacuole of chloride cell; × 6,000.

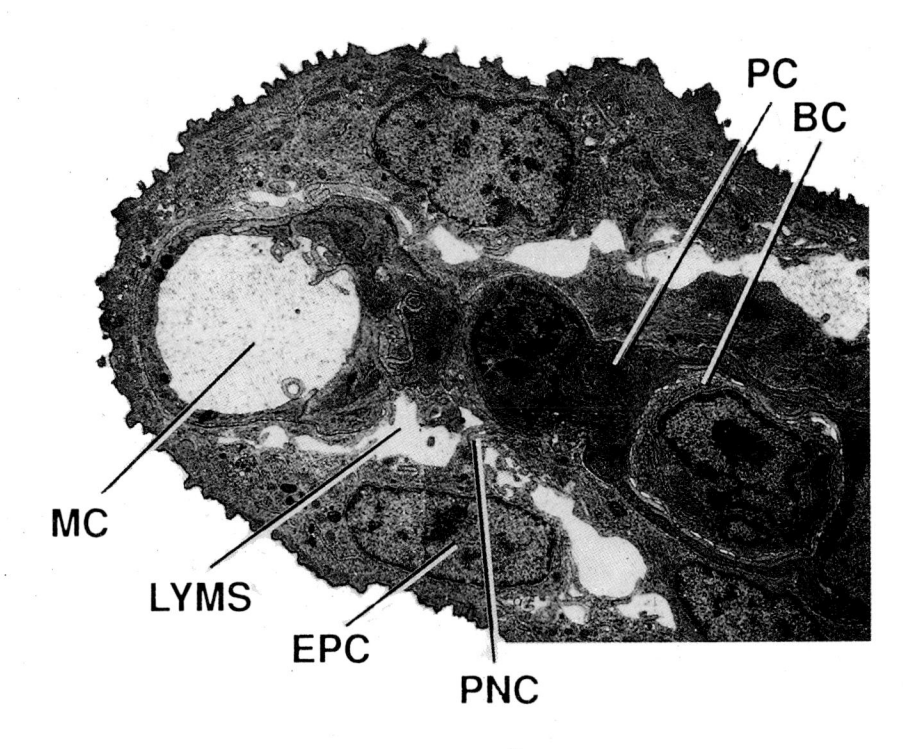

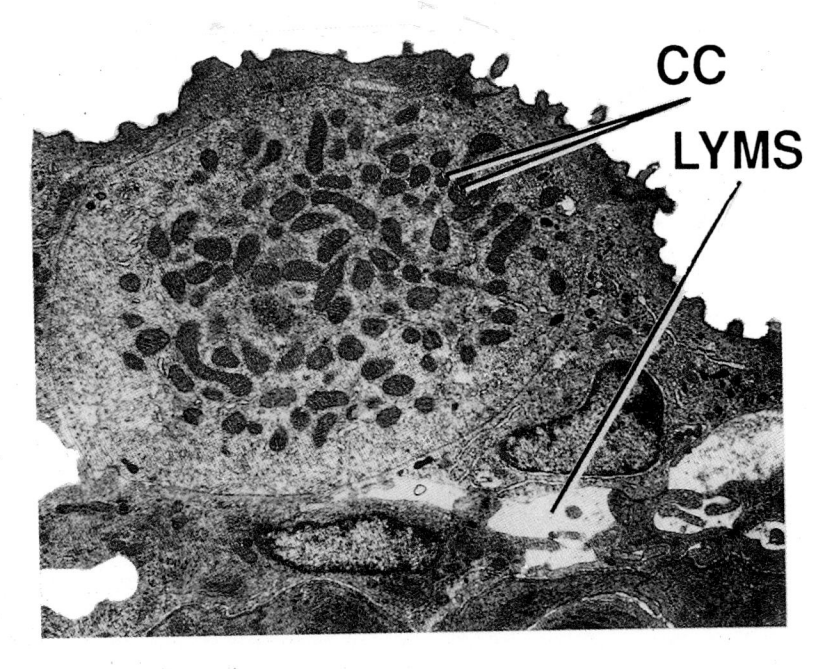

Fig. 13F. Secondary lamella of 96 h in 6 mg/L malathion-exposed fish.
BC—blood channel; CC—chloride cell; LYMS—lymphatic space; PNC—pinocytosis; × 6,000.

an increased number of nucleoli. A few nucleoli were also observed in the cytoplasm of the oocyte. A few atretic oocytes (ATC) were seen but no mature stage IV oocytes were visible (Dutta et al., 1994c).

Clumping of the cytoplasm in the oocytes along with some degeneration in follicular cells (FC) (Fig. 11B) occurred after 24 h of exposure, and 48-h exposure caused more clumping and the cytoplasm became granular. The number of nucleoli increased (Fig. 11C). The degenerative changes were still in process after 72 h. The oocyte shape became deformed with an eccentric nucleus (Fig. 11D). Some oocytes became devoid of nucleus. Oocytes of different stages started to adhere to one another (Fig. 11E). A 96-h exposure showed more degeneration in the oocytes. The nuclear materials of all the oocytes shrank to a smaller lump. Fusion occurred between the oocytes and the follicular epithelium ruptured (NE). A few atretic oocytes were visible (Fig. 11F). Oocytes I were more damaged than oocytes III and II in all the exposure periods.

The radioimmunoassay indicated the estrogen level in control fish to be 96 ng/ml, which came down to 73 ng/L of serum after malathion exposure for 72-h. A 72-h exposure caused the maximum damage to ovarian oocytes (Dutta et al., 1994c).

It is evident from this study that the effects of sublethal exposure of malathion occur after 24 h but are more pronounced after 72 h. Malathion may cause fewer viable eggs. This may have an impact on population dynamics of species in contaminated areas (Dutta et al., 1994c).

Electron microscopic study

Scanning electron microscopic study of gills. Scanning electron microscopy is an important modern tool for revealing the surface ultrastructure of various organs including the gills. The gills of fish are a vulnerable organ to be affected by any toxicant dissolved in water, because fish have to pass a large amount of water over the gills for respiration. Therefore, the present study dealt with the effect of different concentrations of diazinon on the gill morphology of a bluegill fish, *Lepomis macrochirus*. Gill arch and lamellae were taken as a case study. Procedures for SEM and TEM were adopted (with some modifications) from Robinson et al. (1987) and Wischnitzer (1981).

Each gill is composed of (a) inner gill arch with pyramid gill rakers and (b) gill filaments with lamellae (Fig. 12A). The surface of the gill arch is covered by microridged (MRC) epithelial cells amidst which mucous glands are present (MGO) (Fig. 12B).

Treated gill. With 15 µg/L of diazinon exposure, the gill arch cell boundaries are not prominent and the microridges have lost their normal pattern (Fig. 12C). The loss of microridges after fish exposure to different toxicants has also been observed by other authors (Roy et al., 1986; Roy and Munshi, 1991). Some RBCs were visible, suggesting some vascular damage along with some dead cellular debris. Several droplets of oozing material were seen, imparting a pimpled appearance to the microridged cells (EXTR).

With 30 µg/L exposure, only a few ridges are visible on the gill arch epithelium. The ridges developed a pimpled appearance due to extrusion of cytoplasmic materials (Fig. 12D). With 45 µg/L exposure, some amount of normalcy in the microridges of the gill arch is visible. With 60 µg/L exposure, the microridges on the gill arch are swollen with reduced interridge spaces. At several places the ridges are fused (Fig. 12E). With 75 µg/L, the gill arch exhibits cellular extrusion at a few places (Fig. 12F).

It is evident that even the sublethal dose of diazinon affected the gill morphology. Gills have two different reactions to toxicants. The first reaction includes necrosis and rupture and are known as direct deleterious effects (Temmink et al., 1983). These effects are related to dose and are caused by two mechanisms: autolysis induced by the cells' own enzyme and rapid lysis caused by direct action of the toxicant (Mallatt, 1985). The second reaction includes epithelial lifting, fusion, hypertrophy, hyperplasia, and mucus secretion, mucous and chloride cells proliferation. These are defense responses of the fish (Mallatt, 1985; Dutta et al., 1994b).

In this study mostly the second group of changes in the gill were recorded except for a few ruptures. Organic xenobiotics such as pesticides pass quickly through the gills and contaminate other organs through the blood stream. But the gills have the capacity to metabolize and eliminate the xenobiotics. The oozing of some intercellular material from the surface of the cells after exposure to low doses (15 μg/L and 30 μg/L) in the present study may be the eliminating mechanism for the absorbed and metabolized toxicants. Mucus hypersecretion was observed, which is meant for checking the entrance of toxicants. At higher doses (45 μg/L upward) the extrusion became less and cells were found swollen. The ridges of the gill arch fused at 60 μg/L. But at lower doses the ridges looked pimpled and perforated. The pimples of the ridges seem to be vesication of plasma membrane at the tips, which breaks and appears perforated. Similar changes in the microridges were observed by Kudo and Kimura (1984) in the gills with bacterial infection. Lamellar swelling is a defense mechanism as it reduces the water space (interlamellar space). Lamellar swelling seems to be due to epithelial lifting, hyperplasia and formation of lymphatic spaces.

It is evident that fish gills possess many structures which aid them in remaining safe. In case of breakage in one barrier, other defensive responses, such as hyperplasia, lifting of epithelial cells, and thickening of the blood capillary basal lamina, occur. The branchial responses that serve to slow the entry of toxicants may have the undesirable effect of threatening to suffocate the fish (Richmonds and Dutta, 1989; Dutta et al., 1993a). Severe damage, such as necrosis and rupture of gill epithelium and microridges, results in hypoxia and failure in respiration (Prasad, 1991).

This study therefore confirms that pesticides such as diazinon have a damaging effect on the gills, the life sustaining structures of fish, which may cause a difficult condition for survival (Dutta et al., 1994b).

Transmission electron microscopic study of normal and treated gills. Transmission electron microscopy provides a clear and indepth understanding of normal ultrastructures in the gills as well as changes in them.

The normal gill lamellae exhibit well-defined epithelial cells, pillar cells, and central blood vessels as well as a marginal capillary. A very narrow lymphatic space is visible (Fig. 13A). With malathion (4 mg/L) exposure for 24 h, the lymphatic space (LYMS) becomes more apparent between the pillar cell system and the epithelium. The double-layered epithelium becomes slightly disarrayed. There is an upward movement of chloride cells (UMCC) (Fig. 13B). At a 48-h exposure, the outer epithelial cells become stretched like a thin boundary wall. The lymphatic spaces become very enlarged and a large amount of plasma extrude accumulates in them. The chloride cells become enlarged and come into contact with the basal lamina (BL) of the central blood canal (BC). The chloride cells also come into contact with the lymphatic space and the exterior. Several marginal folds (MF) come out from the pillar cell flanges into the lymphatic space. Fragments of some cytoplasmic masses (CM) of cells are visible in the lymphatic spaces (Fig. 13C).

After 72 h of exposure the lamellar epithelium is found in a much distorted condition. The lymphatic space continues to dilate. The epithelium ruptures and many spherical bodies, cytoplasmic organelles (CO), along with some chloride cells packed with large vesicles extrude. This shows extrusion of some cytoplasmic content (Figs. 13D and 13E). The basal lamina (BL) becomes very thick, leading to the constriction of the central blood channels, which can cause vascular stasis. No erythrocytes are visible in the blood channels but leukocytes (LEUC) occur in large number in the lymphatic space (Figs. 13D and 13E).

With 96 h of exposure, the lymphatic space becomes narrower. A large number of chloride cells are visible. Several pinocytotic vesicles are seen in the lymphatic space. Many marginal folds of pillar cell flanges develop pinocytotic vesicles. The central channels remain thick (Fig. 13F).

This study revealed that ultrastructural changes in the gill structures are initiated even at an exposure of 24 h, but the changes become severe after 72 h of exposure. However, at 96 h of exposure the gills undergo a process of restoration wherein reduction in the size of lymphatic spaces and leucocyte population as well as formation of epithelial cells takes place (Dutta et al., 1994d). Changes in the ultrastructures seem to be related to the process of defense. The defensive structural changes include lymphatic space dilation, leukocyte infiltration, vascular stasis, epithelial lifting, swelling, and chloride cells proliferation. There is a direct deleterious effect of the malathion pesticide which includes necrosis and rupture of the branchial epithelium. The deleterious effect is mostly dose-dependent and more often occurs under lethal rather than sublethal conditions (Mallatt, 1985). But in this case even a sublethal dose of malathion caused deleterious effects at 72 h of exposure, with cellular contents emerging from the lamellae (Dutta et al., 1994d). The death of branchial cells and their rupture could be either by autolysis or by rapid lysis, caused by the direct lytic action of toxicants on cell constituents (Abel, 1976).

Proliferation of chloride cells was observed in this study and corroborates the work of Leino et al. (1987), Fischner-Scherl and Hoffmann (1988), and Kudo and Kimura (1983). The chloride cells are connected with the lymphatic spaces and the vascular bed with their inner ends, whereas their outer surface is exposed to the outer environment through apical pits. The chloride cells are involved in metabolizing and excreting the absorbed malathion. For details concerning the function of chloride cells, see Dutta et al. (1994d). Evidently, that might be the main reason for the surge of these cells at 72 h of exposure when damage to the gills was maximum. The appearance of a large number of leukocytes in the lymphatic spaces is related to the increase in the immune system of the affected fish.

TOWARDS A COMPOSITE MODEL

A descriptive model showing changes in structure, physiology, and behavior in fish exposed to pesticides was constructed (Fig. 14). It represents the way malathion and diazinon affect the structures, physiology, and behavior of fish. Circle 1 represents the malathion and the diazinon. Solid lines show the direct effects of malathion and diazinon contamination while broken lines indicate the subsequent effects.

A Composite Model Showing Changes in Structure, Physiology and Behavior in Fish

Model 1

- 1: MALATHION & DIAZINON CONTAMINATION
- 2: HISTOPATHOLOGY IN GILL BY SEM
- 3: HISTOPATHOLOGY IN GILL BY TEM
- 4: HISTOPATHOLOGY IN LIVER
- 5: HISTOPATHOLOGY IN OVARY
- 6: FORMATION OF NEW SERUM PROTEIN BANDS (QUALITATIVE INCREASE)
- 7: QUANTITATIVE INCREASE IN GLOBULIN FRACTION
- 8: FRACTION 2 (ALBUMIN)
- 9: CHANGES IN OPTOMOTOR BEHAVIOR
- 10: CHANGES IN ACETYL-CHOLINESTERASE ACTIVITY

→ Direct Effect of Malathion and Diazinon
⋯▶ Subsequent Effect of Malathion and Diazinon

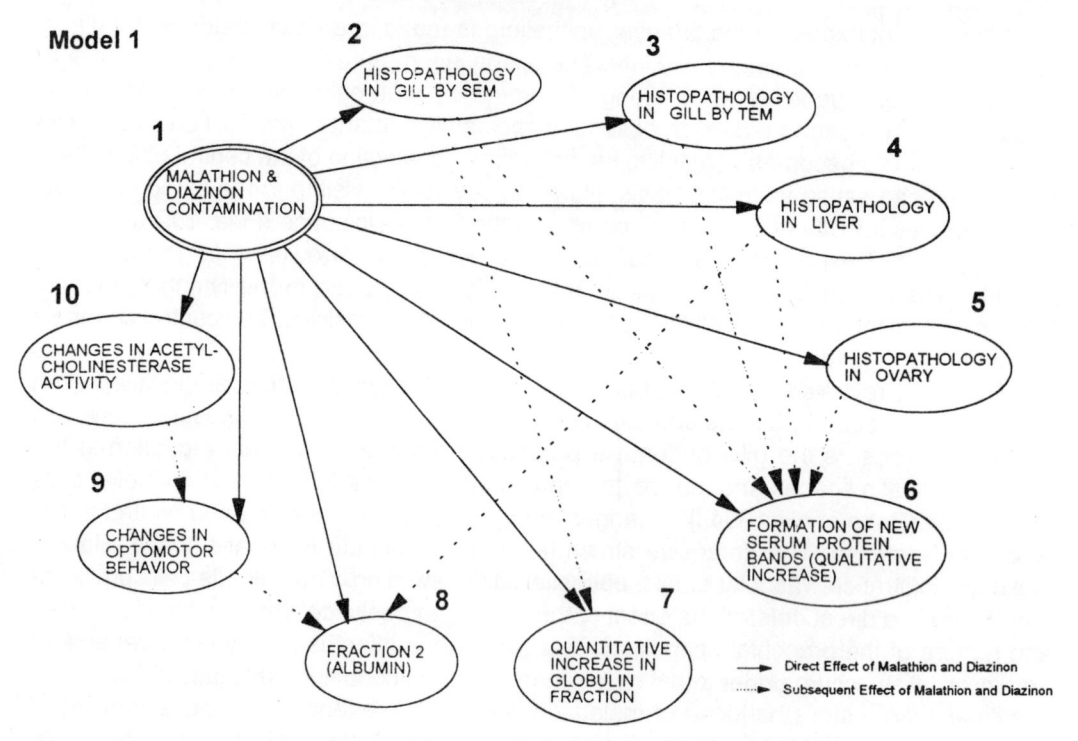

Fig. 14. Represents a model showing changes in structure, physiology, and behavior in fish exposed to pesticides. See text for explanations.

Circle 2 shows some of the results in the structures studied with SEM. The breakdown in microridged epithelial cells, RBCs, and other cellular components are some of the damages which occur in the gill. Circle 3 shows results obtained from TEM, namely, increase in chloride, mucous cells, leukocytes, and macrophages. It also indicates changes in RBCs, blood capillaries, and an increase in lymphatic spaces. Extrusion of cellular components in the gills is also observed. Increase in endoplasmic reticulum, intramitochondrial granules and damage to the mitochondria also occur. Circle 4 represents the histopathology (obtained by light microscopy) in the liver in the form of shrinkage of the nuclei and fatty vacuolation in the hepatocytes. Changes in the hepatocyte diameter also occur. Circle 5 represents the histopathology (obtained by light microscopy) in the ovary in the form of excessive vitellogenesis and yolk formation, complete karyolysis, disappearance of chromatin reticulum of the nucleoli and vacuolation of the cytoplasm. Fewer mature oocytes and several atretic and damaged oocytes in the ovary are also visible. Circle 6 represents the formation of new serum protein bands (qualitative increase). This may be the result of a new protein formation due to the breakdown of red blood cells and other cellular components. Arrows indicate the influence of 1, 2, 3, 4, and 5 on Circle 6. Circle 7 represents the quantitative increase in globulin fraction. This increase may be in response to the immune system. Damage to the epithelial barrier may result in some infections. These infections can induce an immune response which will increase globulin (fraction 1). Arrows indicate the influence from 1 and 2 on Circle 7.

Circle 8 exhibits the formation of fraction 2 (albumin) which showed a trend toward decrease in quantity. Fish exposed to malathion and diazinon are expected to experience severe stress. The decrease in this fraction can be explained on a functional basis. Albumin is thought to have three functions: osmotic regulation of blood volume, an easily available protein reserve, and a transport protein. The hyperactivity caused by these two pesticides may lead to the utilization of this easily accessible protein reserve fraction containing albumin, resulting in a decrease in its quantity. Shrinkage of the nuclei, vacuolation, and breakage in mitochondria and endoplasmic reticulum may result in decreased synthesis of albumin. Arrows indicate the influence from Circles 1, 4 and 9 on Circle 8.

Circle 9 represents changes in the optomotor behavior in fish. Hyperactivity may be observed with certain concentrations of these two pesticides. Hyperactivity is related to the decline in activity of AChE. Optomotor behavior is negatively related to AChE activity. With an increase in the concentration of acetylcholine, fish initially become hyperactive, followed by hypoactivity and lethargy. Lethargic fish are certainly easy prey for any predator and, furthermore, are not able to maintain position when swept downstream. The optomotor function is necessary to escape from predators, search for food, and to maintain position in an aquatic environment. The effect of pesticides on these functions may result in population reduction and changes in community structure and trophic relationships. Arrows indicate the effects of Circles 1 and 10 on Circle 9.

Circle 10 exhibits the changes of the acetylcholinesterase activity. The decline in AChE activity is observeable. Acetylcholine acts as a neurotransmitter and AChE acts as a neuroregulator. Inactivation of this enzyme results in the accumulation of acetylcholine, causing excessive stimulation. It has been established that oxygen analogues of the pesticides are potential inhibitors of AChE in animals. Decrease in brain AChE in animals may cause various physiological and behavioral changes that reduce their survival ability. Arrows from Circle 1 indicate its effects on Circle 10.

This model elicits composite effects including structural, physiological and behavioral changes in fish caused by the pesticides malathion and diazinon. These changes are not isolated as they interact with each other to precipitate the effects generated.

Acknowledgment

This article is the outcome of a research project supported by Smithsonian Institution (USA), no. 0022570000 Appr. 33FT566-00-A20, and was sponsored in India by the University Grants Commission.

The author sincerely thanks Mrs. Linda L. Matz for word processing several drafts of the manuscript. Dr. A. Graham is specially thanked for helping me take the SEM photographs in his laboratory (Department of Biological Sciences, Kent State University, Kent, Ohio, USA). Ms. Jannnette Killius of Northeastern Ohio Universities College of Medicine, Rootstown, Ohio, USA, is thanked for preparation of transmission electronmicrographs. The author also thanks Mr. Larry Ruben, Director of the Kent State University Audio-Visual Center for preparing all the illustrations. Dr. Chelliah Richmonds, of the Case Western Reserve University Medical College, Cleveland, Ohio, USA is specially thanked for granting consent to use diagrams from his dissertation in this article.

References

Abel, P.D. 1976. Toxic action of several lethal concentrations of an anionic detergent on the gills of the brown trout (*Salmo trutta* L.). *J. Fish Biol.* 9:441–446.

Abiola, F.A., P. Houets, and F. Dutta. 1991. Agricultural organophosphate applicators, cholinesterase activity and lipoprotein metabolism. *Bull. Environ. Contam. Toxicol.* 46:351–360.

Ahmad, G. and G.J. Srivastava. 1985. Histopathologic alterations in the liver and skin of a freshwater teleost, *Heteropneustes fossilis* (Bloch), exposed chronically to a sublethal concentration of methylene blue. *Pakistan M. Zool.* 17:239–246.

Anderson, L.O. 1979. Transport proteins. *In* B. Blomback and L.A. Hanson (eds.), *Plasma Proteins*, pp. 44–54. John Wiley and Sons, New York.

Ansari, B.A. and K. Kumar. 1988. Diazinon toxicity: Effect on protein and nucleic acid metabolism in the liver of Zebra fish, *Brachydanio rerio* (Cyprinidae). *Sci. Total Environ.* 76:63–68.

Antwi, L.A.K. 1987. Fish head acetylcholinesterase activity after aerial application of temephos in two rivers in Bukina Faso, West Africa. *Bull. Environ. Contam. Toxicol.* 38:461–466.

Arecchon and J.A. Plumb. 1990. Sublethal Effects of Malathion on Channel Catfish, *Ictalurus punctatus*. *Bull. Environ. Contam. Toxicol.* 44:435–442.

Atchison, G.M., G. Henry, and M.B. Sandheinrich. 1987. Effects of metals on fish behavior: a review. *Environ. Biol. Fishes*, 18:11–25.

Avella, M., J. Berhaut, and M. Bornancin. 1993. Salinity tolerance of two tropical fishes, *Oreochromis aureus* and *O. niloticus*. 1. Biochemical and morphological changes in the gill epithelium. *J. Fish Biol.* 42:243–254.

Baatrup, E. 1991. Structural and functional effects of heavy metals on the nervous system, including sense organs of fish. *Comp. Biochem. Physiol.* 100C(1/2):253–257.

Bardach, J.E. and T.S. Life. 1983. Pollution effects on fish. *In* J.R. Pfaffin and E.N. Ziegler (eds.), *Encyclopedia of Environmental Science and Engineering*. Gordon and Breach Science Publishers, New York (2nd ed.), 893 pp.

Borlakoglu, J.T. and R. Kickuth. 1990. Behavioral changes in *Gammarus pulex* and its significance in the toxicity assessment of very low level of environmental pollutants. *Bull. Environ. Contam. Toxicol.* 45:258–265.

Bouck, G.R. and R.C. Ball. 1966. Influence of capture methods on blood characteristics and mortality in the rainbow trout (*Salmo gairdneri*). *Trans. Amer. Fish. Soc.* 95:170–176.

Brown, J.A., P.H. Johansen, P.W. Colgan, and R.A. Mathers. 1987. Impairment of early feeding behavior of largemouth bass by pentachlorophenol exposure: A preliminary assessment. *Trans. Amer. Fish. Soc.* 116:71–78.

Campbell, G.J. and P.W. Bettoli. 1992. Behavioral reactions of fishes exposed to unbleached Kraft Mill effluent. *Bull. Environ. Contam. Toxicol.* 49:157–164.

Carbett, J.R. 1974. *The Biochemical Mode of Action of Pesticides*. Academic Press, New York.

Carpenter, P.L. 1975. *Immunology and Serology*. W.B. Saunders Company, Philadelphia (3rd ed.), 254 pp.

Casillas, E., M. Myers, and A. Warren. 1983. Relationship of serum chemistry values to liver and kidney histopathology in English sole (*Parophrys vetulus*) after acute exposure to carbon tetrachloride. *Aquatic Toxicol.* 3:61–78.

Cederblad, G. 1979. Plasma protein involved in haem metabolism and in transport of metals, hormones, and vitamins. *In* B. Blomback and L.A. Hanson (eds.), *Plasma Proteins*, pp. 94–117. John Wiley and Sons, New York.

Cook, G.H., J.C. Moore, and D.C. Coppage. 1976. The relationship of malathion and its metabolites to fish poisoning. *Bull. Environ. Contam. Toxicol.* 16:283–290.

Dantzer, R. 1980. Modelés animaux en toxicologie compartmentale. *Sci. Tech. Anim. Lab.* 5:178–187.

DePeyster, A. and W.F. Long. 1993. Fathead minnow optomotor response as a behavioral endpoint in aquatic toxicity testing. *Bull. Environ. Contam. Toxicol.* 51:88–95.

Dhar, M. and K. Gapal. 1991. Neurobehavioral changes in freshwater fish *Channa punctatus* exposed to fenitrothion. *Bull. Environ. Contam. Toxicol.* 47:455–458.

Doggell, S.M. and R.G. Rhodes. 1991. Effects of a Diazinon formulation on unialgal growth rates and phytoplankton diversity. *Bull. Environ. Contam. Toxicol.* 47:36–42.

Doving, K.B. 1991. Assessment of animal behavior as a method to indicate environmental toxicity. *Comp. Biochem. Physiol.* 100C(1/2):247–252.

Dubale, M.S. and P. Shah. 1979. Histopathological lesions induced by Malathion in the liver of *Channa punctatus*. *Ind. J. Exp. Biol.* 17:693–697.

Dutta, H.M. and A.Z. Haghighi. 1986. Methyl mercuric chloride and serum cholesterol level in the bluegill (*Lepomis macrochirus*). *Bull. Environ. Contam. Toxicol.* 36:181–185.

Dutta, H.M., S.B. Lall, and A.Z. Haghighi. 1983. Methyl mercury induced changes in the serum proteins of bluegills. *Lepomis macrochirus. Ohio J. Sci.* 83:119–122.

Dutta, H.M., J. Marcelino, and C. Richmonds. 1992a. Brain acetylcholinesterase activity and optomotor behavior in bluegills, *Lepomis macrochirus*, exposed to diazinon. *Arch. Intern. de Physio., de Biochim. et de Biophys.* 100:331–334.

Dutta, H.M., J.V.V. Dogra, N.K. Singh, P.K. Roy, S.S.T. Nasar, S. Adhikari, J.S.D. Munshi, and C. Richmonds. 1992b. Malathion induced changes in the serum proteins and hematological parameters of an Indian Catfish, *Heteropneustes fossilis* (Bloch). *Bull. Environ. Contam. Toxicol.* 49:91–97.

Dutta, H.M., S.S.T. Nasar, J.S.D. Munshi, and C. Richmonds. 1992c. Malathion induced changes in the optomotor behavior of an Indian carp, *Labeo rohita. Bull. Environ. Contam. Toxicol.* 49:562–568.

Dutta, H.M., C.R. Richmonds, and T. Zeno. 1993a. Effects of diazinon on the gills of bluegill sunfish *Lepomis macrochirus. J. Environ. Toxic. Oncol.* 12(4):219–227.

Dutta, H.M., S. Adhikari, N.K. Singh, P.K. Roy, and J.S.D. Munshi. 1993b. Histopathological changes induced by malathion in the liver of freshwater catfish, *Heteropneustes fossilis* (Bloch). *Bull. Environ. Contam. Toxicol.* 51: 895–900.

Dutta, H.M., S.S.T. Nasar, J.S.D. Munshi, and C. Richmonds. 1994a. Behavioral changes in an air-breathing fish, *Anabas testudineus*, exposed to malathion. *Bull. Environ. Contam. Toxicol.* 52:80–86.

Dutta, H.M., J.S.D. Munshi, P.K. Roy, N.K. Singh, and L. Motz. 1994b. Effect of diazinon on gill of bluegill sunfish, *Lepomis macrochirus*: Observation with scanning electron microscope (submitted).

Dutta, H.M., A. Nath, S. Adhikari, P.K. Roy, N.K. Singh, and J.S. Datta Munshi. 1994c. Effect on ovary of *Heteropneustes fossilis* after short-term exposure to sublethal dose of malathion. *Hydrobiologia* 294:215–218.

Dutta, H.M., J.S.D. Munshi, P.K. Roy and J. Killius. 1994d. Malathion induced changes in the respiratory lamellae of a catfish, *Heteropneustes fossilis*, with particular reference to chloride cells. An electron microscopic study (submitted).

Dutta, H.M., J.S.D. Munshi, G.R. Dutta, N.K. Singh, S. Adhikari, and C.R. Richmonds. 1995. Age-related differences in the inhibition of brain acetylcholinesterase activity of *Heteropneustes fossilis* (Bloch) by malathion. *Comp. Biochem. Physiol.* Vol. IIIA(2):331–334.

Ellman, G.L., K.D. Courtney, V. Andres Jr., and R.M. Featherstone. 1961. A new and rapid colorimetric determination of acetylcholinesterase activity. *Biochem. Pharmacol.* 7:88–95.

Faber, D.S., J.R. Fetcho, and H. Korn. 1989. Neural networks underlying the escape response in goldfish. *Ann. NY Acad. Sci.* 563:11–33.

Fischner-Scherl, T. and R.W. Hoffmann. 1988. Gill morphology of native brown trout, *Salmo trutta fario*, experiencing acute and chronic acidification of a brook in Bavaria, F.R.G. *Dis. Aquat. Org.* 4:43–51.

Frank, R., B.S. Clegg, B.D. Ripley, H.E. Braun. 1987. Investigations of pesticide contamination in rural wells, 1979–1984, Ontario, Canada. *Arch. Environ. Contam. Toxicol.* 16:9–22.

Fujii, Y. and S. Asaka. 1982. Metabolism of diazinon and diazoxon in fish liver preparations. *Bull. Environ. Contam. Toxicol.* 29:453–460.

Galgani, F. and G. Bocquene. 1990. *In vitro* inhibition of acetylcholinesterase from four marine species by organophosphates and carbamates. *Bull. Environ. Contam. Toxicol.* 45:243–249.

Haider, S. and R.M. Inbaraj. 1988. *In vitro* effect of malathion and endosulfan on the LH-induced oocyte maturation in the common carp, *Cyprinus carpio* (L.). *Water, Air and Soil Pollution* 39:27–31.

Hanneman, E.H. 1992. Diisopropylfluorophosphate inhibits acetylcholinesterase activity and disrupts somitogenesis in the zebrafish. *J. Exp. Zool.* 263(1):41–52.

Hartwell, T. and K.B. Doving. 1989. Toxicity versus avoidance response of golden shiner, *Notenigomus orysoleucas. Fish Biol.* 35:447–456.

Heath, A.G. 1987. *Behavior and Nervous System Function in Water Pollution and Fish Physiology*. CRC Press, Inc. Boca Raton, Florida, pp. 181–196.

Henry, M.G. 1984. *The comparative effects of three toxic substances on bluegill behavior*. Ph.D. thesis. Iowa State University, Ames, Iowa.

Hughes, G.M. and J.S. Datta Munshi. 1973. Nature of the air-breathing organs of the Indian fishes *Channa, Amphipnous, Clarias* and *Saccobranchus* as shown by electron microscopy. *J. Zool. Lond.* 170:245–270.

Inesterosa, N.C. and A. Perelman. 1990. Association of acetylcholinesterase with the cell surface. *J. Memb. Biol.* 118:1–9.

Jarvinen, A.W. 1971. *The analysis of blood samples as a potential indicator of hydrogen sulfide pollution.* M.S. thesis. University of Minnesota, Minneapolis, Minn.

Johansen, P., R.A. Mathers, and J.A. Brown. 1987. Effect of exposure to several pentachlorophenol concentrations on growth of young-of-year largemouth bass, *Micropterus salmoides*, with comparisons to other indicators of toxicity. *Bull. Environ. Contam. Toxicol.* 39:379–384.

Kanazawa, Jun. 1978. Bioconcentration ratio of diazinon by freshwater fish and snail. *Bull. Environ. Contam. Toxicol.* 20:613–617.

Kennedy, R.S., R.L. Wilkes, W.P. Dunlap, and L.A. Kuntz. 1987. Development of an automated performance test system for environmental and behavioral toxicology studies. *Percept. Motor Skills* 65:947–962.

Koeman, J.H. 1991. From comparative physiology to toxicological risk assessment. *Comp. Biochem. Physiol.* 100C(1/2):7–10.

Kozlovskaya, V.I. and B.A. Flerov. 1980. Organophosphate pesticides and their danger to aquatic animals. Materials 3rd USSR-USA Symposium.

Kozlovskaya, V.I. and F.L. Mayer, Jr. 1984. Brain acetylcholinesterase and backbone collagen in fish intoxicated with organophosphate pesticides. *Great Lakes Res.* 10(3):261–266.

Kudo, S. and N. Kimura. 1983. Transmission electron microscopic studies on bacterial gill disease in rainbow trout fingerlings. *Jap. J. Ichthyol.* 30(3):247–260.

Kudo, S. and N. Kimura. 1984. Scanning electron microscopic studies on bacterial gill disease in rainbow trout fingerlings. *Jap. J. Ichthyol.* 30:393–403.

Kulshrestha, S.K. and L. Jauhar. 1984. Effects of sublethal dose of Thiodan and Sevin on liver of *Channa striatus*. *Proc. Sem. Eff. Pest Aq. Fau.*, pp. 71–78.

Kumar, S. and S.C. Pant. 1985. Renal pathology in fish (*Puntius conchonius* Ham.) following exposure to acutely lethal and sublethal concentrations of monocrotophos. *Bull. Environ. Contam. Toxicol.* 35:228–233.

Lawless, E.W., R. Von Rumker, and T.L. Ferguson. 1972. *The Pollution Potential in Pesticide Manufacturing.* U.S. Environmental Protection Agency, Washington, D.C.

Lee, E.L. 1969. *Measurement of pesticide toxicity by fish respiration rate.* D.Sc. thesis. Washington University, Sever Institute of Technology, 123 pp.

Leino, R.L., P. Wilkinson, and J.G. Anderson. 1987. Histopathological changes in the gills of pearl dace, *Semotilus margarita*, and fathead minnows, *Pimephales promelas*, from experimentally acidified Canadian lakes. *Can. J. Fish Aquatic. Sci.* 44 (suppl. 2): 126–128.

Maina, J.N. 1990. A study of the morphology of the gills of an extreme alkalinity and hypersomotic adapted teleost, *Oreochromisalcalicus grahami* (Boulenger) with particular emphasis on the ultrastructure of the chloride cells and their modifications with water dilution: an SEM and TEM study. *Anat. Embryo.* 181:83–98.

Malla Reddy, P. 1988. *Effect of synthetic pyrethroids on selected physiological aspects of the common carp, Cyprinus carpio.* Ph.D. thesis, Srikrishna-devaraya University, Anantapur, India.

Mallatt, J. 1985. Fish gill structural changes induced by toxicants and other irritants: A statistical review. *Can. J. Fish. Aquat. Sci.* 42:630–648.

Mandal, P.K. and A.K. Kulshrestha. 1980. Histopathological changes induced by the sublethal sumithion in *Clarias batrachus*. *Linn. Ind. Jr. Exp. Biol.* 18:547–558.

Mani, K. and P.K. Saxena. 1985. Effect of safe concentrations of some pesticides on ovarian recrudescence in the fresh-water murrel, *Channa punctatus* (BL). *Ecotox. Environ. Saf.* 9(3):241–249.

Mantz, W.E., Jr. 1983. *Effects of organophosphate insecticides on aspects of reproduction and survival in small mammal.* Ph.D. thesis, Virginia Polytechnic Institute and State University, Blacksburg, VA 176 pp.

Martinez-Tabche, L., C.I. Galari, M.B. Ramirez, R.A. Morales, and F.C. German. 1994. Parathion effect on acetylcholinesterase from fish through an artificial trophic chain: *Ankistrodesmus falcatusmoina macrocopa–Oreochromis hornorum*. *Bull. Environ. Contam. Toxicol.* 52:360–366.

McLean, S., M. Sameshina, T. Katayama, J. Iwata, C.E. Olney, and K.L. Simpson. 1984. A rapid method to determine microsomal metabolisms of organophosphate pesticides. *Bull. Jap. Soc. Sci. Fish.* 50:1419–1423.

Menezes, M.R. and S.Z. Qasim. 1984. Effects of mercury accumulation on the electrophoretic patterns of the serum haemoglobin and eye lens proteins of *Tilapia mossambica* (Peters). *Wat. Res.* 18:153–161.

Menzel, B.W. 1970. *An electrophoretic analysis of the blood proteins of the subgenus Luxulus (Notropis cyprinidae).* Ph.D. thesis, Cornell University, Ithaca, NY.

Mount, D.I. and C.E. Stephan. 1967. A method for establishing acceptable toxicant limits for fish-malathion and the butoxy-ethanol ester of 2,4-D. *Trans. Amer. Fish. Soc.* 21:185–193.

Moya, M.A., M.E. Fuentes and N.C. Inestrosa. 1991. A comparison of the *Xenopus laevis* oocyte acetylcholinesterase with the muscle and brain enzyme suggests variations at the post-translational level. *Comp. Biochem. Physiol.* (98C):299-305.

Munshi, J.S.D. and R.K. Singh. 1971. Investigation on the effects of insecticides and other chemical substances on the respiratory epithelium of the predatory and weed fishes. *Ind. J. Zootomy* XII(2):127-134.

Munshi, J.S.D. and A. Singh. 1992. Scanning electron microscopic evaluation of effects of low pH on gills of *Channa punctata*. *J. Fish Biol.* 41:83-89.

Murphy, S.D. 1968. Pesticides. In J. Doull, C.D. Klaassen, and M.O. Amdur (eds.), *Caserett and Doull's Toxicology*, pp. 360-408. MacMillan Publishing Co., Inc., New York.

Murugesan, A.G. and M.A. Haniffa. 1992. Histopathological changes in the oocytes of the air-breathing fish *Heteropneustes fossilis* (Bloch) exposed to textile-mill effluent. *Bull. Environ. Contam. Toxicol.* 48:929-936.

Naqvi, S.M. and R. Hawkins. 1988. Toxicity of selected insecticides (Thiodon, Security, Spartan, and Sevin) to mosquito fish, *Gambusia affinis*. *Bull. Environ. Contam. Toxicol.* 40:779-784.

Ney, J.J. and L.L. Smith, Jr. 1976. Serum protein variability in geographically defined bluegill *Lepomis macrochirus* populations. *Trans. Amer. Fish. Soc.* 2:281-289.

O'Brien, R.D., B. Hetnarski, R.K. Tripathi, and G.J. Hart. 1974. Recent studies on acetylcholinesterase inhibition. In G.K. Kohn (ed.), *Mechanisms of Pesticide Action*. American Chemical Society, Washington, D.C.

Ohkawa, K., Y. Tsukada, W. Nunomura, M. Ando, I. Kimura, A. Hara, N. Hibi, and H. Hirai. 1987. Main serum protein of rainbow trout (*Salmo gairdneri*): Its biological properties and significance. *Comp. Biochem. Physiol.* 88B:497-501.

Orsati, S. and P.W. Colgan. 1987. Effects of sulphuric acid exposure on the behavior of largemouth bass, *Micropterus salmoides*. *Govern. Biol. Fishes* 19(2):119-129.

Pavlo, D.D., G.M. Chuiko, Y.V. Serrassimov, and V.D. Tonkoply. 1992. Feeding behavior and brain acetylcholinesterase activity in bream (*Abramis brama* L.) as affected by DDVP, an organophosphorus insecticide. *Comp. Biochem. Physiol.* 103c:563-568.

Pierce, K.V., B.B. McCain, and S.R. Wellings. 1978. Pathology of hepatomas and other liver abnormalities in English sole (*Prophrys vetulus*) from the Duwamish River estuary, Seattle, Washington. *J. Natl. Cancer Inst.* 60:1445-1453.

Plack, P.A., E.R. Skinner, A. Rogie, and A.I. Mitchell. 1979. Distribution of DDT between the lipoproteins of serum protein. *Comp. Biochem. Physiol.* 62c: 119-125.

Powell, M.D., D.J. Speare, and J.F. Burka. 1992. Fixation of mucus on rainbow trout (*Oncorhynchus mykiss* Walbaum) gills for light and electronmicroscopy. *J. Fish Biol.* 41:813-824.

Prasad, M.S. 1991. SEM study on the effects of crude oil on the gills and air-breathing organs of climbing perch, *Anabas testudineus*. *Bull. Environ. Contam. Toxicol.* 47:882-889.

Rand, G.M. 1985. Behavior. In G.M. Petrocelli and S.R. Rand (eds.), *Fundamentals of Aquatic Toxicology*, pp. 221-263. Hemisphere Publishing Corporation, New York.

Reddy, M.S. and K.V.R. Rao. 1988. In vivo recovery of acetylcholinesterase activity from phosphamidon and methylparathion nervous tissue of Penaeid prawn (*Matapenaeus monoceros*). *Bull. Environ. Contam. Toxicol.* 40:752-758.

Reddy, M.S. and K.V.R. Rao. 1990. Influence of salinity on the toxicity of phosphamiden to the estuarine crab, *Cylla scrrata* (Forskal). *Bull. Environ. Contam. Toxicol.* 44:859-864.

Reddy, P.M. and G.H. Philip. 1994. In vivo inhibition of AChE and ATPase activities in the tissues of freshwater fish, *Cyprinus carpio*, exposed to technical grade cypermethrin. *Bull. Environ. Contam. Toxicol.* 52:619-626.

Reddy, P.M., G.H. Philip, and M.D. Bashamohideen. 1992. Regulation of AChE system of freshwater fish, *Cyprinus carpio*, under fenvalerate toxicity. *Bull. Environ. Contam. Toxicol.* 48:18-22.

Richmonds, C. 1989. *Effects of malathion on some physiological, histological and behavioral aspects of bluegill lungfish, Lepomis macrochirus*. Ph.D. thesis, Kent State University, Kent, Ohio.

Richmonds, C. and H.M. Dutta. 1989. Histopathological changes induced by malathion in the gills of bluegill *Lepomis macrochirus*. *Bull. Environ. Contam. Toxicol.* 43:123-130.

Richmonds, C. and H.M. Dutta. 1992a. Variations produced by malathion on the serum protein fractions of bluegill sunfish, *Lepomis macrochirus*. *Comp. Physiol. Biochem.* 102C No. 3:403-406.

Richmonds, C. and H.M. Dutta. 1992b. Effect of malathion on the brain acetylcholinesterase activity of bluegill sunfish, *Lepomis macrochirus*. *Bull. Environ. Contam. Toxicol.* 49:431-435.

Richmonds, C. and H.M. Dutta. 1992c. Effect of malathion on the optomotor behavior of bluegill sunfish, *Lepomis macrochirus*. *Comp. Biochem. Physiol.* 1002C(No. 3):523-526.

Robinson, D.A., V. Ehlert, R. Herken, B. Hermann, F. Mayer, and S.W. Schiirmann (eds.). 1987. *Methods of Preparation for Electron Microscopy.* Springer-Verlag, New York.

Roy, P.K. and J.S.D. Munshi. 1987. Toxicity of technical and commercial grade malathion on *Cirrhinus mrigala* (Ham.), a major carp. *Biol. Bull. India* 9:50–56.

Roy, P.K. and J.S. Datta Munshi. 1991. Malathion induced structural and morphometric changes in gills of a freshwater major carp, *Cirrhinus mrigala* (Ham.). *J. Environ. Biol.* 12(1):79–87.

Roy, P.K., J.S. Datta Munshi and J. Datta Munshi. 1986. Scanning electron microscopic evaluation of effects of saponin on gills of the climbing perch *Anabas testudineus* (Bloch) Anabantidae:Pisces. *Indian J. Exp. Biol.* 24:511–516.

Roy, P.K., J.D. Munshi, and H.M. Dutta. 1990. Effect of saponin extracts on morpho-histology and respiratory physiology of an air-breathing fish *Heteropneustes fossilis* (Bloch). *J. Freshwater Biol.* 2(2):135–145.

Sailatha, D., I.K.A. Sahib, and K.V.R. Rao. 1981. Toxicity of technical and commercial grade malathion to the fish *Tilapia mossambica* (Peters). *Proc. Indian Acad. Sci.* 90:87–92.

Sastry, K.V. and K. Sharma. 1980. Diazinon effect on the activities of brain enzymes from *Ophiocephalus (Channa) punctatus. Bull. Environ. Contam. Toxicol.* 24:326–332.

Scherer, E. and S.E. Harrison. 1979. The optomotor response test. In E. Scherer (ed.), *Toxicity Tests for Freshwater Organisms,* vol. 44, p. 179. Canadian Special Publication of Fisheries and Aquatic Sciences.

Shukla, J.P., M. Banerjee, and K. Pandey. 1987. Deleterious effects of malathion on survivality and growth of the fingerlings of *Channa punctatus* (Bloch.), a freshwater murrel. *Acta Hydrochim. Hydrobiol.* 15(6): 653–657.

Smith, J.R. 1984. Fish neurotoxicity. *Aquatic Toxicol.* 2:107–151.

Spiller, D. 1961. A digest of available information on the insecticide malathion. In R.C. Metcalf (ed.), *Advances in Pest Control Research,* pp. 316–317. Interscience Publishers, Inc., N.Y.

Srivastava, A.K. and A.K. Srivastava. 1988. Effects of aldrin and malathion on blood chloride in the Indian catfish *Heteropneustes fossilis. J. Environ. Biol.* 9(1 suppl): 91–95.

Steele, C.W., D.W. Owens, and A.D. Scarfe. 1990. Attraction of zebrafish, *Brachydanio rerio,* to alanine and its suppression by copper. *J. Fish Biol.* 36:341–352.

Sukumar, A. and P.R. Karpagaganapathy. 1992. Pesticide-induced atresia in ovary of freshwater fish, *Colisa lalia* (Hamilton-Buchanan). *Bull. Environ. Contamin. Toxicol.* 48:457–462.

Tandon, R.S. and A. Dubey. 1983. Toxic effects of two organophosphorus pesticides on fructose-1,6-diphosphate aldolase activity of liver, brain and gills of the freshwater fish *Clarias batrachus. Environ. Pollut.* (series A) 31:1–7.

Temmink, J.P., P. Bouweister, P. deJong, and J. Van der Berg. 1983. An ultrastructural study of chromate induced hyperplasia in the gill of rainbow trout (*Salmo gairdneri*). *Aqua. Toxicol.* 4:165–179.

Terrant, K.A., H.M. Thompson, and A.R. Hardy. 1992. Biochemical and histological effects of the aphicide demeton-s-methyl on house sparrows (*Passer domesticus*) under field conditions. *Bull. Environ. Contam. Toxicol.* 48:360–366.

Wajsbrot, N., A. Gasith, A. Diamant and D.M. Popper. 1993. Chronic toxicity of ammonia to juvenile gillhead seabream *Sparus aurata,* and related histopathological effects. *J. Fish. Biol.* 42:321–328.

Warner, R.E., K.K. Peterson, and U. Borgmann. 1976. Behavioral pathology in fish: a quantitative study of sublethal pesticide toxication. *J. Appl. Ecol.* 3:223.

Wester, P.W. 1991. Histopathological effects of environmental pollutants β-HCH and methyl mercury on reproductive organs in freshwater fish. *Comp. Biochem. Physiol.* 100c(1/2):237–239.

Wester, P.W. and J.H. Canton. 1987. Histopathological study of *Poecilia reticulata* (guppy) after long-term exposure to bis(tri-n-butylin)oxide (TBTO) and di-n-butyltinchloride (DBTC). *Aquat. Toxicol.* 10:143–165.

Wester, P.W. and J.H. Canton. 1991. The usefulness of histopathology in aquatic toxicity studies. *Comp. Biochem. Physiol.* 100(NO. 1/2):115–117.

Wester, P.W., J.H. Canton, and J.A. Ma. 1988. Pathological effects in freshwater fish *Poecilia reticulata* (guppy) and *Oryzias latipes* (medaka) following methyl bromide and sodium bromide exposure. *Aquat. Toxicol.* 12:323–344.

Wester, P.W., J.H. Canton, A.A.J. Vanlersel, E.I. Krajne, and H.A.M.G. Vaessen. 1990. The toxicity of bis(tri-n-butylin)oxide (TBTO) and di-n-butyltindichloride (DBTC) in the small fish species *Oryzias latipes* (medaka) and *Poecilia reticulata* (guppy). *Aquat. Toxicol.* 16:53–72.

Wischnitzer, S. (ed.). 1981. *Introduction to Electron Microscopy*. Pergamon Press, New York.
Ysargil, G.M. and E. Sandri. 1990. Topography and ultrastructure of commissural interneuron that may establish reciprocal inhibitory connections of the mauthner axon in spinal cord of the trench, *Tinca tinca* L. *J. Neurocytol*. 19:111-126.

Author Index

Aasenhaug, B. 184
Abel, P.D. 269
Abiola, F.A. 252
Abrahamsson, T. 52
Abraham, M. 123, 188
Abramowitz, A.A. 161
Achtenberg, H. 185
Adachi, S. 116, 121
Adhikari, S. 89, 251, 252, 253, 254, 257, 259, 266, 267
Agarwala, N. 157
Agey, P.J. 171, 173, 175
Aghajanian, J. 84, 98, 99
Agius, C. 82, 85
Ahlstrom, E.H. 130, 132
Ahmad, G. 266
Aida, K. 80, 88
Ainis, L. 51
Akita, Y.K. 187
Albert, A. 156
Albright, J.T. 180, 185, 186
Alderdice, D.F. 123, 131
Alexander, R.Mc.N. 23, 170, 179, 180, 186, 187
Alexander, S. 193, 199
Allan, F.D. 51
Allen, M.D. 125, 126, 130, 131
Al-Hussaini, A.H. 59, 66
Al-Kadhomiy, N.K. 219, 221, 222, 228, 230
Amanze, D. 135
Ammann, A. 210
Amthauer, R. 84, 88
Anand, K. 152
Anand, T.C. 152
Andersen, P. 176
Anderson, E. 111, 112, 113, 116, 117, 119, 123, 124, 125, 126, 127, 128, 130, 131
Anderson, J.G. 37, 269
Anderson, L.O. 256
Ando, M. 252
Ando, S. 79
Andres, V. Jr. 252, 253
Andrew, W. 59, 60
Andriashev, A.P. 180
Androntico, F. 123, 129
Anker, G.C. 16, 17
Anker, G.G. 2, 3, 17
Ansari, B.A. 250

Antwi, L.P.K. 252
Applegate, L. 66
Arambourg, C. 21, 23
Aranson, S. 52
Arecchon 250, 253
Aristotle 66
Arndt, E.A. 128
Arnfinnson, J. 59
Asakawa, H. 89
Asaka, S. 250
Aseeva, N.L. 104
Assali, N.S. 51, 55
Atchison, G.M. 253
Au, C.Y.W. 88
Avella, M. 254
Azevedo, C. 112, 113, 119, 124
Azzat, A. 113

Baatrup, E. 253
Babu, N. 150
Bac, N. 78
Bahuguna, S.N. 73
Bailey, R.M. 193
Bainbridge, R. 187
Baksi, S.M. 89
Baldridge, H.D. 83
Ball, N.J. 161
Ball, R.C. 255, 257
Banerjee, M. 250
Ban, M. 121
Ban, T. 83, 85
Bara, G. 118, 120, 149, 151, 155, 156, 157, 158, 160
Barber, D.L. 106
Bardach, J.E. 59, 61, 63, 66, 67, 252
Barel, C.D.N. 1, 11, 12, 13, 15, 16, 17, 18, 19
Barets, A. 176
Barnabe, G. 89
Barnes, R.D. 60, 61
Barni, S. 83, 87
Barrington, E.J.W. 59, 65, 66, 71, 96
Barr, W.A. 147, 149, 152, 156, 158
Bashamohideon, M.D. 259
Bassi, R. 123, 129
Bauer, P.R. 26
Beach, A.W. 149
Beadle, L.C. 187
Beams, H.W. 112

Begovac, P.C. 123, 128
Beklemishev, V.N. 61
Belsare, D.K. 97, 149
Belvedere, P.C. 116, 119, 120
Bengelsdorf, H. 80
Bennett, M.B. 222
Bereiter-Hahn, J. 33, 35
Bergman, H.L. 180, 181, 186
Berg, L.S. 23
Berhaut, J. 254
Berkawitz, E.J. 154
Bernard, L.M. 175
Bernocchi, G. 83, 87
Bern, H.A. 147
Berry, P.Y. 68
Bertin, L. 21, 23, 96, 187
Betlger, W.J. 106
Bettoli, P.W. 253
Bhargava, N.H. 157
Bhatia, R. 153
Bhattacharya, S. 114, 116, 117, 118, 121, 122, 123
Bhat, G.K. 150, 151, 152, 155
Bhujle, B.V. 151
Biagianti, S. 78, 105
Biagianti-Risbourg, S. 78, 79, 81, 89, 97, 98, 99, 106
Bielek, E. 106
Bieniarz, K. 116, 119, 120
Bilard, R. 116, 117, 118, 119, 121, 133, 134
Billard, R. 152
Birgi, E. 104
Biswas, N. 219, 222, 228, 230
Black, E.C. 187
Blake, I.H. 68
Blanc-Livni, N. 156, 158
Blaxter, J.H.S. 66, 173, 187, 193
Blenkarn, G.D. 235
Bluntschli, H. 96
Blust, R. 184, 185
Bock, P. 83
Bocquene, G. 252
Bodznick, D. 26
Boglazova, E.K. 113
Boland, E.J. 39, 40
Bond, C.E. 59, 66
Bone, Q. 169, 172, 173, 174
Boreus, I. 52
Borgmann, V. 253
Borg, B. 152
Borlakoglu, J.T. 253
Bornancin, M. 254
Borowy, Z.J. 81
Bosman, G. 120
Bostrom, S.L. 173
Botte, V. 156, 157, 158, 161
Bouche, G. 83
Bouchier, A.D. 95

Bouck, G.R. 255, 257
Bouix, G. 104
Boujard, T. 80
Bouweister, P. 268
Bowmaker, J.K. 3, 10
Bowser, P.R. 89
Boyd, J.D. 52
Boyd, R.B. 38
Bradley, T.M. 81, 85, 87
Brakevett, C.R. 149, 156
Bratberg, B. 89
Braunbeck, T. 82, 84, 88, 89
Braun, H.E. 250
Breder, C.M. 187
Breton, B. 111, 116, 117, 118, 119, 120, 121, 122, 123, 152, 159
Bretschneider, L.H. 149
Brill, R.W. 171, 173, 175
Brinkman, C.R. 51, 55
Brivio, M. 123, 130
Brivio, M.F. 123, 129
Browning, H.C. 147, 155
Brown, C.E. 245
Brown, J.A. 250
Brown, M. 89
Brummett, A.R. 130, 133
Brusle, J. 78, 81, 82, 89
Brusle, S. 111, 123
Bucke, D. 68
Buddington, R.K. 59
Budd, J. 83, 98
Bullock, T.H. 26
Burggren, W. 52
Burka, J.F. 253
Burkhardt-Holm, P. 89
Burnstock, G. 60
Burridge, M. 130
Busson-Mabillot, S. 112, 113, 123, 124, 125, 126, 128, 130, 131,
Byczkowska-Smyk, W. 83, 87, 89, 205
Bye, V.J. 152
Byland, G. 106
Byskov, A.G. 147

Cade, T.J. 193
Callard, I.P. 148, 149, 154, 155, 156, 157, 158
Callard, K.P. 114, 117
Caloianu-Iordachel, M. 124
Cameron, J.J. 37
Campbell, C.M. 152
Campbell, G. 39, 61
Campbell, G.J. 253
Cannon, M.D. 193
Canton, J.H. 253, 255
Carbett, J.R. 260
Carey, F.G. 171

Carslaw, H.S. 238
Carson, J.L. 84, 98, 99
Carter, G.S. 187
Carusone, C. 89
Casillas, E. 254
Castillos, F.A. 133, 134
Cecagno, C. 123, 129
Cederblad, G. 257
Chakrabarte, J. 97
Chambolle, P. 161
Chan, K. 154, 156, 157, 158, 159, 160, 161, 163
Chan, K.C. 113, 117, 118, 119, 120
Chan, S.T.H. 118, 150
Chan, Y. 23
Chao, L.N. 61, 63
Chapman, G.B. 84
Chardon, M. 48
Chaudhry, H.S. 111, 124, 126
Chen, T. 23
Chen, X. 23
Cherr, G.N. 125, 133
Chervinski, J. 184
Chesley, L.C. 71
Chiba, A. 79, 80
Chieffi, G. 112, 148, 149, 154, 155, 156, 157, 158, 161
Chinareva, I.D. 112, 113, 114, 124, 126, 127
Choi, H.Y. 245
Choudhary, D.P. 219, 222, 223, 224, 225, 227, 230
Choudhary, S. 214, 215
Christensen, A.K. 120
Christen, R. 123
Christiansen, H.E. 156, 158
Chuiko, G.M. XXIII, 260
Clarke, A.J. 68
Clarke, W.C. 112, 113, 116, 117, 118, 119, 120, 121
Clark, Jr., W.H. 125, 133
Clark, W.C. 154, 157, 158, 159, 160
Classen, H. 174
Clegg, B.S. 250
Cockell, K.A. 106
Coetzee, D.J. 61
Coe, M.J. 180, 181
Colgan, P.W. 253
Colombo, L. 116, 119, 120
Connes, R. 89, 98, 99
Contini, A. 114
Cook, G.H. 251, 252
Copeland, D.C. 180, 185
Copeland, D.E. 179, 180, 185, 186
Copley, H. 186
Coppage, D.C. 251, 252
Cornelius, C.E. 95
Corning, H.K. 186
Costagnoli, N. 126, 128, 129
Costa, J.R.V. 126, 128, 129

Cotelli, F. 123, 129, 130
Coughlan, D.J. 38
Coulson, J.M. 238
Courtney, K.D. 252, 253
Craig-Bennet, A. 156
Crank, J. 238
Crespo, S. 78, 81, 82
Crim, L.W. 111, 116, 157, 160
Crossman, E.J. 61
Cruz-Hofling, M.A. 113
Cruz-Landin, C. 113

Dale, J.E. 34
Damant, G.C.C. 187
Dam, L. Van. 179, 180
Dandotia, O.P. 219, 222, 223, 224, 225, 227, 230
Daniel, T.L. 38
Dantzer, R. 253
Daoust, B.G. 187
Das, P.K. 26
Datta Munshi, J.S. 21, 25
Datta, N.C. 193
David, H. 98
Davies, A.J. 104
Davison, W. 35, 36, 174
Dawes, B. 59, 66, 68
Dawe, C.J. 106
Day, F. 21
de Beer, M. 4, 6
De Laney, R.G. 48, 49, 51, 52, 53
De Saint Aubain, M:L. 52
de Visser, J. 4, 5, 6, 7, 8, 16, 17
Dehadrai, P.V. 179, 187, 216
Dejours, P. 55, 209
Dekkers, W.J. 219
Delaney, R.G. 48
Denis Donini, S. 123, 129
Denizot, J.P. 26
Denny, M. 199
Denton, E.J. 173, 179, 180, 193
DePeyster, A. 262
Desai, N.S. 71
DeSilva, C. 222
Desportes-Livage, I. 104
Devi, U. 27
DeVries, A.L. 38, 87
Diamant, A. 254
Diamond, J.M. 59
Diaz, J.P. 89, 98, 99
Dijkgraaf, S. 193, 200
Disler, N.N. 193
Divanach, P. 89
Dixit, R.K. 157
Dobbin, C.N. 180
Dodd, J.M. 157
Dogell, S.M. 250

Dogra, J.V.V. 251, 252
Dole, S. 130, 131
Donaldson, E.M. 59, 114, 116, 126, 152
Donato, A. 114
Donovan, M. 113, 123, 124, 125, 130, 131, 133, 134, 135
Dorier, A. 105
Dorn, E. 180, 186
Dorothy, M. 52
Doumen, C. 184, 185
Doving, K.B. 251, 253
Droller, M.J. 124
Dubale, M.S. 253
Dubey, A. 249
Dube, S.C. 205, 206, 219, 222, 223, 224, 225, 226, 227, 230
Dullemeijer, P. 1, 11, 15, 17
Dumont, J.N. 130, 133
Dunel, S. 33
Dunel-Erb, S. 33, 36, 37
Dunlap, W.P. 253
Dunn, J.F. 174
Dutta, F. 252
Dutta, G.R. 259
Dutta, H.M. 11, 38, 59, 60, 63, 65, 68, 71, 73, 89, 251, 252, 253, 254, 255, 256, 257, 258, 259, 262, 266, 267, 268, 269
Dutt, N.H. 150, 151, 152, 155
Duvyene, de Wit, J.J. 149
Dyte, H.M. 114, 116

Eastman, J.T. 38, 87
Ebeling, A. 187
Eddy, F.B. 37
Egami, N. 87
Egginton, S. 175
Ehlert, V. 267
Ekau, W. 131
Eller, L.L. 153
Ellias, H. 80, 98
Elliott, W.M. 100
Ellis, L.C. 100
Ellman, G.L. 252, 253
El-Tantawy, S.A.M. 104, 105
Emel'Yanova, N.G. 112, 113, 123,
Enns, T. 186, 187
Epter, P. 116, 119, 120
Erhardt, H. 112, 115, 119, 123, 124, 125, 127
Eriksson, J.C. 106
Ermolina, N.O. 112
Eurell, J.A. 80, 82, 83
Evans, H.M. 187
Evensen, O. 89
Ezza, A. 113

Faber, D.S. 253
Fahlen, G. 180, 185, 186, 187
Falk-Petersex, J.B. 130, 131
Fange, G. 179, 180, 185, 186, 187
Fange, R. 187
Farrel, A.P. 59
Fasulo, S. 51, 114
Fawcett, D.W. 185
Fay, F.S. 51
Featherstone, R.M. 252, 253
Ferguson, H.W. 82, 83
Ferguson, T.L. 250
Fernandes, M.R. 222
Ferri, S. 84, 85, 98
Fetcho, J.R. 253
Fichter, G.S. 21
Fick, A. 207
Fierstine, H.L. 169
Fige, F.H.J. 53
Fink, S.V. 23, 27
Fink, W.L. 23, 27
Fischner-Scherl, T. 269
Fisher, K.C. 111, 113, 115, 119, 123, 124, 125, 126, 130
Fisher, M.M. 96, 99, 100
Fishman, A.P. 48, 49, 50, 51, 52, 53
Flegler, C. 123, 124, 125, 126
Flerov, B.A. 252
Flik, G. 38
Flood, P.R. 173, 174
Flugel, H. 111, 112, 113, 115, 116, 119, 123, 124, 125, 126, 130
Fomena, A. 104
Fontaine, Y.A. 116
Fostier, A. 111, 116, 117, 118, 119, 120, 121, 122, 123, 152
Frange, R. 68
Franklin, C.E. 35, 36
Frank, R. 250
Frazier, G.M. 89
Friedrich-Freksa, H. 26
Friess, A. 130
Fromm, P.O. 33, 34, 35, 36, 37, 41
Frost, L.A. 130, 131
Fryer, G. 184
Fuentes, M.E. 252
Fujii, Y. 250
Fujita, H. 83, 84, 85
Fusimi, T. 133
Fynn-Aikins, F.K. 80

Gabaeva, N.S. 112, 113
Galari, C.I. 252, 258
Galgani, F. 252
Galis, F. 16
Gannon, B.J. 39

Gasith, A. 254
Gas, N. 83
Geevarghese, C. 63, 68
Gee, J.H. 180
Gehr, P. 34, 208, 210, 211, 212, 215, 224, 225, 228, 230
Gennser, G. 52
George, G.J. 71
George, J.C. 173
German, F.C. 252, 258
Gerzeli, G. 83, 87
Geyer, F. 187
Ghosh, T.K. 25, 34, 39, 40, 41, 49, 211, 212, 213, 215
Gillim, S.W. 120
Gillis, D.J. 87
Gilloteaux, J. 95, 101, 103
Gilloteaux, L.C. 103
Gill, H.S. 170
Ginsburg, A.S. 130, 133, 134
Gloss, S.P. 38
Goetz, F.W. 152, 160
Gokhale, S.V. 149
Goldblatt, P.J. 81, 98, 99
Goldenberg, H. 83
Goldschmidt, T. 4, 5, 6, 7, 8, 16, 19
Gonzalez, G. 78, 81, 82, 83, 89
Goolish, E. M. 180, 186
Gopesh, A. 23, 26, 27
Gorgas, K. 82, 84, 88, 89
Gosline, W.A. 193, 199, 200
Goswami, S.V. 149, 152
Gotting, K.J. 111, 113, 119, 123, 124, 125, 133, 134, 135
Gotting, M.J. 112, 119, 123, 124
Gouder, B.Y.M. 151
Goudswaard, P.C. 6
Gould, M. 135
Graham, E.F. 124
Graham, J.M. 35
Gray, I.E. 205, 218, 222, 227, 230
Gray, J. 187
Greene, C.W. 68
Greenwald, G.S. 114, 116, 122
Greenwood, P.H. 18, 21, 23, 179
Green, C.W. 179
Green, S.L. 187
Greer-Walker, M. 171
Greven. H. 123, 124, 128
Grisham, J.W. 84, 98, 99
Groat, E.P. 123
Groff, J.M. 80
Groman, D.B. 59, 66, 68
Gros, G. 37, 209
Grote, J. 240
Grove, D. 68

Guha, G. 205, 206, 219, 222, 223, 224, 225, 226, 227, 230
Gulliksen, B. 80
Guppy, M. 174
Guraya, S.S. 111, 112, 113, 114, 115, 116, 117, 119, 120, 121, 122, 123, 124, 126, 128, 129, 130, 131, 132, 133, 147, 149, 150, 151, 152, 153, 154, 155, 156, 157, 158, 159

Hacking, M.A. 83, 98
Haensly, W.E. 68, 80, 82, 83
Hafey, S. 113, 130
Haghighi, A.Z. 251, 255
Hagstrom, B.A. 130
Haider, S. 249, 255
Hakim, A. 219, 222, 223, 224, 225, 227, 230
Halder, S. 114, 116, 117, 118, 121, 122, 123,
Hale, P.A. 63
Hall, F.G. 179, 185, 186, 187
Hamazaki, T.S. 123, 128
Hampton, J.A. 81, 97, 98, 99
Handy, R.D. 37
Haniffa, M.A. 255
Hanneman, E.H. XXIII, 261
Hanson, D. 53
Hara, A. 121, 252
Harboe, S.B. 80
Harder, W. 169
Hardy, A.R. 253
Harpole, J.H.Jr. 34, 35
Harrison, S.E. 262
Harris, J.E. 71
Hartwell, T. 253
Hart, G.J. 250, 258
Hart, N.H. 113, 123, 124, 125, 130, 131, 133, 134, 135
Hashimoto, Y. 106
Hasler, A.D. 179
Haslewood, G.A.D. 95, 96
Haug, T. 80
Hawkes, J.W. 113, 124, 125, 128, 130, 132, 133, 134,
Hawkins, R. 253
Hawkins, W.E. 84
Hayar, S.R. 174
Heath, A.G. 253
Hebibi, N. 36, 37, 38
Heinrichs, S.M. 66
Heiser, J.B. 23, 25, 26, 28, 193
Helm, W.T. 63
Helvik, J.V. 123, 129
Henry, G. 253
Henry, M.G. 265
Herken, R. 267
Hermann, B. 267
Hetnarski, B. 250, 258

Heymann, M.A. 51, 53
Hibiya, T. 80, 88
Hibi, N. 252
Hightower, L.E. 89
Hilge, V. 131, 134
Hills, B.A. 235, 236, 240, 244, 245, 246
Hinton, D.E. 81, 82, 97, 98, 99
Hirai, A. 133
Hirai, H. 252
Hiroi, O. 121
Hirose, K. 80, 88, 113, 121, 123, 125, 126
Hisaw, F.L. 148, 149, 155, 156, 157, 161
Hisaw, F.L.Jr. 148, 149, 155, 156, 157
Hoar, W.S. 31, 112, 113, 116, 117, 118, 120, 121, 152, 154, 156, 157, 158, 159, 160, 161, 163
Hoa-Ren, L. 114, 116
Hobe, H. 36, 37
Hochachka, P.W. 171, 173, 174, 175
Hodson, K. 83, 98
Hoffmann, R.W. 130, 269
Holiday, F.G.T. 124
Holmgren, A. 236
Holmgren, S. 52
Honma, Y. 79, 80, 149
Hoogerhoud, R.J.C. 16, 18, 19
Hoppeler, H. 174
Hora, S.L. 180
Horboe, S.B. 99, 100
Hosaja, M. 133
Hosokawa, H. 80
Hosokawa, K. 133
Hossain, A. 59, 63, 65, 68, 69, 70, 71, 73
Hossain, A.M. 59, 63, 68
Hossain, M.A. 59, 60, 63, 68, 71
Hossler, F.E. 33, 34, 35, 36, 38
Houets, P. 252, 253
Howe, E. 130, 131
Hubbs, C.L. 193
Hudson, R.C.L. 176
Hughes, G.M. 25, 31, 34, 35, 171, 173, 174, 175, 205, 206, 208, 210, 211, 212, 215, 216, 218, 219, 222, 223, 224, 225, 226, 227, 228, 229, 230, 235, 236, 237, 240, 245, 246, 254
Huiskamp, R. 88
Hulbert, W.C. 174
Humbert, W. 37
Hung, S.S.O. 80
Hunter, G.A. 114, 116
Hurley, D.A. 111, 113, 115, 119, 123, 124, 125, 126, 130
Hyrtl, C.J. 97

Idler, D.R. 72, 111, 116, 152
Ikeda, M. 52
Iles, T.D. 184
Inagawa, T. 106

Inbaraj, R.M. 249, 255
Inesterosa, N.C. 252
Ishida, J. 71, 72, 73
Ishihara, A. 41
Ishii, K. 88
Ishii, S. 123
Itami, T. 37
Itazawa, Y. 221
Ito, T. 84
Iuchi, I. 123, 128
Ivankov, V.N. 130
Ives, P.J. 68
Iwamatsu, T. 113, 117, 118, 120, 133, 134
Iwama, G.K. 37
Iwasaki, V. 118
Iwasaki, Y. 157, 160
Iwata, J. 250, 253
Iyengar, A. 135
Iyer, V. 123, 131

Jacinski, A. 180
Jackson, D. 35
Jacobs, W. 187
Jacyna, R. 95
Jaeger, J.C. 238
Jaffe, L.A. 135
Jager, S.de. 219
Jalabert, B. 111, 116, 117, 118, 119, 120, 121, 122, 123, 152, 159
Jansson, B.O. 66, 72, 73
Jarvinen, A.W. 251
Jauhar, L. 266
Jensen, J.K.S. 176
Jesen, D.N. 106
Jirge, S.K. 63, 87
Jobling, M. 63
Johansen, K. 47, 53, 180, 184
Johansen, P. 250
Johansson, R.G. 173
Johnson, E.Z. 123
Johnson, R.K. 59, 68
Johnston, I.A. 170, 173, 174, 175
Jollie, L.C. 111, 113, 123, 124, 125
Jollie, W.P. 111, 113, 123, 124, 125
Jones, D.R. 52
Jones, F.R.H. 179, 180, 185, 186, 187, 188
Jong, P.de. 268
Jorgensen, J.M. 26, 193
Juario, J.V. 88

Kafuku, T.T. 66
Kagawa, H. 112, 113, 116, 117, 118, 119, 120, 121, 122, 154, 157, 158, 159, 160
Kaiso, T. 80
Kambegawa, A. 157, 160
Kamoi, I. 71

Kanazawa, Jun. 251
Kanwisher, J. 187
Kapoor, A.S. 193
Kapoor, B.G. 59, 61, 63, 65, 66, 71, 72
Kapoor, C. 112
Karas, R.H. 174
Karkare, S. 95, 101
Karlsson, L. 33, 35
Karpagaganapathy, P.R. 253, 255
Kasahara, N. 121
Katayama, T. 250, 253
Kaur, S. 111, 112, 113, 119, 120, 124, 150, 152, 154, 156, 157, 158, 159
Kazic, D. 106
Kazubski, S.L. 104
Keenan 172
Kelly, T.R. 95, 101
Kemp, N.E. 125, 126, 130, 131
Kendall, M.W. 34, 84
Kennedy, R.S. 253
Kent, G.C. 59, 60
Keshavnath, P. 152
Kessel, R.G. 112
Khandelwal, O.P. 193
Khanna, S.S. 59, 68, 150, 157, 158
Khoo, K.H. 118, 149, 157, 158, 160
Kickuth, R. 253
Kilarski, W. 180
Killius, J. 38, 254, 269
Kimball, D.C. 63
Kimura, I. 252
Kimura, N. 35, 268
King, J.A. 34, 35
King, J.A.C. 38
King, P.E. 111, 112, 113, 124, 125, 126, 127, 129
Kirchinskaja, E.B. 113, 114, 124, 126, 127
Kirsch, R. 37
Kisia, S.M. 171, 172, 173, 174, 175
Kjorsvik, E. 59, 130, 131
Klingstedt, G. 41
Klobukar, N. 83
Knutton, S. 35
Kobayakawa, M. 130
Kobayashi, K. 85
Kobayashi, W. 112, 124, 133, 134, 135
Kock, K.H. 131
Kodama, T. 80
Koeman, J.H. 251
Koga, A. 98
Kojiyama, H. 80
Kon, Y. 106
Korn, H. 253
Korsgaard, B. 80
Kosek, E. 95, 101
Koshtoyanz, C.S. 179
Kovaleva, A.A. 104

Koyama, T. 240
Kozlovskaya, V.I. 250, 252
Ko, W. 95, 101
Kraft, A.V. 111, 112, 124
Krajne, E.I. 253
Kramar, R. 83
Krauskopl, M. 84, 88
Krogh, A. 180, 185, 208, 209
Kruysse, A. 39, 246
Kryvi, H. 59, 173, 174
Kubono, K. 37
Kuchnow, K.P. 130, 133, 134
Kudo, R.R. 104, 105
Kudo, S. 35, 133, 268
Kugel, B. 130
Kulshreshtha, S.K. 266
Kulshrestha, A.K. 266
Kumari, S.D.R. 133
Kumar, K. 250
Kumar, S. 124, 150, 250
Kuntz, L.A. 253
Kunwar, G.K. 219, 222, 227, 230
Kuonen, E. 240
Kuo, C.M. 126, 153
Kvenseth, P.G. 59
Kylstra, J.A. 235, 245

Laale, H.W. 123, 127, 130, 133
Labbe 106
Labhart, P. 63
Ladner, C. 51, 55
Lagler, K.F. 59, 61, 63, 66, 67
Laguesse 106
Lahiri, S. 48, 49, 50, 53
Lall, S.B. 251, 255
Lambert, J.D.G. 88, 116, 117, 118, 120, 150, 151, 155, 157, 158, 159, 160
Lamina, C.L. 123, 129
Lam, T.J. 38, 152, 153, 156, 161, 163
Lance, V.A. 114, 117, 147, 149, 154, 155, 156, 157, 158
Langdon, J.S. 83
Langille, B.L. 52
Lang, I. 150, 152, 155, 158
Lanphier, E.H. 245
Lanphier, E.J. 235
Lantz, R.C. 81, 98, 99
LaRosa, M.L. 123, 129
Larsen, W.J. 113
Lauren, D.J. 38, 81, 98, 99
Laurent, P. 25, 33, 35, 36, 37, 38, 48, 49, 50, 51, 52, 53, 181
Lau, S.R. 66
Lawless, E.W. 250
Lawrence, J.M. 66, 72
Leatherland, J.F. 38, 80, 87

Leavitt, W.W. 118, 121, 123
Lee, E.L. 251
Lehri, G.K. 150, 156, 157, 158
Lehtinen, K.J. 41
Leino, R.L. 37, 106, 269
Leme Dos Santos, H.S. 126, 128, 129
Lenfant, C. 53
Leray, C. 118, 128
Lev, M. 50, 53
Lewis, D.H. 68
Life, T.S. 252
Ligtvoet, W. 6
Lincoln, R.F. 80
Lindsay, G.H. 71
Lissmann, H.W. 26
Livanov, M.N. 179
Livni, B.N. 118
Livni, N. 118, 120
Lock, R.A.C. 37
Lofts, B. 118, 147
Logemann, R.B. 171, 173, 175
Lok, D. 120
Londraville, R.L. 175
Lone, K.P. 80
Long, W.F. 262
Lonning, S. 125
Lopes, R.A. 126, 128, 129
Lorenzini, S. 176
Lotto, W.N. 81
Lougheed, W.M. 81
Low, M.P. 68
Loyning, Y. 176
Luczynski, M. 133
Luling, K.H. 179, 187
Lupo di Prisco, C. 148, 149, 154, 156, 158, 161
Luprano, S. 170
Lutes, P.B. 80
Lutton, C. 98
Lykkeboe, G. 180, 184

Machida, Y. 126
Machin, K.E. 26
Mackay, N.J. 152, 157, 158, 160
MacMurphy, D.M. 51
MacMurphy, T. 52
Maetz, J. 35, 36
Maina, J.N. 181, 215, 222, 228, 230, 254
Maitland, B. 174
Makeeva, A.P. 129
Malla Reddy, P. 260
Mallatt, J. 38, 253, 268, 269
Malmfors, O. 52
Maloiy, G.M.O. 174, 175, 180, 184, 210, 215, 222, 228, 230
Maly, I.P. 81, 83
Mandal, P.K. 266

Mani, K. 150, 153, 253, 255
Manna, P.R. 114, 116, 117, 118, 121, 122, 123
Manner, H.W. 125, 130
Mann, M. 187
Mantz, W.E.Jr. 250, 259
Manwell, R.D. 104
Maple, G. 112, 113, 115, 117, 119, 120, 154, 156, 159
Marcelino, J. 251, 252, 259, 262
Marchiondo, A.A. 106
Marias, J.F.K. 63, 65
Markey, D. 37
Markle, D.F. 130, 131
Markov, K.P. 132
Marshall, N.B. 173, 179, 180, 185, 186, 187, 188
Marshall, W.S. 38
Martineau, D. 89
Martinez-Tabche, L. 252, 258
Martin, N.V. 71
Mascarello, F. 170
Mathers, R.A. 250
Matheus, C.E. 184
Mathews, L.H. 158
Mathews, S.A. 149, 156, 161
Mathieu-Costello, O. 171, 173, 175
Matsusato, T. 133
Mattei, X. 113, 114, 128,
Mattey, D.L. 106
Mattheij, J.A.J. 157
Mattheij, J.A.M. 118
Mattison, A. 180
Maugeri, A. 114
Maurer, F. 51, 53
Mayberry, L.F. 106
Mayer, F. 267
Mayer, F.L.Jr. 250, 252
Ma, J.A. 253, 255
McMahon, B.R. 47, 53, 54
McBee, R.H. 68, 71
McCormick, J.H. 37
McCuskey, P.A. 97
McCuskey, R.S. 81, 97
McFarland, W.N. 23, 25, 26, 28, 193
Mckeown, B.A. 87
Mclean, E. 59
McLean, S. 250, 253
McIlwain, T.D. 33, 35, 36, 38
McMillan, C.R. 149, 156
Mehrota, B.K. 59, 68
Meis, S. 35
Melmon, K.L. 51
Mendelson, J.R. 193, 199
Menezes, M.R. 251
Menzel, B.W. 256
Merz, W.A. 207
Mester, R. 116

Micklem, K.J. 35
Migdalski, E.C. 21
Mikodina, E.V. 129, 130, 131
Miller, J.E. 59, 61, 63, 66, 67
Milsom, W.K. 52
Milton, P. 222
Mishra, A.K. 219, 222, 227, 230
Mitchell, A.I. 256
Mitchell, L.G. 105
Mochizuki, M. 240
Mohsin, S.M. 59, 63, 66, 68
Moluskey, P.A. 81
Mommsen, T.P. 80
Mondolfin, R.M. 34
Mooi, R.D. 130
Moore, J.C. 251, 252
Morales, R.A. 252, 258
Morgan, M. 31, 33, 39, 53, 106
Morgan, S.L. 222
Morimoto, T. 89
Mori, K. 79, 157
Morrison, C.M. 106
Morris, S.M. 180, 185, 186
Morton, J. 60, 63, 65, 68
Mosconi-Bac, N. 79, 82, 84, 88
Moser, G.H. 156
Moser, H.B. 130
Moser, H.G. 132
Motomatsu, K. 85
Motz, L. 38, 254, 368
Mount, D.I. 250
Moya, M.A. 252
Muehleman, C. 125, 130
Mugiya, Y. 41
Mugnaini, E. 80, 99, 100
Muir, B.S. 205, 222
Muller, H. 125
Mullinger, A.M. 26
Munk, O. 180
Munshi, J.S.D. 34, 35, 38, 39, 40, 41, 49, 89, 205, 206, 208, 210, 211, 212, 213, 214, 215, 216, 218, 219, 222, 223, 224, 225, 226, 227, 228, 230, 251, 252, 253, 254, 257, 259, 266, 267, 268, 269
Murakami, T. 39
Murphy, B. 174
Murphy, S.D. 252
Murugesan, A.G. 255
Mwangi, D.K. 210
Myers, G.S. 21, 23
Myers, M. 254

Nagahama, V. 119, 120, 121
Nagahama, Y. 112, 113, 116, 117, 118, 119, 120, 121, 122, 123, 128, 134, 147, 152, 154, 156, 157, 158, 159, 160, 161, 163
Nair, N.B. 126, 133, 150

Nair, N.D. 133
Nair, P. 51
Nair, P.V. 156, 158
Nakamura, J. 119, 120, 121
Nakamura, K. 79
Nakano, E. 128
Nakano, M. 89
Nakashima, S. 113, 133, 134
Nadkarni, V.B. 118, 120, 150, 151, 152, 153, 154, 155
Naqvi, S.M. 253
Narahara, A. 181
Nasar, S.S.T. 251, 252, 253
Nash, C.E. 126
Nath, A. 253, 267
Neaves, W.B. 120
Nelson, G.J. 193, 200
Nelson, J.S. 21, 23, 195
Neumayer, L. 96
Newstead, J.D. 33, 39
Ney, J.J. 255, 256
Ng, T.B. 88, 152
Nicholas, G. 104
Nicholls, T.J. 154, 156, 158, 159
Nicholls, T.O. 112, 113, 115, 117, 119, 120
Nielsen, J.C. 180
Nikolskii, G.V. 21, 23
Nikolsky, G.V. 61, 65
Nilsson, S. 52
Nimi, A.Z. 221, 222
Nopanitaya, W. 84, 98, 99
Northcote, T.G. 65, 180
Nunomura, W. 252

Oates, K. 41
Obara, T. 71
Oberhansli-Weiss, I.M. 51
Odense, P.H. 106
Oeda, H. 116, 121
Ogilive, D. 89, 100
Oguri, M. 79, 80, 82, 83
Ohata, M. 84
Ohkawa, K. 252
Ohta, H. 113, 117, 118, 133, 134
Ohta, T. 113, 117, 120, 133
Oikawa, S. 221, 222
Ojha, J. 25, 34, 39, 40, 41, 211, 212, 213, 215, 216, 218, 219, 222, 223, 225, 227, 228, 230
Okamoto, K. 37
Olney, C.E. 250, 253
Olson, K.R. 25, 31, 32, 33, 34, 35, 36, 37, 38, 39, 40, 41, 49, 210, 212, 213, 215
Olson, L. 52
Olsson, R. 66, 72, 73
Onitako, K. 118
Onozato, H. 117, 118, 119, 120, 154

Oppen-Bernstein, D.O. 123, 129
Orsati, S. 253
Orvig, T. 23
Osanai, K. 130, 133
Osborn, M. 33, 35
Oshima, E. 113
Otten, E. 11, 16
Oura, C. 119
Ourth, D.D. 37
Owens, D.W. 253
Owman, C. 52
O'Brien, R.D. 250, 258

Paganelli, C.V. 245
Pancharatna, M. 151
Panchen, A.L. 25
Pandey, A. 27, 224, 227, 230
Pandey, K. 153, 250
Pandey, K.N. 25
Pandey, S. 161, 163
Pant, M.C. 149
Pant, S.C. 250
Paris, H. 83
Parker, W.N. 48, 53
Part, P. 37
Pasha, K. 65
Passino, D.R.M. 59, 61, 63, 66, 67
Pasternak, C.A. 35
Patil, H.S. 153
Patterson, R.J. 65
Patzner, R.A. 130, 131
Paulo, D.D. XXIII
Pavlo, D.D. 260
Pearse, A.S. 185
Pearse, D.C. 185
Peek, W.D. 96, 99, 100
Pelizaro, M.G. 126, 128, 129
Pelster, B. 179
Perelman, A. 252
Perlmutter, A. 186
Perry, S.F. 35, 36, 37, 38, 185, 187, 222
Peterson, K. 52
Peterson, K.K. 253
Peters, H.M. 111, 112, 124
Peute, J. 88, 118, 120, 150, 151, 154, 155, 157, 158, 159, 160
Pfautch, M. 40
Philipart, R.C. 184
Philip, G.H. 252, 259
Phillips, J.G. 150
Pietra, G.G. 38
Pietri, R. 113, 130, 131
Piiper, J. 238, 245
Piper, J. 222
Pisam, M. 35, 36
Pithawalla, R.B. 133, 134

Plack, P.A. 256
Player, D. 112
Plumb, J.A. 250, 252, 253
Poisson, R. 104
Polder, J.J.W. 149
Pool, C.R. 82, 97, 98
Popper, D.M. 254
Popta, C.M. 187
Potts, W.T.W. 41
Pough, F.H. 23, 25, 26, 28, 193
Powell, M.D. 253
Power, G.G. 245
Prakash, S. 25, 27
Pramoda, S. 151
Prasad, M.S. 219, 222, 254, 268, 269
Price, J.W. 205, 221, 222
Pull, G.A. 171

Qasim, S.Z. 251
Quaglia, A. 82
Quintana, N. 99
Qutob, Z. 179

Rabergh, C.M.I. 106
Rahimullah, M. 59, 63, 66, 68
Rai, B.P. 149, 156, 158
Rajalakshmi, M. 149, 150, 151, 156, 158
Ramadan, A. 113
Ramirez, M.B. 252, 258
Ram, R.N. 153
Randall, D.J. 31, 39, 180, 181, 186
Rand, G.M. 253
Rantin, F.T. 222
Rao, K.V.R. 249, 251, 252
Rao, M.A. 151
Rappaport, A.M. 81
Rastogi, M.G. 193
Rastogi, R.K. 111, 126, 133, 149, 150, 151, 156
Rayner 172
Ray, P.K. 193, 199, 200
Reddy, M.S. 249, 252
Reddy, P.M. 252, 259
Reid, R.C. 240
Reifel, C.W. 59, 63,
Rejte, O.B. 53, 184
Renard, P. 123
Renfro, J.L. 89
Rheinalt, T. 89
Rhodes, R.G. 250
Ribble, D.O. 63, 65
Richardson, J.F. 238
Richard, C.W. 66
Richmonds, C. 38, 251, 252, 253, 254, 255, 256, 257, 258, 259, 260, 261, 262, 263, 264, 265, 266, 268
Richter, C.J.J. 118, 157

Riehl, R. 111, 112, 113, 115, 116, 123, 124, 125, 127, 128, 129, 130, 131, 132, 133, 134, 135
Ripley, B.D. 250
Robertson, D.A. 130
Robertson, J.C. 81, 85, 87
Robertson, J.I. 53
Robertson, O.H. 78, 87
Roberts, B.L. 169, 173
Roberts, R. 112
Robinson, D.A. 267
Rodriguez, E. 84, 88
Rogie, A. 256
Rohsenow, W.M. 245
Romanello, M.G. 170
Romer, A.S. 55, 59, 60
Rooj, N.C. 219, 222, 227, 230
Rosen, D.E. 21, 23, 193, 199
Ross, P. 95
Roth, T.F. 124
Roy, P.K. 38, 49, 89, 208, 210, 211, 212, 213, 219, 221, 222, 224, 226, 227, 230, 251, 252, 253, 254, 257, 266, 267, 268, 269
Roy, S.K. 114, 116, 121, 122
Rubstov, V.F. 131
Ruby, J.R. 33, 35, 36, 38
Rudolph, A.M. 51, 53
Ruwet, J.C. 184

Saez, L. 84, 88
Saharya, R. 97
Sahib, I.K.A. 251
Saidapur, S.K. 120, 147, 150, 151, 152, 153, 154, 155, 161
Sailatha, D. 251
Saito, Y. 87
Sakai, N. 113
Sakano, E. 84, 85
Sakari, N. 113, 133, 134
Saksena, D.N. 157
Salman, N.A. 37
Salmon, C. 116, 121
Sameshina, M. 250, 253
Samuel, M. 157
Sandercock, F.K. 72
Sandheinrich, M.B. 253
Sandri, E. 253
Sand, A. 193
Sanwal, R. 156, 157, 158
Sardet, C. 35, 36
Sargent, P.A. 79, 89, 100
Sarkar, S. 11
Sasse, D. 81, 83
Sastry, K.V. 249, 251
Sastry, V.K. 66, 72
Sathyanesan, A.G. 149, 151, 153, 156, 158
Sato, M. 26

Saupe, M. 186
Saxena, P.K. 111, 124, 150, 152, 153, 157, 253, 255
Scammon, R.E. 96
Scapolo, P.A. 170
Scarfe, A.D. 253
Schar, M. 81, 83
Scheid, P. 179, 245
Scherer, E. 262
Schiirmann, S.W. 267
Schlichting, H. XXII, 240
Schlote, W. 40
Schmehl, M.K. 124
Schneider, H. 179
Schoenfisch, W.H. 235
Scholander, P.F. 179, 180, 186, 187
Schreibman, M.P. 154
Schulte, E. 112, 113, 123, 124, 133, 134, 135
Schunke, M. 84, 88
Schwartzkopff, J. 179
Schwartz, F.J. 106
Schwerdtfeger, W.K. 33, 35
Scott, A.P. 80
Scott, J.R. 130, 133, 134
Scott, W.B. 61
Scripcariu, D. 116
Seal, M. 26
Seal-Prasad, M. 23
Segner, H. 83, 88, 89
Selcer, K.W. 118, 121, 123
Selman, K. 112, 117, 123, 124, 126
Senoo, H. 85
Serrassimov, Y.V. XXIII, 260
Sesso, A. 84, 85
Sfsvhi, D. 116, 121
Shackley, S.E. 111, 112, 113, 124, 125, 126, 127, 129
Shafland, P.L. 66
Shah, P. 253
Shanbag, A.B. 118, 120
Shanbhag, A.B. 150, 151, 152, 155
Sharma, K. 249, 251
Sharma, M.S. 193
Sharma, R.S. 193
Sharma, S.N. 222
Shedadeh, Z.H. 126
Sheldon, H. 113, 114, 123, 126, 127, 128, 130, 132
Shelton, W.L. 132
Shephard, K.L. 37.
Sherwood, T.K. 240, 244, 245
Shimeno, S. 80
Shin, Y.C. 78
Shivers, R.R. 89, 100
Shlaifer, A. 187
Shrivastava, S.S. 150, 156, 158
Shukla, J.P. 250
Shukla, T.K. 153

Shul'Man, B.S. 104, 105
Sidell, B.D. 175
Sidon, E.W. 87, 79, 96, 99, 100
Silva, D.G. 52
Simonneaux, V. 37
Singh, A. 38, 254
Singh, B. 222, 227, 230
Singh, B.N. 212, 216
Singh, B.R. 205, 206, 216, 219, 222, 223, 224, 225, 226, 227, 228, 230
Singh, H. 153
Singh, H.R. 73
Singh, K.P. 23
Singh, M. 25, 27
Singh, N.K. 38, 251, 252, 253, 254, 257, 259, 266, 267, 268, 259
Singh, N.K. 89, 219, 222, 227, 230
Singh, O.N. 219, 222
Singh, R. 219, 222, 227, 230
Singh, R.K. 253
Singh, T.P. 153
Sinha, A.L. 219, 222, 223, 224, 225, 227, 230
Sinha, S.K. 219, 222
Sire, M.F. 82, 98
Sis, R.F. 68
Sjoberg, N.O. 52
Skinner, E.R. 256
Skjerve, E. 89
Sloaf, W. 80
Sloan, R. 89
Smithson, T.R. 25
Smith, H. 59, 61, 63, 65, 66, 71, 72, 73
Smith, H.M. 25
Smith, J.R. 266
Smith, L.L. 255, 256
Smith, M.H. 63, 65
Sobhana, B. 126, 133
Solemdal, P. 125, 130
Sonstegard, R.A. 87
Soto, R. 157
Speare, D.J. 253
Specker, J.L. 119, 120, 121
Speilberg, L. 89
Sperry, D.G. 34
Spille, D. 251
Sponaugle, D.L. 130, 132
Srivastava, A.K. 250
Srivastava, C.B.L. 23, 25, 26, 27, 193
Srivastava, D. 23
Srivastava, G.J. 266
Srivastava, M.D.L. 193
Srivastava, M.P. 21
Srivastava, S. 23, 25
Stahl, P.A. 128
Stanton, M.F. 106
Steele, C.W. 253

Steen, J.B. 179, 180, 186, 187, 246
Steen, J.G. 38, 39
Stehr, C.M. 113, 114, 115, 123, 124, 125, 126, 127, 128, 129, 130, 132, 133, 134
Stenger, A.H. 149
Stenkowski, I.K. 104
Stensio, E. 25
Stephan, C.E. 250
Sterba, G. 125
Sternlieb, I. 99
Stevens, E.D. 169, 171, 173, 175
Stgreba, G. 200
Stickland, N.C. 170
Stoever, H.J. 246
Stolk, A. 149
Storch, V. 82, 84, 88, 89
Stroband, H.W.J. 59, 66, 67
Strussmann, C.A. 88
Sufi, G.B. 157
Sugawara, A. 79
Sugimura, M. 106
Sukumar, A. 253, 255
Sumida, B.Y. 132
Sumpter, J.P. 157
Sunderaj, B.I. 149, 152
Sunderaraj, B.T. 152
Sundnes, G. 186, 187
Sun, J. 113, 117
Suyehiro, Y. 59, 61, 63, 65, 66, 67, 68, 98
Suzuki, K. 121
Suzuki, T. 71
Swan, H. 47
Szabo, T. 26
Szamier, R.B. 26
Szidon, J.P. 50, 53
Szollosi, D. 133, 134, 159

Takahashi, S. 80, 85, 132, 133
Takahashi, Y. 37, 85
Takano, K. 112, 113, 116, 117, 118, 119, 120, 121, 133, 154, 157, 158, 159, 160
Takashima, F. 88
Takeda, M. 80
Tamaoki, B.I. 121
Tamura, S. 83, 85
Tam, P.P.L. 88
Tanahashi, K. 80
Tanaka, Y. 87
Tandon, R.S. 249
Tang, F. 118
Tanuma, Y. 84, 85, 97, 98, 99
Tatsumi, H. 83, 85
Tavolga, W.N. 179
Taxi, J. 49, 52
Taylor, C.P. 210
Tazawa, H. 240

Te Winkel, L.E. 148, 149, 155, 156
Teal, J.M. 171
Temmink, J.P. 268
Teranishi, T. 113, 117, 118, 134
Terrant, K.A. 253
Tesoriero, J.V. 112, 125, 126, 127, 128, 129
Thakur, R.N. 205, 206, 223, 224, 225, 228, 230
Thiaw, O.T. 113, 114, 128
Thompson, H.M. 253
Thorarensen, H. 59
Thurmond, T. 68
Tonkoply, V.D. XXIII, 260
Toor, H.S. 124, 150
Toshimori, K. 112, 115, 116, 119, 123
Totland, G.K. 173, 174
Tovell, P.W.A. 33, 39
Travill, A.A. 59, 63
Travison, P. 71
Treasurer, J.W. 124
Trewavas, E. 180, 181, 188
Tripathi, R.K. 250, 258
Tripathi, S. 23, 25
Tromp-Blom, N. 149
Tsubouchi, H. 85
Tsukada, Y. 252
Tsukahara, J. 123, 130
Tung, H.N. 112
Tyler, A.V. 66, 73
Tytler, P. 187

Ubelaker, J.E. 106
Ueda, H. 121
Uhlmann, E. 174
Ultsch, G.R. 37, 209
Umezawa, S.I. 34
Upadhaya, S.N. 123, 154
Ursin, E. 219, 220, 222
Uspenskaya, A.V. 104

Vaessen, H.A.M.G. 253
Van Bergeijk, W.A. 193, 199
Van Bohemen, C.G. 88
Van den Hurk, R. 118, 120, 150, 151, 154, 155, 157, 158, 159, 160
Van den Kraak, G. 114, 116
Van der Berg, J. 268
Van der Klaauw, C.J. 11
Van der Meeren, T. 59
Van der Meer, H.J. 2, 3, 4, 6, 8, 9, 10, 11, 18
Van Kreijl, C. 80
van Oijen, M.J.P. 6
Van Oordt, P.G.W.J. 88, 117, 118, 120, 157, 158, 160
Van Ree, G.E. 150, 151, 152, 157
Van Rijs, J.H. 38
Van Veen, T. 152

Vancura, M. 125, 130
Vanlersel, A.A.J. 253
Vantue, V. 66
Varanasi, U. 37
Varma, S.K. 120
Varo, M.I. 116
Vasilenko, T.D. 179
Vassilenko, P.D. 179
Va-Ree, G.E. 120
Vejdovsky 105
Verheyen, E. 184, 185
Verighina, I.A. 59, 61, 63, 65, 66, 71, 72
Verigina, I.A. 23
Vernier, J.M. 82, 98
Vethaak, D. 89
Viehberger, G. 101, 106
Vivien, J.H. 149
Vlaming de, V.L. 152, 157
Vogel, V. 40
Vogel, W. 40
Vogel, W.O. 40
Volki, A. 82, 84, 88, 89
Von Hayek, H. 50
Von Rumker, R. 250
Voth, M. 33, 35

Wachtel, A.W. 26
Wagner, H.J. 8
Wajsbrot, N. 254
Wake, K. 85
Wallace, R.A. 112, 117, 123, 124, 126, 128
Walsh, J.M. 111, 116
Walsh, P.J. 181
Walters, V. 169
Waltker, B.T. 123, 129
Wanink, J.H. 6
Wanink, J.J. 19
Wardle, C.S. 173
Warner, R.E. 253
Warren, A. 254
Wassersug, R.J. 34
Wassersug, R.L. 59, 68
Watanabe, T. 89
Watts, S.A. 66, 72
Weatherly, A.H. 170
Webb, P.W. 180
Weber, K. 33, 35
Weibel, E.R. 34, 208, 209, 210, 211, 215, 216, 224, 225, 228, 230, 240
Weil, C. 111, 118, 120, 121, 122, 123
Weis, P. 83
Weitzman, S.N. 23
Welcomme, R.L. 184, 185
Welsch, U. 84, 88
Wendelaar Bonga, S.E. 35, 38
Werneri, R.G 123

Wert, S.E. 113
Westermann, J.E. 106
Wester, P.W. 250, 252, 253, 255
Weston, P. 51, 55
West, G.C. 68, 71
West, J.B. 236
Wexler, B.C. 78, 87
Whalen, W.J. 51
Wietzman, S.N. 21, 23
Wilkes, R.L. 253
Wilkinson, P. 269
Willmer, E.M. 187
Wilson, L.J. 133, 134
Wingstrand, K.G. 52
Winterbottom, R. 130, 170
Wischnitzer, S. 267
Witcomb, D.M. 68
Wittenberg, B.A. 11
Wittenberg, J. 185
Wittenberg, J.B. 11, 185, 186, 187
Witte, F. 4, 5, 6, 7, 8, 16, 19
Witte-Mass, E.L.M. 6
Witt, U. 83
Wodtke, E. 84, 88
Woodland, W.N.F. 180, 186
Wood, C.M. 180, 181, 186
Wooten, R.J. 152
Woo, N.Y.S. 88
Wourms, J.P. 111, 112, 113, 114, 123, 124, 125, 126, 127, 128, 130, 132
Wright, A. 150
Wright, D.E. 34, 106

Wright, P.A. 180, 181, 186
Wu, X. 23

Xavier, F. 147, 155, 161
Xizai, W. 113, 117

Yadava, A.M. 216, 219, 222
Yadava, A.N. 219, 222
Yamagami, K. 123, 128
Yamamoto, K. 98, 99, 100, 117, 118, 119, 120, 123, 149, 154
Yamamoto, M. 87, 112, 113, 116
Yamamoto, T. 98, 99
Yamamoto, T.S. 124, 133, 134, 135
Yamauchi, K. 121
Yamazaki, F. 149
Yaron, Z. 118, 120, 150, 151, 155, 156
Yashou, A. 118
Yasuzumi, F. 112, 115, 116, 123
Yokote, M. 80, 88
Young, G. 116, 117, 119, 121, 122, 157, 160
Youson, J.H. 79, 87, 89, 96, 99, 100
Ysargil, G.M. 253

Zaccone, G. 51
Zeno, T. 38, 253, 256, 265, 266, 268
Zihler, F. 63
Ziswiler, P. 63
Zona, Y. 116, 117, 118, 119, 121, 122, 152
Zuvic, T. 84, 88
Zweers, G. 15, 16

Subject Index

Absorbent cell   XX
Accessory respiratory organs   XVII, 203, 206, 212, 215, 223
Acetylcholine   52
acetylcholinesterase   252, 253
Acoustico lateralis system   193
Acromatization   122
Actinopterygians   23, 25, 27
Adductor muscle   16
Adenyl cyclase   116
Adrenergic plexus   52
Adult   220, 221, 224, 225
Afferent branchial artery   32, 39, 50
Age changes   87
Air sac(=tube)   204, 214, 224
Albumin   256, 257
Amacrine cells   10
Animal pole   133
Anisotropy coefficient   175
Apical cells   102
Aponeurosis   170
Apophysis   133
Asphyxiation   185
Atretic follicles   XXI
Autapomorphies   21, 23, 24, 28

Basal lamina   111, 112, 119, 135
Behavior   250, 253
BHCH   250
Bile canaliculi   84, 85
Bile duct   82, 99
Bile salts   95
Biliary system   XX, 96, 97, 98
Bimodal respiration   56
Biological function   17
Bipolar cells   10
Boundary layer   242, 246
Branchial arches   49, 51
Branchial chamber   213
Branchiostegals   22

Caeca   59
Caecal enzymes   71
Cardiorespiration   XIX
Carotid labyrinth   25
Cephalic canals   199

Chloride cell   33, 35, 36, 37, 38
Chondrosteans   22
Circadian rhythm   8
Clupeomorphan   22
Coeliaco-mesenteric artery   185, 186
Collagen fibre   183, 184
Collagenous connective tissue   183
Color   78
Color vision   10
Complex relationship   88
Composite model   269
Cones   4, 6, 8, 9, 10, 11
Corpus luteum   XXI
Cortex radiatus   123, 124, 129, 132, 133
Counter current principle   237
Cranial nerves   193
Cretaceous   23

Dendritic organ   215, 223
Desmosomes   112
Detoxification   31
Development   96
Diazinon   249, 251, 255
Diffusing capacity   207, 225
Diffusion   238, 239
Diffusion coefficient   239
Diffusion distance   214, 225, 226
Digestive system   XX
Digestive tract diverticulae   60, 61, 66
Dorsal aorta   48, 49, 50, 183
Ductal cells   98
Ductus arteriosus   XIX, 48, 49
Ductus caudally   52
Ductus myocytes   52

Ecological morphology   XVII, XVIII
Efferent branchial artery   32, 40, 50
Egg   XXI
Electron microscopicstudy   267
Electroreceptors   24, 26
Elepomorphan   22
Elopids   23
Embryogenesis   118, 126
Endocrine cells   XX
Energy consideration   246
Environmental effect   73
Eocene   22

Euphysoclistic   185, 186
Euphysoclists   180

Fat storing cells   85
Feeding behaviour   250
Feeding effect   80, 88
Fenestrae   84
Fick's law   XXIII, 239, 240
Film coefficient   242
Follicular epithelium   111, 112, 113, 114, 115, 116, 117, 118, 119, 120, 121, 122, 123, 124, 126, 127, 132, 135
Folliculogenesis   112
Following   263, 264
Functional morphology   XVII, XVIII

Gall bladder   XX, 95, 96, 99
Ganglion   8, 11
Gas exchange   235, 240
Gastrointestinal tract   216
Gas-secreting cells   183, 184, 187
Genome organization   128
Gill   XIX, 203, 205, 210, 218, 235, 236, 254, 265
Gill area   219, 221, 224
Gill dimension   XXII, 236, 238
Gill fan   213, 215, 228
Gnathostomes   25
Gonadotropin   121, 122
Grods   206

Harmonic mean   208
Hepatic architecture   80, 81
Hepatic parenchyma   81
Hepatic stroma   97
Hepatocytes   83, 87
Hepatosomatic index   79
Hepatic lobules   81
Histological findings   68, 69
Histology   96, 100
Histopathology   253, 255
Holosteans   22, 23
Hyoidean hemibranch   48
Hypoxia   184, 185

Immunoglobulin   37
Infraorbital canal   198, 200, 201
Isotropic   206
Ito cells   85

Juvenile   220, 221, 224, 225

Kupffer cell   85

Labyrinthine organ   204, 215, 223, 228
Lamellae   205, 206, 218
Lamina flow   241

Lateral line system   193
Ligamentum   52
Liver   77, 81, 85, 89, 96, 97, 98, 254, 266
Locomotory muscles   XXI, 169, 170, 176

Macrophage   85
Macula adherens   115
Malathion   249, 251, 255
Mandibular canal   198, 199, 201
Mass transfer   238
Melano macrophage centres   82
Merz grid   207, 208
Mesozoic   23
Microcirculation   XIX, 40
Microorganisms   71
Micropyle   XXI, 111, 132, 133, 134, 135
Microretia mirabilia   183
Microridges   34, 35, 36
Microscopic anatomy   68
Microvilli   101, 115, 123, 124, 126, 129, 135
Morphometrics   XVII, 203, 206, 229
Mucous cells   37, 38
Musculus adductor arcus palatini   16
Musculus adductor mandibulae   12, 13, 16, 17, 18
Myogenesis   169
Myoglobin   173
Myomere   169, 170
Myopic position   2
Myosepts   170
Myotomes   169, 170, 171

Neuromasts   193, 195, 198, 199, 200, 201
Newtonian   241
Nutrient pathways   40

Ontogeny   XX, 3, 6, 11, 19, 66
Oocyte   111, 112, 114, 115, 116, 117, 121, 123, 124, 125, 126, 128, 129, 132, 133, 135, 266, 267
Oolemma   115, 123, 127
Opercular chamber   204, 216
Optomotor behaviour   261
Osmoregulation   XIX
Osmoregulatory organ   31
Ostariophysian   22, 23, 28
Osteoglossids   23
Osteoglossomorphan   22
Otophysic   22
Ovary   XXI, 114, 266
Oxygen conduction line   236
Oxygen diffusion   XXI, 174
Oxygen tension   235
Oxygen uptake   229

Paleocene   23
Parasite ultrastructure   105

Parasites   103, 104
Parenchymal cells   100
Pavement cells   33, 34, 35
Pentachlorophenol   250
Pericytes   183
Pesticides   249
Photoreceptors   3
Pharyngeal chamber   204, 212, 223
Phylogenetic lineage   23, 24, 28
Phylogenetic position   XVIII
Phylogeny   XVII, XX
Physiology   71
Physostomes   186, 187, 188
Pink muscle   172
Plasticity   11, 17, 18
Polymorphism   85
Polyspermy   134
Portal triads   81
Post opercular canal   196, 199
Post orbital canal   198
Pre orbital canal   198, 199, 201
Pre opercular canal   195, 196, 199
Pseudobranchial neurosecretory   27
Pulmonary artery   49, 50, 52, 53
Pulmonary vein   48, 49

Recessed cells   35
Red muscle   171, 172, 173, 174, 175
Respiration   235, 238, 240
Respiratory islets   212, 214
Rete chorioidea   11
Rete mirabile   182, 186, 187
Retina   1, 2, 3, 8, 10.11, 12
Retinal resolution   2, 3
Retinomotor response   8
Retractor tentis muscle   2
Reversal   263, 264
Reynold's Analogy   245
Reynold's Number   243, 244
Rodlet cells   105
Rods   8
Rostrocaudal   2

Samonids   23
Scaling   218, 223
Scotopic system   8
Season effect   80, 88
Secretory cells   XX
SEM   254
Sensory apparatus   XXII
Serosa   119
Serum protein   256
Sex effect   80, 87

Shape   78
Sinusoids   84
Skin area   225
Snorkel   48
Space of Disse   84
Species variability   87
Sphincter   182
Steady state   239
Steady supply   238
Stereocilia   26
Steroidogenesis   114, 116, 117, 118, 120, 121, 122, 126
Stress effect   80
Submuscularis layer proper   183
Supraorbital canal   195, 198, 199, 200
Surface cells   35
Swim(air) bladder   XXII, 179, 180, 181, 182, 183, 184, 185, 186, 187, 188, 216
Synapomorphies   22, 24
Synaptic network   52
Systemic circulation   48, 49, 53

Tarbulent flow   241
Temperature   87
Theca   111, 135,
Thyroxine   257
Toxins   105
Transcription   128
Transferrin   256, 257
Translation   128
Treated gill   267
Tunica externa   183

Ultrastructure   97, 98, 99, 102

Vasa-vasorum   50
Vascular papillae   212, 213, 224
Vascularization   79
Vasoactive harmones   40
Vasomotor segments   52, 53, 54
Vena cava   48, 49
Ventricle   48
Vestigial ductus   52
Vitellogenesis   XX, 114, 118, 131

Water velocity   246
Weberian ossicles   22, 23
White muscle   171, 172, 173, 175

Zona material   124, 126
Zona pellucida   111, 112, 114, 115, 123, 124, 125, 126, 127, 128, 129, 130, 131, 132, 133, 134, 135
Zona radiata   123, 124, 129

Systematic Index

Abramis brama XXII, 80, 260
Acacia auriculaeformis 254
Acanthobrama terrae-sanctae 118, 120, 150, 151, 156
Acanthurus blochii 82
Acernia 180
Acipenser 96
Acipenser sturio 104
Acipenser transmontanus 125, 133
Agonus cataphractus 125
Amphipnous cuchia 150, 151, 220
Anabas 215, 220, 221, 223, 224, 231, 254
Anabas testudineus XXII, 34, 211, 215, 217, 219, 220, 221, 222, 223, 224, 225, 226, 227, 230, 254
Anguilla 180
Anguilla anguilla 35, 78, 83, 87, 222
Aphyosemion splendopleur 114
Aplocheilus panchax 195, 196, 197, 198, 199, 200
Apocryptes lanceolatus 216
Arapaima 187
Arius thalassinus 112
Astyanax bimaculatus 129

Bathybates 14
Blennius ocellatus 104
Blennius pholis 125, 126, 128, 222
Boleophthalmus boddaerti 211, 216, 217, 219, 220, 222, 227, 230
Boleophthalmus viridis 216
Botia dario 219, 222
Botia lohachata 222, 227, 230
Brachydanio rerio 118, 120, 124, 133, 134, 150, 151, 152, 157, 158, 159, 160, 249
Brosmius brosme 104

Carany spp. 81
Carassius auratus 80, 84, 89, 98, 118, 120, 121, 129
Carassius carassius 98, 174
Catla catla 217, 219, 222, 227, 230
Centronotus gunellus 104
Cepola 104
Cestracion 104
Cetorhinus 173
Cetorhinus maximus 149, 156, 158
Chaenogobius isaza 132, 133
Chaenogobius spp. 80

Channa 34
Channa gachua XXII, 118, 120, 150, 155, 212, 217, 219, 220, 222, 223, 224, 225, 227, 230
Channa punctatus XXII, 211, 212, 213, 217, 219, 220, 222, 223, 224, 225, 227, 230, 250, 253
Channa striatus XXII, 211, 212, 213, 217, 219, 220, 221, 222, 223, 224, 225, 226, 227, 230, 231
Chanos chanos 88
Chimaera 174
Chloromyxum histolyticum 104
Chloromyxum leydigi 104
Chloromyxum morovi 105
Chloromyxum rosenbushi 104
Chloromyxum trijugum 104
Choloromyxum truttae 104
Chromis chromis 114
Chrysemis scripta 52
Chum salmon 129
Cichlasoma nigrofasciata 120, 125, 126, 128, 130, 131
Ciliata mustela 104
Cirrhina mrigala 193, 211, 213, 217, 219, 220, 221, 222, 224, 227, 230, 254, 255
Clarias 211, 220, 224, 231
Clarias batrachus XXII, 34, 38, 40, 150, 156, 158, 211, 215, 217, 219, 220, 222, 223, 224, 225, 227, 230, 249
Clarias lazera 118
Clarias mossambicus 175, 215, 222, 227, 230
Clupea 104
Clupea harengus 131
Cnidospora XX
Colisa fasciatus 215, 222
Coregonus 180
Coregonus peled 114, 127
Coryphaena hippurus 222
Cottus 179
Crenilabrus cinercus 125
Crenilabrus mediterraneus 125
Crenilabrus melops 125
Ctenolabrus exoletus 125
Ctenolabrus rupestris 125
Cyathopharynx or Chilotilapia (rhodesi) 14
Cynolabias belotti 125
Cynolabias ladigesi 114, 127
Cynolabias melanotaenia 111, 114, 125, 127, 130, 132

298

Cyprinodon variegatus 131
Cyprinus carpio 83, 84, 88, 120, 131, 187, 220, 221, 222, 249, 260

Decimodus 14
Dermogenys pusillus 125, 126
Dicentrarchus labrax 78, 79, 82, 84, 88, 98, 99
Diodon holacanthus 80, 82
Diodon spp. 87
Dipnorhynchus 25
Dorosoma petenense 132
Drepanoepsetla 104

Engraulis encrasicolus 173, 175
Epiplatys spilargyreus 114, 128
Erethmodes 14
Esox americanus 200
Esox lucius 135
Esox niger 200
Etmopterus 174
Eucalia inconstans 149, 156
Eusthenopteron 25
Euthynnus affinis 173

Fundulus 131, 180, 185
Fundulus heteroclitus 112, 117, 125, 126, 130, 186

Gadus callarias 186
Gadus marrhua 125
Gadus merlangus 149
Gadus morhua 83, 129, 171
Gaidropsarus cirratus 104
Galeus 174
Galeus melastomus 171, 173
Gambusia affinis 83
Garra lamta 219, 222, 227, 230
Gasterosteus aculeatus 104, 149, 156
Gasterosteus spinachia 104
Glossogobius giuris 219, 222
Glytothorax nectinopterus 149
Gobio gobio 112, 115, 125, 129, 133
Gobius flavescens 180
Gobius giuris 149, 151, 156, 158
Gobius paganellus 149
Gymnarchus 187

Haplochromis argens 4, 6, 7, 8, 10, 11
Haplochromis elegans 12
Haplochromis heusinkveldi 4, 6, 7, 8
Haplochromis multicolor 120
Haplochromis pyrrhocephalus 4, 6, 7, 8, 10, 11
Haplochromis reginus 4, 6, 7, 8, 10
Haplochromis sauvagei 4, 11
Haplochromis sp. 4
Haplotaxodon(microlepis) 14
Hemibates(stenosoma) 14

Hemichromis 113
Heterandria formosa 115
Heteropneustes 211, 220, 224, 231
Heteropneustes fossilis XXII, 149, 156, 158, 211, 213, 217, 219, 220, 222, 223, 224, 225, 226, 227, 230, 249, 254, 255, 259, 266
Hippocampus brevirostris 104
Hippocampus erectus 124, 125
Hippocampus guttulatus 104
Hippoglossus 104
Hippoglossus hippoglossus 80
Hoplerythrinus unitaeniatus 187
Hoplias malabaricus 222
Hucho hucho 187
Hypoglossoides platessoides 125

Ichthyostega 25
Ictalurus melas 34
Ictalurus punctatus 32, 34, 37, 40, 82, 98, 253
Ictarulus furecatus 104

Katsuwonus pelamis 169, 171, 173, 175, 222

Labeo bata 219, 222
Labeo rohita 211, 221, 224, 227, 230
Lampetra spp. 78
Lebistes reticulatus 115, 125, 149
Lepidocephalichthys guntea 216, 219, 222
Lepidosiren 47, 50, 53, 55
Lepidosiren paradoxa XIX, 49
Lepidosiren spp. 88
Lepomis macrochirus 59, 60, 65, 66, 125, 255, 267
Leptotheca agilis 105
Leptotheca trijugum 104
Leptotheca vikrami 104
Leptothus agilis 104
Lethrinops or Labeotropheus(fuelleborni) 14
Leuciscus idus 82, 83, 84, 88
Limanda limanda 125
Liopsetta 128
Liza 99
Liza spp. 78, 79
Lobochilotes (elabiatus) 14
Lophius 104, 179
Lota 97
Lota lota 83
Lotta 180
Lutianus analis 124
Lutjanus bohar 78, 82, 81

Macrognathus aculeatum 211, 219, 220, 222, 227, 230
Maena vulgasis 104
Merluchius gayi 104
Microcystis aeroginosa XX
Micropogon undulatus 80, 82, 83

Micropterus dolomieu 38, 221, 222
Micropterus salmoides 187, 250
Misgurnus anguillicaudatus 118, 134
Mixidium gasterostei 104
Mixidium kudoi 104
Mollugo pentaphylla 254
Molva vulgaris 104
Monopterus 224, 226
Monopterus albus 118, 150
Monopterus cuchia 34, 211, 212, 223, 224, 225, 227, 230
Morone saxatilis 34, 38
Motella 104
Mugil capito 35, 118
Mugil cephalus 36, 149, 153
Mustelus canis 148, 149, 156, 161
Mutatis mutandis 11
Mystus cavasius 120, 150, 151, 154, 155, 157, 158, 217, 219, 222, 227, 230
Mystus seenghala 149, 156, 158
Mystus tengara 150, 157
Mystus vittatus 217, 219, 222, 227, 230
Myxine glutinosa 99, 100
Naemachilus rupicola 219, 222, 227

Neoceratodus 47, 96
Noemacheilus barbatulus 112, 115, 125, 128, 129, 133
Noemacheilus triangularis 133
Notopterus 187
Notopterus chitala 216
Notopterus notopterus 150, 156, 158, 216
Notothemia gibberifrons 175

Odontesthes bonariensis 88
Oncorhynchus gorbuscha 125, 133
Oncorhynchus keto 133, 134
Oncorhynchus kisutch 120
Oncorhynchus mykiss 78, 80, 81, 82, 84, 106, 129
Oncorhynchus rhodurus 120
Oncorhynchus spp. 87
Ophiocephalus punctatus 149, 193, 249, 251
Opsanus 180
Opsanus tau 220, 222, 227, 230
Oreochromis alcalicus grahami XXII, 180, 181, 182, 183, 184, 185, 186, 187, 188, 254
Oreochromis aureus 254
Oreochromis niloticus 171, 173, 174, 175, 222, 254
Oryzias latipes 87, 118, 125, 126, 127, 129, 131, 134, 250
Oryzias melastigma 195, 196, 197, 198, 199, 200
Osteolepis 25

Parophrys vetulus 254
Periophthalmus lanceolatus 219, 222
Periophthalmus pearsi 216

Periophthalmus schlosseri 219, 222
Periophthalmus vulgaris 211, 216
Petromyzon marinus 79, 87, 99, 100
Pimelodus maculatus 98
Pimephales promelas 37, 125
Platichthys flesus 125, 171
Platichthys stellatus 125
Plecoglossus 112
Plecoglossus altivelis 80, 112, 115, 119, 123, 149
Pleuronectes 104, 179
Pleuronectes flesus 125
Pleuronectes limanda 125
Pleuronectes platessa 125, 126, 149, 156, 158
Pleuronichthys coenosus 125, 132
Poecilia reticulata 117, 120, 150, 151, 155, 195, 196, 197, 198, 199, 200, 250
Pomatomus saltatrix 34
Pomatoschistus minutus 113, 115, 131, 134
Pomoxis sparoides 104
Prionace glauca 171
Protopterus 47, 49, 50, 55
Protopterus aethiopicus XIX, 49
Protopterus spp. 88
Psettodes stellatus 133
Pseudopleuronectes americanus 131
Puntius conchonius 250
Puntius sarana 133

Raia 104
Raja binoculata 148, 149, 156
Raja erinacea 148, 149
Rana 52
Rhamphochromis 14
Rhina 171
Rhinichthys 180
Rhinomugil corsula 217, 219, 222, 227, 230
Rhodeus amerus 149
Ruvettus 173

Salmo fario 187
Salmo fontinalis 185
Salmo gairdneri 98, 99, 101, 118, 120, 129, 150, 151, 154, 157, 158, 159, 160, 169, 171, 175, 222, 227, 230
Salmo irideus 185, 187
Salmo salar 81, 185
Salmo trutta 87, 125, 185
Salvelinus fontinalis 116, 125
Salvelinus leucomaenis 118, 120
Sarda sarda 34
Sardina pilchardus 171
Sarotherodon aureus 118
Scarus spp. 82, 83
Scoliodon 114, 119
Scoliodon sorrokowah 112, 149, 153, 154
Scomber 179

Scomber scomber 104, 120, 149, 151, 156, 158, 160
Scomber scombrus 34, 36, 171, 222
Scorpena scrofa 103
Scylliorhinus canicula 100, 148, 149, 154, 156, 158, 169, 171, 222, 227
Scylliorhinus stellaris 148, 149, 154, 156, 158, 171, 222
Seriola quinqueradiata 80
Serranochromis or Spathodus (marlieri) 14
Serranus cabrilla 78, 81, 82
Sicamugil cascasia 219, 222
Silurus glanis 131
Sorpena scrofa XX
Sparus aurata 78, 254
Sphaeromyxa balbianii 104
Sphaeromyxa exneri 104
Sphaeromyxa gasterostei 104
Sphaeromyxa hollandi 104
Sphaeromyxa incurvata 104
Sphaeromyxa sabrazesi 104
Squalus 96, 173
Squalus acanthias 114, 117, 148, 149, 154, 156, 158
Squalus suckleyi 148, 156
Stizostedian vitreum 222
Syngnathus fuscus 124, 125

Tanganicodus or Tropheus (moorii) 14
Teopedo mormorata 100, 101
Testudo graeca 52
Tetractenos glaber 187
Thunnus albacares 222
Thunnus thynnus 222
Thysanophris japonicus 104
Tilapia 41
Tilapia aurea 118
Tilapia mossambica 87, 251

Tilapia nilotica 117, 120, 150, 151
Tilapia thollori 112
Tinca tinca 175, 222, 227, 230
Tor tor 149, 156
Torpedo ocellata 148, 149, 154, 156
Torpedo sp. 148
Trachurus mediterraneus 118, 120
Trachurus mediterraneus lamellae 34
Trematocara (nigrifrons) 14
Trematomus newnesi 175
Trigloporum lastoviza XX, 101
Tropedo 100, 101, 104
Tropedo marmorata 148, 149, 154, 156, 157, 158, 161
Trutta fario 104
Trygon pastinaca 104

Umbra 187
Unicauda 105
Uranoscopus 102, 105
Uranoscopus sp. XX

Wallago attu 222, 227, 230

Xenentodon cancila 149, 156
Xenochromis (=Perissodus) 14
Xenotis megalotis 104
Xiphophorus helleri 113, 115, 118, 195, 196, 197, 198, 199, 200

Zschokella 104
Zschokella russelli 104
Zschokella sturionis 104
Zschokella sturionis 104
Zeus faber 104